Latin America
& the Caribbean

LANDS AND PEOPLES

third edition

Latin America & the Caribbean

LANDS AND PEOPLES

third edition

DAVID L. CLAWSON
University of New Orleans

McGraw Hill Higher Education

Boston   Burr Ridge, IL   Dubuque, IA   Madison, WI   New York   San Francisco   St. Louis
Bangkok   Bogotá   Caracas   Kuala Lumpur   Lisbon   London   Madrid   Mexico City
Milan   Montreal   New Delhi   Santiago   Seoul   Singapore   Sydney   Taipei   Toronto

LATIN AMERICA & THE CARIBBEAN: LANDS AND PEOPLES
THIRD EDITION

Published by McGraw-Hill, a business unit of The McGraw-Hill Companies, Inc., 1221 Avenue of the Americas, New York, NY 10020. Copyright © 2004, 2000, 1997 by The McGraw-Hill Companies, Inc. All rights reserved. No part of this publication may be reproduced or distributed in any form or by any means, or stored in a database or retrieval system, without the prior written consent of The McGraw-Hill Companies, Inc., including, but not limited to, in any network or other electronic storage or transmission, or broadcast for distance learning.

Some ancillaries, including electronic and print components, may not be available to customers outside the United States.

This book is printed on recycled, acid-free paper containing 10% postconsumer waste.

1 2 3 4 5 6 7 8 9 0 QPD/QPD 0 9 8 7 6 5 4 3

ISBN 0–07–252144–9

Publisher: *Margaret J. Kemp*
Sponsoring editor: *Thomas C. Lyon*
Developmental editor: *Fran Schreiber*
Executive marketing manager: *Lisa L. Gottschalk*
Lead project manager: *Joyce M. Berendes*
Production supervisor: *Sherry L. Kane*
Coordinator of freelance design: *Michelle D. Whitaker*
Cover designer: *Kaye Farmer*
Lead photo research coordinator: *Carrie K. Burger*
Compositor: *Shepherd-Imagineering Media Services, Inc.*
Typeface: *10/12 New Caledonia*
Printer: *Quebecor World Dubuque, IA*

Cover Images
Main and back cover images: *Colonial era Roman Catholic church of Campeche, Mexico, at sunset.*
© *David L. Clawson*
Inset image: *Highland market woman, Saquisilí, Ecuador.* © *Oscar H. Horst*
Bottom image: *Iguaçu (Iguazú) Falls, Paraná River, near the junction of Brazil, Argentina, and Paraguay.*
© *Oscar H. Horst*

The credits section for this book begins on page C-1 and is considered an extension of the copyright page.

Library of Congress Cataloging-in-Publication Data

Clawson, David L. (David Leslie), 1948–
    Latin America & the Caribbean : lands and peoples / David L. Clawson. —
3rd ed.
        p. cm.
    Includes bibliographical references and index.
        ISBN 0–07–252144–9 (hard copy : alk. paper)
            1. Latin America. 2. Caribbean Area. I. Title: Latin America and the
Caribbean. II. Title.

F1408.C62   2004
980—dc21
2002155766

CIP

www.mhhe.com

This book is dedicated to Margaret
and our five children,
Elizabeth, Ernest, Emma, Mary, and John

CONTENTS

Chapter 6
Political Change 114

Chapter 7
Race, Ethnicity, and Social Class 141

Chapter 8
Latin America as a Culture Region 173

Chapter 9
Religion 202

Part III
ECONOMIC PATTERNS 233

Chapter 10
Agriculture and Agrarian Development 234

Chapter 11
Mining, Manufacturing, and Tourism 267

Chapter 12
Urbanization, Population Growth, and Migration 305

Chapter 13
Development and Health 353

LIST OF MAPS

LIST OF TABLES

PREFACE

························

This book has been written primarily as a geography of Latin America and the Caribbean, but is also intended to serve as an interdisciplinary introduction to the region. Latin America and the Caribbean are extremely diverse in their physical attributes. Massive mountain ranges, many of them among the highest and least accessible in the world, are set among enormous lowland plains on the continental mainland while tiny islands dot the Caribbean Sea. Virtually every climate, vegetative association, and soil type found on the earth are present to one degree or another. Cultural diversity also abounds. Economic activities range from traditional slash and burn agriculture to modern, high-technology manufacturing. Almost every racial and ethnic group and political and philosophical persuasion are represented among its peoples. Human settlements vary from remote peasant villages to huge urban agglomerations containing neighborhoods that range from palatial mansions to inner city tenements and peripheral shantytowns.

Yet, beneath this external diversity, there exists a dominant Hispanic cultural heritage whose values and institutions have shaped much of human behavior throughout the region for over 500 years. Although some economic and social subgroups have embraced this heritage more fully than others, it is these values and institutions that have made Latin America a distinct culture realm. This text has been written with the intent of enabling the reader to recognize and appreciate both the diversity and the unity of Latin America and the Caribbean.

ORGANIZATION

The material is organized topically rather than regionally. One of the principal advantages of a thematic approach is its capacity to place current developments in their historical contexts, enabling the reader to know not only what things are like, but how they came to be that way. This understanding, in turn, prepares the student to evaluate the causes and conse-

Case Study 4.1: Biodiversity and the Tropical Rain Forest

The term "biodiversity" is often used to refer to the number of life forms, be they plant or animal, that exist in a given area. Biodiversity is important to the stability of an ecosystem because, all else being equal, the greater the number of species and varieties, the greater the ability of the ecosystem to withstand the loss of some of its members. Various researchers have estimated that somewhere between 4 and 30 million forms of life exist on the earth today and that only 1.5 to 1.8 million of those have been identified to date. As a general rule, the warmer and wetter the environment of a given area, the greater the biodiversity. Two of the most genetically diverse biomes on earth are the coral reefs found in warm, shallow tropical waters and the tropical rain forests.

Scholars believe that approximately half of all life existing on the planet today are associated with the tropical rain forests, which are especially rich in flowering plant and insect life. The Amazon Basin contains the largest surviving stand of rain forest vegetation and may be the most genetically diverse rain forest as well. One indication of the incredible diversity of life in Amazonia is found in two recent studies of a single 5.0 square kilometer (1.93 square mile) section of the Manu Biosphere Reserve of southeastern Peru where researchers identified over 5,000 species of plants and animals. These included 1,147 species of vascular plants, 244 species of mosses and liverworts, 117 species of mammals, 415 species of birds, 128 species of reptiles and amphibians, 210 species of fish, and 2,935 species of insects and spiders (Wilson and Sandoval 1996;

Dallmeir, Kabel, and Foster 1996). As amazing as the findings were, the investigators estimated that the remainder of the reserve contains five to ten times the number of species found in the small study site.

Just as the tropical rain forest exhibits an immense collective array of life, it also varies greatly in its vegetative composition from one microregion to another. These variations can be reflections of minute differences in physical habitats, such as underlying landforms, soil moisture and nutrient levels, rainfall, and exposure to sunlight, as well as diverse human uses of the natural environment (Jain 2000; Mazer 1996). This, in turn, suggests that conservation measures will succeed only if large tracts of land can be set aside as forest preserves. The varied genetic makeup also helps us understand why large-scale monoculture, or the cultivation of a single crop over a wide area, is environmentally incompatible with the rain forest biome. Yet the Amazonian rain forests have been cut and burned recently at a rate equivalent to an area the size of Belgium each year. As a result of the destruction of the Amazonian rain forests, and others in Latin America, Africa, Asia, and Australia, the earth is losing an estimated 4,000 to 6,000 species annually (Wilson 1989).

Why are so many people so concerned about the genetic erosion currently under way? The answers are varied but fall into two general categories. The first is that the genetic erosion results in the loss of invaluable medicinal and otherwise economically useful products (Mittermeir and Konstant 2001; Calderon et al. 2000;

Plotkin 1988). To use medicinal substances as one example, it is estimated that only 12.5 percent of all medicinal drugs potentially available from rain forest plants have been discovered to date (Mendelsohn and Balick 1995). Derivatives from just one of these, the rosy periwinkle, offer a 99 percent chance of remission for victims of lymphocytic leukemia and a 58 percent chance of recovery from Hodgkin's disease. Quinine, an alkaloid derived from the bark of cinchona trees, is used to prevent and treat malaria. Numerous other rain forest substances are used to treat illnesses ranging from heart, eye, and neurological disorders to sickle-cell anemia, Parkinson's disease, and bacterial infections (Posey 2000; Chivian 1997; Wilson 1992). As rain forest species become extinct, we are risking the loss of an incalculable number of medicinal cures and treatments without even realizing what we have forfeited. Other species are of considerable potential worth for use as fuels, fibers, foods, and industrial compounds.

Perhaps an equally compelling argument for maintaining biodiversity is that each species is irreplaceable and, as such, represents a small portion of our earthly heritage—in a sense, the loss of any species is a loss of part of ourselves. In spite of all our technological knowledge, we do not yet have, and may never have, the capacity to create a single new species of life. For these reasons, concerned individuals and institutions from the Amazon Basin nations and elsewhere are endeavoring to develop programs and strategies that will help to stem the loss of biodiversity within the region.

quences of change, be it the loss of biodiversity within lands formerly covered by tropical rain forest or such cultural issues as the plight of street children and the evolving status of women.

The text is designed for maximum flexibility. Written to be readable for beginning students, it is also generously documented with scholarly references for the research needs of advanced undergraduate and graduate students. The individual chapters can be used in all or in part and in any order. The volume is suitable to either single term or multiterm courses.

PEDAGOGY

The focus of this text is on comprehension of concepts, patterns, and issues rather than on memorization of facts and figures. The latter are provided, but they are intended primarily to illustrate the underlying conditions rather than to serve as the focus of the text. Two pedagogical tools that have been employed to accomplish this objective are highlighted key terms and case studies. The maps and photographs are intended to promote a deeper understanding of the Latin American way of life and its regional expressions.

FIGURE 8.10

Literacy rates (percentage) in Latin American and Caribbean nations.
Source: *United Nations Statistical Yearbook 1998* 2001, 73–74.

CHANGES AND ADDITIONS
FOR THIRD EDITION

The text has been thoroughly updated for the third edition. New and updated topics include the devastating El Salvador earthquakes of 2001, the on-going loss of Latin America's irreplaceable biodiversity, the changing economic and cultural parameters of the South American drug trade, the hidden costs of corruption, and the evolving status of women throughout the region. Political conditions for each nation have been updated along with the impacts of International Monetary Fund structural reforms. Patterns of population growth, urban and rural landuse, and international migration are each addressed, as well as human development indicators, social class structure, and the growth of charismatic Protestantism and Catholic liberation theology. Treatments of human health and well-being focus on advances in nutrition, improved water supplies and health care, increasing human life expectancies, and the ominous resurgence of malaria, dengue fever, and other debilitating diseases of the tropical lowlands. The maps and graphs have been expanded and updated to reflect the latest available data, and additional photographs have been added to enable the reader to gain a greater feeling for how the people live.

ACKNOWLEDGMENTS

As in the first two editions, I am indebted to my wife Margaret, who is not only an inspiration but who also functions as typist, editor, and critic. I would also like to acknowledge my Latin Americanist colleagues and students in geography and its allied disciplines for the things they have taught me.

The following have given generously of their slides and photographs for the production of this text:

Delwin L. Clawson

Alexander Coles
Tulane University

Cyrus B. Dawsey
Auburn University

John W. DeWitt
Radford University

Oscar H. Horst
Western Michigan University

Robert L. Layton
Brigham Young University

Charles M. Nissly
Millersville University

Kally Marie Ray Squires

Christoph Stadel
University of Salzburg, Austria

Peter H. Yaukey
University of New Orleans

Thanks is also given to the following individuals who have reviewed one of the editions:

Jorge A. Brea
Central Michigan University

Alexander Coles
Tulane University

William V. Davidson
Louisiana State University

Cyrus B. Dawsey
Auburn University

John W. DeWitt
Radford University

Richard Ethorne
Northern Michigan University

Gary S. Elbow
Texas Tech University

Daniel W. Gade
University of Vermont

Jerry D. Gerlach
Winona State University

Charles Gildersleeve
University of Nebraska at Omaha

Linda Greenow
State University of New York, New Paltz

Charles F. Gritzner
South Dakota State University

Dennis G. Gruwell
a physician practicing in New Orleans, Louisiana

Lawrence M. Harmon

Oscar H. Horst
Western Michigan University

Merrill L. Johnson
University of New Orleans

Gregory V. Jones
Southern Oregon University

Gregory W. Knapp
University of Texas, Austin

Elizabeth M. Larson
Arizona Humanities Council

Joseph S. Leeper
Humbolt State University

James Loucky
Western Washington University

Colin M. MacLachlan
Tulane University

Ines M. Miyares
Hunter College-CUNY

Charles M. Nissly
Millersville University

Samuel M. Otterstrom
Brigham Young University

David J. Robinson
Syracuse University

Kathleen Schroeder
Appalachian State University

Latin America & the Caribbean

LANDS AND PEOPLES

third edition

1

The Changing Face of Latin America and the Caribbean

Latin America and the Caribbean form one of the earth's most important and rapidly changing culture realms. Their lands and peoples, regrettably, are also among the least understood. Our perceptions of Latin America are too often shaped by somewhat sensationalized and fragmentary media accounts of distant crises. A day seldom passes, for example, without media reports of political or military strife. Other accounts carry predictions of impending economic ruin, images of physical devastation, and scenes of intense human poverty and deprivation.

These momentary glimpses, while generally accurate in the limited contexts in which they are presented, often convey unintentionally a distorted picture of the daily life of most Latin American and Caribbean peoples. They also communicate little, if any, appreciation of the almost infinitely diverse physical landscapes of the region. They fail, further, to convey an understanding of the centuries-old struggle for human dignity and development that has been waged against both internal and external exploitation.

It is human nature, perhaps, to generalize and oversimplify what we know little about (Goodwin 2003; Pike 1992; Weber 1988). Misconceptions of Latin America abound. Physically, the region is often assumed to be dominated by vast, impenetrable tropical rain forests or by scorching deserts. Those who live in the technologically advanced, more industrialized nations have often thoughtlessly attributed Latin America's collective underdevelopment to a lack of natural resources. Others have speculated that many Latin Americans are poor because the tropical heat makes them lazy and virtually incapable of sustained physical labor. A surprising number still believe that most Latin Americans are illiterate

Indians living simple lives in tiny farming villages that are ruled over by crude military dictators residing in distant capital cities. Yet another misconception is that virtually all Latin Americans are still traditional Roman Catholics who faithfully engage in reassuring rounds of Masses, fiestas, and sacraments throughout their lives.

One of the reasons for the widespread ignorance of Latin America and the Caribbean is the rapidity and intensity with which change has come to the region. While each of these stereotypes has some basis in fact, each is outdated and inaccurate. Tropical rain forests and deserts, for example, are but two of the many environments found in Latin America and the Caribbean (Figures 1.1, 1.2, and 1.3). Contrary to popular perception, Latin American and Caribbean collective underdevelopment has occurred despite the presence of many of the most richly endowed mineral and agricultural regions on earth (Figures 1.4 and 1.5). Those who believe that most Latin Americans are poor because of climatically induced laziness and lethargy ignore both the relatively mild temperatures that characterize much of the region and the common needs and desires of peoples of all climates, cultures, and races to work hard and to provide adequately for themselves and their loved ones. They further fail to recognize that the peoples of all nations and regions are equally capable and industrious when provided proper health care and educational and employment opportunities.

The stereotype that most of the peoples of Latin America and the Caribbean are poor peasant farmers residing in remote villages is countered by the realization that the region is rapidly becoming one of the most urbanized on earth, with many of the world's largest and most environmentally threatened cities (Figure 1.6).

1

FIGURE 1.1

Tropical rain forest vegetation bordering the Acre River, Amazon Basin, Brazil.

FIGURE 1.2

Desert vegetation in northwestern Mexico.

FIGURE 1.3

Highland environment of the central Andes.

FIGURE 1.4

The Chuquicamata open-pit mine of northern Chile is one of the world's leading sources of copper.

FIGURE 1.5

Bananas awaiting processing on a plantation, Sarapiquí, Costa Rica.

FIGURE 1.6

Mexico City has grown to be the second largest city in the world, with a population that exceeds that of New York City, London, and Paris. It is also one of the most environmentally burdened cities on earth, with health-threatening levels of atmospheric contaminants and solid wastes.

FIGURE 1.7

A small, outdoor market in Bahia, northeastern Brazil.

FIGURE 1.8

Traditional and modern lifestyles are often found side by side in contemporary Latin America. This Anglo American fast food restaurant faces the main plaza in La Ceiba, Honduras.

The popular impression that most Latin American countries are ruled by ruthless military dictators is likewise an anachronism. In reality, most Latin American nations now enjoy democratically elected civilian governments, and this trend is likely to be strengthened over the long term by the steady expansion of an economic middle class committed to egalitarian political values. Finally, the perception that all Latin Americans are religiously active Roman Catholics has never been accurate, given two facts: first, the historical dominance of nominal Roman Catholicism, and second, the unprecedented expansion of fundamentalist Protestantism that is presently modifying the religious landscape.

The forces that are transforming the traditional face of Latin America and the Caribbean have altered some areas more than others. There remain many regions where life appears to be much the same as it has been for centuries (Figure 1.7). Indeed, even in those areas most affected by change, there are many, especially of the older generations, whose lives appear to outsiders to have been altered but little. To comprehend the unevenness of change is to better understand why Latin America and the Caribbean is a region whose cultural diversity and complexity match the diversity and complexity of its physical landscapes (Figure 1.8). It helps to explain how donkeys and automobiles can be double-parked on narrow colonial streets lined by both traditional outdoor markets and modern commercial establishments. It helps to explain why barefooted Indian men dressed in loose-fitting, off-white, knee-length trousers and shirts can be seen cutting or "mowing" the grass along paved superhighways with *machetes,* or long knives. It further helps us understand the vast gulf that persists not only between urban and rural zones but between the various ethnic and socioeconomic subgroups that make up the people of the region.

In order to fully comprehend the uneven development of Latin America, it is necessary to recognize that change has almost always occurred within a context of wasted human resources. A tradition of rigid social stratification, which was found among the indigenous peoples of the region and which was reinforced by conquering Europeans, has resulted historically in the exclusion of a large majority of the population from opportunities for personal economic advancement and political initiative. Land tenure inequalities have discouraged the development of an economic middle class and lowered agricultural production. Education of the masses has been neglected, thereby limiting industrial and technological development. Political oligarchies have oppressed the poor and misappropriated vast amounts of desperately needed national revenues. Graft and corruption have been accepted as normal behavior at all levels, with consequent widespread waste and fraud. As if these internal abuses were not enough, foreign nations, both colonial and neocolonial, have exploited the resources of the region while leaving far too little in return. Contemplating the ramifications of these abuses, many observers wonder how any nation or people can realistically expect to fully develop when the great majority of its citizens are precluded, by circumstances beyond their control, from achieving their full individual potential.

As Latin America and the Caribbean begin the twenty-first century, serious challenges abound. Foreign debt and trade imbalances continue to plague most nations and inflation is eroding the living levels of the people. Illicit narcotic drugs have become the most profitable export crop of many regions, with the wealth derived from drugs financing shadow governments that threaten newly won democratic gains. The unprecedented destruction of tropical rain forests, with the at-

FIGURE 1.10

Soil erosion and landslides are yet another major environmental threat to the Latin American peoples. Enormous gullies have formed on many mountain slopes in the highlands of central Ecuador.

FIGURE 1.9

Tropical deforestation, seen here in the northcentral Brazilian state of Pará, continues to be one of Latin America's most serious environmental challenges.

FIGURE 1.11

This squatter settlement on the outskirts of San José, Costa Rica, is situated within sight of new high-rise apartment buildings.

tendant losses of native plant and animal life, have combined with ever-accelerating levels of soil erosion and some of the world's most severe cases of urban air and water pollution to create calamitous levels of environmental degradation (Figures 1.9 and 1.10). Poverty, rapid population growth, and uncontrolled urbanization have contributed to the creation of massive shanty-towns and the rise of a socially and politically unstable urban underclass (Figure 1.11).

As difficult as these and other challenges facing the region may be, it is important to realize that they are occurring in the midst of unprecedented progress. Never before, for example, have so many Latin Americans been so well fed, clothed, and housed. Never before have so many had access to quality medical care and education. Never before have so many been so free to believe and speak and do as they wish.

It is essential, then, as we study Latin America and the Caribbean, that we view those conditions that presently characterize the region from a long-term historical perspective (Sauer 1941; Robinson 1989; West 1998). When we do so, we may conclude, as Dickens did in his classic *A Tale of Two Cities*, of late-eighteenth-century England, that Latin Americans are experiencing "the best of times" and "the worst of times." Just as the England of Dickens' age was being transformed socially

and economically in the early stages of the industrial revolution, so too the lifestyles and values of ever-increasing numbers of Latin Americans are being deeply altered—in some ways for better and in others for worse—by economic and social modernization and development. While both success and failure will be experienced along the way, it is to be hoped that the good will ultimately outweigh the bad and that life will continue to improve for the majority of the people.

GEOGRAPHY AND REGIONAL STUDIES

Geography, as a scientific and academic field of study, had its Western beginnings among the ancient Greeks, who combined the prefix *geo,* meaning "earth," with the word *graphia,* meaning "to write about or describe." Geography, then, in its fullest sense, was virtually all-encompassing, and included the study of both the physical and cultural phenomena that characterize the earth's surface. A term that we frequently use today to describe the all-encompassing nature of geographic inquiry is **holistic.** Just as the historian combines, from a temporal perspective, what may initially appear to be unrelated social, political, economic, and physical circumstances to paint as complete a picture as possible of a given period's events, the geographer integrates and synthesizes diverse physical, social, political, and economic conditions in order to analyze all or part of the earth from a **spatial perspective.** In both history and geography, the lists of individual phenomena—be they dates or names or features or products—are not as important as the study of the **interrelationships** of the parts and the singular insights that can be gained therefrom.

In an age when the peoples of the earth are becoming increasingly interdependent, a knowledge of the lands and peoples of Latin America and the Caribbean is essential to the political and economic development of the world as a whole. Geography, with its integrative, holistic perspective, occupies a prominent place in foreign area studies. Indeed, numerous geographers (Sauer 1925; Whittlesey 1954; Pattison 1964; Parsons 1964; Lewis 1985; Martin and James 1993; Abler 1993; National Research Council 1997) have recognized regional studies as one of the principal interests of the discipline. In creating regional syntheses, which spatially integrate diverse topical phenomena, geographers make some of their most distinctive scholarly contributions.

In assessing the inner character of a region, the geographer must come to know not only what the people are like but how and why they came to be as they are. Hart (1982, 25–26) has observed that acquiring this deeper level of understanding requires a knowledge of a people's values. Values influence both the ways in which different individuals or groups of people perceive their physical environments and the manner in which they attempt to modify those environments. Just as originally distinct groups of people will develop over time a common culture as they acquire shared values, diverse physical environments can be molded by their occupants into a common cultural landscape that eventually becomes, as it were, "a medal struck in the likeness of a people" (Vidal de la Blache 1903, 8).

The ultimate objective of this text is to increase our understanding of, and appreciation for, the Latin American and Caribbean peoples and the landscapes struck in their likenesses. Hart (1982, 21) has noted that, in a very real sense, regions are "subjective artistic devices" that "must be shaped to fit the hand of the individual user." This is as true in attempting to delimit Latin America as it is in endeavoring to define such vague but commonly used regional terms as the "Middle East" or the "Far East." While disagreement will and should probably always exist among scholars about exactly where to draw the borders, or even what name to use, in defining a region (Allen, Massey, and Cochrane 1998), most geographers would agree with Hart (1982, 22) that "understanding is more important than classification, and the core usually is more important than the fringes." They would further endorse Fenneman's (1914, 87) classic definition of a region as an area that allows the "largest possible number of general statements before details and exceptions are taken up."

DELIMITING LATIN AMERICA AND THE CARIBBEAN

In attempting to delineate Latin America, it is important that we focus first on the core. Because Latin America extends spatially over such a vast area, it is impossible to identify a physical core in the sense of a common climate or landform. Latin America, if it exists at all, must be seen as a cultural entity. Recognizing this, we next ask, what binds Latin America together culturally? The answer is a common Latin, or Roman, heritage. Speaking literally, then, Latin America consists of those parts of the American continents whose core culture is derived from the ancient Roman culture of the Mediterranean Basin. The Mediterranean countries that most clearly contributed a Roman culture to Latin America and the Caribbean are Spain and Portugal and, to a lesser degree, France.

Utilizing this definition, virtually all observers would agree that the core of Latin America extends from Mexico on the north southward across the continental mainland to southernmost South America. While the core is relatively easy to define, two fringe areas exist whose Latin or Hispanic culture is less dominant (Figure 1.12). The first is the southwestern border region of the United States. Much of this area was settled by Spaniards in the seventeenth and eighteenth centuries and continues to have a significant and, in many respects, growing Hispanic presence today (Torrans 2000; Smith 1999; Martínez 1994; Carlson 1990; Nostrand 1992 and 1970; Weber 1992; Wilson and Mather 1990; Arreola 1987). In this volume, however, this border region is not included within Latin America, owing to the area's political status as part of the

FIGURE 1.12

Cultural core and fringe areas of Latin America.

FIGURE 1.13

English, Asian, and African cultural influences are all evident in this scene from Belize City, Belize.

Anglo-dominated United States of America and to the economic and social characteristics of the region that resemble more closely those of the technologically advanced industrialized nations than those of the Latin American and other developing countries.[1]

The second cultural fringe area is the Caribbean realm, which consists of the Caribbean Islands, Belize, the easternmost rimlands of Central America, and the northern coastal plains of Guyana, Suriname, and French Guiana (Augelli 1962). Although the Caribbean was dominated in the early colonial era by Spain, most of the islands, as well as the Guiana coast of northern South America, subsequently fell to opposing European powers as Spanish military influence waned. The Caribbean Islands and rimlands have thus evolved into a culturally diverse region that includes English, Dutch, French, Asian, and American elements, as well as Spanish, all fused onto an African base (Figure 1.13). Because, with the exceptions of Cuba, the Dominican Republic, and Puerto Rico, traditional Hispanic values are largely missing from many of the Caribbean Islands, most scholars do not consider them to be a part of Latin America. Although not all of the Caribbean realm is necessarily a part of Latin America in the fullest cultural sense, this volume studies the Caribbean with Latin America owing to the close historical ties between

the regions as well as to the shared economic and social challenges facing their peoples.

LATIN AMERICAN AND CARIBBEAN SUBREGIONS

Latin America and the Caribbean are studied most frequently as two physical subregions, Middle America and South America (Figure 1.14). **Middle America** is a highly diverse and fragmented area that can be divided into three units. The first is Mexico, a relative giant, which accounts for 57 percent of the total population and 72 percent of the landmass of Middle America (Table 1.1). The second physical unit is **Central America,** which consists of seven small nations: Belize, Guatemala, El Salvador, Honduras, Nicaragua, Costa Rica, and Panama. These countries, which are situated south of Mexico and west and north of South America, range in size from the area of Massachusetts to that of Michigan and support one-fifth of the Middle American population.

The third physical unit of Middle America is the **Caribbean Islands,** which in turn are often subdivided into the Greater and Lesser Antilles. The largest of the **Greater Antilles** is Cuba, containing almost half the landmass of all the Caribbean Islands combined and a far greater share of the relatively level, easily farmed land. Hispaniola is the second largest island of the Caribbean and is the only one divided into two nations, French-speaking Haiti on the west and the Spanish-speaking Dominican Republic on the east. Jamaica and Puerto Rico make up the remaining members of the Greater Antilles, which collectively dominate the northern Caribbean Sea and jointly account for 90 percent of both the land and the people of the Caribbean. The **Lesser Antilles** consist of hundreds of tiny islands, islets, and cays found primarily along the eastern rim of the Caribbean (Figure 1.15). The terms "Leeward Islands" and "Windward Islands" evolved in the colonial era as British administrative designations for the northern (excluding the Bahamas) and southern islands, respectively, but have no climatological or physiographic basis. The complex cultural makeup of the Lesser Antilles and the Bahama Islands is evidenced by their present division into sixteen distinct political entities.

The second, and by far the largest, physical subregion of Latin America is the continent of **South America,** which consists politically of twelve independent nations and French Guiana, a department of France. These range in size from mighty Brazil, which ranks fifth in size and in population among the nations of the world, to the three Guianas, whose combined population of 1.6 million is considerably less than that of either greater Atlanta, Phoenix, or St. Louis. The nations of South America are equally varied in their physical and cultural characteristics. It is to the physical diversity of Latin America and the Caribbean that we will first turn our attention.

[1] Social scientists and economists have not succeeded in developing consensus on what terms to use when referring to multinational groupings based on economic and social indicators. Modern versus traditional, industrialized versus less industrialized, developed versus developing, and First World versus Third World are just a few of the many term pairs that have been in vogue at various times (Porter and Sheppard 1998; Simpson 1994; Dickenson et al. 1991). All are judgmental to some degree and are, therefore, potentially offensive to certain interest groups. The use of these and related terms throughout this volume is necessitated by the absence of acceptable substitutes but in no way implies the cultural, social, or moral superiority of one group when compared to the other.

FIGURE 1.14

Latin American and Caribbean subregions.

TABLE 1.1

Nations and Territories of Middle and South America

NATION OR TERRITORY	Population (millions)	AREA Square Miles (thousands)	Square Kilometers (thousands)	Capital City
Middle America				
Mexico	103.4	756.2	1,958.2	Mexico City
Central America				
Belize	0.3	9.0	23.0	Belmopan
Costa Rica	3.8	19.7	51.1	San José
El Salvador	6.4	8.2	21.4	San Salvador
Guatemala	13.3	42.0	108.9	Guatemala City
Honduras	6.6	43.3	112.1	Tegucigalpa
Nicaragua	5.0	49.4	127.9	Managua
Panama	2.9	29.8	77.1	Panama City
Caribbean Islands				
Anguilla	0.01	0.1	0.2	The Valley
Antigua and Barbuda	0.07	0.2	0.4	St. Johns
Aruba	0.07	0.1	0.2	Oranjestad
Bahamas	0.3	5.4	13.9	Nassau
Barbados	0.3	0.2	0.4	Bridgetown
Cayman Islands	0.04	0.1	0.3	George Town
Cuba	11.2	44.2	114.5	Havana
Dominica	0.07	0.3	0.8	Roseau
Dominican Republic	8.7	18.7	48.4	Santo Domingo
Grenada	0.1	0.1	0.3	St. George's
Guadeloupe	0.4	0.7	1.7	Basse-Terre
Haiti	7.1	10.7	27.8	Port-au-Prince
Jamaica	2.7	4.4	11.4	Kingston
Martinique	0.4	0.4	1.1	Fort-de-France
Montserrat	0.01	0.04	0.1	Plymouth
Netherlands Antilles	0.2	0.3	0.8	Willemstad
Puerto Rico and the U.S. Virgin Islands	4.0	3.6	9.1	San Juan
St. Christopher (Kitts)-Nevis	0.04	0.1	0.3	Basseterre
St. Lucia	0.2	0.2	0.6	Castries
St. Vincent and the Grenadines	0.1	0.2	0.4	Kingstown
Trinidad and Tobago	1.2	2.0	5.1	Port-of-Spain
South America				
Argentina	37.8	1,073.1	2,780.1	Buenos Aires
Bolivia	8.5	424.2	1,098.6	Sucre/La Paz
Brazil	176.0	3,286.5	8,512.0	Brasília
Chile	15.5	284.5	736.9	Santiago
Colombia	41.0	440.8	1,141.8	Bogotá
Ecuador	13.5	104.5	270.7	Quito
French Guiana	0.2	32.3	83.5	Cayenne
Guyana	0.7	83.0	215.0	Georgetown
Paraguay	5.9	157.0	406.8	Asunción
Peru	28.0	496.1	1,285.2	Lima
Suriname	0.4	63.2	163.8	Paramaribo
Uruguay	3.4	72.2	186.9	Montevideo
Venezuela	24.3	352.1	912.1	Caracas

Sources: *The Statesman's Yearbook 2002*, 2001; *The World Almanac and Book of Facts 2001*, 2001; U.S. Census Bureau International Programs Center, 2002. www.census.gov/cgi-bin/ipc/idbrank.pl

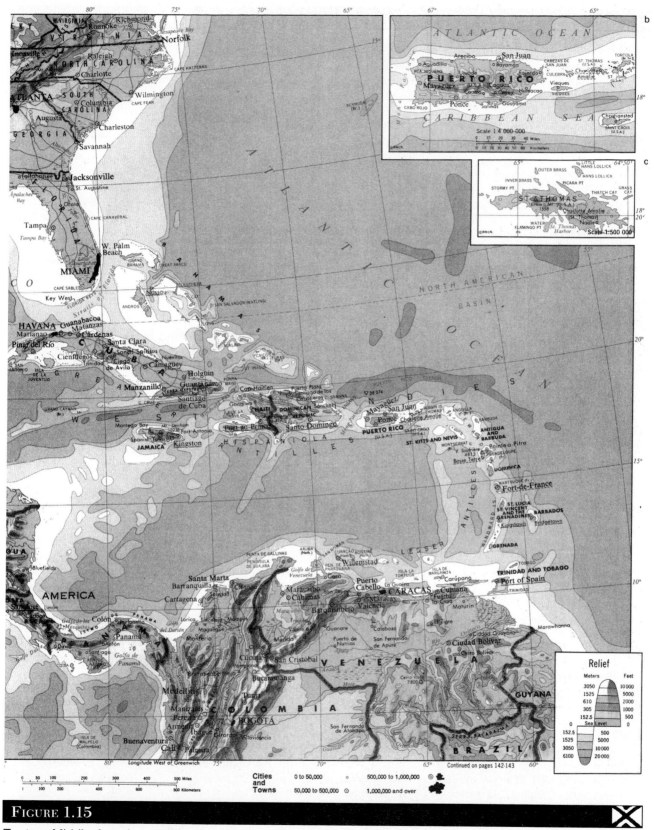

Continued on pages 142-143

FIGURE 1.15

Eastern Middle America and the Caribbean Islands. Map from *Goode's World Atlas,* 20th edition © 2000 by Rand McNally & Company.

KEY TERMS

holistic 6
spatial perspective 6
interrelationships 6

Middle America 8
Central America 8
Caribbean Islands 8

Greater Antilles 8
Lesser Antilles 8
South America 8

SUGGESTED READINGS

Abler, Ronald F. 1993. "Desiderata for Geography: An Institutional View from the United States." In *The Challenge for Geography: A Changing World: A Changing Discipline*, ed. R. J. Johnson, 215–238. Oxford: Blackwell.

Allen, John, Doreen Massey, and Alan Cochrane. 1998. *Rethinking the Region*. London: Routledge.

Arreola, Daniel D. 1987. "The Mexican American Cultural Capital." *Geographical Review* 77:17–34.

Augelli, John P. 1962. "The Rimland-Mainland Concept of Culture Areas in Middle America." *Annals of the Association of American Geographers* 52:119–129.

Carlson, Alvar W. 1990. *The Spanish-American Homeland: Four Centuries in New Mexico's Río Arriba*. Baltimore: Johns Hopkins University Press.

Dickenson, J.P., et al. 1991. *A Geography of the Third World*. London: Routledge.

Fenneman, Nevin M. 1914. "Physiographic Boundaries Within the United States." *Annals of the Association of American Geographers* 4:84–134.

Goodwin, Paul B., Jr. 2003. "Latin America: Myth and Reality." In *Global Studies: Latin America*, ed. Paul B. Goodwin, Jr., 3–6. Guilford: Dushkin/McGraw-Hill.

Hart, John Fraser. 1982. "The Highest Form of the Geographer's Art." *Annals of the Association of American Geographers* 72:1–29.

Lewis, Pierce. 1985. "Beyond Description." *Annals of the Association of American Geographers* 75:465–478.

Martin, Geoffrey J., and Preston E. James. 1993. *All Possible Worlds: A History of Geographical Ideas*, 3rd ed. New York: Wiley.

Martínez, Oscar J. 1994. *Border People: Life and Society in the U.S.–Mexico Borderlands*. Tucson: University of Arizona Press.

National Research Council. 1997. *Rediscovering Geography: New Relevance for Science and Society*. Washington, DC: National Academy Press.

Nostrand, Richard L. 1970. "The Hispanic-American Borderland: Delimitation of an American Culture Region." *Annals of the Association of American Geographers* 60:638–661.

_____. 1992. *The Hispano Homeland*. Norman: University of Oklahoma Press.

Parsons, James L. 1964. "The Contributions of Geography to Latin American Studies." In *Social Science Research on Latin America*, ed. Charles Wagley, 33–85. New York: Columbia University Press.

Pattison, William D. 1964. "The Four Traditions of Geography." *Journal of Geography* 63:211–216.

Pike, Fredrick B. 1992. *The United States and Latin America: Myths and Stereotypes of Civilization and Nature*. Austin: University of Texas Press.

Porter, Philip W., and Eric S. Sheppard. 1998. *A World of Difference: Society, Nature, and Development*. New York: Guilford.

Robinson, David. 1989. "Latin America." In *Geography in America*, eds. Gary L. Gaile and Cort J. Willmott, 488–505. Columbus: Merrill.

Sauer, Carl O. 1941. "Foreward to Historical Geography." *Annals of the Association of American Geographers* 31:1–24.

_____. 1925. "The Morphology of Landscape." *University of California Publications in Geography* 2:19–53.

Simpson, E. S. 1994. *The Developing World: An Introduction*, 2nd ed. Essex: Longman.

Smith, Jeffrey S. 1999. "Anglo Intrusion on the Old Sangre de Cristo Land Grant." *Professional Geographer* 51:170–183.

The Statesman's Yearbook 2002. 2001. Ed. Barry Turner. New York: Palgrave.

The World Almanac and Book of Facts 2001. 2001. Mahwah, NJ: World Almanac Education Group, Inc.

Torrans, Thomas. 2000. *Forging the Tortilla Curtain: Cultural Drift and Change along the United States–Mexico Border from the Spanish Era to the Present*. Fort Worth: Texas Christian University Press.

Vidal de la Blache, Paul. 1903. *Tableau de la Géographie de la France*. Paris: Hachette.

Weber, David J. 1988. *Myth and the History of the Hispanic Southwest*. Albuquerque: University of New Mexico Press.

———. 1992. *The Spanish Frontier in North America*. New Haven: Yale University Press.

West, Robert C. 1998. *Latin American Geography: Historical-Geographical Essays, 1941–1998*. Baton Rouge: Department of Geography and Anthropology Geoscience and Man Publications, Volume 35.

Whittlesey, Derwent S. 1954. "The Regional Concept and the Regional Method." In *American Geography: Inventory and Prospect*, eds. Preston E. James and Clarence F. Jones, 19–68. Syracuse: Syracuse University Press.

Wilson, James R., and Cotton Mather. 1990. "Photo Essay: The Rio Grande Borderland." *Journal of Cultural Geography* 10 (2): 66–98.

ELECTRONIC SOURCE

U.S. Census Bureau International Programs Center. www.census.gov/cgi-bin/ipc/idbrank.pl

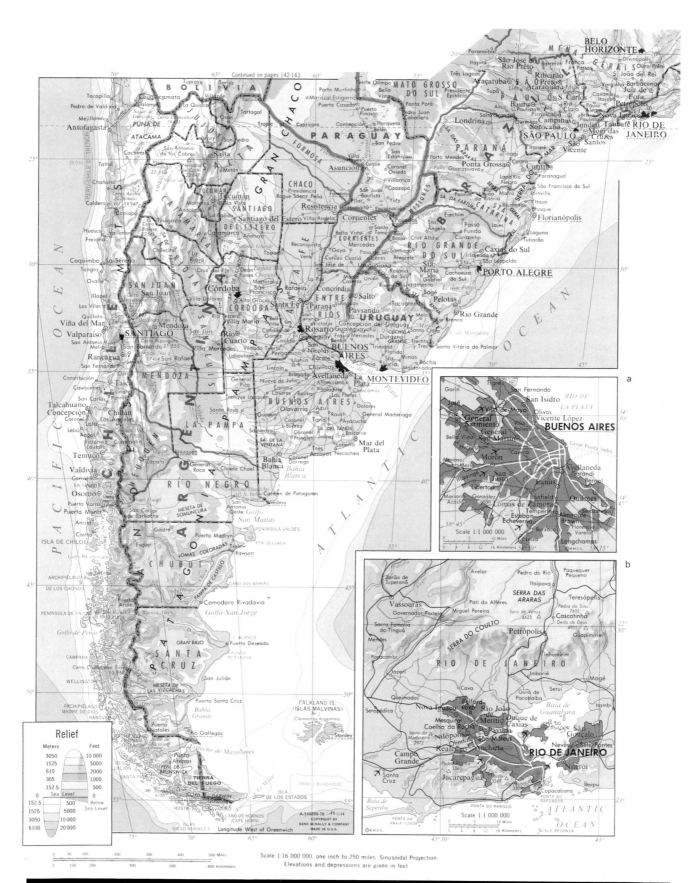

PLATE 1   Southern South America.

Map from *Goode's World Atlas*, **20th Edition** © 2000 by Rand McNally, R. L. 95-S-177.

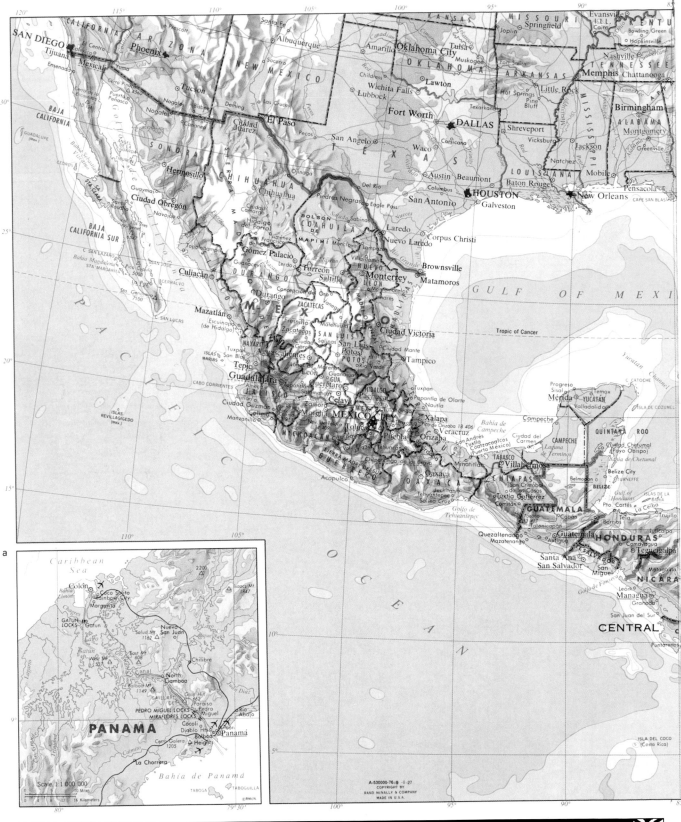

PLATE 4   Western Middle America.

Map from *Goode's World Atlas,* **20th Edition © 2000 by Rand McNally, R. L. 95-S-177.**

Part I

PHYSICAL GEOGRAPHY

Latin America and the Caribbean are endowed with great scenic beauty and physical diversity. Their landscapes include towering snowcapped volcanoes, powerful rivers, endangered tropical rain forests, and some of the world's most extensive grasslands. They are also home to beautiful white sand beaches and amazingly clear, blue-green tropical waters where brilliantly colored fish, plants, and coral can be seen as though they were but a few feet away. Even the dry, desolate deserts offer, to those who come to know them well, quiet hidden beauty within their intricate and delicate ecosystems.

The physical landscapes of Latin America and the Caribbean may be studied from a number of different perspectives. We might begin by asking where various natural phenomena are found, then analyze the forces that influence their distribution and occurrence. Once the individual phenomena are understood, we are prepared to study their interrelationships within natural systems or regions and, finally, to assess the human utilization of those regions.

We might begin an analysis of the tropical rain forests of Latin America and the Caribbean, for example, by studying how their present distribution is related to the rain forest climate. A knowledge of the climate would, in turn, prepare us to understand the unique structure and composition of the forest vegetation and the fragile soils that nourish it. Only then would we be able to recognize which development strategies are environmentally sustainable and worthy of support, and which are not.

While a knowledge of the physical environments of Latin America is useful for its own sake, perhaps its greatest value is to enable us to appreciate the context within which the region's cultural and economic development and underdevelopment have occurred. Our study of the physical geography of Latin America and the Caribbean is divided into three chapters. Chapter 2 presents an overview of the principal landforms and their mineral characteristics, chapter 3 explains weather patterns and their relationships to landuse and settlement, and chapter 4 describes the natural regions of Latin America and assesses their development potentials.

2

Landform Regions of Latin America and the Caribbean

The physical diversity of Latin America and the Caribbean is nowhere more evident than in the landforms that have served as the various stages upon which the human experiences have unfolded. The landform, or physiographic, regions can most easily be understood when seen as parts of three structural zones: the eastern highlands located entirely within South America, the central lowlands of South America, and the western alpine system that extends through South America, Central America, Mexico, and the Caribbean Islands (Figure 2.1).

THE EASTERN HIGHLANDS

Most of eastern South America consists of ancient igneous and metamorphic rocks. These appear at the earth's surface as three great upland regions: the Brazilian Highlands, the Guiana Highlands, and the Patagonian Plateau (Figure 2.2).

The **Brazilian Highlands** form the heart of the nation of Brazil and cover over one-third of its territory. They extend from near Uruguay and Paraguay in the far south nearly to the Amazon River in the north (Color plate 3). Although generally referred to as highlands, a more appropriate designation might be uplands, for these old, deeply eroded landscapes averaging, for the most part, 600 to 900 meters (2,000 to 3,000 feet) above sea level.

The Brazilian Highlands can best be understood when viewed as three distinct subregions. The first, and historically most significant, is the Great Escarpment, which rises abruptly from the narrow, often swampy, Atlantic coastal plain to elevations that exceed, in some areas, 2,700 meters (9,000 feet). The escarpment histori-

cally has functioned as a major barrier to human occupation of the interior regions to the west, and resulted in Brazil's initial settlements being concentrated along a narrow strip of coastal plains extending from Ceará southward to Rio de Janeiro. The escarpment appears, from much of the coastline, as a single mountain range that is known by a variety of local names, the most common of which is the **Serra do Mar.** Occasionally, as in Rio de Janeiro's famed Guanabara Bay, outliers of the escarpment extend into the sea as islands (Figure 2.3). At about 29° S latitude in southern Brazil, the escarpment turns west, forming a mountain range called the **Serra Geral** at the southern edge of the Paraná Plateau.

Some of the world's richest mineral deposits are found within the weathered crystalline rocks that dominate the escarpment and the interior hill country to the west. Settlers were first attracted to the Brazilian mining state of Minas Gerais by reports of gold and diamonds. While many of these have been exhausted, newer discoveries include world-class deposits of titanium, manganese, chromium, tungsten, molybdenum, and other industrial minerals. Some of the world's largest deposits of high-grade iron ore have been discovered both in Minas Gerais and in distant inland reaches of the Mato Grosso Plateau.

The second subregion of the Brazilian Highlands consists of a large interior zone where sedimentary rocks have been deposited over the old crystalline base. In the semiarid north, the sedimentary strata are composed of layers of limestone and sandstone into which the São Francisco River and its tributaries have cut deep valleys, forming a severely eroded backland known as the *sertão.* In the more humid southern portion of the interior sedimentary zone, the **Paraná Plateau** has been

FIGURE 2.1

Structural zones of South America. Copyright: Hammond World Atlas Corporation, NJ., Lic. No: 12450

FIGURE 2.2

Physiographic provinces of South America. Copyright: Hammond World Atlas Corporation, NJ.,
Lic. No: 12450

FIGURE 2.3

Rio de Janeiro's physical setting among coastal outliers of the Serra do Mar is one of the most striking of the world's major urban centers.

FIGURE 2.4

Rolling hills of the Paraná Plateau in Paraná state.

built through the deposition of layers of volcanic lava, some of which have weathered into the fertile *terra roxa* soils that have formed the agricultural heartland of Brazil, first through coffee production and more recently through soybeans, sugar cane, fruits, and vegetables. The Paraná Plateau has a softer relief than the *sertão*, with the Paraná River and its tributaries occupying broad valleys that drain fingerlike upland areas (Figure 2.4). Along the western margins of the plateau, the rivers cascade over the edges of the lava tableland forming spectacular waterfalls. The most famous, Iguaçu Falls (Iguazú in Spanish), consists of a 6.4-kilometer (4.0-mile)-wide arc of 275 cataracts situated near the junction of Brazil, Argentina, and Paraguay (Figure 2.5). Numerous other waterfalls have formed along both the middle Paraná River and the lower São Francisco River, where stream channels depart the interior uplands and enter the surrounding lowland plains. The falls have limited the navigational value of the waterways, which have been dammed at numerous sites for the development of hydroelectric power plants and irrigated agriculture.

The third, westernmost, section of the Brazilian Highlands consists of the vast **Mato Grosso Plateau** and associated features. Most of the plateau is covered by sedimentary strata that in many areas have been eroded to expose the older, underlying crystalline rock. The result is a mature, deeply weathered landscape of rolling hills interspersed with higher, more resistant structures known locally as either *serras* (mountain ranges) or *chapadas* (plateaus).

The uplands of the Mato Grosso Plateau are drained by many of the principal tributaries of the Amazon. These streams, major rivers in their own right, flow in a mostly northerly direction out of the interior highlands and down onto the Amazon floodplain. They include, from east to west, the Tocantins and its

FIGURE 2.5

A portion of Iguaçu Falls, which have formed as the Paraná River flows over the western edge of the Paraná Plateau.

principal tributary, the Araguaia; the Xingú; the Tapajós; and the Madeira.

Brazilians have long viewed the Mato Grosso uplands as a frontier region of virtually unlimited economic potential. The founding in 1960 of Brasília, a new national capital, on the eastern margins of the Mato Grosso Plateau symbolized the determination of the Brazilian people to occupy and develop their western territories (Figure 2.6).

While agriculture has sustained in a modest way the economy of the ever-advancing western frontier, the activity of greatest commercial value to the Brazilian Highlands as a whole has historically proven to be mining. The opening of the Mato Grosso Plateau to mining in recent decades has often followed remarkably closely the sequence of events that began in the eastern escarpment centuries before. Settlers are initially attracted to new areas by reports of gold and diamond deposits, deposits that soon play out. These are followed by discoveries of enormous reserves of basic industrial ores, especially iron ore, aluminum ore or bauxite, tin ore or cassiterite, nickel, and manganese. Minerals vital to more technologically advanced industries, including titanium, zirconium, beryllium, chromium, and tungsten, are subsequently developed. The diverse and seemingly

inexhaustible mineral base of the Brazilian Highlands, within both the eastern escarpment and the Mato Grosso Plateau, has enabled Brazil to become a major exporter of industrial ores and, just as significantly, to develop a growing domestic manufacturing base.

The second major division of South America's eastern highlands is the **Guiana Highlands.** In a physiographic sense, the Guiana Highlands are simply a large northern outlier of the Brazilian Highlands, separated by a great depression through which the Amazon River flows. Like the Brazilian Highlands to the south, the Guiana Highlands consist of an old metamorphic and igneous core over which, in many areas, have been deposited younger sedimentary strata. Resistant sandstone caps are common in the central and southernmost reaches of the highlands, where they have given a distinctive flat-topped, plateaulike appearance to many of the mountains, known locally as *tepuis* (Figure 2.7). A few peaks, such as Mt. Roraima at the junction of Brazil, Venezuela, and Guyana, exceed 2,700 meters (9,000 feet) in elevation, but most are considerably lower. As with the São Francisco and Paraná rivers in Brazil, the interior streams drop off the harder elevated formations onto the bordering lowlands, forming numerous waterfalls. The highest waterfall in the world is Venezuela's Angel Falls, which descends 979 meters (3,212 feet)

FIGURE 2.6

The Mato Grosso Plateau of central Brazil with the capital city of Brasília in the foreground.

FIGURE 2.7

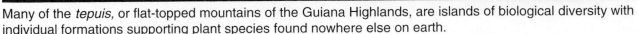

Many of the *tepuis,* or flat-topped mountains of the Guiana Highlands, are islands of biological diversity with individual formations supporting plant species found nowhere else on earth.

after flowing over the edge of a quartzite-capped *tepuy* (Figure 2.8). With the exception of dams in Venezuela, the hydroelectric potential of the Guiana Highlands remains largely untapped.

Many parts of the central and southern Guiana Highlands are so inaccessible that they have yet to be intensively surveyed, much less developed in an economic sense. Northernmost Brazil and southernmost Venezuela, Guyana, Suriname, and French Guiana constitute, in this regard, one of the earth's last remaining natural and cultural preserves. The same cannot be said, however, of the northern Guiana Highlands which, like the Brazilian Highlands, are proving to be one of the richest mining regions of the world. The development pattern once again tends to be scattered fields of gold and diamonds attracting some initial settlement, only to be overshadowed by the subsequent exploitation of massive iron ore and bauxite deposits. Commercial agriculture in Guyana, Suriname, and French Guiana is found almost exclusively along the northern coastal plain. Here fertile alluvial soils have been built up through the flooding of the Essequibo, Courantyne, and Maroni rivers as well as through the deposits of Amazonian sediments that are carried by the prevailing ocean currents from the mouth of the river northwestward along the northern continental shelf.

The southernmost of the eastern highland regions of South America is the semiarid **Patagonian Plateau,** which extends from the Río Negro, situated at the southern edge of the pampa, southward some 1,600 kilometers (1,000 miles) to eastern Tierra del Fuego, the "land of fire," which protrudes into the icy waters of Antarctica. Patagonia consists of a series of sedimentary plateaus that have formed over older granitic shield materials. The plateaus rise abruptly above the Atlantic coast and progress westward in steplike fashion, reaching elevations of over 1,500 meters (5,000 feet) along their western margins where they are separated from the Andes by a north-south-running trough known as the Pre-Andean Depression (Figure 2.9).

A half-dozen rivers, which are fed by Andean snowmelt and remnant glacial lakes, have cut deep valleys, called *bajos,* into the sandstone and lava strata of the plateaus on their way to the Atlantic. These sheltered, well-watered valleys afford a degree of protection from the never-ending winds and sustain the limited agricultural development of the region. What few people the area supports are found primarily in small port cities, most of which are situated where the rivers meet the sea. The largest of these, Comodoro Rivadavia, has also benefitted from its proximity to modest

FIGURE 2.8

Angel Falls are the world's highest. The volume of flow fluctuates greatly between the wet and dry seasons.

FIGURE 2.9

The Patagonian Plateau consists of a series of smaller plateaus that rise one above another westward from the southern Atlantic Ocean toward the Andes Mountains. Visible in this scene is the western edge of the plateau with the Andes rising in the distance.

petroleum and natural gas deposits. Other small fossil fuel reserves have been developed in Neuquén and Río Negro provinces in northern Patagonia and in Tierra del Fuego, but the region is devoid of industrial ores.

THE CENTRAL LOWLANDS

The second great structural zone of South America consists of a series of interior plains that extend the length of the continent, from Venezuela on the north to central Argentina on the south. This expanse, which stretches across some fifty-two degrees of latitude, is known collectively as the central lowlands and comprises four distinct physiographic provinces: the Llanos, the Amazon Plain, the Gran Chaco, and the Pampa (Color plate 2).

The northernmost of these, the **Llanos** or Orinoco River Plains of present-day Venezuela and Colombia, occupies an immense geologic trough or depression that existed formerly as an arm of the sea. The trough gradually filled, first with thousands of meters of marine sediments, and later with alluvial debris carried down by rivers and streams originating in the adjoining highlands (Brunnschweiler 1972). Because the rate of erosion has been far greater in the younger, higher, and more geologically active Andes than in the older, more resistant Guiana Highlands, the Llanos reach to their highest elevations along the foothills of the Andes where braided river valleys are separated by intervening alluvial terraces of moderate relief. The land then descends ever so gradually toward the base of the Guiana Highlands where the Orinoco River follows a 2,600-kilometer (1,600-mile) course to the sea.

The Llanos are one of the most monotonous and featureless regions in all of Latin America. The land is so level that the average rate of descent over the easternmost 1,200 kilometers (750 miles) is only 10.2 centimeters per 1.61 kilometers (four inches per mile) (Crist and Leahy 1969). This is not to suggest that the Llanos are completely without relief, for smaller feeder streams have left natural levees or ridges scattered across the countryside. The difference in elevation between these ridges and the remaining land is extremely minute, however, often only a meter or less, and does little to alter the pervasive flatness of the region (Figure 2.10).

The principal mineral resource of the Llanos is petroleum, which was trapped anciently in weak subterranean folds and faults within the underlying marine sediments. Reserves have proven, to date, to be far more extensive in the lower, eastern Llanos of Venezuela than in the western Colombian sector.

The usefulness of the Orinoco for navigation and, by extension, for promoting the economic development of the region generally, has been limited by the shallow depth of the lower channel and by rapids and waterfalls further upstream. Dredging in the 1950s opened the river to ships of twelve-meter (forty-foot) draft as far in-

FIGURE 2.10

The principal landuse of the Llanos is cattle ranching.

land as Ciudad Guayana, and smaller vessels regularly reach the Maipures cataract, some 1,500 kilometers (900 miles) from the sea. Except for the Ciudad Guayana–Ciudad Bolívar area, which is examined in chapter 11, the Llanos remain one of the most remote and least developed regions of Latin America.

As the Llanos bend southward in eastern Colombia, the drainage of the Andes and Guiana Highlands ultimately shifts into the **Amazon Plain.** This transition is so gradual in places that one stream, the Casiquiare River in southernmost Venezuela, actually divides and feeds tributaries of both the Orinoco to the north and the Amazon to the south.

The mere mention of the name Amazon evokes images of a region so vast in size and development potential as to almost defy description. Over 1,100 tributaries drain an area holding a third of the world's remaining tropical rain forests. Approximately 20 percent of the freshwater discharge of the entire earth (more than the combined discharge volume of the world's next eight largest rivers) flows through the lower channel at a rate of 175,000 cubic meters per second (McKnight and Hess 2002). From Manaus downstream, the river varies in width from 3.1 to 14.2 kilometers (1.9 to 8.8 miles) at flood stage and its mouth is so large that it contains an island, the Ilha de Marajó, which is as large as Switzerland (Figure 2.11).

The headwaters of the Amazon extend almost 6,500 kilometers (4,000 miles) inland to the heart of the Peruvian Andes. Here, close to the Pacific Ocean, the upper Marañón River flows northward through a spectacular valley for several hundred kilometers before entering the western Amazonian plains through a deep gorge known as the Pongo de Manseriche. The river then assumes a generally eastward course, receiving at frequent intervals various tributaries that dissect the eastern foothills of the Andes into large protruding uplands.

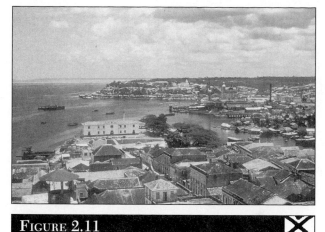

FIGURE 2.11

The Rio Negro at Manaus, Brazil.

FIGURE 2.12

Homes on and at the base of the bluff of the Amazon River at Iquitos, Peru. The lower row of structures consists of houseboats, which will float upon the river in seasons of high water.

At Iquitos, 3,200 kilometers (2,000 miles) upriver in eastern Peru, the Amazon, or Marañón as it is still called, is deep enough to accommodate vessels of four-meter (fourteen-foot) draft. Once the river enters Brazilian territory, its name changes to the Rio Solimões and the channel gradually deepens. By the time it reaches the great inland port of Manaus, 1,600 kilometers (1,000 miles) from the Atlantic, transoceanic passenger liners and large cargo ships can be accommodated. From Manaus eastward the name of the river changes to Amazonas, and its depth and volume continue to increase as it is joined by the huge tributaries draining the Brazilian and Guiana Highlands.

For much of its middle and lower course, the river flows between relatively steep-sided bluffs that stand some three to nine meters (ten to thirty feet) above the mean water level. Amazonian towns and villages have traditionally been built on these bluffs for flood protection (Denevan 1996; Figure 2.12). These communities, which typically are situated slightly downstream from the junctions of the major tributary streams with the Amazon, historically have dominated river trade in the basin and have become the major commercial and urban centers of the region (Figure 2.13). These include Belém, which is situated at the mouth of the Amazon downstream from the Tocantins junction; Santarém at the Tapajós-Amazon confluence; and Itacoatiara and Manaus near the mouths of the Madeira and Negro, respectively.

The dominant color of the water of the tributaries is a function of the composition of the surface formations being drained and the levels of dissolved organic matter and mineral sediments. It ranges from the yellow of the Madeira to the greenish hues of the Tapajós and the black of the Negro. The waters of the Xingú and Tocantins are relatively clear.

Although the channel of the Amazon is navigable to Iquitos, the tributary streams are blocked by waterfalls that have formed where the rivers descend off the Brazil-

FIGURE 2.13

Transportation within the Amazon Basin has centered historically on waterways rather than on land-based highways. These farmers are transporting produce to a market town between Leticia and Manaus.

ian and Guiana Highlands and onto the softer floodplain deposits below. They therefore offer, like many of the major inland waterways of the continent, considerable potential for hydroelectric energy development, but they do little to facilitate modern transportation within the basin. Efforts to link Amazonia with the more developed areas of eastern Brazil have focused primarily on the Trans-Amazonian Highway network (chapter 11) but also include dredging the Araguaia and the Tocantins rivers to connect the central plateau with Belém.

The southern limit of the Amazon drainage basin is found in northern Bolivia where the Beni and Mamoré rivers drain a large lowland region called the **Llanos de Mojos** before joining the Madeira near the westernmost reaches of the Mato Grosso Plateau. East of the Llanos de Mojos lies the **Chiquitos Plateau,** a moderately elevated region of low, undulating sedimentary hills situated virtually at the geographical center of South America (Figure 2.2). The Chiquitos Plateau forms the north-south drainage divide of the continent.

South of the plateau lies the third of South America's great interior lowland regions—the **Gran Chaco** (Color plate 1). The physiographic evolution of the Chaco is similar in many ways to its northern counterpart, the Llanos of Venezuela and Colombia. Like the Llanos, the Chaco is believed to have formed originally as a shallow Atlantic inlet between the Brazilian Highlands to the east and the younger, more tectonically active Andean chains to the west. Because the Andean ranges were higher and more easily eroded, the ensuing alluvial deposition was greater on the western margins of the basin, with the result that the Chaco slopes steadily down from the Andean foothills to the base of the Paraná Plateau. There the Paraguay River drains southward, eventually joining the Paraná River coming down from the eastern highlands to form the Río de la Plata.

The Chaco is one of the most physically forbidding regions of all Latin America (Jones 1977). Only three permanent streams, the Pilcomayo, the Bermejo, and the Salado, survive the loss of water to the porous surface to discharge their sediments into the Paraguay River. It stretches the truth somewhat to describe even these three streams as permanent, given their pattern of shifting channels in response to the filling of former courses with debris brought down by the seasonal floodwaters.

The Paraguay River exhibits similar limitations. Subject to tremendous seasonal fluctuations in volume, it consists in its upper course of a maze of twisting, winding channels draining a great inland swamp known as the **Gran Pantanal.** Here the streams are so serpentine that the actual travel distance from one point to another is often twice or three times the official estimate. Furthermore, one never knows from one year to the next whether a given channel will lead to one's destination or will simply end in an uncharted marsh. Even agriculture is affected by the unpredictable rivers. In a region too sparsely settled to justify the cost of permanent irrigation systems, farmers often sow their crops in shallow depressions called *bañados,* whose locations shift from year to year.

The Chaco continues into northern Argentina where the Río Dulce discharges onto a marshy, saline plain north of Córdoba and Santa Fé. South of this line the poverty of the Gran Chaco gives way to one of Latin America's most productive and prosperous agricultural regions, the Argentine Pampa.

The **Pampa** is a piedmont alluvial plain on which thick layers of fine-grained, wind-borne soil particles, called loess, have been deposited. These have been enriched over time by the addition of volcanic ash from Andean peaks to the west and of large quantities of or-

FIGURE 2.14

Recently harvested land in the southeastern Pampa. Visible in the background are hills belonging to the Sierra de la Ventana.

ganic matter from the native grasses, resulting in the formation of prime prairie soils. The soils are deepest in the eastern portions of the Pampa, which are characterized by numerous marshes that have formed as the east-flowing surface streams empty into low depressions called *esteros* or *cañados.* The southeastern Pampa also contains two small, crystalline mountain ranges known as the Sierra de la Ventana and the Sierra del Tandil.

In its physical endowment, then, the Pampa is a region of unsurpassed agricultural potential. Imagine a flat, almost never-ending plain roughly the size of Texas blessed with adequate rainfall, moderate temperatures, and rich topsoil almost 300 meters (1,000 feet) deep in places (Figure 2.14). That is the Pampa, the heart of Argentina, whose agricultural abundance has sustained the growth of one of Latin America's greatest cities, Buenos Aires, at the head of the Río de la Plata.

THE WESTERN ALPINE SYSTEM

The third major structural zone of Latin America is the western alpine system, which consists of the South American Andean ranges and the highlands of Central America, Mexico, and the Caribbean Islands. Like the Rockies, its North American counterpart, Latin America's western alpine system is part of the circum-Pacific **Ring of Fire,** a series of geologically young and extremely active mountain ranges whose ongoing development is a by-product of the collision and subduction of segments of the earth's crust.

These segments, called **plates,** are shifting laterally in response to seafloor spreading, which is driven by the circulation of huge convective cells of molten rock, or magma, deep within the earth (Figure 2.15).

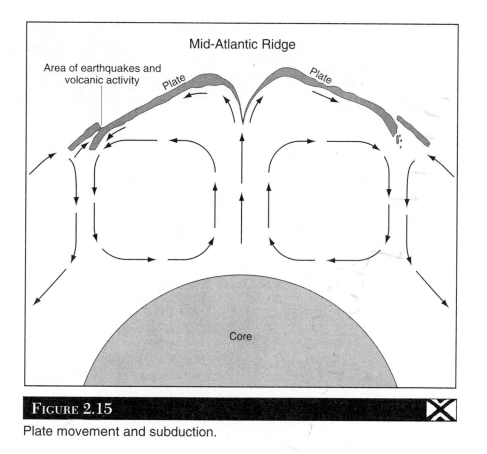

Mid-Atlantic Ridge

Area of earthquakes and volcanic activity

Plate

Plate

Core

FIGURE 2.15

Plate movement and subduction.

As the top of one plate is forced beneath the bottom of another, tension between the plates builds up until the plates suddenly lurch or slip into new positions. If the movement is great enough, it will be felt as an earthquake at the earth's surface. Numerous aftershocks are common as the crustal material above the movement area readjusts its position through settling. Meanwhile, new mountain ranges are likely to form along the contact zone between the plates as the crustal materials are compressed, folded, and faulted. Pent-up magma may also be released, often with great violence, forming new volcanoes or increasing the size of existing ones.

The geological activity of Latin America's western alpine system results from the interactions of seven tectonic plates (Figure 2.16). Most of North and South America belong to the North and South American Plates, both of which are moving westward in response to seafloor spreading along the Mid-Atlantic Ridge. As the North American Plate spreads westward, it pushes up in the area of the Gulf of California against the eastern margin of the Pacific Plate, forming the mountains of northwestern Mexico (Wallace 1992). Meanwhile, the Caribbean Plate, which is drifting slowly toward the southwest, is itself being overtaken along its eastern margin by the more rapidly moving South and North American Plates, resulting in frequent but mostly minor

volcanic eruptions and earthquakes along the inner arc of the Lesser Antilles (Jansma 2000; Shephard 1989; Pascal and Lambert 1988). Simultaneously, the small Cocos Plate is advancing rapidly toward the northeast, which causes it to override the southern edge of the North American Plate and the western side of the Caribbean Plate. As the Caribbean Plate is forced beneath the Cocos Plate, a zone of intense geologic instability manifests itself in western Central America and southern Mexico, where catastrophic earthquakes occur on an average of once every ten to twelve years (Suter et al. 2001; Coates 1997; Sealey 1992, 11). The Andes were formed in response to the convergence of the westward-moving South American Plate and the eastward-shifting Nazca and Antarctic Plates, which occupy most of the southeastern Pacific Basin (Taboada 2000; Spikings et al. 2000; Gutscher 2000).

Recent Earthquake and Volcanic Activity

One of the most active areas in recent times for volcanic activity and earthquakes has been western Central America, especially Nicaragua, El Salvador, and Guatemala (Macias et al. 2000; McCann and Pennington 1990). Guatemala has thirty-two volcanoes nestled within its western highlands and Nicaragua has twenty-eight along a 290-kilometer (180-mile) fracture zone

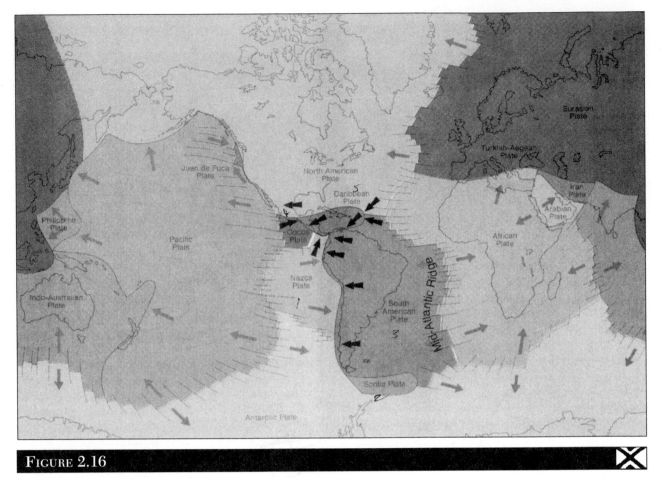

FIGURE 2.16

Tectonic plates of Latin America.

(Incer 1987). El Salvador experienced a devastating series of earthquakes in January and February of 2001 (Case Study 2.1). Other especially unstable regions include the mountainous areas of central and southern Mexico, the inner arc islands of the Lesser Antilles, the northern Andes of Colombia and Ecuador, and the southern Andes from southern Peru southward (Table 2.1) (Figure 2.17).

The short-term damage inflicted by the earthquakes and volcanoes extends far beyond the official mortality figures (Table 2.2). The earthquake that struck Managua, Nicaragua's capital city, on December 23, 1972, for example, is estimated to have killed 10,000 persons. In addition to that, however, more than 20,000 people were injured and over half the population of 400,000 was left homeless. The death and injury figures are often greatest if the earthquake hits at night when most of the people are sleeping inside their homes. The likelihood of roofs collapsing on people in these situations is such a concern that building codes in some areas require that sheet metal or other lightweight material be used for roofing rather than the traditional heavy tiles.

While the immediate effects of these natural disasters are catastrophic, it should also be pointed out that once the volcanic lava and ash have weathered, they form some of the finest soils on earth. The moderate temperatures afforded by the highlands are another long-term benefit of the mountain-building process. For these and other reasons, the western alpine system has historically supported some of the densest populations in Latin America. Those who reside within sight of the beautiful volcanoes know that their lives are both sustained by their fertile soils and threatened by the eruptions that can occur almost without warning (Reading, Thompson, and Millington 1995; Dorbath, Cisternas, and Dorbath 1990; Oliver-Smith 1986).

Physiographic Regions of the Andes

Although the physiography of the Andes is extremely complex, the collision of the South American and Nazca plates has given a common composition and structure to the entire chain. The core of the system consists of igneous rocks that, in some areas, have been overlaid by sedimentary strata and that, in other regions, have

Case Study 2.1: The El Salvador Earthquakes

On Saturday, January 13, 2001, at approximately 11:35 in the morning, a powerful earthquake, whose epicenter was situated some 105 kilometers (sixty-five miles) off the Pacific coast, struck the small nation of El Salvador. Although earthquakes are extremely common to Central America, the strength of this quake, which measured 7.6 on the Richter Scale, combined with its estimated depth of sixty kilometers (thirty-seven miles), resulted in one of the most devastating natural disasters to strike the country in modern history. Perhaps the most dramatic consequence of the quake was the triggering of a massive landslide that roared down a largely denuded hillside in the Las Colinas neighborhood of Santa Tecla, a western suburb of the capital city of San Salvador. The displaced earth buried some 800 to 900 middle-class homes, killing an estimated 300 persons. Although most of San Salvador itself was spared severe damage, hundreds of smaller towns and villages in outlying areas became virtually uninhabitable, with 50 to 90 percent of their buildings being leveled or rendered structurally unsound. Extensive damage also occurred to numerous sections of rural roads and highways, and great numbers of highly productive coffee farms were lost, with some of them being swallowed up entirely by gaping cracks and holes that formed in the earth.

Despite their grief, the Salvadoran people began at once to rebuild. A National Emergency Committee (COEN) was formed by President Francisco Flores and, together with the Salvadoran Armed Forces, coordinated relief and rehabilitation efforts. Offers of assistance poured in from sympathetic nations and international agencies, and within days many temporary shelters had been erected and transportation and utility services restored. Incredibly, however, a second major quake, measuring 6.6 on the Richter Scale, struck on February 13, with an epicenter just thirty kilometers (nineteen miles) east of San Salvador.

Still reeling from the effects of the first disaster, the country's suffering was seemingly compounded many times over by the second quake. In addition to immense physical damage, the densely populated nation was soon confronted with widespread outbreaks of gastrointestinal and respiratory illnesses associated with contaminated water supplies and prolonged exposure to outdoor elements. As serious as the physical illnesses were, they were matched or exceeded by unprecedented levels of mental stress. Great numbers of persons, traumatized by the thousands of aftershocks, resorted to sleeping in streets and other outdoor areas for fear of their homes collapsing upon them during the night. Countless others, exhausted by the ongoing ordeal, developed nervous ticks, diarrhea, skin rashes, psychosomatic pains, and other symptoms of emotional disorders. Other costs of the quakes included the interruption of classes at universities and schools and the closing of many businesses for extended periods of time.

Because so much of the destruction occurred in remote, isolated rural areas, the enormity of the damage did not become fully known until months after the quakes struck. By late 2001, however, final casualty estimates indicated that between 1,100 and 1,200 persons had died, many of them buried alive, with another 8,000 injured. Approximately 150,000 homes were lost and an additional 185,000 damaged. Some 130 rural communities were totally destroyed. In sum, approximately 1.6 million persons (one-quarter of the entire population) were affected, and it was believed that reconstruction costs would eventually rise to $2.8 billion, a figure equal to approximately half the nation's annual economic output.

In the aftermath of the quakes, considerable thought has been given to what can and should be done to prepare for future seismic events that are almost sure to come. Recommendations that have been offered include more effective controls on the cutting of forest vegetation in areas underlain by fault lines and more comprehensive urban planning designed to minimize human settlement in the most vulnerable regions. Building codes and disaster mitigation and emergency response plans can also be strengthened. To these ends, the Salvadoran government announced in late 2001 the creation of a new institute, to be called the *Sistema Nacional de Estudios Territoriales (SNET)* (National System of Territorial Studies). Patterned after a highly effective Nicaraguan counterpart agency, called the Territorial Studies Institute (INETER), SNET hopes to provide the geographical research needed for future Salvadoran planning efforts. While earthquakes cannot be prevented, it is to be hoped that these and other measures will help to minimize the enormous harm and suffering that inevitably result.

Sources: United Nations Office for the Coordination of Humanitarian Affairs (OCHA) Relief Web: www.reliefweb.int/w/rwb.nsf; Programa de las Naciones Unidas para el Desarrollo: www.terremotoelsalvador.org.sv/; *Martinez 2001.*

TABLE 2.1

Selected Recent Volcanic Eruptions in Latin America

REGION/COUNTRY	NAME OF VOLCANO	HEIGHT OF VOLCANO IN METERS	DATE OF MOST RECENT MAJOR ERUPTION
South America			
Chile	Lascar	5,990	1991
	Guallatiri	6,060	1987
Ecuador	Sangay	5,230	1988
	Cotopaxi	5,897	1975
Colombia	Purace	4,756	1977
	Ruíz	5,400	1989
	Galeras	4,266	1993
Central America			
Costa Rica	Arenal	1,657	1993
	Irazú	3,432	1992
	Poás	2,722	1992
Nicaragua	Masaya	635	1993
	El Viejo	1,780	1987
	Concepción	1,556	1986
	Las Pilas	1,071	1954
El Salvador	Izalco	2,362	1966
	San Miguel	2,132	1986
Guatemala	Santa María	3,768	1989
	Fuego	3,958	1991
	Tacana	4,090	1988
Mexico			
	Popocatépetl	5,503	2001
	Colima	4,265	1991
	El Chichón	2,225	1983
	Paricutín	2,808	1943
Caribbean			
Montserrat	Soufriere Hills	942	2001
St. Vincent	Soufrière	1,219	1979
Martinique	Mt. Pelée	1,398	1902

Sources: Smithsonian National Museum of Natural History Global Volcanism Program: www.volcano.si.edu/grp/; West 1964, 75; Sealey 1992, 20.

FIGURE 2.17

Earthquake-damaged building near Mexico City's central plaza.

penetrated older, preexisting sedimentary materials. These igneous and sedimentary materials have subsequently been uplifted and folded into great, parallel, north-south-trending ranges and valleys and remnant plateaus (Gregory-Wodzicki 2000; Bock et al. 2000) (Figure 2.18). The mountainous landscape has been modified further through widespread vulcanism and alpine glaciation. The result is one of the most physically impressive and geologically active regions in all the world.

The Andes begin as relatively low mountains in western Tierra del Fuego and the bordering Pacific islands of southernmost Chile (Figure 2.19). It was here, in one of the stormiest places on earth, that Ferdinand Magellan, a Portuguese mariner sailing for Spain, finally

TABLE 2.2

Recent Catastrophic Earthquakes in Latin America

DATE OF EARTHQUAKE	LOCATION	NUMBER OF DEATHS	MAGNITUDE (RICHTER SCALE)
2001	El Salvador	1,150	7.6
2001	Peru	102	8.1
1999	Colombia	938	6.0
1997	Venezuela	82	6.9
1994	Colombia	1,000	6.8
1991	Costa Rica–Panama	11	7.4
1990	Peru	115	6.3
1988	Guatemala	unknown	6.2
1987	Colombia–Ecuador	+4,000	7.3
1985	Mexico	9,500	8.1
1985	Chile	146	7.8
1983	Colombia	250	5.5
1979	Colombia–Ecuador	800	7.9
1976	Guatemala	22,778	7.5
1972	Nicaragua	10,000	6.2
1970	Peru	66,794	7.7
1960	Chile	5,000	8.3

Sources: *The World Almanac and Book of Facts 2002*, 2002, 188; Incer 1987, 19–25; Sealey 1992, 14.

FIGURE 2.18

This sharply ridged, snowcapped peak in southern Peru is expressive of the intense geological folding, uplift, and glaciation that have been associated with the formation of many of the Andean ranges.

FIGURE 2.19

The Andes begin in Tierra del Fuego, in the southernmost portion of South America, as relatively low snow-covered mountains that rise up out of the cold Antarctic waters.

found a sea route in 1520 through the strait that came to bear his name. Even in these southernmost reaches of the continent, the volcanic peaks rise above 2,400 meters (8,000 feet), as if to prepare the traveler for the more massive landforms to come.

As one moves northward, the height of the mountains continues to increase. Elevations of 3,400 to 4,000 meters (11,000 to 13,000 feet) are common throughout southern Chile. The countryside in this region is similar in some respects to that of coastal Norway and the Pacific coast of Canada and Alaska, with innumerable rocky, coastal islands interspersed among magnificent fjords, or drowned glacial valleys, which conduct the cold Pacific waters far inland along transverse fault lines (Figure 2.20).

The folding of the Andes becomes evident at around 43° S latitude where a great interior valley appears to the west of the principal Andean ranges (Color plate I). The valley, which continues northward to the Peruvian border, is separated from the Pacific by low plateaus whose

FIGURE 2.20

The southern Chilean coastline is characterized by numerous spectacular fjords, or drowned glacial valleys.

remnant hills form terraced coastal ranges that descend to steeply cliffed coastlines. The southernmost portion of the interior valley is dotted with scenic glacial lakes and is called the Chilean Lake District (Figure 2.21).

The great alluvium and ash-filled middle **Central Valley** is the heart of Chile. Irrigated commercial agricultural crops, many of which are destined for Anglo-American winter markets, dominate the highly urbanized region. Some of the world's largest nitrate deposits are worked in a series of arid basins that have formed in the northern Atacama portion of the interior valley, between the coastal ranges and huge alluvial fans that have built up along the western base of the Andes. The central and northern Chilean Andes also contain large copper reserves that have contributed significantly to the economic development of the country.

The Andes themselves increase rapidly in height in central Chile and neighboring Argentina. Mt. Aconcagua, at 7,021 meters (23,035 feet) the highest peak in the Western hemisphere, guards the 4.0-kilometer (2.5-mile-high) Uspallata Pass, which links Chile's principal port of Valparaíso on the Pacific coast with Mendoza, Argentina and the distant Pampas beyond.

FIGURE 2.21

Lake Villarica is situated on the northern end of the Chilean Lake District, which consists of a series of beautiful lakes nestled against the Andes in the southernmost section of the Central Valley.

FIGURE 2.22

The Andean region of northern Argentina, Bolivia, and Peru is characterized by high intermontane plateaus called *punas,* many of which contain small lakes and salt flats.

From central Chile northward through Bolivia, Peru, and Ecuador, the Andes contain many of the highest human-occupied regions on earth. As the width of the Andes increases in northern Chile and Argentina, high interior plateaus, called *punas,* appear between the volcanic peaks. Many of the *punas* are characterized by interior drainage where seasonal meltwaters from snow and ice collect and evaporate in the bottomlands to form salt flats called *salares* (Figure 2.22).

Once the mountains enter southern Bolivia, they divide into two distinct ranges, the Cordillera Real or Oriental on the east and the Cordillera Occidental on the west. Between them lies the **Altiplano,** the largest of the great Andean intermontane plateaus. The northern Altiplano is dominated by Lake Titicaca, South America's largest freshwater lake and, at 3,812 meters (12,505 feet) above sea level, the world's highest navigable freshwater lake (Figure 2.23). The lake, which is shared by Peru and Bolivia, was considered by the ancient Inca to be the navel, or center, of the universe. It continues today to sustain traditional Aymara fishermen and agriculturists who market their surplus in the nearby urban centers.

FIGURE 2.23

Lake Titicaca is the heart of the Bolivian and Peruvian Altiplano.

The greatest of these is La Paz, the seat of government of Bolivia, which is situated in a deep chasm at the western base of the towering, snowcapped Cordillera Real.

Lake Titicaca drains southeastward through the Desaguadero River, which loses much of its water to the coarse alluvial and glacial debris that forms the surface of the Altiplano. What water remains is eventually discharged into a broad, shallow depression forming Lake Poopó, a Pleistocene, or Ice Age, remnant whose depth rarely surpasses 2.5–3.0 meters (eight to ten feet). Because the volume of water entering Lake Poopó seldom exceeds the amount lost to infiltration and evaporation, the lake has no steady outlet nor means of cleansing itself and has become saline. Occasionally, flood waters from Lake Titicaca will cause Lake Poopó to spill out onto a neighboring salt flat known as the Salar de Coipasa which in turn is bordered on its south by the vast Salar de Uyuni.

Although the population of Bolivia has concentrated historically on the relatively temperate northern Altiplano, much of the nation's wealth has come from silver and tin mines situated at very high elevations on the windswept, treeless slopes of the Cordillera Oriental. These cold, desolate reaches contrast markedly with the fertile, mid-elevation agricultural valleys of Cochabamba, Sucre, and Tarija and with the luxuriant tropical vegetation of the lower, eastern flanks of the Cordillera. The northern portion of the Cordillera is drained by numerous tributaries of the Rio Madeira. This area, known as the Bolivian *Yungas,* is among the least accessible in all of Latin America and historically has functioned as one of the principal coca-producing regions of the continent.

In southern Peru, the Cordillera Real eventually curves to the northwest and joins with the Cordillera Occidental in a jumbled mass of mountains known as the Vilcanota Knot (Bowman 1916). From there, the mineral-rich ranges divide again as they continue northwestward through Peru paralleling the Pacific coastline. The western range, the Cordillera Occidental, feeds a number of short streams that sustain life on the dry, narrow coastal plains and hills that border the Pacific (Figure 2.24). East of the Cordillera Occidental lies a *puna*-like region of interior basins and valleys, beyond which appears the eastern *sierra,* or Cordillera Oriental (Figure 2.25). The latter has been dissected into incredibly deep, rock-walled canyons by the Marañón, or upper Amazon River, and its tributaries. These include the Huallaga and the Ucayali, with the latter being fed in turn by the Apurimac and the Urubamba (Figure 2.26). The severely eroded eastern Andean slopes of Peru are referred to as the *montaña* and consist of densely forested foothills and plains that drain eastward into the vast Amazon Basin.

As the Andes continue northward into Ecuador, the highlands narrow considerably and are soon reduced to two parallel volcanic ranges that are separated only by a narrow trough, or intermontane valley, called the **Calle-**

FIGURE 2.24

Irrigated farmland on the arid coastal plain of southern Peru.

FIGURE 2.25

The high snowcapped Andes near Cuzco, Peru.

jón Andino. Within the Callejón are ten fertile basins lying between 2,100 and 3,000 meters (7,000 and 10,000 feet) above sea level (Figure 2.27). The ranges are once again known as the western or Occidental and eastern or Oriental cordilleras and include Cotopaxi and Chimborazo, both of which exceed 5,800 meters (19,000 feet) in height. Cotopaxi is one of Latin America's most geologically active volcanoes.

The cool, temperate Ecuadorian basins have long supported dense human populations, the largest of which is centered in Quito, the nation's capital. The agricultural productivity of the basins is attributable in large measure to the rich ash and alluvial soils that have formed from volcanic eruptions and from sediments brought down by the mountain streams. These same streams have cut deep gullies into the porous soils before breaking through the basin rims and building massive alluvial fans along the western flanks of the

The Urubamba River, shown here in southern Peru, is one of several streams that flow through deep valleys within the Cordillera Oriental.

The Basin of Riobamba is one of ten large valleys forming Ecuador's Callejón Andino. Visible in the background is Mt. Chimborazo.

Cordillera Occidental. The land then levels out in central Ecuador onto the broad Guayas River floodplain, which drains southward to Guayaquil, being separated from the Pacific by the remnants of a low coastal plateau (Figure 2.28). The eastern Andean slopes and adjacent lowlands, known collectively as the *Oriente,* drain into the Amazon Basin and are the source of most of Ecuador's petroleum reserves.

In southern Colombia, the Andes gather into a second knot, or core, in the area of Pasto, and then emerge as three distinct, parallel ranges, the Western, Central, and Eastern Cordilleras (Abele 1992; Parsons 1982) (Color plate 2) (Figure 2.29). The Andes and the associated Caribbean and Pacific coastal lowlands form the economic heart of Colombia. The distant easternmost two-thirds of the nation consists of the as yet only marginally integrated Llanos and Amazonian floodplains.

Two great north-south-trending valleys separate the cordilleras. The westernmost valley is occupied primarily by the Cauca River, which emerges in a volcanic ash plateau near Popayán on the northern flank of the Pasto Knot (Crist 1952). It then meanders northward across a fertile, relatively broad floor that has been raised to 900 meters (3,000 feet) above sea level through the accumulation of alluvial materials from the severely eroded highlands. The Cauca Valley as a physiographic region ends at Cartago, where the river enters a maturely dissected, crystalline upland known as the Antioquia Plateau. The low, rounded hills of Antioquia eventually give way to the Caribbean lowlands, which appear at the northern limits of the Western and Central Cordilleras.

The larger of Colombia's valleys is occupied by the Magdalena River, which flows northward, parallel to the Cauca, between the Central and Eastern Cordilleras. Although the headwaters of the stream reach far into

FIGURE 2.28

Ecuador's Guayas River Basin experiences widespread flooding during the annual rainy season.

FIGURE 2.29

Much of the rugged Cordillera Central of Colombia is sparsely settled with tiny subsistence farmsteads scattered among the towering mountain slopes and intervening valleys.

southern Colombia, the greatest historical significance of the system has been to provide an outlet for the products of Bogotá and other cities of the Eastern Cordillera that are linked to the river by rail and highway. Downstream from the rapids at Honda, the Magdalena Valley broadens steadily before the waterway turns westward and, joined by the César, Cauca, and San Jorge rivers, flows through a marshy alluvial plain that is subject to widespread seasonal inundations. The entire system eventually discharges into the Caribbean in the vicinity of Cartagena and Barranquilla through a network of ill-defined distributaries.

The Eastern Cordillera, lying east of the Magdalena Valley, merits special mention owing to its prominence within Colombia and to its relationship to the Venezuelan Andes. Human settlement within the Eastern Cordillera has concentrated on a series of relatively isolated intermontane basins, which descend gradually in elevation as the chain extends northeastward toward the Venezuelan border. The moderate temperatures and fertile, well-watered soils of these

FIGURE 2.30

Most of Venezuela's population has resided historically in a series of fertile Andean valleys that extend in a chain from west to east parallel to the Caribbean coast. Pictured here are the valley and city of Mérida.

highland valleys have attracted dense populations since the pre-Columbian era.

The largest of the basins, called Cundinamarca, is situated at over 2,400 meters (8,000 feet) above sea level roughly halfway between Ecuador and Venezuela. Bogotá, Colombia's capital and principal city, was founded here in 1538. Northeast of Bogotá are additional high basins occupied by Sogamoso, Tunja, and Chiquinquirá, while a third, lower cluster of basins includes the expanding tropical agricultural centers of Bucaramanga and Cúcuta near the Venezuelan border (Eidt 1968).

The Eastern Cordillera divides in the vicinity of Bucaramanga, with the westernmost arm continuing almost due north toward the Caribbean under the name of the Sierra de Perijá. A granite outlier called the Sierra Nevada de Santa Marta rises to elevations of over 5,800 meters (19,000 feet) along the coast and descends on its northeastern slopes into a parched, windswept coal-mining and drug-running area known as the Guajira Peninsula.

Meanwhile, the remaining arm of Colombia's Eastern Cordillera pushes northeastward into Venezuela as the Cordillera de Mérida, or the Sierra Nevada. Part of the sierra curves around the eastern rim of the oil-rich **Maracaibo Basin,** which drains through shallow Lake Maracaibo into the Gulf of Venezuela, an inlet of the southern Caribbean Sea. The remaining ranges continue on toward the Caribbean coastline, housing the bulk of Venezuela's population in a series of beautiful, relatively low mountain valleys. The largest city within these valleys is Caracas, which is located in a narrow trough some 900 meters (3,000 feet) above sea level. West of Caracas lies the fertile basin of Valencia and, further west, the basins of Barquisimeto, Mérida, and San Cristóbal (Figure 2.30). North of these basins, the sierra degenerates into a zone of low ranges that continue as far east as the Paria Peninsula. They then dip beneath the sea before reappearing in the southeastern Caribbean as the islands of Trinidad and Tobago.

Physiographic Regions of Central America

Central America, Mexico, and the Caribbean Islands represent a northern extension of South America's geologically young western alpine system. The low coastal ranges and interior rivers of the Darién region of eastern Panama, for example, are physically related to the Serranía de Baudó and the Atrato River Basin of the Chocó area of northwestern Colombia.

The highly urbanized, cosmopolitan lowlands of Panama's central canal zone, or *Area Canalera*, end abruptly with the appearance of the magnificent **Central American Volcanic Axis,** which parallels the Pacific coastal plains from western Panama northward to Chiapas state in southern Mexico (Figure 2.31). With the exception of the 3,600-meter (12,000-foot)-high Talamanca range of southwestern Costa Rica, which is an exposed mass of granitic rock, the interior highland region of Central America consists largely of densely populated, ash-filled valleys nestled among emerging volcanoes, most of which range from 2,100 to 3,600 meters (7,000 to 12,000 feet) above sea level. The largest of the basins is the Meseta Central of Costa Rica (Figure 2.33). Here rich volcanic soils have sustained some of Latin America's most productive coffee farms, called *fincas*, many of which are controlled by wealthy families of European heritage who reside in San José and neighboring urban centers (Figure 2.34).

The Volcanic Axis continues into northwestern Costa Rica as the Cordillera de Guanacaste, which ends at the Nicaraguan border. At that point, the high mountains withdraw from along the Pacific coast of Nicaragua in favor of numerous smaller volcanoes in the 900- to 1,800-meter (3,000- to 6,000-foot) range. The dominant structural feature of western Nicaragua is a faulted, down-dropped valley, or graben, which extends from the Gulf of Fonseca in the northwestern corner of the country southeastward to Costa Rica and then eastward along the international border to the Caribbean (Figure 2.32). The valley has been filled with ash and alluvium to an average width of eighty-one kilometers (fifty miles) and is occupied by Central America's largest freshwater lakes, Lake Managua and Lake Nicaragua. The former drains into the latter, which empties into the Caribbean by way of the San Juan River. The western Nicaraguan lowlands historically have supported one of Central America's densest populations (Figure 2.35). Its potential to become one of Latin America's most prosperous regions has never been realized, however, owing in part to the nation's turbulent political history and to repeated devastation from earthquakes, volcanic eruptions, and hurricanes.

North and west of the Gulf of Fonseca, the Volcanic Axis appears in El Salvador in the form of low mountains and upland basins, which have for centuries nourished a dense rural population. Guatemala and the neighboring Mexican state of Chiapas are dominated by two sets of adjacent mountain ranges, one of which is an extension of the Central American Volcanic Axis that parallels the Pacific coast (Figure 2.36). The other is a series of ranges that extend into Honduras and central Nicaragua before arching gently toward the northeast and the Caribbean Sea. Scattered throughout are picturesque highland basins, many occupied in part by scenic lakes.

The final physiographic provinces of Central America are the coastal plains. These are relatively narrow and arid on the Pacific side. They are much wider along the Caribbean coast, where broad, swampy river valleys have formed from Belize to northern Costa Rica in response to the heavy rains that drench the east-facing, windward slopes of the interior highlands. With the exception of a few favored areas, such as the central Panamanian lowlands and the Ulúa River Valley of northern Honduras, the eastern coastal plains have remained sparsely settled and physically isolated from the more densely populated interior highlands of Central America.

Physiographic Regions of Mexico and the Caribbean

Mexico, like Central America, consists of an interior highland core bordered on the east and west by coastal lowlands. A fundamental difference between the two areas, however, is Mexico's sheer size, which exceeds by many times the combined surface area of all the Central American states.

Mexico begins physiographically in a low, hilly depression west of Chiapas called the isthmus of Tehuantepec (Color plate 4). Although the isthmus is the narrowest and one of the lowest parts of the country, its isolation from Mexico City and lack of dense indigenous populations have resulted in its remaining less developed to this day in spite of the presence of a transcontinental railroad, highway, and oil pipeline.

The isolation of Tehuantepec is attributable, in large measure, to the rugged **Southern Highlands** that, along with the equally desolate Balsas River Basin, separate the isthmus from the Mesa Central. It is said that the Spanish conquistador Hernán Cortés was once asked what New Spain, or Mexico, looked like. Cortés, who had allocated to himself vast areas of southern Mexico, is alleged to have picked up a sheet of paper and wadded it violently in his hands. Then, holding up the crumpled ball, he replied, "This is New Spain."

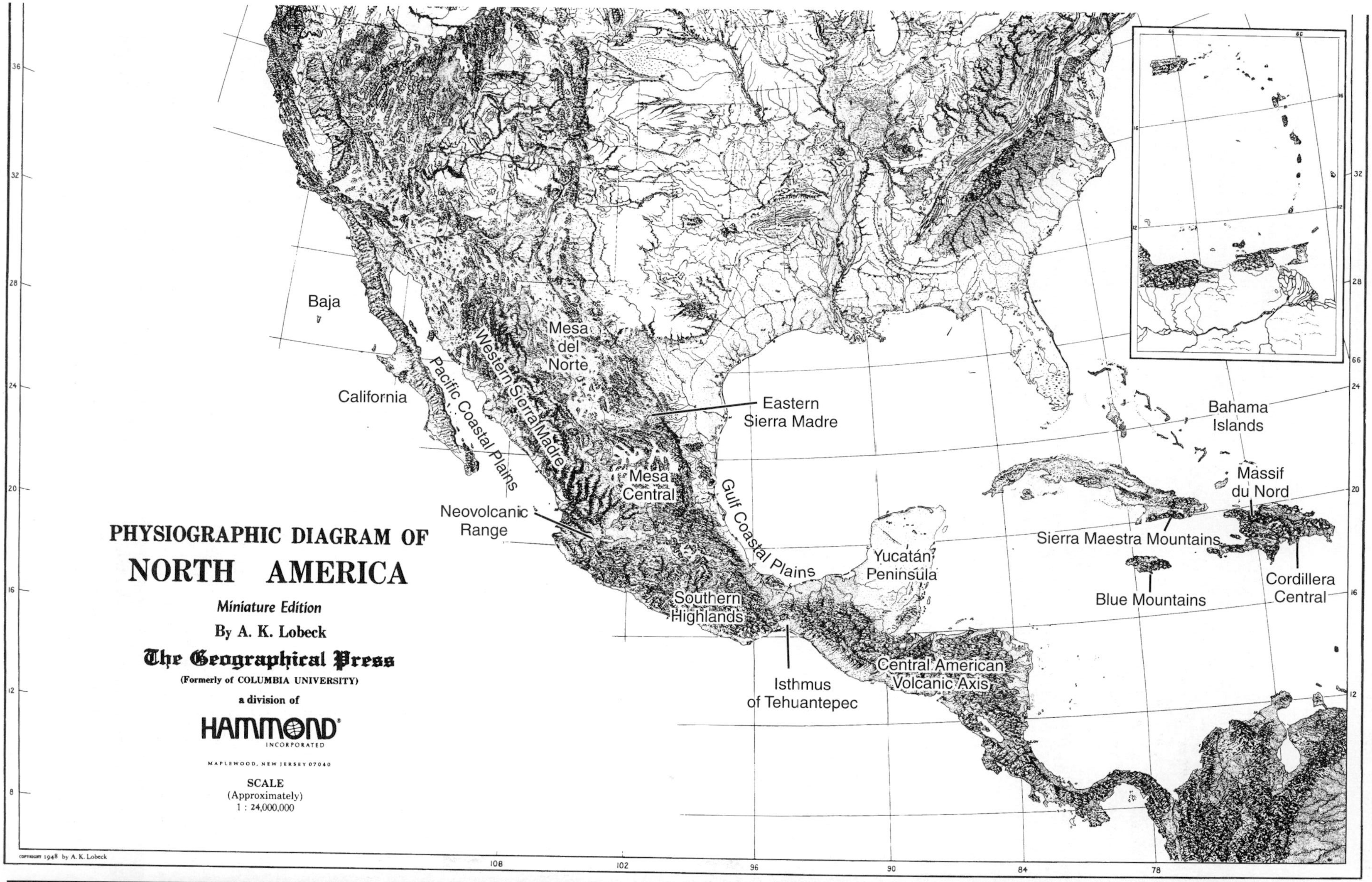
PHYSIOGRAPHIC DIAGRAM OF
NORTH AMERICA
Miniature Edition
By A. K. Lobeck
The Geographical Press
(Formerly of COLUMBIA UNIVERSITY)
a division of
HAMMOND
INCORPORATED
MAPLEWOOD, NEW JERSEY 07040
SCALE
(Approximately)
1 : 24,000,000
copyright 1948 by A. K. Lobeck
Baja
California
Pacific Coastal Plains
Western Sierra Madre
Mesa del Norte
Eastern Sierra Madre
Mesa Central
Neovolcanic Range
Gulf Coastal Plains
Southern Highlands
Isthmus of Tehuantepec
Yucatán Peninsula
Central American Volcanic Axis
Bahama Islands
Massif du Nord
Sierra Maestra Mountains
Blue Mountains
Cordillera Central
36
32
28
24
20
16
12
8
108
102
96
90
84
78
66

FIGURE 2.31

Physiographic diagram (provinces) of North America. Copyright: Hammond World Atlas Corporation, NJ., Lic. No: 12450

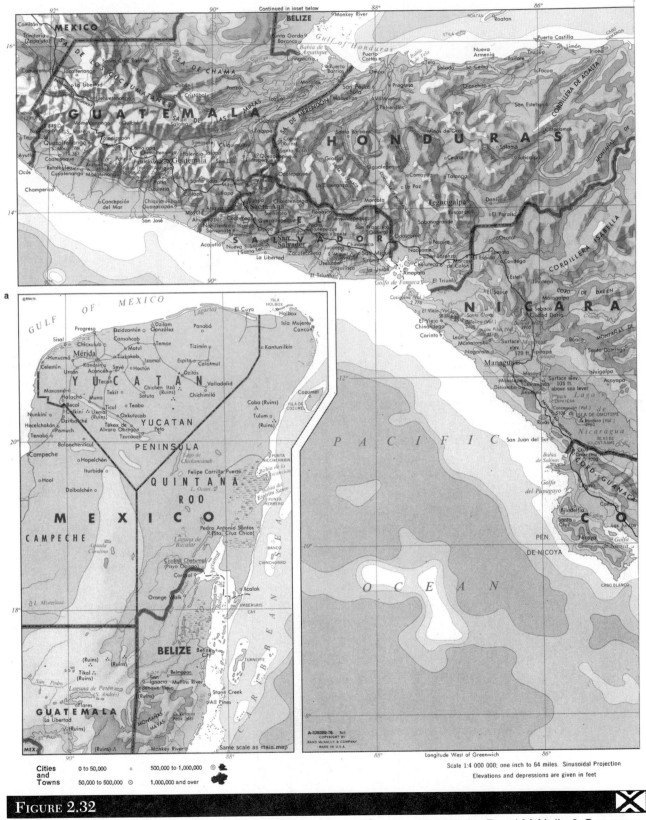

Scale 1:4 000 000; one inch to 64 miles. Sinusoidal Projection

Elevations and depressions are given in feet

Longitude West of Greenwich

FIGURE 2.32

Western Central America. Map from *Goode's World Atlas,* 20th edition © 2000 by Rand McNally & Company.

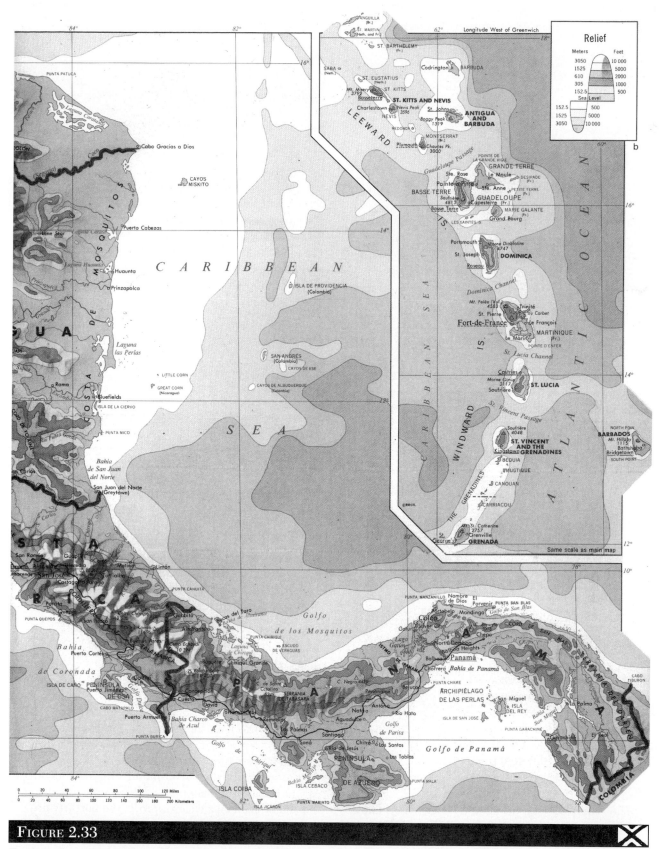

FIGURE 2.33

Eastern Central America. Map from *Goode's World Atlas*, 20th edition © 2000 by Rand McNally & Company.

FIGURE 2.34

San José is beautifully situated in the heart of the volcanic Costa Rican Meseta Central.

FIGURE 2.35

The fertile volcanic soils of the western Nicaraguan lowlands have sustained for centuries one of Central America's densest populations.

FIGURE 2.36

Western Guatemala is dominated by the volcanoes and basins of the Central American Volcanic Axis. Shown here is the Chimaltenango Valley and the nearby Mt. Fuego volcano.

FIGURE 2.37

The Southern Highlands of Mexico are a region of intense rural poverty.

Those familiar with the Southern Highlands know that physically, nothing has changed. With the exception of the broad faulted valley occupied by Oaxaca City, almost all of Oaxaca and Guerrero states are a jumbled crystalline mass of remote mountains and valleys linked by some of the poorest roads in all of Latin America. This extreme ruggedness has hindered rural development and enabled many of the native Indian peoples to continue relatively traditional lifestyles in tiny remote villages (Figure 2.37).

The poverty and isolation of the Southern Highlands are broken at approximately 19° N latitude by a series of spectacular volcanoes and associated high intermontane basins. The **Neovolcanic Range** begins on the east with the 5,700-meter (18,701-foot)-high Mt. Orizaba, or Citlaltépetl, whose snowcapped summit can be seen on a clear day from the port city of Veracruz (Figure 2.38). It then continues across the country almost due west to the Pico de Colima, which rises to 4,265 meters (13,993 feet) near the Pacific Ocean. Beneath the volcanoes are high, broad basins, including the Valley of Mexico, whose volcanic soils and cool

Mt. Orizaba, the eastern terminus of Mexico's Neovolcanic Range, is the highest peak in Middle America.

climates have attracted since preconquest times some of the densest populations in all of Latin America.

North of the Neovolcanic Range, Mexico consists of a high central plateau framed by two ranges, the Western and Eastern Sierra Madre. The plateau itself is a huge tilted block that slopes northward from 2,100 to 2,400 meters (7,000 to 8,000 feet) along its southern margin, where it adjoins the transverse Neovolcanic Range, to just 900 meters (3,000 feet) or less along the border with Texas, New Mexico, and Arizona. Because the plateau is surrounded on three sides by mountain ranges, most of the drainage is toward lower interior basins, called *bolsones*, where the water is quickly lost to soil infiltration and evaporation, leaving salty, alkaline deposits at the surface. Scattered across the **Mesa del Norte,** the arid northern reaches of the plateau, are smaller, north-south-trending block ranges. While seldom exceeding 900 to 1,200 meters (3,000 to 4,000 feet) in height, they frequently act as water catchment areas whose substrata can be tapped by wells to support small isolated settlements dependent upon irrigated agriculture and ranching (Figure 2.39).

Faulted block mountain ranges of the Mesa del Norte, northern Mexico.

FIGURE 2.40

Mexico's Mesa Central is a region of great industrial cities and traditional peasant villages where maize and maguey are cultivated much as they have been for hundreds of years.

FIGURE 2.41

The high, arid valleys of Mexico's Western Sierra Madre range are reminiscent of the Bolivian and Peruvian Altiplano. Shown is the Fresnillo Valley of the state of Zacatecas.

The southern portion of the central plateau, often referred to as the **Mesa Central,** experiences greater rainfall and more moderate temperatures than the Mesa del Norte. The Mesa Central supports several of Mexico's major cities as well as traditional rural zones characterized by scattered peasant villages set among abandoned remnants of pre-Revolution haciendas. Small herds of sheep or goats can often be seen grazing amidst fields of maize and maguey, a large spiny plant that yields fibers and a fermented drink called *pulque* (Figure 2.40).

As one travels westward across the central plateau, the volcanic **Western Sierra Madre** range appears as a barren series of low mountains rising some 600 to 900 meters (2,000 to 3,000 feet) above the high arid plain. It was here that the Spaniards discovered, in the mid-sixteenth century, some of the world's richest silver deposits, following a north-south line along the eastern base of the sierra. A number of towns, including Guanajuato, Zacatecas, and Durango, were founded in adjoining high valleys whose cold, sparsely vegetated landscapes are reminiscent of the Andean Altiplano (Figure 2.41).

The true ruggedness of the Western Sierra Madre can be appreciated only when viewed from the narrow coastal plains that separate the mountains from the Gulf of California. Seen from the plains, the mountains form a massive 3,050- to 3,650-meter (10,000- to 12,000-foot)-high barrier that offers little more than an occasional stream or river in support of irrigated agriculture. The Gulf of California itself is a drowned graben, or faulted down-dropped block of the earth's crust, which formerly extended into the Imperial and

Coachella valleys of southern California and the Colorado River Delta. The desolate Baja California Peninsula is a horst, the uplifted counterpart of the Gulf of California, and slopes westward across volcanic landscapes toward the Pacific Ocean.

The **Eastern Sierra Madre,** which marks the eastern limits of the central plateau, consists of a series of folded limestone ridges that spreads northward from the eastern terminus of the Neovolcanic Range to Monterrey, beyond which it gradually blends with the basin and range country of the Mesa del Norte. Rising abruptly out of the Gulf coastal plain to elevations exceeding 3,050 meters (10,000 feet) in many areas, the sierra long served not only as a formidable barrier to access to the interior but also as a climatic and cultural divide. The easternmost ranges of the sierra receive torrential tropical downpours during the rainy season, with the result that numerous rivers have carved deep, steep-sided valleys, called *barrancas,* into the verdant landscapes. The dry leeward slopes, on the other hand, are subject to violent wind and dust storms and, like the Western Sierra Madre, support highland ranching and silver mining activities.

The Mexican Gulf coastal plains are, on the average, considerably wider than those of the Pacific coast, are subject to severe river flooding, and tend to have brown, muddy coastlines with numerous marshes, mangrove swamps, and tidal lagoons (Figure 2.42). The largest of the marsh and swamp areas is in central and eastern Tabasco and neighboring Campeche states where the Usumacinta and Grijalva rivers discharge vast quantities of sediments brought down from the interior highlands of Chiapas and Guatemala.

FIGURE 2.42

Mangrove swamps and tidal lagoons along the Mexican Gulf coastal plains.

FIGURE 2.43

The moist, interior floor of this central Yucatán *cenote* has been planted to bananas and palms.

FIGURE 2.44

Sinkhole development on the limestone plains of Yucatán and Cuba.

East and north of Tabasco the underlying strata of the Gulf coastal plains change to limestone and the plains expand to encompass virtually all of the **Yucatán Peninsula.** The northern peninsula is one of the flattest regions in Latin America, and features that appear from a distance to be low, rounded hills turn out, as often as not, to be ancient overgrown Mayan pyramids. The southern peninsula includes the Petén region of northern Guatemala, the nation of Belize, as well as portions of the Mexican state of Campeche, and is characterized by limestone hills that give way to a crystalline outcrop known in Belize as the Maya Mountains.

The entire peninsula is a superb example of karst topography, which is associated worldwide with limestone surfaces. Two of the most characteristic indicators of karst topography are the absence of surface streams and the presence of sinkholes or, as they are called in Yucatán, *cenotes* (Figure 2.43). The sinkholes occur naturally as the rainwater reacts beneath the surface with the carbon in the limestone substrata to form carbonic acid. The acid ultimately dissolves the limestone, forming huge underground caverns that appear as craters after the collapse of the surface (Figure 2.44). Many of the *cenotes* are fully or partially filled with

FIGURE 2.45

Cuba is blessed with an abundance of fertile soils derived from an underlying limestone platform. This farm is located on the outskirts of the city of Trinidad in the central part of the island.

water from the fluctuating water table and historically have served as the principal sources of groundwater in the northern peninsula (Veni 1990; West 1964).

The limestone platform of the Yucatán Peninsula extends far out into the Gulf of Mexico as a shallow continental shelf known as the Campeche Banks. The banks have long served as one of Mexico's leading fishing and shrimping grounds and took on added significance in the mid-1970s with the development of major petroleum reserves. The platform reemerges east of the Yucatán Channel as the rolling plains of western and central Cuba (Figure 2.45). It then continues eastward to form the Bahamas, a chain of approximately 700 low-lying islands and countless smaller cays and rocks interspersed among some of the most beautiful coral reefs in the tropical world. The warm, shallow, crystal-clear waters of the islands provide recreational diving and sport fishing opportunities.

Just as the limestone platform extends eastward from Yucatán into the Caribbean, so the ranges of northern Central America reappear as the principal highlands of the Greater Antilles. These include the Sierra Maestra and Baracoa Mountains of southeastern Cuba, the Massif du Nord of northern Haiti, the Cordillera Central and the Cordillera Septentrional of the Dominican Republic, and the Cordillera Central of Puerto Rico and the Virgin Islands (Figure 2.46). Similarly, the Cayman Islands, the Blue Mountains of Jamaica, and the mountainous southwestern peninsula of Haiti are eastern extensions of the crystalline ranges of Honduras and Nicaragua.

The origin of the Lesser Antilles can be traced to two volcanic arcs that have risen from the ocean floor.

FIGURE 2.46

The interior of the Dominican Republic is dominated by high and rugged mountains.

The older of these is the **outer arc,** which includes the islands of Anguilla, St. Martin, St. Barthélemy, Barbuda, Antigua, Marie Galante, and eastern Guadaloupe (Grand Terre) (Figure 2.47). Following their initial formation, the outer arc volcanoes became dormant and were submerged beneath the sea. When they subsequently reemerged, they were capped with layers of porous limestone and lacked the relief necessary to induce significant rainfall from the prevailing trade winds. Today they appear as tiny, impoverished specks of arid land along the outer margins of the Caribbean Sea, too dry and mineral poor to support the development of agricultural and mining sectors.

The younger, inner islands, which extend from Saba, St. Eustatius, and St. Kitts southward to Grenada, have mountainous cores, and receive abundant moisture on their windward slopes. The heavy rains and fertile volcanic soils support a lush tropical vegetation that, together with scenic beaches, attracts great numbers of tourists annually. The **inner arc** islands,

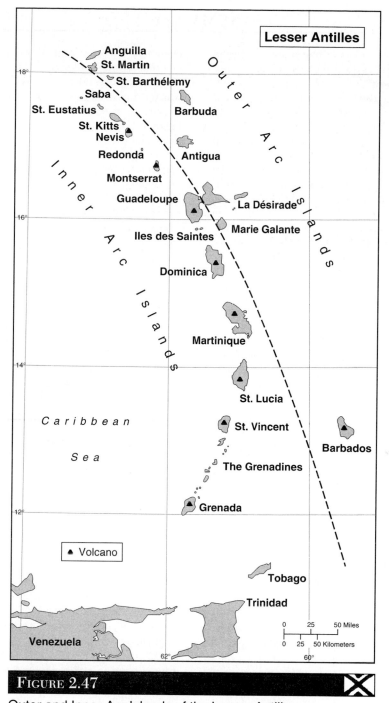

FIGURE 2.47

Outer and Inner Arc Islands of the Lesser Antilles.

as was noted previously, are also subject to the periodic eruptions and earthquakes characteristic of volcanically active regions. Two of the inner arc islands that have experienced severe damage from volcanic activity during the past century are Martinique and Montserrat. The 1902 eruption of Mt. Pelée killed over 28,000 persons and destroyed the city of St. Pierre on the island of Martinique (Smith and Roobol 1990). Similarly, lava flows and ash emissions from the Soufriere Hills volcano, which began erupting in the mid-1990s, resulted in the forced evacuation of approximately two-thirds of Montserrat's 12,000 residents and in the relocation of the capital city of Plymouth (Williams 1997).

SUMMARY

Although Latin America and the Caribbean are among the largest and most physically diverse regions on earth, their landforms can be grouped into three basic structural divisions. The eastern highlands, including the Brazilian and Guiana highlands, are composed of ancient igneous and metamorphic cores that have been eroded into extensive areas of moderate relief. Their soils are generally poor, but they are rich in mineral ores. The second structural division consists of the interior lowlands, which are dominated by the Orinoco, Amazon, and Río de la Plata rivers and their tributaries. Economic development in many of these regions has been hampered by low population densities, inadequate transportation networks, disease, and sheer distance. The quality of their soils varies widely, as does the presence of mineral resources. Although isolated and remote, they include many of Latin America's remaining frontier zones and will likely come under increasing development pressures. The final structural division is the western alpine system, which extends from southern Chile to northern Mexico and the Greater and Lesser Antilles. This physiologically complex region includes numerous highland basins and valleys, whose inhabitants are favored with moderate climates, fertile soils, and adequate moisture. Catastrophic volcanic eruptions and earthquakes are sobering reminders that this system is still developing and is one of the most geologically unstable regions on earth.

The physical diversity of Latin America and the Caribbean exists not only in their landforms but also in their weather and climate patterns. We will turn next to an analysis of these patterns and the forces that sustain them.

KEY TERMS

Brazilian Highlands 14
Serra do Mar 14
Serra Geral 14
sertão 14
Paraná Plateau 14
Mato Grosso Plateau 17
Guiana Highlands 18
Patagonian Plateau 19
Llanos 20
Amazon Plain 20
Llanos de Mojos 21

Chiquitos Plateau 21
Gran Chaco 22
Gran Pantanal 22
Pampa 22
Ring of Fire 22
plates 22
Central Valley of Chile 28
Altiplano 29
Callejón Andino 30
Maracaibo Basin 33

Central American Volcanic Axis 34
Southern Highlands 34
Neovolcanic Range 38
Mesa del Norte 39
Mesa Central 40
Western Sierra Madre 40
Eastern Sierra Madre 40
Yucatán Peninsula 41
outer arc 42
inner arc 42

SUGGESTED READINGS

Abele, Gerhard. 1992. "Landforms and Climate on the Western Slope of the Andes." *Zeitschrift für Geomorphologie* Supplemental Bulletin 84:1–11.

Bock, B., et al. 2000. "Tracing Coastal Evolution in the Southern Central Andes from Late Precambrian to Permian with Geochemical and Nd and Pb Isotope Data." *Journal of Geology* 108:515–535.

Bowman, Isaiah. 1916. *The Andes of Southern Peru.* New York: American Geographical Society.

Brunnschweiler, Dieter. 1972. *The Llanos Frontier of Colombia.* East Lansing: Michigan State University Latin American Studies Center Monograph No. 9.

Coates, Anthony G. 1997. "The Forging of Central America." In *Central America: A Natural and Cultural History,* ed. Anthony G. Coates, 1–37. New Haven: Yale University Press.

Crist, Raymond E. 1952. *The Cauca Valley, Colombia: Land Tenure and Landuse.* Baltimore: Waverly Press.

Crist, Raymond E., and Edward P. Leahy. 1969. *Venezuela: Search for a Middle Ground.* New York: Van Nostrand Reinhold.

Denevan, William M. 1996. "A Bluff Model of Riverine Settlement in Prehistoric Amazonia." *Annals of the Association of American Geographers* 86:654–681.

Dorbath, L., A. Cisternas, and C. Dorbath. 1990. "Assessment of the Size of Large and Great Historical Earthquakes in Peru." *Seismological Society of America Bulletin* 80 (3): 551–576.

Eidt, Robert C. 1968. "Some Comments on the Geomorphology of Highland Basins in the Cordillera Oriental of Colombia." *Revista Geográfica* 68 (June): 141–156.

Gregory-Wodzicki, Kathryn M. 2000. "Uplift History of the Central and Northern Andes: A Review." *Geological Society of America Bulletin* 112:1091–1105.

Gutscher, Mark-André. 2000. "Geodynamics of Flat Subduction: Seismicity and Tomographic Constraints from the Andean Margin." *Tectonics* 19:814–833.

Hudson, John C., and Edward B. Espenshade, Jr., eds. 2000. *Goode's World Atlas,* 20th ed. Chicago: Rand McNally.

Incer, Jaime. 1987. "Nature's Red-Hot Caldrons." *Americas* 39:19–25.

Jansma, Pamela E., et al. 2000. "Neotectonics of Puerto Rico and the Virgin Islands, Northeastern Caribbean, from GPS Geodesy." *Tectonics* 19:1021–1037.

Jones, Tristan. 1977. "Incredible Voyage." *Quest* 2 (May/June): 5–10, 113–116.

Lobeck, A. K. 1975. *Physiographic Diagram of North America.* Maplewood, N.J.: Geographical Press.

Macias, J. L., et al. 2000. "Late Holocene Pelean-Style Eruption at Tacana Volcano, Mexico and Guatemala: Past, Present, and Future Hazards." *Geological Society of America Bulletin* 112:1234–1249.

Macpherson, John. 1980. *Caribbean Lands,* 4th ed. Trinidad: Longman Caribbean.

Martínez, Moisés. 2001. "A. L. se prepara enfrentar 'El Niño.'" *La Prensa de Managua, Nicaragua.* 29 de Noviembre, A-16.

Martinson, Tom L. 1997. "Physical Environments of Latin America." In *Latin America and the Caribbean: A Systematic and Regional Survey,* 3rd ed., eds. Brian W. Blouet and Olwyn M. Blouet, 11–44. New York: Wiley.

McCann, W. R., and W. D. Pennington. 1990. "Seismicity, Large Earthquakes, and the Margin of the Caribbean Plate." In *The Caribbean Region,* Vol. H. of *Geology of North America,* eds. Gabriel Dengo and J. E. Case, 291–306. Boulder, Colo.: Geological Society of America.

McKnight, Tom L., and Darrel Hess. 2002. *Physical Geography: A Landscape Appreciation,* 7th ed. Upper Saddle River, NJ: Prentice-Hall.

Oliver-Smith, Anthony. 1986. *The Martyred City: Death and Rebirth in the Andes.* Albuquerque: University of New Mexico Press.

Parsons, James J. 1982. "The Northern Andean Environment." *Mountain Research and Development* 2:253–262.

Pascal, B., and J. Lambert. 1988. "Subduction and Seismic Hazard in the Northern Lesser Antilles: Revision of Historical Seismicity." *Bulletin of the Seismological Society of America* 78:1965–1983.

Reading, Alison J., Russell D. Thompson, and Andrew C. Millington. 1995. *Humid Tropical Environments.* Oxford: Blackwell.

Sauer, Carl O. 1963. "Geography of South America." In *Handbook of South American Indians.* Vol. 6, ed. Julian H. Steward, 319–344. Washington, DC: Bureau of American Ethnology Bulletin No. 143, Smithsonian Institution.

Sealey, Neil. 1992. *Caribbean World: A Complete Geography.* Cambridge: Cambridge University Press.

Shephard, J. B. 1989. "Eruptions, Eruption Precursors and Related Phenomena in the Lesser Antilles." In *Volcanic Hazards: Proceedings in Vulcanology 1,* ed. J. H. Latter, 292–311. Berlin: Springer-Verlag.

Smith, Alan L., and M. John Roobol. 1990. *Mt. Peleé, Martinique: A Study of an Active Island-Arc Volcano.* Boulder, Colo.: Geological Society of America.

Smith, Guy-Harold. 1963. *Physiographic Diagram of South America.* Maplewood, NJ: Geographical Press.

Spikings, Richard A., et al. 2000. "Low-Temperature Thermochronology of the Northern Cordillera Real, Ecuador: Tectonic Insights from Zircon and Apatite Fission Track Analysis." *Tectonics* 19:649–668.

Suter, Max, et al. 2001. "Quaternary Intra-Arc Extension in the Central Trans-Mexican Volcanic Belt." *Geological Society of America Bulletin* 113:693–703.

Taboada, Alfredo. 2000. "Geodynamics of the Northern Andes: Subductions and Intracontinental Deformation (Colombia)." *Tectonics* 19:787–813.

The World Almanac and Book of Facts 2002. 2002. New York: World Almanac Education Group, Inc.

Thomas, M. F. 1994. *Geomorphology in the Tropics.* Chicester: Wiley.

Veni, George. 1990. "Maya Utilization of Karst Groundwater Resources." *Environmental Geology and Water Sciences* 16 (1): 63–66.

Wallace, Paul, et al. 1992. "Vulcanism and Tectonism in Western Mexico: A Contrast of Style and Substance." *Geology* 20:625–628.

West, Robert C. 1964. "Surface Configuration and Associated Geology of Middle America." In *Natural Environment and Early Cultures: Handbook of Middle American Indians.* Vol. 1, ed. Robert C. West, 33–83. Austin: University of Texas Press.

West, Robert C., and John P. Augelli. 1989. *Middle America: Its Lands and Peoples.* 3rd ed. Englewood Cliffs, NJ: Prentice-Hall.

Williams, A. R. 1997. "Montserrat: Under the Volcano." *National Geographic* 192: 58–75.

ELECTRONIC SOURCES

Programa de las Naciones Unidas para el Desarrollo: www.terremotoelsalvador.org.sv/

Smithsonian National Museum of Natural History Global Vulcanism Program: www.volcano.si.edu/gvp/

United Nations Office for the Coordination of Humanitarian Affairs (OCHA) Relief Web: www.reliefweb.int/w/rwb.nsf

3

Weather and Climate

When Latin America and especially the Caribbean are mentioned in casual conversation and the popular media, images of lush tropical paradises often come to mind. Listeners may picture themselves sunbathing on a beautiful white sand beach with graceful coconut palms swaying gently overhead, or traveling through a dense, almost impenetrable, tropical rain forest. Because these scenes are so exotic and appealing to many, especially to those enduring seemingly interminable midlatitude winters, some outsiders assume that all of Latin America is this way. Yet, as has been noted, Latin America is an extremely large and physiographically complex region. As such, it experiences, to some degree, almost every climate found on the earth.

Our principal objective in this chapter is to come to understand the nature and location of the weather patterns and climates of Latin America and the Caribbean. This can best be accomplished through an overview of the principal physical controls of the two most prominent aspects of climate: temperature and precipitation. Once we appreciate how these basic controls interact with each other, we will better comprehend the environmental and economic capacities and limitations of each climatic region.

It should be understood that the interactions of these forces of nature are extremely complex and that a simple explanation does not always exist for the presence of each of the infinite number of microenvironments found throughout the region. An old Latin American saying, *No hay reglas fijas,* meaning "firm or fixed rules do not exist," applies as often to the physical milieu as it does to the cultural.

CONTROLS OF CLIMATE

Temperature

The temperatures associated with any given location in Latin America are a function primarily of the interactions of three factors: altitude, ocean currents, and latitude.

ALTITUDE

Because so much of Latin America is situated relatively close to the equator, where the length of day and night and daily temperatures vary little throughout the year, the best predictor of temperature is altitude, or elevation above sea level. Although rates vary, depending on whether the air is stationary or moving and on whether precipitation is occurring or not, the atmospheric temperature decreases on the average 6.4° C per 1,000 meters (3.5° F for every 1,000 feet) of increased elevation.

One of the most significant consequences of the correlation between elevation and temperature is the existence in Latin America of **altitudinal life zones** (Troll 1968; Tosi and Voertman 1964; Holdridge 1967) (Figure 3.1). In studying these zones, we must keep in mind that physical environments often change gradually, rather than abruptly, and that the dividing elevations between the zones are simply human-created averages that may or may not apply exactly to a specific locale (Stadel 1986; Gade 1975; Brush 1977). We should also remember that the zones reach higher elevations in the low-latitude regions close to the equator and extend to lower elevations away from the equator.

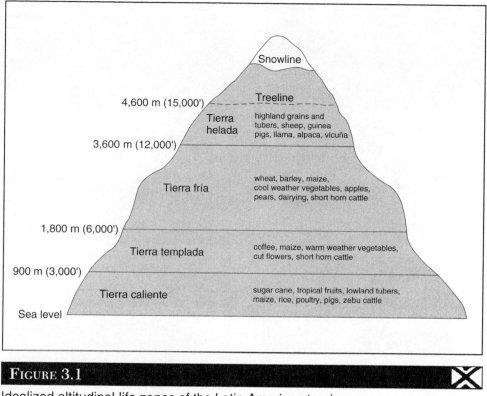

FIGURE 3.1

Idealized altitudinal life zones of the Latin American tropics.

CALIENTE

With these qualifications, it can be said that most of Latin America, from sea level to approximately 900 meters (3,000 feet), is dominated by *tierra caliente*, or the hot lands. Because freezing temperatures are never approached, the *tierra caliente* is associated with the cultivation of perennial, cold-sensitive, so-called tropical crops that cannot be grown commercially in the midlatitude countries. The dominant of these, when measured in terms of acreage, is often sugar cane. The *tierra caliente* also produces a range of bamboos and palms, including the coconut, which graces countless coastlines, and the increasingly valuable African oil palm (Judziewicz et al. 1999; Horst 1997; Jones 1995). Other food crops common to the *tierra caliente* include bananas, mangoes, papaya, citrus, pineapple and other fruits, as well as cacao or chocolate and a diversity of melons and lowland tubers. The most significant of the tubers are manioc, sweet potatoes, yams, and dasheen—a New World member of the taro family with large, elephant ear-shaped leaves (Commission on International Relations 1975) (Figures 3.2, 3.3, 3.4, and 3.5). Corn or maize, rice, and assorted legumes, including beans and peanuts, are also cultivated widely for domestic consumption. Pigs and poultry, being scavengers and capable of keeping themselves alive in the midst of poverty, are found wandering about al-

most every rural household. The *tierra caliente* thus includes regions of great agricultural abundance as well as large tracts of land that historically have remained sparsely settled and relatively underdeveloped owing mostly to the persistence of malaria, yellow fever, and a host of other debilitating illnesses (chapter 13). Another principal landuse of the *tierra caliente*, centuries old, is extensive ranching, which is associated today with ever-increasing infusions of *Bos indicus* (Zebu or Brahma) cattle. The population of the *tierra caliente* tends to be concentrated in port cities, which relay goods and people passing between the interior urban centers and the outside world.

TEMPLADA

As the elevation increases, the incessant heat and humidity of the lowlands fade into a zone of year-round mild temperatures called the *tierra templada*, or temperate lands. Many outside observers have described these regions, which extend to approximately 1,800 meters (6,000 feet) above sea level, as lands of eternal spring. In reality, they are far more temperate than midlatitude springs, which are often characterized by violent climatic extremes. The *tierra templada* historically has supported numerous population clusters in Latin America whose settlers were attracted by relatively healthy and comfortable physical environments. The principal commercial crop of these regions

FIGURE 3.2

Coconut palms are found throughout the hot coastal zones of Latin America and the Caribbean, as well as in many inland areas of the *tierra caliente.* The plant yields a number of valuable products, including construction materials and fibers as well as foods for eating and drinking.

FIGURE 3.3

Papaya plant and fruit. Papaya, a product of the *tierra caliente,* is one of the most widely consumed fruits in Latin America and the Caribbean but is rarely exported owing to its highly perishable nature when ripe and also to its susceptibility to disease when grown under large-scale commercial conditions.

has been coffee, but fresh-cut flowers and warm-weather vegetables, including tomatoes, cucumbers, bell peppers, and green beans are increasingly common (Figure 3.6). In terms of agriculture, the *tierra templada* is often a transition zone, supporting sugar cane, bananas, and other lowland crops in its lower reaches and cool-weather products along its upper margins.

The third life zone is the *tierra fría,* or cold lands, which extends in many regions to 3,600 meters (12,000 feet) above sea level. Although the word *fría* translates literally as "cold," it is probably more accurate to call these the "cool lands," for bitter temperatures and hard freezes are common only in the upper

reaches of this zone. By wearing warm clothing, Europeans and Indians alike have traditionally found the *tierra fría* zones much to their liking. In fact, many of the great population centers of Latin America, including Mexico City, Bogotá, and Quito, are found within this realm. Midlatitude crops dominate, including maize, wheat, barley, apples, pears, potatoes and other tubers, and cool-weather vegetables such as broccoli, cabbage, onions, and carrots (Figures 3.7 and 3.8). Cattle ranching and dairying are also common, with European shorthorn (*Bos taurus*) breeds predominating (Figure 3.9).

Caliente

FIGURE 3.4 ✗

Cassava, also called manioc or yuca, is the most widely cultivated tuber of the Latin American lowlands, owing to its high yields and environmental hardiness.

Caliente

FIGURE 3.5 ✗

The mango, one of the most popular fruits of the Latin American lowlands, is native to the Asian tropics and grows on a large evergreen tree that may bear hundreds of fruits each year.

Where the mountains and valleys continue to rise above the *tierra fría,* one passes into the highest of Latin America's altitudinal life zones, the *tierra helada* or "frozen lands." The extent to which they are truly frozen remains, of course, a function of eleva-

FIGURE 3.6 ✗

Coffee is the principal commercial crop of the *tierra templada.* The beans are obtained from small berries that individually ripen to a dark red on the bush or tree.

FIGURE 3.7 ✗

Apples intercropped with maize and beans in the *tierra fría.*

tion. Cold-weather agriculture is still possible at the lower reaches of the *tierra helada,* and in the Andes focuses largely on native grains and tubers. Included in the former are quinoa, amaranth, and cañihua, while indigenous root crops include two species of potato, ulluco, oca, añu, and the parsnip-like arracacha (Popenoe et al. 1989; Wilson 1990; Stegemann, Majino, and Schmiediche 1988; Gade 1970) (Figure 3.10). Sheep, llamas, alpaca, and vicuña are herded for meat, fiber, and dung (Figure 3.11). Many Andean households also raise domesticated guinea

Cool-weather vegetables, such as these potatoes growing on the upper slopes of the Irazú Volcano of Costa Rica, are frequently cultivated in the Latin American highlands as commercial crops destined for nearby urban markets.

Animal grazing is widespread within the *tierra fría,* where year-round cool temperatures, rainfall, and snowmelt sustain the growth of pastures.

Andean potatoes and oca being sold at market.

Llamas grazing in the *tierra helada* of southern Peru.

pigs, which are allowed to roam freely about the packed dirt floors of the home until selected for the family stew (Morales 1995).

Depending upon local meteorological conditions, the absolute upper limits of crop cultivation are reached in the 4,000 to 4,600-meter (13,000 to 15,000-foot) range, where one will encounter the treeline and, ultimately, a couple thousand feet above that, the snowline (Figure 3.12). The snowline, or the lower limit of snow cover, is seldom a fixed elevation, but rather advances down the slopes during the wet season and recedes as the snow cover melts during the dry months.

As would be expected, human populations are sparse within the *tierra helada.* The one area of relatively dense settlement is the Altiplano of Bolivia and Peru, although even here, La Paz and Cuzco, the largest population centers, are found at 3,658 and 3,380 meters (12,001 and 11,089 feet) respectively, in the transition zone between the *tierra fría* and the *tierra helada.* Scattered miners and herders can be found living above 4,600 meters (15,000 feet), but they are few in number and subject to a variety of high-altitude physiological disorders (chapter 13).

Altitude thus serves as the leading control of temperature in the Latin American tropics. It is a remarkable experience to ascend a great volcano from its lower to upper slopes. Not only does one pass through each of the agricultural zones described above, but the natural vegetation, if it remains, can change from the diverse hardwoods of the rain forest into a zone of

FIGURE 3.12

Treeline and snowline on the upper slopes of Popocatépetl in the Neovolcanic Range of southern Mexico.

needleleaf conifers within the *tierra fría*, to the alpine meadows and tundra vegetation of the high *páramos* of the *tierra helada*. It is possible to encounter on a single mountain almost every major climate type found in the world, as well as innumerable microenvironments. The environmental diversity of Latin America has had a great impact, in turn, on the economic development of the region, which as a whole presents special challenges to large-scale agricultural technologies.

OCEAN CURRENTS

While local temperatures are most generally a function of altitude, many of the broad regional temperature patterns of Latin America are a reflection of the prevailing **ocean currents.** These currents are far larger and deeper than the rivers of the continents and are classified as either cold or warm. Rather than indicating exact water temperature, these designations are comparative terms. A cold current is simply one whose temperature is colder than the water of the region into which it is moving. Cold currents thus move from polar or high-latitude areas into the equatorial zones.

Conversely, warm currents bring water from equatorial regions into areas whose seas would otherwise be colder.

Because the currents are so large, they have a profound impact on the temperature of the adjoining coastal landmasses. Cold currents produce adjacent land temperatures far cooler than would be expected on the basis of altitude or latitude. Cold and warm ocean currents also affect atmospheric humidity and precipitation on land. Because cold air cannot hold as much water vapor as warm air, the cold currents of Latin America not only lower the temperature of the adjoining land surfaces but also contribute to the presence of arid desert conditions. Warm currents, on the other hand, bring humid, muggy conditions to the *tierra caliente.*

Latin America is influenced by three cold currents, each partly responsible for the presence of a cool desert region (Figure 3.13). The northernmost of these is the California Current, which flows southward along the Baja California peninsula and is a major influence on the aridity of northwestern Mexico.

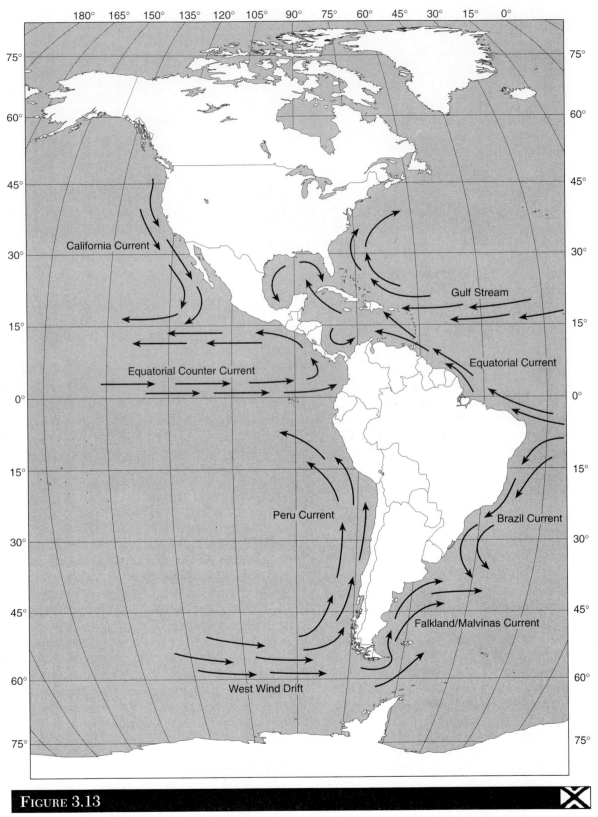

FIGURE 3.13

Latin American ocean currents.

The other two cold currents both derive from the Pacific West Wind Drift off the coast of Antarctica. Part of this water flows past Tierra del Fuego and bends northward into the south Atlantic where it is named the Falkland, or Malvinas, Current after the small islands over which Great Britain and Argentina warred in the 1980s. The Falkland Current dominates the Argentine coast as far north as the Gulf of San Matías, helping bequeath to Patagonia its cold, semi-desert climate. North of the gulf, the Falkland Current curves eastward into the Atlantic, marking the transition to the warm humid lands of the Pampa.

The largest of Latin America's cold currents is the Peru (formerly Humboldt) Current, which extends along the west coast of South America from southern Chile to the Peru-Ecuador border. Interestingly, water temperature is colder off the coast of central Peru than it is off the coast of northern Chile, owing to the upwelling off Peru of extremely cold water from great depths below. Regardless of these minor variations, the current both cools and dries the adjoining land and is largely responsible for a cool desert coastal environment that is temporarily altered on average every three to four years when warm waters from the western Pacific Basin override the colder Peru Current. This condition, known as an **El Niño–Southern Oscillation (ENSO)** event, may bring torrential downpours, widespread flooding, and severe environmental and economic disruption to these otherwise arid regions (McGregor and Nieuwolt 1998; Ropelewski and Halpert 1987; Waylen and Caviedes 1986).

Warm currents dominate the remaining coastal regions of Latin America. These include the Pacific North Equatorial Counter Current, which flows past the west coast of southern Mexico, Central America, and northern South America; the Brazil Current, which extends from northeast Brazil southward to the Argentine Pampa; and the Atlantic North Equatorial Current and the Gulf Stream, which are responsible for the warmth and humidity of the Atlantic coast of northern South America and the greater Caribbean Basin.

LATITUDE

A third control of temperature in Latin America is **latitude.** The general principle operative here is that seasonal temperature fluctuations tend to increase as one moves away from the equator. This happens primarily because areas on the equator experience twelve hours of sunlight and twelve hours of darkness every day of the year, whereas other areas have longer days during the summer and shorter days during the winter. As a result, the greatest summer-winter temperature extremes occur in those parts of Latin America furthest from the equator, while a remarkable, almost monotonous, constancy of both daily and seasonal temperatures characterizes those regions bordering the equator. The highest temperatures in Latin America are thus found in the regions bordering the Tropics of Cancer (23 1/2°N) and Capricorn (23 1/2°S) rather than in those situated close to the equator. If an area experiences alternating wet and dry seasons, the very hottest month generally will occur at the end of the dry season when the sun's direct rays are close to reaching their highest point in the sky, but before the thunderstorms and clouds of the rainy season have begun to moderate somewhat the intensity of the daily heat buildup.

Mexico, the Caribbean, and southern South America are far enough from the equator to make them subject, in their respective winters, to the invasion of polar cold fronts. These are given the descriptive name *nortes,* or "northers," in Mexico and Central America. They bring gentle rain followed by moderate frosts and brilliantly clear, crisp days to the high valleys of southern Mexico and northern Central America (Figure 3.14). Likewise, Antarctic winter fronts, called *pamperos,* sweep northward across the Patagonian Plateau and Argentine Pampa bringing extended periods of cold, damp, drizzly weather. Periodically, light snows and frost occur as far northward as the Paraná Plateau of southern Brazil. The relatively cool and dry air masses may even be strong enough to push up the Gran Chaco between the Andes and the Brazilian Highlands as far northward as Paraguay and eastern Bolivia and from there on into the southwestern Amazon Basin. There the winds, called *friajes* in Peru and *friagem* in Brazil, bring brief but welcome respites from the heat and humidity that would otherwise prevail. Even though the polar cold fronts do not bring excessively low temperatures, as measured simply on a thermometer, the combination of winter cold and humidity often causes great suffering to those who have little warm clothing to wear and who frequently feel colder inside their unheated homes than outside. The frosts also restrict the poleward expansion of the cultivation of cold-sensitive crops, particularly citrus in Mexico and coffee in Brazil.

Precipitation

As with temperature, precipitation patterns in Latin America are primarily an expression of the interrelationships of three physical controls. These controls are continentality, shifting atmospheric pressure belts, and the prevailing winds and orographic effects.

FIGURE 3.14

One of the effects of polar cold fronts, called *nortes* in Mexico and Central America, is the occurrence during the winter months of extended periods of low clouds, fog, and misty rains. This scene is of Tegucigalpa, Honduras.

CONTINENTALITY

One control of precipitation is continentality, or distance from the sea. All else being equal, the further an area is from the ocean, the drier its climate is likely to be. From a macroscale perspective, it is apparent that Latin America's generally narrow, elongated shape has resulted in little continentally-induced aridity, especially when compared to the interior heartland regions of Anglo America, Asia, Africa, and Australia. The only Latin American regions whose dryness is associated to any significant extent with continental location are the Mesa del Norte of north central Mexico and the *puna brava* of the Andean highlands of northwestern Argentina and southern Bolivia. Interestingly, the most continental, or inland, part of Latin America is the heart of the Amazon Basin in central Brazil, which receives large amounts of rainfall.

ATMOSPHERIC PRESSURE BELTS

The key to understanding the major seasonal precipitation patterns of Latin America is knowing the nature and movement of the earth's great atmospheric pressure belts (Christopherson 2002; Strahler and Strahler 2000). These belts, which are depicted in Figure 3.15, are the product of both thermal and nonthermal, or dynamic, forces, and they extend east to west around the earth in huge cells or bands that average some thirty degrees of latitudinal width.

Because warm air is less dense than cold air, a zone of low atmospheric pressure, called the **equatorial low,** is associated with the perpetually hot equatorial regions, extending out to approximately 15 degrees north and south latitude. Poleward of the equatorial low pressure belt, in both the Southern and Northern Hemispheres, are dynamically induced high pressure cells, commonly referred to as the **subtropical highs.**

FIGURE 3.15

Atmospheric pressure belts and idealized seasonal precipitation in Latin America.

These highs, which are centered at approximately 30 degrees north and south latitude, are bordered in turn on their poleward sides by the **subpolar low** pressure belts that adjoin the thermally induced Antarctic and Arctic **polar highs.**

The simplest way to understand the relationship between atmospheric pressure and precipitation is to remember that high pressure is almost always associated with dry, or arid, atmospheric conditions. Conversely, low atmospheric pressure is generally associated with wet atmospheric conditions.

As we apply these basic principles to the precipitation characteristics of the great pressure zones, what emerges is that both the equatorial low and the subpolar low pressure belts are associated with high-precipitation atmospheric conditions, while the subtropical high and polar high pressure cells produce arid conditions. It is the north-south movement or shifting of these belts, in response to the annual migration of the direct rays of the sun, that most accounts for the seasonality of precipitation in Latin America.

Excluding certain local circumstances that produce occasional exceptions to the general patterns, we find that the equatorial regions of South America are rainy throughout most of the year. This is because these areas do not escape for long the influence of the equatorial low pressure belt during the annual passage of the sun's direct rays between the Tropic of Cancer and the Tropic of Capricorn.

If we move poleward some ten to twenty degrees away from the equator, however, we enter extensive areas in both South and Middle America that experience wet summers and dry winters. The relative length and timing of these seasons are a function of the seasonal shifts of the equatorial low and subtropical high pressure belts. Taking Central America, southern Mexico, and the Caribbean as an example, we find that the rainy season generally starts sometime in May with the arrival of the northern edge of the equatorial low. As the low settles in over the region, the heaviest rains occur from June through September, with October usually a tapering-off month. By November, the direct rays of the sun have shifted deep into southern South America, and the Northern Hemispheric subtropical high is influencing most of Middle America, bringing a prolonged low-sun dry season that is not broken until the arrival the following May of the equatorial low.

In the Southern Hemisphere, the precipitation controls are identical, only with the seasons reversed. Throughout much of Brazil and eastern Bolivia the rainy season begins in October with the arrival of the equatorial low. The season reaches its peak from November through March, after which the dry season begins with the growing influence of the subtropical high. Because subtropical high pressure cells are found both north and south of the equator, far more of the Latin American tropics is dominated by a wet summer and dry winter climate than by a year-round wet climate.

Spanish and English terminology for the seasons can cause confusion. In a climatic sense, summer, or *verano* in Spanish, is that part of the year when the sun's rays are most directly overhead, that is, the high-sun period. Likewise, winter or *invierno* is the low-sun period. Owing, however, to the fact that the Mediterranean climate found in coastal Spain and Portugal is characterized by dry summers and wet winters, many Latin Americans refer to the dry low-sun season as *verano* and the wet high-sun months as *invierno* or winter. Also one often hears, in Venezuela, Colombia, and Central America, the term *veranillo,* or little summer, used in reference to a short dry season that falls between two peaks of summer rainfall.

If we continue into the 30 to 40-degree latitude range in each hemisphere, we encounter an area in central Chile and another in extreme northwestern Mexico that experience a summer dry season influenced by the subtropical high and a rainy winter dominated by the subpolar low. Southernmost Chile, on the other hand, never escapes the stormy, rainy influences of the subpolar low.

PREVAILING WINDS AND OROGRAPHIC EFFECTS

The third major control of precipitation patterns in Latin America is called the **orographic effect.** The prefix *oro–* means mountain-related, and orographic precipitation refers to rain or snow that falls as the result of a moisture-carrying air mass being driven up a mountain slope and cooled to the point where condensation occurs.

The dominant winds of Latin America, as in other parts of the earth, are driven by pressure differences between two neighboring regions. The basic operative principle is that air moves from areas of high pressure to areas of low pressure. The greater the difference in atmospheric air pressure, or pressure gradient, the stronger the winds will be.

The huge subtropical high pressure cells noted in our previous discussion are centered at 30 degrees north and south latitude in the Atlantic and Pacific Oceans. If the earth were not rotating on its axis, air, or wind, would blow north and south out of those cells toward the equatorial low and subpolar low pressure belts (Figure 3.16). The spinning or rotation of the earth on its axis, however, deflects the moving air and thereby sets in motion the dominant winds of Latin America (Figure 3.17).

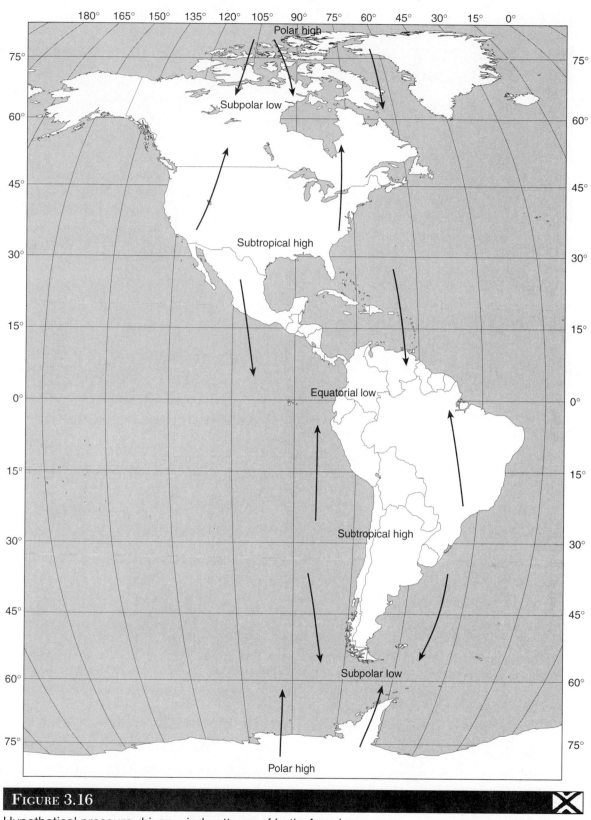

FIGURE 3.16

Hypothetical pressure-driven wind patterns of Latin America.

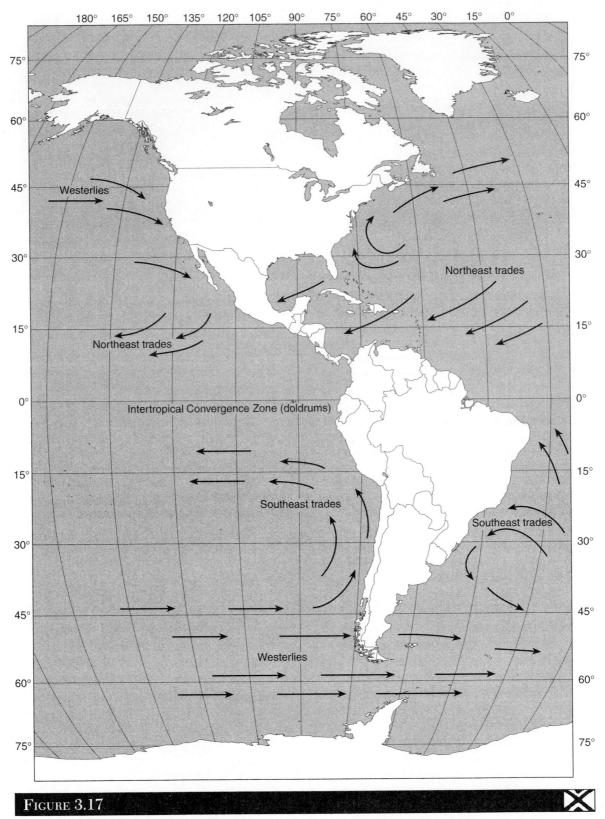

FIGURE 3.17

Actual prevailing winds of Latin America.

These winds, which are named for the direction from which they come, can be divided into two groups. The **westerlies** emanate from the poleward side of the subtropical high pressure cells and become strong west-to-east winds between 40 degrees and 70 degrees latitude. These winds, which control the movement of frontal storms in much of Canada and the United States, do not extend southward enough to significantly affect Mexico and the Caribbean. The greatest impact of the westerlies within Latin America is in southern Chile, where they generate incessant winds that beat against the southern Andes. The winds and associated ocean currents are so strong that it took Magellan thirty-eight days to traverse the dangerous 580-kilometer (360-mile) strait between the continental mainland and Tierra del Fuego that now bears his name. These latitudes subsequently became known to mariners as the "roaring forties," "furious fifties," and "screaming sixties."

The second group of winds originates from the equatorward sides of the subtropical high pressure cells and is deflected westward while moving steadily toward the equator. These are the **northeast trades** of the Northern Hemisphere and the **southeast trades** of the Southern Hemisphere (Figure 3.18). Unlike the powerful westerlies, which bring considerable climatic hardship to the inhabitants of southern Chile, the trade winds historically have contributed much to the physical comfort and economic development of the peoples of the greater Caribbean Basin. They bring refreshing sea breezes and formerly sustained welcome wind-driven trade and commerce, the latter being the basis for the name "trade winds."

As the northeastern and southeastern trades converge toward the equator, their momentum diminishes owing to the increased distance from the high pressure cells and to the increased diameter of the earth as one approaches the equator. The effect is to create a region within seven to eight degrees either side of the equator that has no prevailing winds, and is called the **Intertropical Convergence Zone,** or the **doldrums.** The absence of prevailing winds, combined with the dominance of low atmospheric pressure, which allows the air to rise, has resulted in convective thunderstorm precipitation dominating within these regions.

Elsewhere in Latin America, however, there are numerous examples on both a small, local scale and a large, regional scale of great climatic contrasts created by the collision of the prevailing trades or westerlies with elevated land surfaces. One of the most visually striking examples of contrasting orographic precipitation and drought occurs on those Caribbean

FIGURE 3.18

One expression of the prevailing northeast trade winds of the Caribbean island of Aruba is the almost horizontal growth habit of some of the island's trees.

islands high enough to cause the northeast trades to be lifted up and cooled (Figure 3.19). Because the warm, moisture-laden trade winds flow along a northeast-to-southwest path, torrential downpours occur on the windward, northeastern slopes of the mountainous cores of the islands. Conversely, as the former rain-bearing winds descend the leeward side of the interior highlands, their temperatures rise, resulting in a drying of the air mass and the formation of rainshadow deserts on the southwestern portions of the islands. Puerto Rico is an excellent example of the orographic precipitation patterns characteristic of most Caribbean islands. Average rainfall declines from over 150 centimeters (60 inches) at San Juan on the northeast coast (and considerably more on the windward slopes of the nearby interior mountains) to about 90 centimeters (36 inches) at Ponce on the southern plain just a few kilometers away. This same pattern prevails on the islands of the Lesser Antilles, with the exception of those belonging to the low, drought-stricken outer volcanic arc whose more gentle relief and limestone surfaces conspire both to minimize rainfall and to reduce the presence of surface streams and lakes.

The identical pattern of wet, lushly vegetated, windward regions abutting dry, desertlike leeward zones occurs throughout the Latin American mainland to a far greater extent than is commonly recognized (Table 3.1). A spectacular example in Mexico is the contrast between the luxuriant, often cloud-enveloped uplands of the east-facing escarpment of the southern Mesa Central and the parched, windswept high plateau country to the west

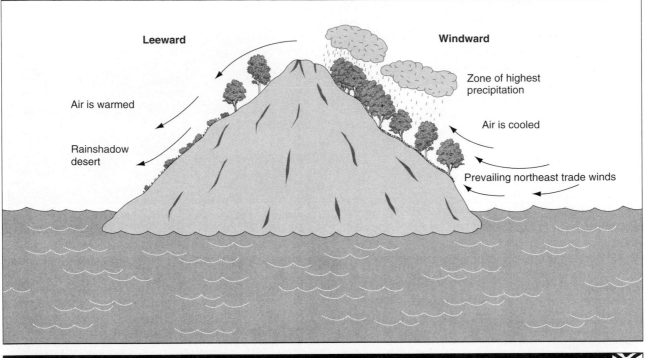

FIGURE 3.19

Orographic precipitation and drought on a Caribbean island.

TABLE 3.1

Orographic Precipitation Regions of Latin America (all figures in centimeters)

Region	WINDWARD STATION		LEEWARD STATION	
	Name	Annual Precip.	Name	Annual Precip.
Puerto Rico	San Juan	154	Ponce	91
Jamaica	Port San Antonio	348	Kingston	80
Southeastern Mexico	Córdoba	229	Tehuacán	48
Panama	Colón	325	Balboa Heights (Panama City)	174
Northwestern Colombia	Buenaventura	394	Cali	97
Tierra del Fuego	Evangelistas Is.	302	Punta Arenas	49
Northeastern Brazil	Fortaleza	163	Quixeramobim	85
Western Amazon Basin/Pacific Coast	Constancia	305	Guayaquil	99

Sources: Köppen and Geiger 1936; Kendrew 1953; Crist 1952; Hiraoka 1989.

FIGURE 3.20

Luxuriantly vegetated, cloud-enveloped windward slopes of the east-facing escarpment of the Mexican Mesa Central.

(Figures 3.20 and 3.21). In Central America, most stations along the eastern Caribbean plains, from Nicaragua southward through Panama, receive from 380 to 635 centimeters (150 to 250 inches) of rain per year, while parts of the leeward Pacific coastal plains average 75 to 150 centimeters (30 to 60 inches).

The Pacific coast of South America contains two of the most intense orographic rainfall zones in all Latin America. The first is the western slopes of the Serranía de Baudó of northwestern Colombia and the neighboring mountains of the Darién of eastern Panama, where winds associated with the Equatorial Counter Current release upwards of 760 centimeters (300 inches) of rain annually in one of the rainiest regions on earth. The second is in southern Chile, where the westerlies lash the deeply fjorded coastline in a manner reminiscent of the winter storms of northwestern Europe. They then descend the leeward slopes of the Argentine Andes, creating a great rainshadow that extends from

FIGURE 3.21

Goats grazing in the semiarid leeward zone of the eastern Mexican Mesa Central.

southernmost Patagonia northward to Mendoza and San Juan. Other large-scale orographic zones in eastern and central South America include the Great Escarpment of the Brazilian Highlands, which separates the well-watered Atlantic coast of Brazil from the semiarid interior backlands of the Northeast, and the *Yungas* and *montaña* regions of northeastern Bolivia and eastern Peru. Although situated on the eastern slopes of the Andes, these latter regions contribute to the uplifting of moist Atlantic winds originating thousands of kilometers to the east, and receive considerably greater rainfall than the central Amazonian plain (Barry and Chorley 1998). In this context, the west coast deserts of Peru can be viewed in part as an arid rainshadow region of the central Andes.

The orographic effect thus plays an extremely prominent role in the precipitation patterns of Latin America, on both a macro- and a microscale. One of the most physically striking experiences available in Latin America and the Caribbean is to pass within a matter of minutes from tropical rain forest to desert vegetation on opposite sides of a single mountain. The presence of many different climates within a small area is yet another expression of the intense physical diversity of the region.

HURRICANES

In addition to the normal weather patterns, which can be explained largely as a function of the controls of climate, there develop in the warm waters of the northern Atlantic and Pacific Oceans storms of extraordinary magnitude whose frequency and paths vary greatly from one year to another. These systems are called **hurricanes** in the Atlantic and Caribbean and *chubascos* along the west coast of Mexico and northern Central America. Hurricanes are tropical cyclones, or low pressure systems, and their development often, but not always, occurs on the poleward side of the Intertropical Convergence Zone, which gradually overspreads the region from the south during the spring and summer months. Many hurricanes develop from low-level atmospheric easterly waves that originate as far away as western Africa and then advance westward across the Atlantic within the northeast trades (Elsner and Kara 1999; Caviedes 1991; Gray 1990 and 1978; Pielke 1990). Only a few of these waves develop into hurricanes, however, with the Caribbean experiencing an average of six to ten storms each year (Landsea 2001; Reading 1990; Neumann, Jarvinen, and Pike 1990; Millas 1968). Another condition required for hurricane formation is an ocean water surface mean temperature of at least 26.7° C (80° F). This is necessary because hurricanes

derive their energy primarily from the latent heat within the ocean water itself, heat that is released in great amounts as water vapor flows into the storm, rises, cools, and condenses. Hurricanes are not known to develop between 5° north and 5° south latitude; consequently, the southern Antilles, Venezuela, Panama, and Costa Rica experience far fewer storms than northern Middle America and the United States.

The simultaneous presence of the Intertropical Convergence Zone and ocean water temperatures of 26.7° C (80° F) or above is seldom reached in the Caribbean until May, and the hurricane season officially begins on June 1. Even so, the first part of the season is usually quiet, and approximately 80 percent of the storms occur during the months of August, September, and October.

The World Meteorological Organization has established three categories of tropical cyclones, based upon the strength or speed of the highest sustained winds. The lowest level is that of a tropical depression, which is upgraded to the status of tropical storm once the sustained winds reach 63 kilometers (39 miles) per hour. Not all tropical storms continue to grow in intensity, but those that do are classified as hurricanes when the sustained winds are 119 kilometers (74 miles) per hour or greater. Beyond that point, hurricanes are classified by categories of intensity, with Category One being the weakest and Category Five reserved for the strongest (Table 3.2). It should be noted that gusts within the hurricane commonly exceed the sustained wind speeds by 25 to 50 percent, meaning, for example, that a Category Three storm with 193-kilometer (120-miles)-per-hour sustained winds would probably generate gusts in the range of 241 to 290 kilometers (150 to 180 miles) per hour.

The arrival of a major hurricane is preceded by beautiful warm, clear weather. As the storm approaches, however, bands of high cirrus clouds can be seen extending out hundreds of kilometers from the eye. These will give way to eerie, totally overcast skies, which precede the arrival of squall lines bringing intermittent gusts and rains. All preparations for the storm must be completed by this time, for relentless winds and torrential downpours, often exceeding thirty centimeters (one foot) of rain, accompany the passage of the system. Palms are bent almost parallel to the ground, massive trees can be uprooted, and homes, power lines, water lines, highways, and crops are all destroyed within minutes by the winds, floods, and mudslides. Although inland districts are subject to major damage, the loss of human life and property is generally greater along the coastline where a storm surge up to 7.6 meters (twenty-five feet) high may sweep across low-lying areas (Figure 3.22).

TABLE 3.2

Rating Scales of Tropical Cyclones

CLASSIFICATION	SUSTAINED WIND SPEED (kilometers per hour)	PROBABLE STORM IMPACTS
Tropical Depression	<63	Moderate winds and thundershowers.
Tropical Storm	63–118	Strong winds, heavy rain, minor damage to poorly constructed structures.
Category One Hurricane	119–153	Moderate damage to trees, crops, and poorly constructed structures; low-lying roads may be flooded and water and electrical supplies disrupted.
Category Two Hurricane	154–177	Most coastal roads under water; widespread roof, window, and door damage to well-constructed structures.
Category Three Hurricane	178–209	Large trees uprooted; serious coastal flooding and erosion; near total destruction of poorly constructed buildings.
Category Four Hurricane	210–249	Roofs blown off well-constructed buildings; widespread inland flooding from storm surge.
Category Five Hurricane	>249	Near total destruction of well-constructed buildings, roads, and bridges.

Sources: Simpson 1971; Case 1990.

FIGURE 3.22

Hurricanes are capable of washing large watercraft, cars, and even homes considerable distances inland. This scene is from the U.S. Virgin Islands.

Hurricanes historically have ranked with earthquakes and volcanic eruptions as the greatest natural disasters of the greater Caribbean Basin. A given location, such as Hispaniola, can expect to be struck over and over again with the passage of time (Figure 3.23). While not all hurricanes are major storms, many are and it is not uncommon for a single tropical cyclone to cause the loss of thousands of human lives and billions of dollars in property damage (Table 3.3). Hurricane Gilbert, which struck Jamaica, the Cayman Islands, and Mexico in September 1988, was the most powerful in recorded history. Hurricane Mitch brought immense destruction and suffering to Central America in October 1998 (Figure 3.24) (Case Study 3.1). One positive consequence of hurricanes, from a hydrologic perspective, is the replenishment of groundwater reservoirs. This, however, is of no consolation to the residents of the affected areas, whose economic and social reconstruction often takes years to complete.

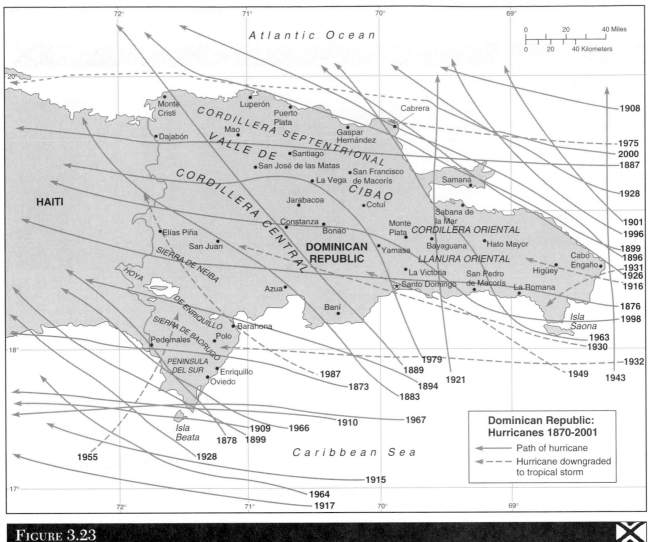

FIGURE 3.23

Dominican Republic hurricanes: 1870–2001.

Redrawn and updated with permission from Oscar H. Horst, "Climate and the 'Encounter' in the Dominican Republic," *Journal of Geography*, Vol. 91, Sept./Oct. 1992, p. 208.

FIGURE 3.24

A temporary bridge built over a section of road destroyed by Hurricane Mitch. This area is part of the interior Honduran highway connecting Tegucigalpa to Valle de Angeles.

TABLE 3.3

Recent Major Caribbean Hurricanes

HURRICANE	YEAR	AREAS MOST AFFECTED	DAMAGE
Unnamed	1931	Belize City, British Honduras	Major flooding and winds killed 1,000 people, approximately 2 percent of the national population.
Unnamed	1955	Chetumal City, Yucatán Peninsula	282-km sustained winds; widespread damage to coastline and inland forests.
Hattie	1961	Belize City, British Honduras	Storm surge inundated areas almost 2 km inland, prompting work to begin on Belmopan, a new capital city 80 km inland, the following year.
Flora	1963	Tobago, Haiti; central Cuba	Total of 2.5 million people affected throughout the Caribbean Basin; 6,000 lives lost.
Fifi	1974	Caribbean coast of Honduras	17,000 lives lost; widespread flooding of banana plantation lands.
Frederick	1979	Dominican Republic	Extensive infrastructural and crop damage.
Allen	1980	Saint Vincent	Destroyed almost the entire banana crop, along with most roads.
Gilbert	1988	Jamaica, Cayman Islands, Mexico	Most powerful hurricane ever recorded in Western Hemisphere; over $10 billion in property damage affecting approximately 6 million people.
Hugo	1989	Leeward Islands, Puerto Rico, Virgin Islands, southeastern United States	Inflicted severe damage in the Caribbean before striking the United States mainland; caused 504 deaths.
Luis	1995	Leeward Islands	Destroyed the airport and harbor of St. Bart's; inflicted over $240 million in damage to St. Kitts-Nevis and washed 1,000 yachts and houseboats across piers and beaches on St. Maarten.
Georges	1998	Northern Lesser Antilles, Puerto Rico, Dominican Republic, Haiti, Cuba, southeastern United States	Damaged three-fourths of the homes on St. Kitts and left over 100,000 homeless in the Dominican Republic; inflicted major losses to banana and other plantation crops of the Greater Antilles.
Mitch	1998	Honduras, Nicaragua, El Salvador, Guatemala	Possibly the worst natural disaster of the twentieth century in the Western Hemisphere; over $6 billion in damages, including the loss of most of the infrastructure of Honduras and Nicaragua; over 10,000 confirmed deaths with another 10,000 missing; over 3 million left homeless.
Iris	2001	Belize	Struck southern Belize as a small but intense Category Four hurricane that generated a 5.5 meter (18 foot) tidal surge. Several communities lost 80–90 percent of their buildings and over 13,000 persons were left homeless.

Sources: Dixon 1991; Eyre 1989; Case 1990; Macpherson 1980; James and Minkel 1986; *Britannica Book of the Year,* annual; National Hurricane Center, National Oceanographic and Atmospheric Administration: http://www.nhc.noaa.gov

Case Study 3.1:   Hurricane Mitch

On Wednesday, October 21, 1998, a weak tropical depression formed from an easterly wave located in the western Caribbean Sea about 580 kilometers (360 miles) south of Kingston, Jamaica. Wandering slightly toward the west, then southward, and later northward almost back to the point where it had originated, the somewhat disorganized system initially caused little alarm. Over the next several days, however, it gradually strengthened as it continued on a mostly northerly track, achieving hurricane status on Saturday, October 24. At that point, the hurricane seemingly exploded in strength and began to move westward toward the Central American coast.

By the following Monday afternoon, the central atmospheric pressure in Hurricane Mitch, as the system had been named, had dropped to 906 millibars, making it one of the most intense Atlantic cyclones in recorded history. Sustained wind speeds briefly reached 290 kilometers (180 miles) per hour as the storm passed over Swan Island and then steadily lessened as it drifted toward northern Honduras. Had Mitch maintained a steady forward movement, the ensuing damage likely would have been relatively light, but it stalled off the coast for three days. As the torrential rains, often totaling over ten centimeters (four inches) per hour, fell without letup over the interior mountain regions of Honduras, rivers quickly overflowed their banks, causing unprecedented flooding. Tens of thousands of residents along the north coast survived only by climbing to the tops of nearby low-lying hills, by waiting out the devastating floods on the roofs of buildings, or by clinging to the upper branches of trees. In Teguci-

galpa, the national capital, water from the normally shallow Choluteca River surged through downtown streets destroying homes and businesses alike.

Finally, on Thursday, October 29, Mitch assumed a forward motion and moved inland over central Honduras, all the while continuing to bring almost constant downpours to the affected region, which now also included Guatemala, El Salvador, and Nicaragua. The following day, the United States National Weather Service issued a bulletin warning that the weakened tropical storm could still bring an additional thirty-eight to sixty-four centimeters (fifteen to twenty-five inches) of rain in northern Central America. In Nicaragua, one of the effects of the unrelenting rains was to cause a substantial rise in the levels of many lakes found within the craters of dormant volcanoes. On Friday, October 30, the increased pressure created by the sudden growth of the crater lake of Casitas Volcano, located 90 kilometers (55 miles) northwest of Managua, caused a huge hole to abruptly open in the side of the mountain. As the resulting wall of mud and water rushed down the slope without warning, ten villages were engulfed, killing thousands of persons. By Sunday, November 1, the center of Mitch had moved westward across southern Guatemala, where it brought still more heavy rains and severe flooding. Near the Chiapas border, the system, now downgraded to a tropical depression, assumed a northward path across Mexico, emerging briefly in the Bay of Campeche before continuing northeastward across western Yucatán on its way to southern Florida and the colder waters of the northern Atlantic Ocean.

With the end of the storm, the stunned survivors were confronted with a panorama of almost unimaginable ruin and devastation. Six days of ceaseless downpours had inflicted economic losses estimated at over $6 billion. Over 3 million people, approximately one-tenth of Central America's population, were left homeless. In Honduras, over 60 percent of the nation's infrastructure, including 92 percent of its bridges, was destroyed, leaving most of the cities as functional islands for weeks thereafter. The banana and other commercial plantation crops of the northern coastal plains were a near total loss, but widespread famine was averted owing to the survival of most of the corn, rice, beans and other staple crops planted on the small, better-drained fields of the interior highlands. Nicaraguan authorities estimated that over 70 percent of the nation's roads were impassable immediately following the passage of the storm, leaving hundreds of thousands of rural residents without access to adequate food, water, or health care. Many of these migrated as desperate refugees into Managua in the weeks that followed. Damage in Guatemala and El Salvador was less severe, yet the former lost an estimated 95 percent of its banana crop and a third of its beef cattle. El Salvador lost much of its cotton and coffee plantings. Confirmed deaths in the four countries totaled well above 10,000 with another 10,000 or more reported missing and likely to be declared dead at a later time. In the aftermath of the storm, additional hundreds of people died from the spread of life-threatening diseases, including cholera, malaria, diarrhea, classic and hemorrhagic dengue fever, and leptospirosis.

continued

Case Study 3.1:   Hurricane Mitch *continued*

(Leptospirosis is transmitted by contact with water contaminated by waste products of rodents and causes liver and kidney failure.) In truth, the exact number of persons killed by Mitch will probably never be known.

Even with the massive international relief efforts mounted in the aftermath of the storm, it is now clear that Central America's recovery from Hurricane Mitch will require years, and in some areas possibly decades, to complete. In addition to the losses described earlier, other long-term impacts include a prolonged disruption of schooling for many youth, widespread unemployment owing to the closing of businesses and industries, and chronic physical limitations and emotional trauma in the lives of many of its victims. The storm will surely be viewed historically as one of the worst natural disasters ever to strike the Western Hemisphere.

SUMMARY

The weather and climate patterns of Latin America are an expression of the complex interactions of numerous climatic controls. The temperatures experienced at a given location are most influenced by three factors: altitude, ocean currents, and latitude. All else being equal, the higher the altitude or elevation of an area, the colder the temperatures will be. Four altitudinal life zones can be distinguished in Latin America: *tierra caliente, tierra templada, tierra fría,* and *tierra helada.* Each is associated with the cultivation of crops and raising of animals specifically adapted to the host environment. The cold ocean currents are largely responsible for Latin America's three cool desert regions, while the warm ocean currents provide the moisture and warmth necessary to sustain the region's tropical rain forest and savanna biomes. Most of Latin America is close to the equator, but southern Argentina and Chile and northern Mexico are far enough away to be subject to polar air masses during their respective winters.

The precipitation patterns of Latin America are controlled by continentality, atmospheric pressure belts, and the orographic effect. Of these, the shifting of the low and high pressure belts is most responsible for the seasonality of precipitation. The interplay between prevailing winds and elevated landmasses accounts for dramatic variations in annual rainfall totals between locations extremely close to one another. Hurricanes, or tropical cyclones, occur in both the Caribbean and Pacific waters of Middle America, often inflicting great loss of human life and property.

One of the most appealing aspects of Latin America is that it offers almost every climate found on the earth, with radically different climates frequently found within short distances of each other. Climate is the major determinant of vegetation type, and climate, vegetation, and relief and parent rock material are the principal influences in the development of the various soils. Climate, vegetation, and soils are, in turn, the most significant components of natural regions or biomes, the nature and distribution of which are the focus of chapter 4.

KEY TERMS

altitudinal life zones 46
ocean currents 51
El Niño–Southern Oscillation
(ENSO) 53
latitude 53
equatorial low 54

subtropical highs 54
subpolar low 56
polar highs 56
orographic effect 56
westerlies 59

northeast trades 59
southeast trades 59
Intertropical Convergence Zone 59
doldrums 59
hurricanes 62

SUGGESTED READINGS

Barry, Roger G., and Richard J. Chorley. 1998. *Atmosphere, Weather, and Climate.* 7th ed. London: Routledge.

Britannica Book of the Year. (annual). Chicago: Encyclopedia Britannica.

Brush, Stephen B. 1977. *Mountain, Field, and Family: The Economy and Human Ecology of an Andean Valley.* Philadelphia: University of Pennsylvania Press.

Case, Bob. 1990. "Hurricanes: Strong Storms Out of Africa." *Weatherwise* 43:23–29.

Caviedes, César N. 1991. "Five Hundred Years of Hurricanes in the Caribbean: Their Relationship with Global Climatic Variables." *GeoJournal* 23:301–310.

Christopherson, Robert W. 2002. *Geosystems: An Introduction to Physical Geography.* 4th ed. Upper Saddle River, NJ: Prentice-Hall.

Commission on International Relations. 1975. *Underexploited Plants with Promising Economic Value.* Washington, DC: National Academy of Sciences.

Crist, Raymond E. 1952. *The Cauca Valley, Colombia.* Baltimore: Waverly Press.

Dixon, Clifton. 1991. "Yucatan After the Wind: Human and Environmental Impact of Hurricane Gilbert in the Central Yucatan Peninsula." *GeoJournal* 23:337–345.

Elsner, James B., and A. Biro Kara. 1999. *Hurricanes of the North Atlantic: Climate and Society.* New York: Oxford University Press.

Eyre, L. Alan. 1989. "Hurricane Gilbert: Caribbean Record-Breaker." *Weather* 44:160–164.

Gade, Daniel W. 1970. "Ethnobotany of Cañihua (*Chenopodium pallidicaule*), Rustic Seed Crop of the Altiplano." *Economic Botany* 24:55–61.

———. 1975. *Plants, Man and the Land in the Vilcanota Valley of Peru.* The Hague: Dr. W. Junk N. V. Publishers.

Gray, William M. 1978. "Hurricanes: Their Formation, Structure and Likely Role in Tropical Circulation." In *Meteorology Over the Tropical Oceans,* ed. D. B. Shaw, pp. 163–172. Bracknell, England: Royal Meteorological Society.

———. 1990. "Strong Association between West African Rainfall and U.S. Landfall of Intense Hurricanes." *Science* 249:1251–1256.

Hiraoka, Mario. 1989. "Ribereños Changing Economic Patterns in the Peruvian Amazon." *Journal of Cultural Geography* 9 (2):103–119.

Holdridge, L. R. 1967. *Life Zone Ecology.* San José, Costa Rica: Tropical Science Center.

Horst, Oscar H. 1992. "Climate and the 'Encounter' in the Dominican Republic." *Journal of Geography* 91:205–210.

———. 1997. "The Utility of Palms in the Cultural Landscape of the Dominican Republic." *Principes* 41(1):15–28.

James, Preston E., and C. W. Minkel. 1986. *Latin America.* 5th ed. New York: Wiley.

Jones, David L. 1995. *Palms Throughout the World.* Washington, DC: Smithsonian Institution Press.

Judziewicz, Emmet J., et al. 1999. *American Bamboos.* Washington, DC: Smithsonian Institution Press.

Kendrew, Wilford G. 1953. *The Climates of the Continents.* 5th ed. Oxford: Clarendon Press.

Köppen, W., and R. Geiger, eds. 1930–1936. *Handbuch der Klimatologie.* Vol. 2, Parts G, H, I, and J. Berlin: Gebruder Borntraeger.

Landsea, Christopher W. 2001. "FAQ: Hurricanes, Typhoons, and Tropical Cyclones. Part E: Tropical Cyclone Records." http://www.aoml.noaa.gov/hrd/tcfag/tcfagE.html

Macpherson, John. 1980. *Caribbean Lands.* 4th ed. Trinidad: Longman Caribbean.

McGregor, Glen R., and Simon Nieuwolt. 1998. *Tropical Climatology: An Introduction to the Climates of the Low Latitudes.* Chichester: Wiley.

Millas, Jose C. 1968. *Hurricanes of the Caribbean and Adjacent Regions, 1492–1800.* Miami, Fla.: Academy of the Arts and Sciences of the Americas.

Morales, Edmundo. 1995. *The Guinea Pig: Healing, Food, and Ritual in the Andes.* Tucson: University of Arizona Press.

Neumann, C. J., B. R. Jarvinen, and A. C. Pike. 1990. *Tropical Cyclones of the North Atlantic Ocean, 1871–1986.* Washington, DC: United States Department of Commerce (NOAA).

Pielke, R. A. 1990. *The Hurricane.* London: Routledge.

Popenoe, Hugh, et al. 1989. *Lost Crops of the Incas: Little Known Plants of the Andes with Promise of Worldwide Cultivation.* Washington, DC: Agency for International Development.

Reading, Alison, J. 1990. "Caribbean Tropical Storm Activity over the Past Four Centuries." *International Journal of Climatology* 10:365–376.

Riehl, H. 1979. *Climate and Weather in the Tropics.* London: Academic Press.

Ropelewski, C. F., and M. S. Halpert. 1987. "Global and Regional Scale Precipitation Patterns Associated with the El Niño/Southern Oscillation." *Monthly Weather Review* 115:1606–1626.

Simpson, R. H. 1971. "A Proposed Scale for Ranking Hurricanes by Intensity." *Minutes of the Eighth National Oceanic and Atmospheric Administration National Weather Service Hurricane Conference.* Miami, Fla.

Stadel, Christoph. 1986. "Del Valle al Monte: Altitudinal Patterns of Agricultural Activities in the Patate-Pelileo Area of Ecuador." *Mountain Research and Development* 6:53–64.

Stegemann, H., S. Majino, and P. Schmiediche. 1988. "Biochemical Differentiation of Clones of Oca. (*Oxalis tuberosa,* Oxalidaceae) by Their Tuber Proteins and the Properties of These Proteins." *Economic Botany* 42:37–44.

Strahler, Alan H., and Arthur N. Strahler. 2000. *Introducing Physical Geography.* 2nd ed. New York: Wiley.

Tosi, Joseph, and R. F. Voertman. 1964. "Some Environmental Factors in the Economic Development of the Tropics." *Economic Geography* 40:189–205.

Troll, Carl. 1968. "Geo-ecology of the Mountainous Regions of the Tropical Americas." *Colloquium Geographicum* 9:1–223.

Waylen, Peter R., and César N. Caviedes. 1986. "El Niño and Annual Floods on the North Peruvian Littoral." *Journal of Hydrology* 89:141–156.

Wilson, H. D. 1990. "Quinua and Relatives (*Chenopodium* sect. *Chenopodium* subsect. Cellulata)." *Economic Botany* 44 (3S):92–110.

ELECTRONIC SOURCE

National Hurricane Center, National Oceanographic and Atmospheric Administration: www.nhc.noaa.gov

4

Natural Regions

Having studied the most prominent landforms and major weather and climate controls, we are now prepared to develop an overview of the principal natural regions of Latin America and the Caribbean. Each of these regions is characterized by a common climate and associated vegetation and soil. As we analyze their physical attributes, we will also come to understand more clearly their economic potentials and limitations and the capacity of each to contribute to the long-term economic development of the nations of which they are a part.

TROPICAL RAIN FOREST

The first, and in many ways the least understood, of the natural regions of Latin America is the tropical rain forest. Although the rain forest and closely related tropical monsoon environments do not dominate lowland Latin America and the Caribbean, they are widely distributed throughout these lands (Figure 4.1). The most extensive zone of tropical rain forest is found in the central and western Amazon Basin of Brazil, Bolivia, Peru, Ecuador, and Colombia. Other smaller, yet significant, areas include the Atlantic coast of Brazil from the city of Salvador southward to Vitória and again from Rio de Janeiro to Santos, the Guianas, and the Pacific coasts of northern Ecuador, Colombia, and eastern Panama. From Panama, the great forests extend northward along the Caribbean lowlands of Central America to Belize, where they extend inland into the Petén region of northern Guatemala before eventually crossing into Mexico, where they end in southern Veracruz and Tabasco states. Impressive rain forests

are also found on portions of the northeastern highlands of Jamaica, Hispaniola, Puerto Rico, and the more elevated windward slopes of the Lesser Antilles (Figure 4.2).

Climatically, Latin America's rain forest regions either are wet throughout the year or have such a short dry period that the vegetative growth is unaffected by drought. The formation of soils in these regions is also influenced by the constantly high temperatures. Shown in Table 4.1 are monthly climatic summaries for three representative rain forest stations. Manaus, Brazil, is the dominant city of the central Amazon Basin; Iquitos, Peru, is the major urban center of the upper Amazon Basin; and Greytown (San Juan del Norte), Nicaragua, is a small fishing community along the Central American Caribbean lowlands.

Notice that in none of these stations is the hottest month of the year more than a few degrees warmer than the coldest. In reality, the tropical rain forest regions experience a far greater temperature range between the heat of the midafternoon and the cool of the predawn hours than they do from one season to the next. For this reason, nighttime is often said to be the "winter" of the tropics. It is at this time that many of the animals are most active (Kricher 1989; Gilmore 1963).

Notice also that annual rainfall totals are substantial in the rain forest regions and that even the drier months receive some precipitation. Typically, most of the rainfall in the tropics is of the convective thunderstorm type. Huge cumulus clouds will begin to form around noon or 1:00 P.M. in response to uneven heating of the humid lowlands, and around 2:00 to 3:00 P.M. a torrential downpour may occur that breaks the heat of

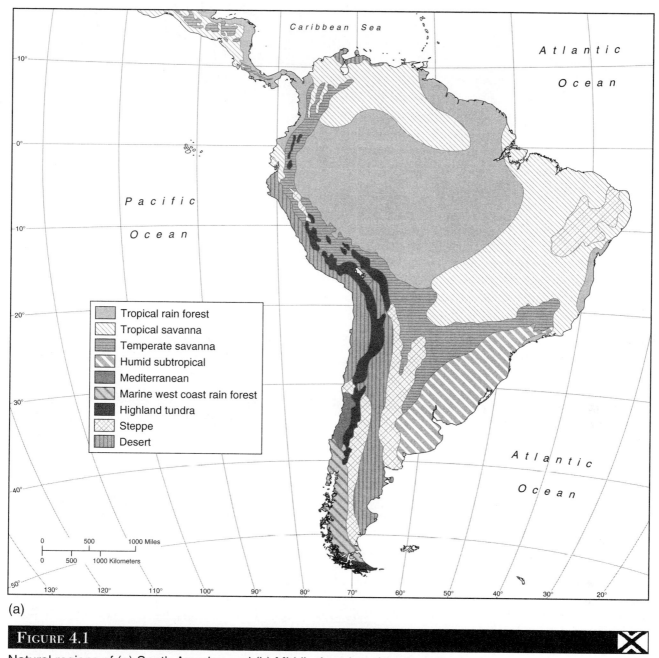

(a)

FIGURE 4.1

Natural regions of (a) South America and (b) Middle America. (*continued*)

the day. Much of the population traditionally goes home for lunch and, if possible, a *siesta* during the hot early afternoon. The stores and businesses generally reopen in the late afternoon and service their clientele into the evening hours.

The plants within the rain forest grow continuously throughout the year in response to the constant heat and abundant soil moisture. As they compete for sunlight, the trees grow to considerable heights and

eventually the mature forest is left with four distinct layers of vegetation (Leigh, Rand, and Windsor 1996; Gentry 1990; Richards 1952) (Figure 4.3). The highest of these are the towering forest giants, whose crowns attain heights of 40 to 50 meters (120 to 150 feet) above the forest floor. These trees frequently have buttressed roots for support. Beneath them is a second layer of trees that achieve heights of 20 to 25 meters (60 to 80 feet). Below these is the third layer whose

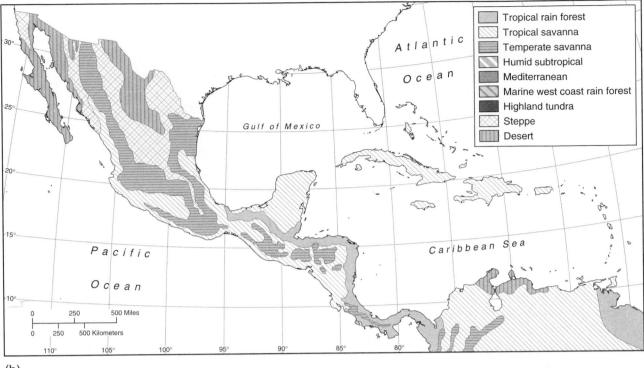

(b)

FIGURE 4.1 *continued*

Natural regions of (a) South America and (b) Middle America.

TABLE 4.1

Tropical Rain Forest Biomes

STATION	ALTITUDE (meters)	*	Jan	Feb	Mar	Apr	May	Jun	Jul	Aug	Sep	Oct	Nov	Dec	YEAR
Manaus	44	T	26.6	26.7	26.5	26.6	26.7	26.7	27.0	27.6	28.2	28.2	27.9	27.0	27.2
		R	24.9	23.4	24.9	21.8	17.8	9.7	5.8	3.6	5.1	10.4	15.5	19.8	182.7
Iquitos	100	T	25.3	25.7	24.6	25.0	24.2	23.5	23.4	24.6	24.6	25.1	25.8	25.5	24.8
		R	25.9	24.9	31.0	16.5	25.4	18.8	16.8	11.7	22.1	18.3	21.3	29.2	261.9
Greytown	sea level	T	25.3	25.4	26.0	27.1	27.1	26.7	26.2	26.3	26.9	26.8	25.8	25.3	26.2
		R	59.2	28.7	16.5	29.0	51.8	58.9	87.4	69.3	44.2	50.8	92.7	70.6	659.1

°T = Mean monthly temperature in degrees Celsius

°R = Mean monthly rainfall in centimeters

Sources: Koeppe and De Long 1958, 324–327; Kendrew 1953, 522–527; Köppen and Geiger 1930–1936.

FIGURE 4.2

Many trees of the tropical rain forest, such as those shown here in northeastern Jamaica, support their great height through roots that resemble stilts or buttresses.

branches reach a "mere" 9 to 12 meters (30 to 40 feet) high. These may be young saplings or species adapted to the darkness of this vertical niche. Finally, along the forest floor are found mosses and shade-loving vines, some of which have been adapted as houseplants for midlatitude homes and offices. A certain class of woody vines, called **lianas,** seek out the trunks of trees to wrap themselves around and may eventually attain lengths of 180 to 250 meters (600 to 800 feet) as they climb to the upper stories and descend back toward the ground in an explosion of vegetative growth.

To understand the rain forest as an ecosystem, we must realize that, owing to the three distinct layers of trees above, very little sunlight—generally less than 1 percent—penetrates to the forest floor. This means that the focus of life in the rain forest is above the ground rather than on it. It is in the trees where the arboreal animals reside, such as monkeys, sloths, jaguars, and snakes, as well as the seemingly endless array of brightly colored tropical birds and insects (Figure 4.4). The forest floor is dark and damp, but contrary to popular perception, has little vegetative and animal life, and is generally easy to move about in.

In addition to the darkness and relative stillness of the forest floor, another attribute of the rain forest is its unequaled biodiversity, often hundreds of species of plants in a single hectare of land (Thorington et al. 1996; Terborgh 1992; Hecht and Cockburn 1989) (Case Study 4.1). One of the most valuable functions of this genetic diversity is to enable the forest ecosystem to withstand the endless onslaughts from the equally large number of diseases and pests that threaten the vegetation.

For centuries, Westerners have been under the misconception that any region that could sustain such luxuriant plant growth must be endowed with tremendously fertile soils that, were the forest to be cleared, could be used to produce vast amounts of food or fiber for humankind. History is littered with failed rain forest development schemes.

What outsiders have generally failed to appreciate is not only the nature of the rain forest but also that of the fragile tropical **oxisol** and **ultisol** soils that lie beneath it. These soils, also called **latosols,** ferralsols, and acrisols, are deeply leached from the year-round precipitation and are generally low in plant nutrients and extremely acidic (Kellman and Tackaberry 1997; Sanchez and Benites 1987; Wambecke 1978). The only way that the plants of the forest survive, then, is to literally live off themselves (Jordan 1985). Most of the nutrients that are found in these lateritic soils are derived from the dead leaves and other fallen litter of the evergreen vegetation above. This organic material is quickly broken down by fungi and other microorganisms at the surface and carried down into the thin topsoils where the tree roots capture the nutrients and recycle them back into the canopies above.

Rather than being a robust, inherently fertile biome, then, the tropical rain forest in reality is one of the most fragile and delicately balanced on earth, and one whose impressive biomass output is more a function of the climate than of the soils that lie beneath. Traditional farmers have adapted to these physical constraints by practicing small-scale polyculture and other environmentally sustainable strategies compatible with the rain forest environment that will be addressed in more detail in chapter 10.

The problems in rain forest agriculture occur primarily when these principles are ignored in large-scale

FIGURE 4.3

Vegetative structure of the tropical rain forest.

FIGURE 4.4

Baby parrots being nursed by a Shipibo Indian in the Amazonian rain forest of eastern Peru. The young birds had fallen from their nest. Animal life in the tropical rain forest is centered on the branches of trees growing high above the forest floor rather than on the ground itself.

Case Study 4.1: Biodiversity and the Tropical Rain Forest

The term "biodiversity" is often used to refer to the number of life forms, be they plant or animal, that exist in a given area. Biodiversity is important to the stability of an ecosystem because, all else being equal, the greater the number of species and varieties, the greater the ability of the ecosystem to withstand the loss of some of its members. Various researchers have estimated that somewhere between 4 and 30 million forms of life exist on the earth today and that only 1.5 to 1.8 million of those have been identified to date. As a general rule, the warmer and wetter the environment of a given area, the greater the biodiversity. Two of the most genetically diverse biomes on earth are the coral reefs found in warm, shallow tropical waters and the tropical rain forests.

Scholars believe that approximately half of all life forms existing on the planet today are associated with the tropical rain forests, which are especially rich in flowering plant and insect life. The Amazon Basin contains the largest surviving stand of rain forest vegetation and may be the most genetically diverse rain forest as well. One indication of the incredible diversity of life in Amazonia is found in two recent studies of a single 5.0 square kilometer (1.93 square mile) section of the Manu Biosphere Reserve of southeastern Peru where researchers identified over 5,000 species of plants and animals. These included 1,147 species of vascular plants, 244 species of mosses and liverworts, 117 species of mammals, 415 species of birds, 128 species of reptiles and amphibians, 210 species of fish, and 2,935 species of insects and spiders (Wilson and Sandoval 1996;

Dallmeir, Kabel, and Foster 1996). As amazing as the findings were, the investigators estimated that the remainder of the reserve contains five to ten times the number of species found in the small study site.

Just as the tropical rain forest exhibits an immense collective array of life, it also varies greatly in its vegetative composition from one microregion to another. These variations can be reflections of minute differences in physical habitats, such as underlying landforms, soil moisture and nutrient levels, rainfall, and exposure to sunlight, as well as diverse human uses of the natural environment (Jain 2000; Mazer 1996). This, in turn, suggests that conservation measures will succeed only if large tracts of land can be set aside as forest preserves. The varied genetic makeup also helps us understand why large-scale monoculture, or the cultivation of a single crop over a wide area, is environmentally incompatible with the rain forest biome. Yet the Amazonian rain forests have been cut and burned recently at a rate equivalent to an area the size of Belgium each year. As a result of the destruction of the Amazonian rain forests, and others in Latin America, Africa, Asia, and Australia, the earth is losing an estimated 4,000 to 6,000 species annually (Wilson 1989).

Why are so many people so concerned about the genetic erosion currently under way? The answers are varied but fall into two general categories. The first is that the genetic erosion results in the loss of invaluable medicinal and otherwise economically useful products (Mittermeir and Konstant 2001; Calderon et al. 2000;

Plotkin 1988). To use medicinal substances as one example, it is estimated that only 12.5 percent of all medicinal drugs potentially available from rain forest plants have been discovered to date (Mendelsohn and Balick 1995). Derivatives from just one of these, the rosy periwinkle, offer a 99 percent chance of remission for victims of lymphocytic leukemia and a 58 percent chance of recovery from Hodgkin's disease. Quinine, an alkaloid derived from the bark of cinchona trees, is used to prevent and treat malaria. Numerous other rain forest substances are used to treat illnesses ranging from heart, eye, and neurological disorders to sickle-cell anemia, Parkinson's disease, and bacterial infections (Posey 2000; Chivian 1997; Wilson 1992). As rain forest species become extinct, we are risking the loss of an incalculable number of medicinal cures and treatments without even realizing what we have forfeited. Other species are of considerable potential worth for use as fuels, fibers, foods, and industrial compounds.

Perhaps an equally compelling argument for maintaining biodiversity is that each species is irreplaceable and, as such, represents a small portion of our earthly heritage—in a sense, the loss of any species is a loss of part of ourselves. In spite of all our technological knowledge, we do not yet have, and may never have, the capacity to create a single new species of life. For these reasons, concerned individuals and institutions from the Amazon Basin nations and elsewhere are endeavoring to develop programs and strategies that will help to stem the loss of biodiversity within the region.

monocultural schemes. Consider that these projects almost invariably depend on mechanized agriculture. Equipment is brought in and the forest is bulldozed and planted to a given crop. Initially, the plantings do well, but after a year or so the residual soil nutrients are exhausted and the only way to keep the crops growing is to add large amounts of chemical fertilizers. The expensive synthetic fertilizers ultimately lower the soil pH even further, kill off most of the beneficial soil microorganisms, and may reduce considerably the profits of the operators. Meanwhile, having temporarily reduced nature's diversity, the farmers find insects and diseases building up in ever-increasing numbers and are forced to resort to more and more frequent applications of chemical insecticides and disease-control compounds, resulting again in accelerated environmental degradation and reduced profit margins.

This is not to suggest that commercial agriculture can never be successfully sustained over the long haul in the rain forest regions of Latin America. For one thing, not all rain forest areas have poor lateritic soils (Reading, Thompson, and Millington 1995; Amazonia Without Myths 1992; Committee on Tropical Soils 1972). Soils along alluvial floodplains or in volcanically active regions, for example, are usually much more fertile. Secondly, even given the natural constraints discussed above, industrial inputs exist that, if used in sufficient quantities, usually make monoculture possible. The problem, however, is that it is usually not environmentally desirable to destroy rain forest for low-productivity ranching and not economically profitable to pursue high-input monoculture (Mattos and Uhl 1994; Furley 1994). Rain forest development planning is thus focusing more and more on the economic and ecological advantages of preserving the rain forest in its natural state or on devising landuse models that allow the environmental balance of the rain forest to be replaced with an equally sustainable system (Place 2001; Hall 2000; Smith 1999; Browder 1989).

TROPICAL AND TEMPERATE SAVANNA

In a climatic sense, the term "savanna" refers to an area that experiences wet summers and dry winters. If the temperatures remain warm throughout the year, as is the case in the *tierra caliente* and the lower reaches of the *tierra templada,* the area is said to have a **tropical savanna,** or tropical wet and dry, climate. If we are dealing with an upland wet and dry region, it is designated as a **temperate savanna** climate. In either case, the distinguishing feature of the savanna biomes of Latin America is the seasonal extremes of precipitation.

The tropical savannas of Latin America and the Caribbean extend almost in a great semicircle along the drier margins of the tropical rain forests (see Figure 4.1). This includes the northern portion of the Gran Chaco, most of the central and eastern Brazilian Highlands, the Guiana Highlands, and the Orinoco and Magdalena river basins. Lowland savannas also cover portions of the Pacific coasts of Central America and southern Mexico, the central Mexican gulf coast, and the Greater and Lesser Antilles. Temperate savannas extend from the Eastern and Western Sierra Madres of Mexico southward through the uplands of Central America to the mountains of western Panama. They then reappear in the cordilleras of Colombia and continue southward through the Ecuadorian and Peruvian Andes into eastern Bolivia, and from there spread eastward into the Paraná Plateau country of southern Brazil.

Table 4.2 contains representative climatic data for tropical and temperate savanna stations in both the Southern and Northern Hemispheres. Goiás, a regional marketing center on the Brazilian uplands northwest of Brasília, and Havana, Cuba, are both examples of tropical wet and dry climates; while Cochabamba, Bolivia, and Guatemala City are found within the temperate savanna zones.

Notice that the average temperatures of each station vary according to altitude and distance from the equator. The timing and relative length of the wet and dry seasons also vary, depending on the hemisphere and location relative to the shifting pressure belts. What all four cities have in common, then, is the pattern of summer rains and winter droughts.

Just as considerable variation exists within the climate patterns of the savanna regions, so too a number of vegetation associations are found. These are an expression not only of climatic differences but of varying local soil moisture balances. The classic savanna vegetation type is defined as tall grasses—many two to three meters (eight to ten feet) or higher—interspersed with scattered trees, with dense galeria forest growing along the banks of the seasonally flooded streams and rivers (Figures 4.5 and 4.6). While this vegetation type does exist in Latin America, it is not widespread and may, in some cases, be more a product of repeated burning by humans than of the local interactions of climate and soil. Much of the Latin American tropical wet and dry area consists of forests of scrubby deciduous trees that drop their leaves during the dry season and scattered palms that gradually give way to cacti and other succulents at the drier margins (Robichaux and Yetman 2000; Furley and Ratter 1988; Cole 1960) (Figure 4.7). The temperate montane savannas, on the other hand, are often characterized by mixed hardwood and softwood

TABLE 4.2 ❌

Tropical and Temperate Savanna Biomes

STATION	ALTITUDE (meters)	*	Jan	Feb	Mar	Apr	May	Jun	Jul	Aug	Sep	Oct	Nov	Dec	YEAR
Goiás	520	T	23.5	23.8	24.1	24.4	23.7	22.4	22.4	24.0	25.6	25.4	24.5	23.7	23.9
		R	31.8	25.2	25.9	11.7	1.0	0.8	0.0	0.8	5.8	13.5	23.9	24.1	164.5
Havana	24	T	21.9	22.1	23.2	24.6	25.9	26.9	27.6	27.6	27.1	26.1	23.9	22.6	24.9
		R	7.4	4.6	4.8	5.6	12.2	16.3	12.2	13.7	14.7	16.8	7.9	6.1	122.3
Cochabamba	2,575	T	18.8	18.5	17.3	16.9	15.6	14.0	15.3	16.3	17.7	19.7	20.0	19.0	17.3
		R	10.4	9.7	6.1	1.0	1.0	0.8	0.5	0.5	1.8	1.5	3.3	9.9	46.5
Guatemala City	1,490	T	16.5	17.3	18.8	19.0	20.8	19.7	19.3	19.3	19.3	18.4	17.5	16.3	18.5
		R	0.8	0.5	1.3	3.3	14.0	29.7	19.8	19.8	23.6	16.8	2.3	0.5	132.4

°T = Mean monthly temperature in degrees Celsius
°R = Mean monthly rainfall in centimeters

Sources: Boucher 1975, 70; Kendrew 1953, 522–527; Robinson 1967, 20; Köppen and Geiger 1930–1936; Koeppe and De Long 1958, 324–327.

FIGURE 4.5 ❌

Tall grass and galeria forest savanna vegetation on the Venezuelan Llanos during the dry season.

FIGURE 4.6 ❌

Flooded savanna vegetation on the Venezuelan Llanos during the rainy season.

forests in Middle America and by broadleaf evergreen forests in South America (Figure 4.8).

Soil conditions are equally variable. Large sections of lowland tropical savanna have reddish-yellow oxisols and ultisols, and agriculture in these areas faces many of the same challenges as were discussed for the tropical rain forest zones. Other lowland areas have **vertisols,** soils whose high clay content causes them

to expand during the rainy season and shrink and crack severely during the dry months (Haase 1992; Montgomery 1988; Medina and Silva 1990). In these districts, plant- and tree-based agriculture is sometimes limited by the tendency of the expanding or shrinking soil to damage the root structure of the crops. Conversely, animal grazing is encouraged since most pasture grasses withstand root tearing. Other

FIGURE 4.7

Scattered, deciduous trees dominate this form of savanna vegetation. Called *campo cerrado*, it is found in much of eastern and central Brazil. Also visible above the new grass that is emerging after burning are two large termite mounds.

FIGURE 4.8

Softwood forest vegetation of the temperate upland savanna of central Mexico.

areas, especially portions of the highland temperate savanna, are favored with exceptionally fertile volcanic soils. Ultimately, the greatest single limiting factor to increased agricultural productivity in the savanna regions of Latin America is the seasonal fluctuation in water supplies. Given adequate irrigation and/or drainage system development, many wet and dry zones are capable of greatly increased food output.

DESERT

Although seen by many as barren and inhospitable, Latin America's deserts are, in their own way, fragile, intricately balanced ecosystems of surprising economic potential. The true deserts of Latin America fall into three broad climatic subtypes (Turner, Bowers, and Burgess 1995; West 1993; Meigs 1966; Jaeger 1957). The first, found in northern Mexico and along a great longitudinal band extending from southern Bolivia southward to southern Patagonia, is characterized by continental temperature controls (Figure 4.1). The second, consisting of the coastal deserts of northern Chile, Peru, and Baja California, is regulated largely by cold, offshore Pacific currents. Finally, a small strip of hot coastal desert occurs within the Guajira Peninsula and the neighboring lowlands of northwestern Venezuela, where the warm Caribbean trade winds are not lifted enough to induce precipitation.

Shown in Table 4.3 are representative climatic data from each of these three zones. Ahome, a small farming community along the Fuerte River near Los Mochis in northern Sinaloa state, is typical of the desert communities of northern Mexico. La Guaira is a Caribbean port north of Caracas, Venezuela. Iquique is situated in the heart of the bone-dry Atacama Desert of northern Chile.

Perhaps the most notable of the patterns depicted in the Ahome data is the large annual temperature range, which can include winter lows in the 4.0 to 7.0°C (40 to 45°F) range and midsummer highs above 43°C (110°F). The average annual temperature of La Guaira, on the other hand, is about three degrees greater than that of Ahome, owing not to higher summer readings but rather to a far more constant temperature pattern throughout the year. Finally, Iquique, despite being situated six degrees closer to the equator than Ahome, has an average annual temperature five degrees cooler than that of its northern counterpart owing to the influence of the cold Peru Current. The other remarkable aspect of Iquique's weather is the almost total absence of rainfall throughout the year. This, too, is attributable to the cold offshore waters, which have made the Atacama Desert the driest on earth.

The vegetation associated with each of Latin America's desert regions is **xerophytic,** meaning that it is adapted to drought conditions. The adaptations take numerous forms. Where groundwater supplies will support them, small leguminous or pod-bearing trees and shrubs, including acacias and larreas in Patagonia and palo verde, mesquite, and creosote in northern Mexico, reduce water loss by maintaining extremely small leaves. Prickly pear, saguaro, and other cacti, for the same reason, bear thorns rather than

TABLE 4.3

Desert Biomes

STATION	ALTITUDE (meters)	*	Jan	Feb	Mar	Apr	May	Jun	Jul	Aug	Sep	Oct	Nov	Dec	YEAR
Ahome	34–85	T	17.3	18.0	19.9	22.1	25.0	27.9	30.4	30.0	29.5	26.5	21.9	17.8	23.9
		R	0.0	0.3	0.5	0.0	0.0	1.0	4.1	8.1	8.1	0.8	1.3	6.6	30.8
La Guaira	sea level	T	25.8	25.8	26.3	26.8	27.3	27.6	27.3	28.1	28.3	28.1	27.5	26.0	27.0
		R	1.3	0.5	2.0	0.5	1.5	2.3	2.5	2.8	3.1	4.1	4.1	3.8	28.5
Iquique	9	T	21.7	21.7	20.6	18.4	17.3	16.7	15.6	16.1	17.2	17.8	19.4	20.6	18.5
		R	0.3	0.0	0.0	0.0	0.3	0.0	1.8	0.8	0.5	0.3	0.0	0.0	4.0

°T = Mean monthly temperature in degrees Celsius
°R = Mean monthly rainfall in centimeters

Sources: Kendrew 1953, 522–527; Köppen and Geiger 1930–1936; Miller 1938, 248–249.

FIGURE 4.9

Desert vegetation of Jujuy province, northwestern Argentina.

FIGURE 4.10

Desert vegetation growing within Mexico's Mesa del Norte.

leaves, while all forms of plants, including grasses, space themselves widely and send out wide lateral roots in order to maximize the amount of water captured during the infrequent, but often intense, rainstorms (Figures 4.9 and 4.10).

Yet another aspect of xerophytic vegetation is the capacity of the seeds of many herbaceous plants to lie dormant on the desert floor for extended periods of time before sprouting once a rain arrives (Turner 1990). The sudden appearance of tender new plants, together with the renewed growth and/or flowering of the existing vegetation, gives the desert the capacity to transform itself dramatically from brown to green in a matter of days. When the rains end, however, the desert quickly fades to its former browns and grays.

Where little or no rain falls, as with the Atacama and along the coast of Peru, one would expect to find no vegetation at all. This, indeed, is the case along the coastal plain and within some interior regions of northern Chile (Figure 4.11). Along the southern Peruvian and northern Chilean coasts, however, a thick fog known as the *garúa* forms between June and October (Cereceda and Schemenauer 1991) (Figure 4.12). As it drifts inland onto the Andean foothills, it sometimes provides sufficient moisture, without precipitation occurring, for the growth of patches of grasses and shrubby plants referred to as the *loma*. Water droplets dripping from the shrubs (fog drip) help sustain this vegetation.

FIGURE 4.11

Much of the Atacama Desert of northern Chile and southern Peru is entirely devoid of vegetation.

FIGURE 4.12

Garúa fog along the coast of southern Peru.

Desert soils, called **aridisols,** are in many ways almost the opposite of the oxisols and ultisols of the rain forests. Rather than losing minerals and nutrients through the process of leaching, as occurs in the humid tropics, aridisols are usually very fertile, owing to their high levels of mineral salts, many of which are plant nutrients. These minerals or salts occur naturally within the earth's bedrock, and minute quantities will dissolve in any rainwater or melted snows that flow out of the highland catchment basins and down onto the floors of the desert valleys where they will be absorbed into the topsoil.

The danger, of course, is that too much of a good thing may be harmful. If a given area is shaped like a basin and the amount of groundwater entering the area cannot sustain a perennial outflowing stream or river,

the pH or alkalinity of the basin will gradually increase as the inflowing waters evaporate and leave behind their salty minerals. This process can ultimately lead to such severe **salinization** of the topsoil that agriculture becomes impossible.

Unfortunately, every time a farmer irrigates his fields, he is mimicking the basin drainage pattern and thus contributing to the salinization process. And yet without irrigation, intensive agriculture is usually not possible in arid regions. This, then, is the dilemma. Worldwide, yields from irrigated agriculture average two to four times per hectare the output of unirrigated lands. This is because of the inherent fertility of desert soils, the nutrients supplied by irrigation water, and also because of fewer losses from rain damage, plant diseases, and pests. Yet numerous irrigation schemes in Latin America and elsewhere have failed after twenty to thirty years because of uncontrolled soil salinization.

The desert regions are thus an enigma. On the one hand, they are among the least developed and most sparsely settled of all of Latin America's natural regions. On the other hand, they have the potential to become some of the most productive and prosperous if irrigation can be provided and properly managed. While groundwater supplies in these arid zones are not unlimited, far more could be done to use them in an environmentally responsible manner. The development of the deserts, like that of Latin America's other natural regions, has thus been limited more by cultural and economic constraints than by physical ones.

STEPPE

Nature seldom changes so abruptly that one passes immediately from humid to arid zones with nothing in between. Rather, what we find are transitional, semiarid regions separating the deserts from the humid environments. Geographers call these semiarid lands **steppes.** They are found in Latin America on the margins of the true deserts (Figure 4.1). This includes much of western Patagonia abutting the humid Andes, portions of the western Pampa and the eastern Gran Chaco, and large areas of the central and northeastern Mexican Plateau. A final steppe region is found in the *sertão*, or backlands, of northeastern Brazil. Shown in Table 4.4 are climatic data for two communities representative of the steppe biomes: Chihuahua, a regional industrial and commercial center of northcentral Mexico, and Quixeramobim, a small interior town of Ceará state in northeastern Brazil.

The common feature of the steppe climates of both Chihuahua and Quixeramobim is a total annual rainfall that is greater than that of the true deserts but less than that of neighboring humid regions. Chihuahua's higher altitude and colder temperatures mean that it has a lower

TABLE 4.4 ✖

Steppe Biomes

STATION	ALTITUDE (meters)	*	Jan	Feb	Mar	Apr	May	Jun	Jul	Aug	Sep	Oct	Nov	Dec	YEAR
Chihuahua	1,423	T	10.0	11.9	15.6	19.1	23.0	26.0	24.8	24.0	21.9	18.2	13.0	9.6	18.1
		R	0.5	1.0	0.7	0.5	0.5	4.3	9.1	9.4	8.4	2.3	1.3	1.0	39.0
Quixeramobim	207	T	28.3	27.7	27.1	26.9	26.4	26.2	26.4	27.1	27.8	28.3	28.5	28.6	27.4
		R	7.9	8.9	14.5	12.2	9.4	3.8	2.0	1.0	0.3	0.0	0.5	2.8	63.3

°T = Mean monthly temperature in degrees Celsius
°R = Mean monthly rainfall in centimeters

Sources: Koeppe and De Long 1958, 324–327; Köppen and Geiger 1930–1936; Kendrew 1953, 522–527.

annual evapotranspiration rate than Quixeramobim and therefore requires less precipitation to be classified as semiarid. The transitional quality of the steppe climate is illustrated by the fact that both areas have humid regions bordering them to the east. In Mexico, for example, annual rainfall totals average 75 to 100 centimeters (30 to 40 inches) along the upper gulf coast, while 150 to 200 centimeter (60- to 80-inch) totals are common along the Brazilian coast from São Luís to Recife.

Many of the steppe regions of the mid to high latitudes, such as the Great Plains of the United States and Canada and parts of the Ukraine, are known for their short grass vegetation. The Latin American steppes, however, are dominated either by a mixture of scattered shrubs and sparse grasses or by forests of scrubby, often thorny, shrubs and trees, whose height and density are a reflection of long-term groundwater patterns (Uhl et al. 1982; Veblen and Lorenz 1988; Vivó Escoto 1964). In western Patagonia, this vegetation is called the *monte.* The steppe of the Gran Chaco is dominated by different species of the quebracho tree, a source of tannin, fuel, and timber. The vegetation of the Brazilian backlands, called the *caatinga,* is a mixture of palms, low deciduous trees, and cacti. The Mexican steppes are characterized by various combinations of grasses, cacti, yucca, and low shrubs (Figures 4.13 and 4.14).

Soils within the steppe regions are for the most part excellent, having neither the excess salts nor the extreme acidity of the more arid and humid zones. Owing primarily to the lack of economic resources, the steppelands of Latin America have been used mostly for extensive animal grazing and food gathering. Many of them, such as the Brazilian *sertão,* have experienced severe environmental degradation from repeated human-induced burning, the effects of which are compounded by severe periodic droughts of several years' duration (Hastenrath 1990; Mechoso, Lyons, and Spahr 1990; Ward and Folland 1991).

FIGURE 4.13 ✖

Steppe vegetation of western Patagonia.

FIGURE 4.14 ✖

Steppe vegetation of northwestern Mexico.

TABLE 4.5

Humid Subtropical Biomes

STATION	ALTITUDE (meters)	*	Jan	Feb	Mar	Apr	May	Jun	Jul	Aug	Sep	Oct	Nov	Dec	YEAR
Buenos Aires	25	T	23.3	22.8	20.4	16.3	12.8	9.8	9.4	10.6	12.8	15.5	18.8	21.6	16.1
		R	7.6	6.4	11.7	7.6	7.1	6.9	5.6	6.1	7.6	9.1	7.1	9.9	92.7
Curitiba	908	T	20.4	21.1	19.3	16.8	13.7	12.2	12.5	13.5	14.6	16.1	18.0	19.6	16.5
		R	16.8	16.0	11.2	7.9	10.2	10.2	6.4	8.1	12.5	14.0	12.7	14.0	140.0

°T = Mean monthly temperature in degrees Celsius
°R = Mean monthly rainfall in centimeters

Sources: Kendrew 1953, 522–527; Köppen and Geiger 1930–1936; Koeppe and De Long 1958, 324–327.

HUMID SUBTROPICAL

One of the most economically productive and densely settled of the natural regions of Latin America is a humid subtropical zone that extends from eastern Argentina northeastward through Uruguay and on into southeastern Brazil (Figure 4.1). As its name implies, this region is characterized by moderate temperatures and precipitation throughout the year. Perhaps even more importantly, with respect to economic productivity, it is endowed with two of the most fertile soil belts in all of Latin America. These are the **mollisols** or rendzinas, formerly called chernozems and chestnuts, of the Pampa of Argentina and Uruguay and the volcanic soils of the Paraná Plateau of southern Brazil.

Presented in Table 4.5 are monthly and annual climate summaries for two humid subtropical stations. Buenos Aires is the commercial and manufacturing heart of the Pampa, and Curitiba is the principal urban center of the uplands of Paraná state in southern Brazil.

The moderate, temperate nature of the humid subtropical biome is evident in the precipitation figures, which show that no month at either station averages less than five centimeters (two inches) or more than eighteen centimeters (seven inches). The annual temperature regime is also mild, with every month of the year averaging from 9° to 23°C (upper 40s to low 70s F).

The natural vegetation grades from the tall prairie grasses of the Pampa to a mixed forest-grassland cover in Uruguay and Rio Grande do Sul in southern Brazil (Figure 4.15). To the north, Santa Catarina and Paraná were once mostly forested, with broadleaf species prevailing in the sheltered river valleys and conifers predominating on the higher plateau country. Unfortunately, little of the natural vegetation remains.

FIGURE 4.15

Tall grasses and broadleaf trees characterize the more humid portions of the Argentine Pampa. The rows of wooden boxes visible in the foreground are beehives.

In the late nineteenth and early twentieth centuries, great numbers of Europeans, especially Italians and Germans, settled the Pampa and southern Brazil. They were attracted to the region by its temperate climate and fertile soils and, most of all, by the hope of building new and better lives for themselves in the New World. They brought with them industrial technologies and urban lifestyles that have contributed to relatively high collective living levels. And yet this area, like Latin America as a whole, has fallen far short of achieving its full potential. The cultural, political, and economic factors contributing to its continued underdevelopment will be addressed in later chapters.

TABLE 4.6 ✕

Mediterranean Biomes

STATION	ALTITUDE (meters)	*	Jan	Feb	Mar	Apr	May	Jun	Jul	Aug	Sep	Oct	Nov	Dec	YEAR
Santiago	520	T	20.4	19.5	16.9	13.7	10.6	7.6	7.9	9.2	11.0	13.8	16.8	19.2	13.9
		R	0.0	0.3	0.5	1.5	5.8	8.1	8.6	6.1	3.1	1.5	0.5	0.5	36.5
San Fernando	335	T	19.9	18.9	16.4	12.9	10.0	6.9	7.3	8.4	10.2	13.2	16.0	18.6	13.2
		R	0.3	0.0	1.0	3.1	15.5	20.8	11.7	8.1	7.1	2.3	2.5	0.8	73.2

°T = Mean monthly temperature in degrees Celsius
°R = Mean monthly rainfall in centimeters

Sources: Koeppe and De Long 1958, 324–327; Köppen and Geiger 1930–1936; Kendrew 1953, 522–527.

MEDITERRANEAN

The mediterranean biome occurs in a tiny corner of northwestern Baja California and in central Chile from approximately 32° to 38° south latitude (Figure 4.1). The physical characteristics of the highly significant Chilean region are unlike those of any other part of Latin America. Shown in Table 4.6 are climatic summaries for two cities in the mediterranean region of Chile. Santiago, the national capital, is situated in the drier northern zone, while San Fernando is a smaller regional center some eighty miles to the south.

Two traits stand out from the mediterranean data. One is the very mild temperatures, attributable both to the proximity of the region to the cold waters of the north-flowing Peru Current and to distance from the equator. The second is the marked seasonality of precipitation—essentially the opposite of that of the temperate savanna. Notice that San Fernando's total annual rainfall is twice that of Santiago's. This is a reflection of the overall Chilean pattern of precipitation totals increasing with distance from the equator. Yet both stations receive over 85 percent of their yearly precipitation during the five-month winter that lasts from May through September. The remaining seven months bring droughtlike conditions that are somewhat alleviated by runoff of the winter snows from the towering Andes to the east.

Chaparral, the natural vegetation of mediterranean Chile, consists of woody shrubs and low evergreen trees capable of withstanding almost seven consecutive months without rain. The soils of the region are mostly **alfisols** (also called lixisols), with some clayey vertisols prevailing in scattered poorly drained bottomlands. The dominant alfisols tend to be a little low in organic matter but overall rank among the finest

agricultural soils in Latin America, having neither the acidity of the humid oxisols nor the potential salinity of the desert aridisols. The seasonal water deficits are largely compensated for by the numerous mountain streams that support abundant harvests of midlatitude grains, fruits, and vegetables.

Mediterranean Chile is thus the heart of the Chilean nation, housing the majority of its people, agriculture, and industry. It is one of Latin America's most favored natural regions. The only major disadvantage in its natural endowment is simply its limited area. History clearly teaches us, however, that social and political conditions are often more important than size in sustaining or limiting the economic development of nations.

MARINE WEST COAST RAIN FOREST

South of the Bío-Bío River, Chile's mediterranean biome gives way to a cold and stormy land that is characterized primarily by a marine west coast climate and, in the far south by subpolar conditions. We use as representative stations, in Table 4.7, Puerto Montt, a small settlement on the Gulf of Ancud in the northern portion of the region, and Evangelistas Island, near the Strait of Magellan in the far south.

Several patterns are evident from the data. One is that the region has a relatively small temperature range for its latitude: the difference between the warmest and coldest months is only 7.6°C (13.6°F) at Puerto Montt and decreases to just 4.7°C (8.5°F) at Evangelistas Island. This is attributable to the almost perpetually overcast skies and also to the penetration of marine influences far inland through the magnificent assemblage of rocky offshore islands and drowned glacial valleys.

TABLE 4.7

Marine West Coast Rain Forest Biomes

STATION	ALTITUDE (meters)	*	Jan	Feb	Mar	Apr	May	Jun	Jul	Aug	Sep	Oct	Nov	Dec	YEAR
Puerto Montt	10	T	15.3	14.5	13.3	11.3	9.8	7.7	7.7	7.8	8.5	10.6	12.0	13.9	11.0
		R	11.7	11.2	15.0	18.8	26.9	25.4	27.4	23.6	16.0	14.0	14.0	13.7	217.7
Evangelistas Island	55	T	8.6	8.6	8.4	7.2	5.6	4.6	3.9	4.4	4.8	5.7	6.2	7.5	6.3
		R	32.5	22.6	31.2	29.5	22.4	21.6	22.6	21.8	19.6	22.9	26.2	24.4	297.3

°T = Mean monthly temperature in degrees Celsius

°R = Mean monthly rainfall in centimeters

Sources: Koeppe and De Long 1958, 324–327; Köppen and Geiger 1930–1936; Miller 1938, 248–249.

A second impression is that the temperatures, while certainly cool year-round, do not appear to be exceptionally cold, especially for latitudes ranging from the low 40s to the middle 50s F. What these figures fail to convey, however, are the bitterly cold sensible temperatures brought on by the merciless winds and high humidities. Perhaps the most difficult aspect of the biting cold is that there is little if any relief, with the storms and seas beating against the battered islands and inlets throughout the year.

A third feature of the climate is that it is one of the wettest in all of Latin America. Notice that a modified mediterranean precipitation regime is evident in the northern districts (Figure 4.16). Puerto Montt, for example, receives almost two-thirds of its annual rainfall during the low-sun months of April through September. The difference between Puerto Montt and San Fernando and other mediterranean stations, however, is that the latter have a pronounced summer dry season, while Puerto Montt receives over eleven centimeters (four inches) of rain every month of the year. In contrast, seasonality of precipitation is entirely absent at Evangelistas Island, with September, the driest month, averaging almost twenty centimeters (eight inches).

The heavy rains and cool temperatures support a dense vegetative cover that might best be described as a temperate rain forest. Although not characterized by the four vertical layers of the tropical rain forest, the temperate rain forest is just as impressive in its own right, with a lush green mat of mosses, grasses, ferns, and vines thriving as an understory to the mixed broadleaf and coniferous canopy growing overhead. The forest fades on the coldest, southernmost islands into stunted tundra shrubs and lichens. It is difficult to identify a dominant soil type because glacial and alluvial erosion have left much of the steeply sloped region

with bedrock exposed at the surface. Where soils have formed in the more favored sites, the processes of **podzolization** and gleization have left them leached, acidic, and lacking in nutrients.

Economic development throughout the region has been and will likely continue to be limited by the harsh climate and the shortage of level, arable land. Southern Chile is not devoid of resources, however, with fishing, forestry, sheep grazing, and mining each offering hope for economic gains in the years to come (Figure 4.17).

HIGHLAND

The highlands, the last of Latin America's natural regions, are also the most complex owing to the fact that the temperature, precipitation, vegetation, and soil patterns frequently change quickly from one locale to another on the basis of differences in altitude, exposure, and relief (Allan, Knapp, and Stadel 1988). The relationship between altitude and temperature is evident from data shown in Table 4.8 for stations situated within each of the four altitudinal life zones discussed in chapter 3. Guayaquil, the largest city of Ecuador, is a port and as such experiences temperatures common to the *tierra caliente*. Mérida is an old colonial city of the *tierra templada* in the highlands of western Venezuela. Quito, the capital of Ecuador, is situated in a basin just south of the equator within the *tierra fría*. Cerro de Pasco is a mining town in the *tierra helada* zone of the central Peruvian Andes.

In addition to the steady drop in temperatures as one ascends from the *tierra caliente* to the *tierra helada*, there is a remarkable constancy of temperature from one month to the next at each of the four stations, with a variance from 2.8°C (5.0°F) at Guayaquil to a mere 0.2°C (0.4°F) at Quito. This is attributable to the proximity of all the communities to the equator.

FIGURE 4.16

The northern portion of Chile's marine west coast climate region is characterized by a modified mediterranean precipitation pattern that sustains prosperous grain, fruit, and dairy farming. This scene is from the city of Osorno.

FIGURE 4.17

Sheep grazing on formerly forested lands near Cerro Castillo, southern Chile.

TABLE 4.8

Highland Biomes

STATION	ALTITUDE (meters)	*	Jan	Feb	Mar	Apr	May	Jun	Jul	Aug	Sep	Oct	Nov	Dec	YEAR
Guayaquil	12	T	26.3	26.3	26.5	26.9	26.0	25.2	24.1	24.5	25.1	24.8	25.8	26.8	25.6
		R	24.6	26.7	18.8	13.5	5.3	2.0	1.0	0.0	0.5	1.0	0.8	4.8	99.0
Mérida	1,640	T	17.8	18.4	18.6	19.2	19.2	18.9	18.9	19.3	19.1	18.9	18.4	17.9	18.7
		R	6.6	3.8	9.7	17.0	28.5	17.5	10.2	14.5	16.0	25.2	20.3	8.4	177.7
Quito	2,850	T	12.6	12.5	12.5	12.5	12.6	12.6	12.5	12.6	12.7	12.6	12.5	12.6	12.5
		R	8.1	9.9	12.2	17.8	11.7	3.8	2.8	5.6	6.6	9.9	10.2	9.1	107.7
Cerro de Pasco	4,350	T	6.7	6.2	6.7	6.7	5.9	5.0	4.7	4.9	5.0	5.4	5.7	5.8	5.7
		R	11.7	11.4	9.1	8.6	5.8	2.3	2.8	3.1	7.1	8.4	8.6	9.4	88.3

°T = Mean monthly temperature in degrees Celsius

°R = Mean monthly rainfall in centimeters

Sources: Kendrew 1953, 522–527; Koeppe and De Long 1958, 324–327; Köppen and Geiger 1930–1936; Miller 1938, 248–249.

Because the lowest three life zones fall within the rain forests, savannas, deserts, and other natural regions presented previously, the remainder of this discussion will focus on the perpetually cold lands of the *tierra helada*. It is impossible to depict on the map every isolated occurrence of highland or *tierra helada* environment, but the principal areas of widespread distribution extend along the high Andes, from central Chile and Argentina northward through Bolivia and central Peru. Additional areas are found in highland Ecuador and Colombia (Figure 4.1).

The soils of these regions vary as widely as the temperature, rainfall, and parent material. However, they can be placed generally into the **inceptisol** and **entisol** orders, meaning that they are, for the most part, in the early stages of development. This is attributable to the cold temperatures which slow the rate of bedrock decomposition and soil formation generally, to mass wasting of existing soils on many of the steeper mountain slopes, and to the frequent additions of volcanic ash and other ejecta that have not fully weathered. The vegetation in Middle America is composed

FIGURE 4.18 ✕

Puna vegetation being grazed by cattle on the high Altiplano of southern Peru.

FIGURE 4.19 ✕

The *frailejón* adds beauty and tranquility to the cold *páramo* landscapes of the northern Andean highlands.

of herbaceous plants that resemble those of the treeless North American arctic tundra. Andean high vegetation can be divided into two broad subtypes—*puna* and *páramo.* The **puna,** consisting mostly of ichu and other bunch grasses and herbs, dominates the drier southern Andes from Peru southward; whereas the **páramo,** composed of the same bunch grasses plus prickly shrubs and mosses, prevails in the more humid northern Andes (Figure 4.18). A plant common to the latter is the *frailejón,* a giant dandelion whose creamy white, velvety "flower" adds a bright note to the otherwise somber landscape (Figure 4.19).

The human carrying capacity of these highland regions is limited by the cold, by the short frost-free growing seasons, and, in many areas, by shortages of water. Residence within excessively high altitudes also presents unique health challenges (chapter 13). Having acknowledged these limiting factors, one must also recognize that the lower reaches of the highlands have, for millennia, supported some of Latin America's densest human populations. Agriculture and animal grazing are possible, soils permitting, up to 4,000 to 4,300 meters (13,000 to 14,000 feet) in most areas, and mining remains a significant economic component. So the highlands likely will continue to play a prominent role in the development of the Andean nations for generations to come.

SUMMARY

Latin America is endowed with one of the most diverse and abundant physical resource bases of any region on earth. Included within its boundaries are vast rain forests and savannas, fragile deserts and fertile prairies, temperate uplands, and highland tundras. Each natural region has its disadvantages with respect to certain types of economic activities; at the same time, each is especially favored for other endeavors.

Latin America's collective resource base is probably neither better nor worse than any of the earth's other culture areas. Anglo America, for instance, includes not only some of the world's richest prairies but also some of the largest tracts of little-used desert and arctic tundra. Several of the great industrial nations of northern Europe have physical environments similar to southern Chile's. Japan has built one of the world's largest economies on one of the most meager land bases on earth.

What the world's most developed regions and nations have in common is not more favorable climates, or more fertile soils, or richer mineral deposits, but rather a historic commitment to the development of the greatest resource of any region—its human population. Latin America's past and future development has been and will surely continue to be shaped primarily not by its physical resource base but by internal and external influences that impact the fulfillment of human potential. Part 2 of this volume traces, from a thematic perspective, the evolution of the social, political, and cultural values and institutions that have impacted the human utilization of Latin America's physical environments.

KEY TERMS

lianas 73	xerophytic 78	chaparral 83
oxisols 73	*garúa* 79	alfisols 83
ultisols 73	*loma* 79	podzolization 84
latosols 73	aridisols 80	inceptisol 85
tropical savanna 76	salinization 80	entisol 85
temperate savanna 76	steppes 80	*puna* 86
vertisols 77	mollisols 82	*páramo* 86

SUGGESTED READINGS

Allan, Nigel J. R., Gregory W. Knapp, and Christoph Stadel. 1988. *Human Impacts on Mountains.* Savage, Md.: Roman and Littlefield.

Amazonia Without Myths. 1992. Washington, D.C.: Inter-American Development Bank.

Boucher, Keith. 1975. *Global Climate.* New York: Wiley.

Browder, John O., ed. 1989. *Fragile Lands of Latin America: Strategies for Sustainable Development.* Boulder, Colo.: Westview Press.

Calderon, Angela I., et al. 2000. "Forest Plot as a Tool to Demonstrate the Pharmaceutical Potential of Plants in a Tropical Forest of Panama." *Economic Botany* 54:278–294.

Cereceda, Pilar, and Robert S. Schemenauer. 1991. "The Occurrence of Fog in Chile." *Journal of Applied Meteorology* 30:1097–1105.

Chivian, Eric. 1997. "Global Environmental Degradation and Biodiversity Loss: Implications for Human Health." In *Biodiversity and Human Health,* eds. Francesca Grifo and Joshua Rosenthal. Washington, D.C.: Island Press.

Cole, M. M. 1960. "Cerrado, Caatinga and Pantanal: The Distribution and Origin of the Savanna Vegetation of Brazil." *Geographical Journal* 126:168–179.

Committee on Tropical Soils. 1972. *Soils of the Humid Tropics.* Washington, D.C.: National Academy of Sciences.

Dallmeir, Francisco, Margo Kabel, and Robin B. Foster. 1996. "Floristic Composition, Diversity, Mortality, and Recruitment on Different Substrates: Lowland Tropical Forest, Pakitza, Río Manu, Peru." In *Manu: The Biodiversity of Southeastern Peru,* eds. Don E. Wilson and Abelardo Sandoval, 61–88. Washington DC: Smithsonian Institution Press.

Eidt, Robert C. 1968. "The Climatology of South America." In *Biogeography and Ecology in South America,* eds. E. J. Fittkau et al., 54–81. The Hague: Dr. W. Junk N. V. Publishers.

Furley, Peter A., ed. 1994. *The Forest Frontier: Settlement and Change in Brazilian Roraima.* New York: Routledge.

Furley, Peter A., and James A. Ratter. 1988. "Soil Resources and Plant Communities of the Central Brazilian Cerrado and Their Development." *Journal of Biogeography* 15:97–108.

Gentry, Alwyn H., ed. 1990. *Four Neotropical Rainforests.* New Haven, Conn.: Yale University Press.

Gilmore, Raymond M. 1963. "Fauna and Ethnozoology of South America." In *Handbook of South American Indians.* Vol. 6, ed. Julian H. Steward, 345–464. Washington, D.C.: Smithsonian Institution Bureau of American Ethnology Bulletin No. 143.

Haase, R. 1992. "Physical and Chemical Properties of Savanna Soils in Northern Bolivia." *Catena* 19:119–134.

Hall, Anthony, ed. 2000. *Amazonia at the Crossroads: The Challenge of Sustainable Development.* London: University of London Institute of Latin American Studies.

Hastenrath, Stefan. 1990. "Prediction of Northeast Brazil Rainfall Anomalies." *Journal of Climate* 3:893–904.

Hecht, Susanna B., and Alexander Cockburn. 1989. *The Fate of the Forest: Developers, Destroyers, and Defenders of the Amazon.* New York: Verso.

Jaeger, Edmund C. 1957. *The North American Deserts.* Stanford: Stanford University Press.

Jain, S. K. 2000. "Human Aspects of Plant Diversity." *Economic Botany* 54:459–470.

Jordan, Carl F. 1985. *Nutrient Cycling in Tropical Forest Ecosystems—Principles and Their Application in Management and Conservation.* New York: Wiley.

Kellman, Martin, and Rosanne Tackaberry. 1997. *Tropical Environments: The Functioning and Management of Tropical Ecosystems.* London: Routledge.

Kendrew, Wilfred G. 1953. *The Climates of the Continents.* 4th ed. London: Oxford University Press.

Koeppe, Clarence E., and George C. De Long. 1958. *Weather and Climate.* New York: McGraw-Hill.

Köppen, W., and R. Geiger, eds. 1930–1936. *Handbuch der Klimatologie.* Vol. 2, Parts G, H, I, and J. Berlin: Gebruder Borntraeger.

Kricher, John C. 1989. *A Neotropical Companion: An Introduction to the Animals, Plants, and Ecosystems of the New World Tropics.* Princeton, N.J.: Princeton University Press.

Leigh, Egbert G., Jr., A. Stanley Rand, and Donald M. Windsor, eds. 1996. *The Ecology of a Tropical Forest.* 2nd ed. Washington, DC: Smithsonian Institution Press.

Mattos, Marli Maria, and Christopher Uhl. 1994. "Economic and Ecological Perspectives on Ranching in the Eastern Amazon." *World Development* 22(2):145–158.

Mazer, Susan J. 1996. "Floristic Composition, Soil Quality, Litter Accumulation, and Decomposition in Terra Firme and Floodplain Habitats Near Pakitza, Peru." In *Manu: The Biodiversity of Southeastern Peru*, eds. Don E. Wilson and Abelardo Sandoval, 89–125. Washington, DC: Smithsonian Institution Press.

McGregor, Glenn R., and Simon Nieuwolt. 1998. *Tropical Climatology: An Introduction to the Climates of the Low Latitudes.* Chichester: Wiley.

Mechoso, Carlos R., Steven W. Lyons, and Joseph A. Spahr. 1990. "The Impact of Sea Surface Temperature Anomalies on the Rainfall of Northeast Brazil." *Journal of Climate* 3:812–826.

Medina, Ernesto, and Juan F. Silva. 1990. "Savannas of Northern South America: A Steady State Regulated by Water-Fire Interactions on a Background of Low Nutrient Availability." *Journal of Biogeography* 17:403–413.

Meigs, Peveril. 1966. *Geography of Coastal Deserts.* Paris: UNESCO.

Mendelsohn, Robert, and Michael J. Balick. 1995. "The Value of Undiscovered Pharmaceuticals in Tropical Forests." *Economic Botany* 49:223–228.

Miller, A. Austin. 1938. *Climatology.* 2nd ed. London: Methuen.

Mittermeir, Russell A., and William R. Konstant. 2001. "Biodiversity Conservation: Global Priorities, Trends, and the Outlook for the Future." In *Footprints in the Jungle: Natural Resource Industries, Infrastructure, and Biodiversity Conservation*, eds. Ian A. Bowles and Glenn T. Prickett, 9–28. New York: Oxford University Press.

Montgomery, R. F. 1988. "Some Characteristics of Moist Savanna Soils and Constraints on Development with Particular Reference to Brazil and Nigeria." *Journal of Biogeography* 15:11–18.

Nieuwolt, S. 1982. *Tropical Climatology: An Introduction to the Climates of the Low Latitudes.* Chichester, England: Wiley.

Place, Susan E., ed. 2001. *Tropical Rainforests: Latin American Nature and Society in Transition*, revised and updated edition. Wilmington, Del.: Scholarly Resources.

Plotkin, M. J. 1988. "Conservation, Ethnobotany, and the Search for New Jungle Medicines: Pharmacognosy Comes of Age . . . Again." *Pharmacotherapy* 8:257–262.

Posey, Darrell Addison. 2000. "Biodiversity, Genetic Resources, and Indigenous Peoples in Amazonia: (Re)Discovering the Wealth of Traditional Resources of Native Amazonians." In *Amazonia at the Crossroads: The Challenge of Sustainable Development*, ed. Anthony Hall, 188–204. London: University of London Institute for Latin American Studies.

Reading, Alison J., Russell D. Thompson, and Andrew C. Millington. 1995. *Humid Tropical Environments.* Oxford: Blackwell.

Richards, Paul W. 1952. *The Tropical Rain Forest.* London: Cambridge University Press.

Robichaux, Robert H., and David A. Yetman, eds. 2000. *The Tropical Deciduous Forest of Alamos: Biodiversity of a Threatened Ecosystem in Mexico.* Tucson: University of Arizona Press.

Robinson, Harry. 1967. *Latin America: A Geographical Survey.* New York: Frederick A. Praeger.

Sanchez, P. A., and J. R. Benites. 1987. "Low-Impact Cropping for Acid Soils of the Humid Tropics." *Science* 238:1521–1527.

Sauer, Carl O. 1963. "Geography of South America." In *Handbook of South American Indians.* Vol. 6, ed. Julian H. Steward, 319–344. Washington, D. C.: Smithsonian Institution Bureau of American Ethnology Bulletin No. 143.

Smith, Nigel J. H. 1999. *The Amazon River Forest: A Natural History of Plants, Animals, and People.* New York: Oxford University Press.

Terborgh, John. 1992. *Diversity and the Tropical Rain Forest.* New York: Scientific American Library, distrib. W. H. Freeman.

Thorington, Richard W., Jr., et al. 1996. "Distribution of Trees on Barro Colorado Island: A Five-Hectare Sample." In *The Ecology of a Tropical Forest*, 2nd ed., eds. Egbert G. Leigh, Jr., A. Stanley Rand, and Donald M. Windsor. Washington, DC: Smithsonian Institution Press.

Turner, Raymond M. 1990. "Long-term Vegetation Change in a Fully Protected Sonoran Desert Site." *Ecology* 71:464–477.

Turner, Raymond M., Janice Bowers, and Tony L. Burgess. 1995. *Sonoran Desert Plants: An Ecological Atlas.* Tucson: University of Arizona Press.

Uhl, Christopher et al. 1982. "Ecosystem Recovery in Amazon Caatinga Forest after Cutting, Cutting and Burning and Bulldozer Clearing Treatments." *Oikos* 38:313–320.

Veblen, Thomas L., and Diane C. Lorenz. 1988. "Recent Vegetational Changes along the Forest/Steppe Ecotone of Northern Patagonia." *Annals of the Association of American Geographers* 78:93–111.

Vivó Escoto, Jorge A. 1964. "Weather and Climate of Mexico and Central America." In *Handbook of Middle American Indians.* Vol. 6, ed. Robert C. West, 187–215. Austin: University of Texas Press.

Wambecke, A. van. 1978. "Properties and Potentials of Soils in the Amazon Basin." *Interciencia* 3:233–242.

Ward, M. Neil, and Chris K. Folland. 1991. "Prediction of Seasonal Rainfall in the North Nordeste of Brazil Using Eigenvectors of Sea-Surface Temperature. *International Journal of Climatology* 11:711–743.

West, Robert C. 1993. *Sonora: Its Geographical Personality.* Austin: University of Texas Press.

Wilson, Don E., and Abelardo Sandoval, eds. 1996. *Manu: The Biodiversity of Southeastern Peru.* Washington, DC: Smithsonian Institution Press.

Wilson, Edward O. 1992. *The Diversity of Life.* Cambridge: Belknap Press.

———. 1989. "Threats to Biodiversity." *Scientific American* 261 (September 3):108–116.

Part II
CULTURAL PATTERNS

We have learned in chapters 2, 3, and 4 that Latin America consists of diverse natural environments, each endowed with physical characteristics that are conducive to some economic activities while less well suited to others. The physical diversity of Latin America is itself evidence that the unity of the region lies in its shared cultural attributes, that is to say, in a common human legacy. It has been the utilization or, alternatively, the underutilization of Latin America's greatest resource—its people—that has influenced most profoundly the economic development and underdevelopment of the region.

In Part 2 we turn to an analysis of the cultural geography of Latin America. We begin in chapter 5 with a historical overview of Latin America's common Iberian, or Latin, heritage, which from the time of the Conquest has governed the social, economic, and political development of the region. Chapter 6 consists of brief national political summaries that should help us to understand the unique political traditions of each country and to place current political developments in their broader historical contexts. Chapter 7 introduces the non-Hispanic peoples of Latin America and summarizes present racial and ethnic relationships and their impact on social and economic development. This is followed, in chapter 8, by a description of the dominant Latin American ethos, or worldview, as it expresses itself in the individual's concept of self and his or her place in society and the universe. Part 2 concludes in chapter 9 with a study of the role of religion, both institutional and personal, in Latin American society, and of the forces that are presently altering the religious geography and, by extension, the traditional culture of much of the region.

Iberian Heritage, Conquest, and Institutions

To appreciate the nature of modern Latin American society and the cultural roots of the economic underdevelopment of much of the region, we must first understand Latin America's Iberian heritage. Of particular importance is the manner in which Roman and Moorish values and culture traits were assimilated by the Spaniards and Portuguese and subsequently transferred to the New World, where they formed the basis of many of Latin America's most pervasive social and economic institutions.

LATIN AMERICA'S ROMAN HERITAGE

Roman settlement of the **Iberian Peninsula,** which is present-day Spain and Portugal, was accomplished in part by soldiers who were awarded huge land grants, called *latifundios,* as compensation for previous military conquests in outlying portions of the Roman Empire. The new estate owners, who came to be called *latifundistas,* attempted to duplicate the exploitative social and economic structures that had come to characterize Roman culture in the Italian heartland. There society had become extremely class conscious and rigidly stratified, with a small, elite upper class monopolizing the wealth generated by an ever-decreasing number of poor freemen and an ever-increasing number of slaves. The proud *latifundistas* did everything in their power to avoid farming and other forms of manual labor, which they associated with lower-class standing. The poor freemen, however, had no alternative but to toil on tiny plots, or *minifundios,* to produce crops of winter grains and vegetables to feed their families and perhaps sell a small surplus in the markets of

the rapidly growing urban centers (Dyson 1992; Frayn 1979; Duncan-Jones 1974).

The *latifundistas,* for their part, valued their landholdings primarily as a symbol of upper-class status. Although the aristocracy hoped to profit from the operations of the great estates, most were so secure financially that they felt little pressure to work the land intensively. In reality, the most important "product" of the *latifundio* was social prestige.

The preeminent social function of the *latifundio* was reinforced by the Roman love of urban life. Partly owing to their Greek intellectual heritage, upper-class Romans came to view residence in a city as an absolute prerequisite of the good life. Indeed, the present English-language words "city," "civility," and "civilization" all derive from the common Latin root of *civilitas,* suggesting that to the Roman elites, rural folk were uncivilized, that is, people of inferior culture and worth. The material comforts and social interactions so valued by the aristocracy were available only in the urban centers.

This, however, presented a potential dilemma to the wealthy. On the one hand, the highest levels of upper-class standing could be maintained only through ownership of a *latifundio.* On the other hand, societal norms required the elites to reside in the cities, which were often great distances from the remote, isolated estates. These outwardly conflicting expectations were reconciled through the practice of **absentee land ownership,** in which the upper-crust estate owners resided throughout the year in palatial urban mansions, leaving only occasionally to inspect the operations of their distant *latifundios.* There they would be received by the hired overseer and the native, lower-class peasants, who would line up meekly to pay homage to the all-powerful **patrón,** or master. The *latifundistas,* for their part, recognized the vast social and economic gulf that separated them from the fiefs and, out of a sense of **noblesse oblige,** bestowed from time to time small favors upon the peasants. As Roman Catholicism diffused throughout the empire, *noblesse oblige* often came to express itself through the *compadrazgo,* or godparenthood, system in which the *patrón* would agree to act as the sponsor or godfather of the children of the poor. This practice had the effect of institutionalizing in a religious context the rigid social stratification of the people.

Another outgrowth of absentee ownership was an emphasis on cattle grazing as the preferred form of landuse on the great estates. Cattle grazing necessitated little if any investment in improving the land itself; perhaps even more importantly, cattle ranching required far less human labor than did food crop farming, and therefore demanded less administrative oversight and involvement on the part of the urban-oriented *latifundista.* To the Roman aristocracy of Iberia, the term "cattle" meant not only beef or milk animals of the *Bos taurus* family, but any large, hoofed animal, including sheep, goats, donkeys, and horses. Because horses were expensive to breed and maintain and were kept primarily for riding and warring rather than as a source of food for the common people, the use of horses became a symbol of power and wealth and, ultimately, of upper-class standing. For example, the Spanish word for horse is *caballo,* which in turn is the root of *caballero. Caballero* came to have a double meaning. One is that of a horseman or one who rides a horse. However, *caballero* also translates as a gentleman or a person of proper upbringing. In other words, in the classic Iberian and, subsequently, Latin American contexts, only upper-class people who owned horses could be gentlemen or members of proper society. The masses of peasants, who walked, were not people equal in worth to those who rode horses. This fact was reemphasized in the Spanish word *hidalgo,* another expression for a gentleman. *Hidalgo* originally evolved as a contraction of *hijo de algo* (*alguien*), meaning the son of someone important. Thus, with the exception of the clergy, upper-class standing in Iberian societies came to require that a person own a large landed estate, reside in the city, and have been born into a proper family.

A final condition of social prominence in Roman Iberia came to be membership in, and at least outward compliance with the teachings of, the Roman Catholic Church, which emerged as a dominant political and economic force within the empire from the fourth century onward. To an extent, the influence of the Church lessened conflicts that would have otherwise emerged among the elites over inheritance of the great landed estates. Owing to the Roman custom of **primogeniture,** wherein the eldest living son of a family received for his inheritance the entire estate, the second and subsequent children had no legal claim on the *latifundio* and consequently little prospect for maintenance of the ostentatious lifestyles to which they had become accustomed. Many, therefore, chose the priesthood as a career, for ascension into the upper levels of the ecclesiastical hierarchy brought with it the potential for control of vast lands and wealth. Those of the disinherited sons who decided not to become priests had, as perhaps their final option, the opportunity to join the military, through which they hoped to receive, as compensation for their conquests, *latifundios* of their own that would enable them to live as the great *caballeros.*

With the death of the Emperor Constantine in 337, the Roman Empire was divided into western and eastern spheres and never regained its former grandeur. Iberia, situated on the margin of the western empire, was largely cut off from the Roman heartland. In the ensuing centuries, the peninsula was invaded by a series of Gothic tribes, who while leaving considerable Teutonic blood in Catalonia and other northern regions, did little to alter the entrenched Roman values and beliefs that ultimately became the bases of land and society in the New World.

ISLAM IN IBERIA

By the sixth and seventh centuries A.D., Iberia and the remaining Roman culture realm were settling into a largely feudalistic way of life dominated both politically and economically by the Christian Church. Meanwhile there emerged in the distant Arabian peninsula a belief system that would soon greatly alter the material and social life of those who would eventually colonize the New World. We refer, of course, to Islam, which within a century of the prophet Mohammed's first revelation in 610 had become the dominant faith of the desert peoples of the Middle East and North Africa.

In 711, mounted Arab-Berber peoples known as **Moors,** bent on spreading the Islamic gospel, crossed the Straits of Gibraltar and, sweeping northward across Iberia and the Pyrenees Mountains, advanced toward Paris. Simultaneously, other Muslim forces were pushing up the Danube River Plain of eastern Europe with the goal of taking Vienna and the alpine strongholds beyond. The invaders were repulsed in pivotal battles near Vienna and at Tours in central France and retreated southward, where, in the case of Spain and Portugal, they were to remain for almost 800 years, leading to the French saying that "Africa begins at the Pyrenees."

The Christians who refused to convert to Islam either fled or were driven underground and eventually came to be concentrated in northern Iberia. Moorish

FIGURE 5.1

The dual Roman and Moorish heritage of medieval Iberia is reflected in the arches of this ruined colonial-era Roman Catholic church of Antigua, Guatemala.

FIGURE 5.2

The colonial aqueduct of Morelia, Mexico, is a New World manifestation of a hydrologic technology originally developed by the Romans and subsequently expanded by the Moors in medieval Iberia.

settlement centered in the south. As the Christians regrouped and waited for their strength to increase, Iberia entered the golden age of Islam, which from the ninth through the thirteenth centuries produced the highest levels of intellectual and material achievement in the Western world. At the same time that the Christian lands of northern Europe were slipping into the so-called Dark Ages, Moorish, Jewish, and Egyptian scholars at the great Iberian Moslem universities at Córdoba and Toledo were achieving significant advances in mathematics, science, and the humanities. With the completion of the military conquest of Iberia, the caliph's court was established at Córdoba and an era of Islamic enlightenment began.

So extensive were the Moorish contributions to the Iberian and subsequent Latin American cultures that it is estimated that approximately one-third of modern Spanish is of Arabic derivation. Examples include most words beginning with the Arabic article *al,* such as *algodón* (cotton), *alfombra* (a floor carpet), and *alameda* (a grove of trees or a public walk or mall). Many of the more common expressions used in everyday Latin American conversation, such as *Ojalá* ("I hope so" or, more literally, "I would to God" [Allah]) and *Esta es su casa* (literally "This is your home," meaning please be comfortable in this home) can be traced to Arab values of homage to deity and hospitality to strangers.

As would be expected, the influence of the ruling Muslims is also evident in governmental terms such as *alcalde* (mayor or justice of the peace), weights and measures, and science and medicine. The urban landscape was enriched with numerous fountains, and many homes and public buildings reflected graceful

Arab styles and motifs (Glick 1979; Parsons 1962; Lane-Poole 1967) (Figure 5.1).

Many of the most significant Arab contributions to the Iberian cultural landscape occurred in the field of agriculture. Numerous crops that were to play prominent roles in Latin America were introduced from far-flung Islamic lands in Indochina, Indonesia, and the Philippines. These included cotton, indigo, saffron, and citrus from South Asia; sugar cane, bananas, and rice from Southeast Asia; and coffee, alfalfa, and melons from Southwest Asia. Magnificent Roman-era aqueducts were rebuilt, and in some instances, new water transport systems were constructed to supply growing urban populations and beautiful gardens and farms formed from lands used previously only for extensive animal grazing (Butzer 1988; Butzer et al. 1985; Crist 1957) (Figure 5.2). So pervasive was the Moorish influence on diet and food preparation that the names of many of Spain's most common foods reflect Iberia's dual cultural heritage. The word *aceite,* for example, widely used in Latin America today as a generic term for any vegetable oil, refers literally to olive oil and is a derivative of *acetuna,* the Arab word for olive fruits. Yet the Spaniards chose to retain the Latin word *olivo,* which originally meant olive fruits, as the word for an olive tree.

SPANISH RESPONSES TO ISLAMIC RULE

In addition to shaping many of the future economic patterns of Latin America, the Moorish occupation of Iberia led also to lasting changes in the attitudes and behaviors of the Spanish and Portuguese Christians who settled the New World. One consequence was a change in the attitudes of the Iberians toward peoples of darker skin color. In contrast to earlier conditions within the Roman empire, the ruling class during the Moorish era was darker-skinned than the subject peoples. To the Spaniards and the Portuguese, this meant that one of the most effective means of achieving upward socioeconomic mobility was to marry a brown-skinned woman. Racial mixing, or miscegenation, which was to become such a taboo among the English Puritans who settled Anglo America, would never be a troublesome issue to the Spanish and Portuguese colonists in Latin America.

Related to racial mixing was the impact of the Moorish occupation on Iberian familial role models and behavior. The Arabs brought to Iberia a faith that permitted the practice of polygamy and that held females in extremely low regard. Not only were wives to be totally subservient to their husbands, but their legal standing was little more than chattel, to be acquired and discarded at the pleasure of the man. While sexual promiscuity is certainly present in all cultures and times, it is probable that Moorish values and behaviors, reinforced by the traditional upper-class Roman pursuit of pleasure, contributed to the acceptance within Spanish and Portuguese societies of a double sexual standard and to the development of the *machismo* complex, which will be addressed in further detail in chapter 8. At the time of the European conquests of the New World, tremendous differences in attitude existed between the pious, family-oriented New England Puritans and the adventurous Spaniards. These differences, in turn, were to have a significant impact on race relations, social class structure, education, and economic development in the Americas.

A third way in which the Moorish occupation of Iberia modified the Spanish and Portuguese cultures was the Arab tradition of charismatic, authoritarian rule. In contrast to the Roman and Greek penchant for democratic political processes (albeit only among those privileged to be citizens), the Moors were accustomed to rule by strongmen who did not hesitate to use physical force to impose their will. This, combined with the long centuries of resistance by the Iberian Christians to Moorish decrees, contributed to the development within the Spanish and Portuguese societies of a general willingness to place one's personal political ambitions above the rule of law. The tradition of authoritarian rule also contributed to the development within Spain and Portugal, and later within Latin America, of bloated bureaucracies staffed through the patronage of the strongman and his subordinates. It also promoted a tolerance of political graft and dishonesty as a means of circumventing bureaucratic inertia and arbitrary and capricious actions by government officials. These traits again contrasted sharply with the English tradition of respect for constitutional processes and the concept of government service as a public trust based on merit.

Yet another outgrowth of the Moorish occupation was the blurring within Spanish and Portuguese societies of the distinction between church and state. Of course, the political dominance of the Church and its representatives had been a reality within both the Islamic and Christian worlds prior to the Arab invasions of Iberia. What was new therefore was not so much the unification of church and state per se, although these were certainly integrated closely, but the degree to which the Moorish occupation shaped perceptions of cultural identity.

Religious allegiance and ethnic nationalism came to be indistinguishable in Iberian society. Because the hated Moors were both foreigners and infidels, membership in and commitment to the Roman Catholic Church came to be viewed as the litmus test of all true Spaniards. Because Spaniards were, by definition, Catholics, non-Catholics or disloyal Catholics were presumed to be either foreigners or traitors and therefore enemies of the state. Preoccupation with church membership came to be so great in the socially stratified Iberian societies that eventually applicants for secular or ecclesiastical office had to prove that they had no Moorish, Jewish, or heretical Christians in their ancestry for four generations prior to themselves. Those who could not demonstrate the purity of their bloodline, called **la limpieza de la sangre,** were assumed to be descendants of "new Christians," that is, non-Christians who converted for reasons of political or economic expediency, and who were therefore considered unworthy of holding office. By the same logic, the terms *Judío* and *Luterano*, meaning, respectively, Jew and Lutheran, that is, Protestant, became synonymous for "foreigner" in colonial Latin America, where captured English pirates were as likely to be tried for the crime of their improper church membership as for their thievery or murders.

Although the melding of church and state occurred throughout Iberia, Moorish influences were weaker and of shorter duration in Portugal, which achieved independence in the twelfth century. In Spain, religious fanaticism became more deeply entrenched in the Spanish character. It was in Spain, therefore, that the principal language of the Christians, called Castellano (the language of Castile), became *"la lengua Cristiana"* (the Christian language), and it was the Spanish monarchs, rather than the Portuguese, who acquired the title *"los reyes Católicos"* (the Catholic kings).

On the relation of church and state, again it is instructive to contrast the consequences of differing Spanish and English values at the time of their arrival in the New World. The Puritans and other Anglo colonists settled America after the Protestant Reformation of Europe—indeed, they came in pursuit of religious liberty and freedom of expression. While extremely strict and not always tolerant of unorthodox religious behavior, they nonetheless soon established a strong tradition of separation of church and state that protected and nourished diversity of thought. The Spaniards, on the other hand, failed to separate church and state in the New World and endeavored to stifle dissent of all kinds in colonial Latin America. So intensely negative were their feelings toward members of other faiths that they even attempted, through the office of the Inquisition, to prevent works published in Protestant nations on economic and other nonreligious topics from entering the region. As a result, though Latin America had been colonized by Spain and Portugal—two of the most economically advanced nations in Europe during the Middle Ages—it remained largely outside the Industrial Revolution and economically and socially fell further and further behind the Anglo-American New England colonies during the colonial era. As we will learn in chapter 9, full freedom of expression, whether religious or political, came to most of Latin America only in the twentieth century.

THE SPANISH RECONQUEST

With the passage of time in Iberia, the Christians who had taken refuge in northern Spain achieved greater strength, and from the eleventh century onward they began winning their military engagements with the Moors. These wars, which are known collectively as **The Reconquest,** or *La Reconquista,* corresponded with a gradual, worldwide decline in the fortunes of the Islamic world. Included in that decline was the sacking by the Mongols of Baghdad in 1258 and the subsequent rise to power within Islam of conservative fundamentalist movements that rejected many of the scientific advances of earlier ages.

As the centuries passed, the Christian armies drove the Moors further and further south, replacing the once-flourishing Arab farms with migrant herds of sheep and cattle. Generation after generation of Spaniards passed away, filled with a crusading fever that was exceeded only by the dream of becoming *caballeros* on the reconquered lands. The Papacy, ever so appreciative of the wars being waged in the name of Christ, and of the wealth flowing to the Church from the acquisition of former Muslim lands and mosques, reciprocated by granting to the Spanish monarchs two far-reaching privileges that were to continue with the Crown during the conquest of the Americas. The first, called the **Royal Patronage,** or *Patronato Real,* was the authority given to the kings to approve or disapprove the appointment of all clergy in reconquered lands. The second authorized the Crown to collect and dispense the church tithe within the newly won territories. Taken together, these extraordinary concessions did much to politicize the institutional Church in colonial Latin America, and for over 400 years the Church sided consistently with conservative forces opposed to social and political reform.

By the fifteenth century, the Moors had retreated to their last remaining southern strongholds in Andalusia; Portugal was independent and exploring the coast of Africa in search of a sea route to India; and Spain was being unified through the joining of the kingdoms of Castile and Aragon. All of Iberia, it seemed, was on a war footing, poised to make further conquests for God and the king and, most importantly, for personal fame and riches. It is historically significant that the final battle between the Moors and Christians was fought at Granada in southern Spain in January of 1492, just eight months prior to Columbus's departure on his voyage of discovery. We can scarcely imagine the force with which the news of the finding of a new world struck the Spanish people, who after centuries of crusading had suddenly found themselves with no one left to conquer and, if of lower-class standing or disinherited through the practice of primogeniture, no means of achieving upper-class respectability. How thrilling it must have been to learn that even greater riches awaited those with the will to undertake the military and spiritual conquests of America.

THE CONQUEST AND SETTLEMENT OF LATIN AMERICA

The Spaniards and Portuguese who conquered the New World were a mixed lot composed primarily of men of lower-class standing and the second sons (*segundos*) of upper-class families who were willing to do almost anything to achieve the aristocratic lifestyles denied them in Iberia. Their purposes in coming to the New World centered on the so-called "three Gs": gold, God, and glory. As Bernal Díaz del Castillo (1989), who chronicled Cortés's conquest of Mexico, observed: "We came here to serve God . . . and also to get rich."

There is no doubt that the thoughts of most of the **conquistadores** centered far more on getting rich than on serving God. Not only did the *conquistadores* covet riches, they wanted them immediately. Their obsession with instant wealth predisposed them to undertake the most arduous expeditions conceivable in search of treasure, both real and imagined. And yet, having noted this, we must not overlook the fact that some of the explorers were men of sincere Christian belief, and even those who were not were accompanied by clerics committed to the propagation of the holy faith.

Of all the heroic figures of the age, none exhibited more openly the dual medieval character traits of piety and personal ambition than Christopher Columbus himself. Recognized by his peers as one of the most capable navigators of his time, Columbus's pleas for financial backing for his proposed westward voyage to the Orient were rejected by both the Portuguese and Spanish courts—that is, until he confided to the confessor of Queen Isabella that he intended to use his future wealth to fund an army that would liberate Jerusalem from the infidel Muslims.

Having used the confessor to persuade Isabella "La Católica" to finance the venture, Columbus set sail in August 1492, armed with an abiding faith that God would sustain him and with the promise that he would be appointed viceroy of the lands he discovered and receive 10 percent of all the gold found therein. Driven by a crusading fervor that compelled him to push on despite a near mutiny by his crew, Columbus landed somewhere in the eastern or central Bahamas, possibly San Salvador (Watling Island) or Samana Cay (Henige 1991; Deagan 1992; Marden 1986). After exploring the northern Caribbean for three months, during which time he discovered Cuba and Hispaniola, Columbus returned in triumph to Spain, where he was proclaimed "Admiral of the Ocean" by Isabella (Figure 5.3).

For the previous century Portugal had been extending its influence southward along the west coast of Africa under the direction of Prince Henry the Navigator and his successors. Now growing restless, Portugal's rulers asked Pope Alexander VI to decide to whom the newly found lands rightfully belonged. The Pope, acting in his role as vicar of Christ, issued a Bull of Demarcation that led to the signing in 1494 of the **Treaty of Tordesillas.** The document established as the boundary between the Portuguese and Spanish spheres a line 370 leagues west of the Cape Verde Islands, or approximately 46° west longitude (see Figure 5.8). Those lands discovered to the east of the line would belong to Portugal and those to the west to Spain.

Columbus was quickly outfitted for a second and far more grandiose expedition, which extended from 1493 to 1496 and included visits to Jamaica and several of the islands of the Lesser Antilles (Figure 5.4). This was followed by a third voyage, from 1498 to 1500, which included an exploration of the Caribbean coast of Venezuela. During his fourth and final voyage, from 1502 to 1504, Columbus desperately probed the inlets of the eastern coastline of Central America from Honduras southward to Panama, hoping against hope that he would somehow succeed in penetrating what

he firmly believed to be outlying islands off the coast of Asia. He died in Valladolid shortly after his return, melancholy and despondent and largely ignored by the far less idealistic generation of explorers who rushed to enrich themselves in the newfound lands (Pérez-Mallaína 1998).

The initial focus of Spanish settlement in the New World was the island of Hispaniola, or "little Spain." It was here in December of 1492 that Columbus established a garrison named La Navidad at a point along the northwest coast where his largest ship was wrecked on a reef. When he returned during his second voyage, the admiral found the fort burned and all the men dead, apparently killed by Arawak Indians. He then spent much of the next two years trying to establish a settlement called La Isabela on a coastal site further to the east, but that attempt also failed. The first permanently occupied Spanish community was founded in 1496 at Santo Domingo along the southern coast of the island. Soon thousands of Spaniards were flocking to the colony in search of gold, bringing with them, almost as afterthoughts, the values and institutions and material culture of fifteenth- and sixteenth-century Spain (Sauer 1966; Gibson 1968).

Much to their disappointment, most of the adventurers found little precious ore. Dreading the thought of having to settle down as farmers, they were soon vying with one another for the opportunity to move on to other islands or to the mainland, where they hoped their luck would improve. One such individual was Vasco Nuñez de Balboa, who was so determined to flee Hispaniola in response to reports of gold and pearls in the Darién that he arranged for his friends to smuggle him off the island in a provision cask. He then proceeded to Panama where, in 1513, he crossed the isthmus and discovered the Pacific Ocean (Figure 5.5). In what was a preview of the ruthlessness with which the Spaniards were to treat one another and the Indians, Balboa was beheaded the following year by his jealous father-in-law, Pedro Arias de Ávila (Pedrarias), upon de Ávila's arrival as governor of the region. Meanwhile, so many others were fleeing Hispaniola that the Crown soon passed laws calling for terribly dire punishments, such as the loss of an arm or a leg, for those who were caught attempting to leave without authorization.

The threatened penalties had little effect, however. Within two decades of the founding of Santo Domingo, Alonso de Ojeda was exploring the Caribbean coast of Colombia in search of the Dorado, the fabled land of gold, and Ponce de León had left Puerto Rico for Florida seeking the elusive Fountain of Youth. Cuba, too, had been settled and, under the cruel rule of Governor Diego Velásquez, was surpassing Hispaniola as the most prominent of the American colonies.

FIGURE 5.3

Columbus' voyages to the New World.

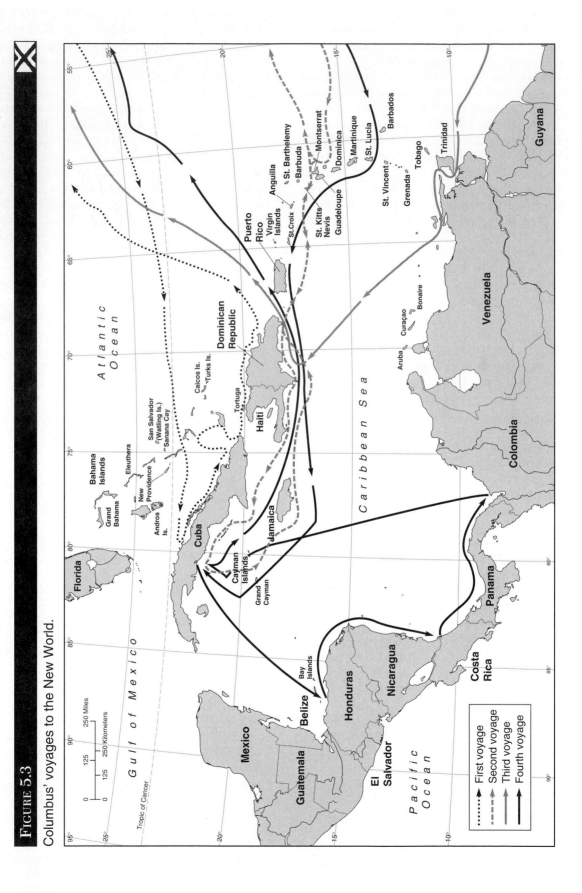

FIGURE 5.4

Discovery Bay, Jamaica, was discovered by
Columbus on his second voyage to the New World.

FIGURE 5.6

La Antigua, Veracruz; the site of the landing of
Cortés and the beginning of the Spanish conquest
of Mexico.

FIGURE 5.5

This monument to Balboa is placed so as to face
the Pacific Ocean in Panama City, Panama.

Mexico and Central America

It was from Cuba that the conquest of Mexico was carried out. The first Spaniard to arrive on the mainland was Francisco Hernández de Córdoba, who, in response to persistent rumors of a rich and powerful Indian state to the west, set sail in 1517. Crossing the narrow channel between Cuba and Mexico, some say in a hurricane,

Hernández de Córdoba arrived off the coast of the Yucatán peninsula, where he saw, in the Maya ruins, evidence of a great civilization. Hernández de Córdoba's expedition led to another the following year under the command of Juan de Grijalva, who explored the coastline as far west as Veracruz. It was Grijalva's voyage that prompted Governor Velásquez to colonize the mainland. Velásquez chose, as leader of the enterprise, Hernando Cortés, an extremely ambitious and energetic man who, prior to settling down in Cuba to a planter's life of gambling and women, had rejected an offer of an *encomienda* in Hispaniola with the disdainful comment that he had not come to America to work as a peasant.

Cortés set sail for Yucatán in February 1519 with a motley band of 500 men and a few horses, cannons, and guns. After skirting the gulf coast, Cortés disembarked at a point near Veracruz where the Antigua River empties into the sea, burned his ships to prevent any of his men from fleeing, and began his march into the interior (Figure 5.6). Pursuing a strategy of divide and conquer, Cortés struck alliances with many of the tribes that had been exploited so severely by the Aztecs, and ruthlessly slaughtered those who refused to join with him. He was greatly aided in his communications with the Indians by a Spaniard named Jerónimo de Aguilar, who had learned the Maya language as a captive of the Maya in Yucatán, and by an Aztec Indian princess named Malintzin, or Marina, who knew both Nahuatl, the Aztec language, and Maya. Marina had lived as a prisoner of

PROFECIA·DEL·GRAN·SACERDOTE·QUETZALCOATL·ANUNCIANDO·SU·REGRESO·EL·AÑO·UNO·CAÑA·

FIGURE 5.7

A mural depicting the promised return of Quetzalcoatl from the east. The work depicts the oral traditions of modern Tlaxcalan Indians and is found in the state capitol in Tlaxcala, Mexico.

the Maya in Tabasco before being presented, along with another twenty or so Indian women, as a gift to Cortés and his officers. She was instrumental in securing the allegiance of a number of tribes.

As valuable as were the swords and guns and horses of the Spaniards, perhaps their greatest weapon was the Aztec legend of having been visited in the distant past by a powerful, fair-skinned, bearded god named **Quetzalcoatl,** who had established peace and prosperity among the ancestors and, upon leaving, had promised to return one day from the east (Figure 5.7). As reports of the advances of Cortés and his tiny band reached Moctezuma in the great highland capital city of Tenochtitlán, the superstitious Aztec ruler vacillated, alternately sending gifts to the Spaniards in the misguided hope that they would leave and devising plans to try to kill them. None of the latter succeeded, however, and Moctezuma finally opted in desperation for ap-

peasement. As Cortés and his men climbed the divide between the towering volcanoes of Popocatépetl and Ixtaccíhuatl, and gazed down on the valley of Mexico with its great pyramids set among ordered fields and villages and waterways, they were greeted as emissaries of Quetzalcoatl by the representatives of Moctezuma, who knelt and kissed the ground at the Spaniards' feet. They were then escorted past hundreds of thousands of Indians to the entrance to the city, where they were received with great pomp by Moctezuma.

We can scarcely imagine the feelings of awe and anticipation that surely swept over Cortés when, after having struggled so hard to achieve his goal, he realized that he stood on the verge of conquering one of the wealthiest and most powerful kingdoms on earth. The thrill and amazement were tempered, however, with the realization that he and his men could be massacred in an instant. He quickly moved, therefore, to imprison Moctezuma and the other Aztec chiefs, and for a period of time he tried to rule through them. Terrible battles then followed in which both sides suffered major losses, but by August of 1521 Tenochtitlán had been leveled, and the Spaniards, many of them married to Aztec princesses, ruled over a kingdom whose population numbered in the millions. Although a new set of masters now governed from the valley of Mexico, the subservient relationships of the countless Indian peasants to their rulers would continue and, indeed, intensify in the years to come.

The Spanish conquest of the Aztecs was scarcely completed before new expeditions were dispatched from Mexico City, which rose from the rubble of Tenochtitlán. Francisco de Coronado traversed much of what today is northern Mexico and the southwestern United States, seeking in vain for the riches of the seven lost cities of Cíbola. Pedro de Alvarado, a member of Cortés's inner circle, traveled southward through the Mixtec and Zapotec settlements of Oaxaca and established himself as governor of Guatemala (Figure 5.8). Cortés himself undertook the conquest of Honduras for fear of losing it to a rival and then returned to Spain, where he died. Others, under the direction of Pedrarias, advanced from Panama northward into Nicaragua and Costa Rica but found neither gold nor Indians in large quantities.

South America

Rumors persisted, however, of an Indian kingdom in the southern continent whose size and wealth dwarfed even that of the Aztecs. In 1531, Francisco Pizarro, a former swineherd and member of Balboa's expedition to Panama, landed at San Miguel in northern Peru, having previously explored the Pacific coasts of Colombia and Ecuador, where he had captured some Indians who

FIGURE 5.8

Exploration and conquest of Latin America.

would later serve as interpreters. In a manner remarkably similar to the strategies employed a decade earlier by Cortés, Pizarro then moved inland with fewer than 200 men toward the highland stronghold of Cajamarca, where Inca Atahualpa awaited him with an army of over 30,000 (Prescott 1848). As with Cortés, Pizarro's admittance into the heart of the empire was facilitated by the legend of a bearded, fair-skinned god, called **Viracocha** by the Incas, who was believed to have introduced law and civilization to the forefathers and then departed, promising to return either to renew the world or to destroy it.

Like Cortés, Pizarro encountered a divided kingdom, not in the sense of tribal rivalries, as was the case in Mexico or New Spain, but rather a civil war raging between Inca Atahualpa and his half-brother, Inca Huáscar. Arriving in Cajamarca, Pizarro staged a surprise attack on Atahualpa, taking him alive as Cortés had Moctezuma. After releasing the chief in return for a huge ransom, Pizarro then imprisoned Atahualpa a second time and sentenced him to death by burning at the stake for having previously killed Huáscar. In response to Atahualpa's pleas for mercy, the heartless Pizarro commuted the sentence to death by hanging in consideration of the Indian's agreeing to be baptized a Christian. By November of 1533, Pizarro's forces had occupied without opposition the Inca capital of Cuzco (Figure 5.9). Two years later they founded Lima along the coast, from which point the Spaniards and their Indian wives plotted for control of an empire that stretched from central Chile to northern Ecuador.

Once Peru was conquered, most of the remainder of South America fell quickly into Spanish and Portuguese hands (Figure 5.10). The Atacama region of northern Chile was first penetrated in 1535 by Pizarro's bitter rival, Diego de Almagro. Almagro later rebelled against Pizarro, who had Almagro strangled in 1538. Pizarro himself was then killed by Almagro's son in 1541, the same year that Pedro de Valdivia reached the Central Valley of Chile and founded Santiago.

Ecuador was first colonized by Sebastián de Benalcázar, who left Pizarro in 1533 to subdue the northernmost Inca province. After having "founded" Quito in 1534 on the site of the previous Indian administrative center, he pushed northward in the eternal search for El Dorado into southern Colombia, where he established Popayán in 1536. Santa Marta and Cartagena had been founded previously along the Caribbean coast of Colombia in 1525 and 1533, respectively, and Bogotá was established in 1538 by Gonzalo Jiménez de Quesada, who had led an expedition up the Magdalena River valley to the Chibcha Indian settlements in the basins of the Cordillera Oriental. As Jiménez de Quesada was preparing to leave, he was joined in the Cundinamarca Basin by Benalcázar, who had come from Popayán in search of gold, and by one Nicolás Feder-

FIGURE 5.9

The construction of buildings of European design on top of ancient Inca structures in Cuzco was both a physical and symbolic expression of the Spanish conquest of the native peoples.

mann. Federmann had come from the western Venezuelan Andes as a representative of the German banking house of Welser, which since 1528 had been attempting to colonize Venezuela under an arrangement with the Spanish Crown (see Figure 5.8). Jiménez de Quesada's claim was subsequently upheld, and many of the basins of the Venezuelan Andes were settled by Spaniards in the mid-1500s. Jiménez de Quesada may also have been the first European to penetrate the Colombian Llanos, leading a party into the headwaters of the Orinoco River from which only 25 of 500 men returned alive.

Meanwhile, Gonzalo Pizarro, who had fought with his brother Francisco in the conquest of Peru, was appointed governor of Quito in 1539. Two years later he undertook an expedition into Ecuador's Oriente region east of Quito. Finding himself short of supplies, Pizarro instructed his next-in-command, Francisco de Orellana,

FIGURE 5.10

Ruins of Machu Picchu, which served as a personal recreation site of the Inca ruler Huayna Capac.

to lead an advance party of fifty men in search of provisions. Upon reaching the junction of the Napo and Marañón Rivers in an area just east of present-day Iquitos, Orellana deserted Pizarro and sailed all the way down the Amazon. Arriving at its mouth in August of 1542, he named the river after the mythical race of female warriors in recognition of a tribe of women whom he claimed to have encountered along the way.

The Atlantic coast of Brazil had been first discovered by the Spanish navigator Vicente Yáñez Pinzón, who owing to the Treaty of Tordesillas made no effort to occupy the region, which was clearly east of the line of demarcation. Portugal's claim to Brazil dated from April of 1500, when Pedro Álvares Cabral, either intentionally or unintentionally, came upon the coast of Bahia after having sailed far to the west of Africa on a voyage to India (Russell-Wood 1998). The first serious Portuguese attempt to settle Brazil did not come until 1533, however, when the region was divided into huge land grants called **capitanias.** The *capitanias* extended from the Atlantic coast westward into the uncharted interior and were given to individual noble-

men (Figure 5.11). Although most of these failed in the short term, a centralized government was in place by 1549 in response to threats from the French and Spaniards, and by 1554 settlements extended along the coast from Pernambuco southward to Santos and inland to São Paulo.

Argentina was first colonized by Spain in 1536, when Pedro de Mendoza founded Buenos Aires on the south shore of the Río de la Plata (see Figure 5.8). Because the nomadic Pampean Indians were hard to control and the settlement site offered little in the way of immediate riches, it was abandoned the following year in favor of Asunción, Paraguay. Asunción was situated far upriver in the midst of the semisedentary Guaraní Indians, whose labor was relatively easy to exploit. It was from Asunción that most of the lower Paraguay and Paraná river valleys were subsequently settled, which led in turn to the reoccupation of Buenos Aires in 1580. The eastern Andean piedmont cities of Argentina, from Tucumán southward to San Luís, were founded in the latter half of the sixteenth century, largely by colonists arriving from central Chile and the Bolivian Altiplano.

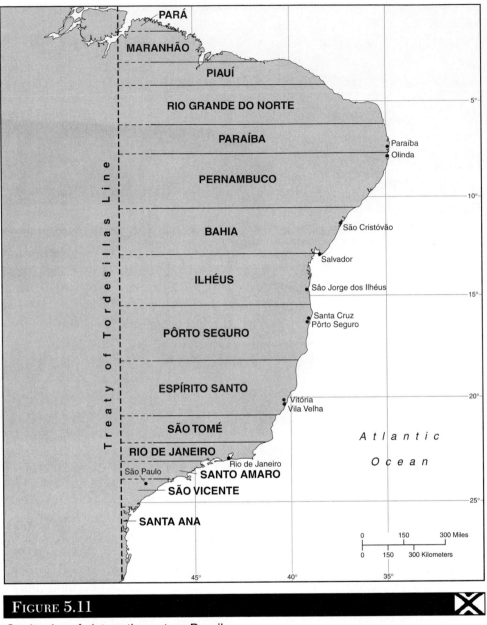

FIGURE 5.11

Capitanias of sixteenth-century Brazil.

THE COLONIAL ERA

As the initial rush of exploration and conquest wound down in Latin America, the Spanish Crown sought to replace the chaotic and often violent rule of the *conquistadores* with a more orderly and responsive governmental infrastructure. Accordingly, in 1524, a number of the most prominent and capable men in the kingdom were appointed to a newly formed body named the **Supreme Council of the Indies,** which was given complete authority to regulate every facet of life in the New World, be it secular or spiritual. Serving as the highest representatives of both the council and the king in the New World were the viceroys, who governed in an authoritarian manner two vast administrative units known as **viceroyalties.** The sprawling Viceroyalty of New Spain consisted of Mexico, which included the present southwestern United States, and also Central America, the Caribbean, and the distant Philippine Islands. The Viceroyalty of Peru initially encompassed virtually all of Spanish South America except for Venezuela. The difficulties inherent in governing such immense and generally isolated territories resulted in the two original units' being divided into four in the eighteenth century. The vague and often erroneous boundaries of the viceroyalties and their subunits eventually came to form the frequently contested outlines of the modern Latin American nations (Figure 5.12).

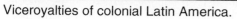

FIGURE 5.12

Viceroyalties of colonial Latin America.

The wealthiest and most powerful of the late eighteenth-century South American viceroyalties was Peru, which consisted of most of modern-day Peru and small, neighboring portions of Bolivia and Brazil. The Viceroyalty of Peru was administered from Lima, the leading city in all the Spanish empire. Most of Bolivia, as well as Paraguay, Uruguay, and Argentina, belonged to the Viceroyalty of La Plata, which was administered first from Asunción and later from Buenos Aires. Ecuador, Colombia, and Venezuela formed the Viceroyalty of New Granada, whose capital was Bogotá. Mexico and the Caribbean Islands composed the Viceroyalty of New Spain, with its capital of Mexico City. Chile and Guatemala (Central America) were designated as captaincy-generals in recognition of their isolation, but reported to the viceroys of Peru and New Spain, respectively. Brazil was classified as a single viceroyalty by the Spanish government from 1580 to 1640, during which time Spain and Portugal were united, and continued to be administered as a single entity thereafter under Portuguese control. The original *capitanias* evolved into some of the present-day states of Latin America's largest country.

The Mercantile System

Responsibility for regulating the economic life of the colonies was delegated by the Council of the Indies to a body known as the **House of Trade,** or *Casa de Contratación.* The House of Trade, in turn, was responsible for the implementation of the large and complex set of laws that made up the Spanish **mercantile system.**

The overarching objective of the mercantile system was to enrich the Spanish monarchs to the fullest extent possible. To that end, the colonies were initially prohibited from trading either directly with one another or with foreign nations. This meant, for example, that if a merchant in Mexico wished to send a ship full of goods to Colombia, he would first have to route the vessel from Veracruz back to Spain and then have the commodities cross the ocean a second time for delivery in Cartagena or Santa Marta. The reason for this policy, of course, was the Spanish desire to levee various sales and import-export taxes on all products produced in the New World. Included among the taxes on precious ores was the **Quinto Real,** or Royal Fifth—a flat 20 percent take on all metals mined in the colonies.

As a means of enforcing the trade restrictions, all cargoes were required to be carried on Spanish-registered vessels, which were supposed to depart annually from Seville or, later, Cádiz. To protect the gold- and silver-laden galleons on their return passage to Spain, the ships were allowed to travel only in convoys, called *flotas,* and a great network of harbor fortresses was constructed in the principal port cities, including Havana, San Juan, Cartagena, and Veracruz (Figure 5.13). The mercantile system

FIGURE 5.13

The old harbor fortress of San Juan de Ulua at Veracruz, Mexico, is a remnant of the colonial mercantile system.

also forced the colonies into unequal, dependent trade relationships by prohibiting the manufacture or cultivation in the New World of any product or crop whose production might threaten the economy of Spain. Included in this category were most industrial products as well as the growing of grapes, wheat, olives, and sheep raising.

The mercantile system retarded the economic development of the Spanish colonies in a number of ways. Perhaps the most obvious consequence was the crippling of industry and commerce, which were already laboring under the extreme handicap of being excluded from technologies originating in the industrializing Protestant nations of northern Europe. Beyond the visible ramifications, however, the system promoted widespread contraband trade and disrespect for law, and contributed to the ever-growing reservoir of ill will between American-born Spaniards, who felt discriminated against, and their European-born associates.

The Encomienda System

While the mercantile system was contributing to the progressive underdevelopment of the Spanish and Portuguese colonies during the sixteenth, seventeenth, and eighteenth centuries, equally lasting damage to the fabric of human society was being inflicted through the **encomienda system,** one of the principal mechanisms through which the European aristocracy exploited the native American Indians. The term *encomienda* came from the Spanish verb *encomendar,* meaning "to entrust," and was derived from the medieval European manorial system in which the serfs were forced to entrust themselves as vassals into the care and keeping of the feudal lord, who in return required tribute and personal services from his subjects.

The *encomienda* system originally was embraced by the Council of the Indies as a means of protecting the Indians from the ravages of the *conquistadores,* many of whom, it was feared, would otherwise enslave and kill vast numbers of the natives. Under its basic provisions, the Indians were gathered, if need be, into new settlements called *congregaciones* or *reducciones,* where they were informed, in a language that few if any understood, that they had entrusted themselves to their new Spanish masters. In exchange for Indian tribute and labor, the European colonists were obligated to "civilize" the natives by teaching them the Spanish language and the Christian faith (Lovell 1990). In theory, the *encomienda* was not a land grant but rather a labor relationship that was valid only for "two or three generations" (*dos o tres vidas*), following which time responsibility for the care of the Indians would revert to the Crown.

In practice, however, the *encomenderos* seldom took seriously their responsibility to protect the Indians, and finding themselves at last in a position to enrich themselves at the expense of others, made the Indians virtual slaves (Simpson 1966; Chevalier 1965). Alarmed at the terrible abuses and the resultant collapse of the native populations that he was witnessing, a priest named Bartolomé de las Casas began in the 1520s to petition the Council of the Indies to abolish the *encomienda* system (de las Casas 1974 and 1992; Clendinnen 1987). His efforts were finally rewarded in 1542, when the council, headed by Fray García de Loaisa, a Dominican Cardinal, replaced the *encomienda* with the New Laws, which mandated the **repartimiento system.** Under the latter, the Spaniards were forbidden to require labor or tribute from individual Indians, but were allowed to demand group services of entire Indian villages whose leaders determined individual work assignments. From the midsixteenth century on, then, the term *encomendero* evolved into a largely honorary title whose decline mirrored that of the rural Indian populations and the agrarian sector generally. With the opening in the later sixteenth century of the fabulous silver mines of Bolivia, Peru, and Mexico, Spanish-Indian labor relationships were governed through most of the colonial era by the *repartimiento* system. In the case of the Andean region, this was frequently viewed by the Indians as an extension of the ancient Incan *mita* system, wherein the ruling Incas had exacted labor from their subject villages.

While the *encomienda* and *repartimiento* systems were intended to protect the native Indian peoples, clearly they failed to do so (Lynch 1992). No amount of enlightened legislation in Spain could alter the attitudes of the New World colonists, whose behavior was reflected in the saying *"Se obedece pero no se cumple"* ("one obeys the letter of the law but not the spirit"). Ironically, a consequence of de las Casas' successful campaign to legislate protection for the Indians was

the introduction of black African peoples as slaves into the plantation regions of the labor-starved American lowlands, whose previous Indian occupants were rapidly dying out (chapter 7). De las Casas' heart-rending written accounts of the sufferings of the Indians at the hands of their cruel Spanish masters were used by England and other Protestant rivals of Spain to propagate a misconception that came to be known as the **Black Legend,** or *La Leyenda Negra,* in which the Spaniards were accused of intentionally killing as many Indians as they could. In reality, it was the English in Anglo America who felt the only good Indians were dead ones, while the Spanish lifestyle required the use of Indians as serfs.

However well intentioned were the *encomienda* and *repartimiento* systems, their ultimate impact was the segmentation of Latin American society into a small, elite aristocracy and a huge underclass. With the establishment by the mid-1500s of a thriving African slave trade, together with the mercantile and *encomienda-repartimiento* systems, the institutions that were to influence most profoundly the economic and social character of Latin America during the remaining two and one-half centuries of colonial rule and beyond were in place.

Decline of Spanish Influence in the Greater Caribbean Basin

Although the basic features of Latin American society were to remain relatively constant throughout the colonial period, two forces were at work that led eventually to the collapse of the Spanish and Portuguese empires. The first was the gradual neglect by Spain of its Caribbean Island possessions, whose importance diminished in the eyes of the Spaniards in proportion to the volume of gold and silver extracted from the highland mining centers of the continental mainland. As Spanish interest and involvement in the Caribbean waned, England, France, and the Netherlands, undeterred by the Treaty of Tordesillas, moved to establish bases throughout the region in the late 1500s. The initial incursions usually took the form of semiofficial support for pirates, such as Sir Francis Drake and his kinsman John Hawkins (Esquemeling 1967). Another prominent raider, operating nearly a century after Drake and Hawkins, was Henry Morgan. In addition to plundering Spanish vessels on the high seas, Morgan bartered cargoes of slaves and manufactured goods with the industry-starved Spanish colonies—all in direct defiance, of course, of the Spanish mercantile regulations prohibiting such trade (Figure 5.14).

The success of the British, French, and Dutch buccaneers soon led both to the formation of West India companies and to the colonization of islands and areas not effectively controlled by the Spaniards. Bermuda

FIGURE 5.14

The ruins of *Panamá Viejo,* the first site of Panama City, which was sacked in 1671 by the pirate Henry Morgan.

FIGURE 5.15

An English-style town clock near the center of Ocho Rios, Jamaica.

was taken by the English in 1612 and was soon joined in the British empire by Barbados and a number of the smaller islands of the northern Lesser Antilles, including St. Kitts, Nevis, Barbuda, Antigua, and Montserrat. Meanwhile, France was occupying Guadeloupe, Martinique, and Marie Galante in the central Lesser Antilles, and the Dutch were gaining control of Saba, St. Eustatius, and St. Martin to the east of Puerto Rico, and Aruba, Bonaire, and Curacao—the so-called ABC islands—off the western coast of Venezuela. Even Denmark got into the act, colonizing St. John and St. Thomas in the Virgin Islands in the 1670s.

While the Lesser Antilles were falling one by one from Spanish control, the British, French, and Dutch were also colonizing the Caribbean coasts of Central America and the northern and eastern coasts of South America. The British were especially effective in occupying the remote eastern lowlands of Central America where, in the mid-to-late seventeenth century, they established settlements from Belize (formerly British Honduras) southward along the Honduran and Nicaraguan coastlines (Woodward 1999; Richardson 1992). The cultural legacy of these efforts includes the widespread use of a modified English among coastal residents today and numerous individuals with Spanish given names and English surnames. The legacy is also manifested in the use of English names for geographic places, such as the town of Bluefields, Nicaragua, and Little and Great Corn Islands off the Nicaraguan coastline. The Dutch were so successful in their attempts to establish a presence along the Atlantic coast of Brazil that they actually ruled northeastern Brazil from 1630 until 1654, when they were overthrown by an army composed primarily of men from the São Paulo region to the south. The following year,

Spain lost Jamaica to the British, and in 1697 the western third of Hispaniola fell to France (Figure 5.15). This colony, which ultimately became Haiti, the French named Saint-Dominique. One consequence of these changes on the human geography of the Caribbean islands was the development over time of mixed cultures. Present-day evidences of this pattern include the widespread use of Spanish names in English-speaking Trinidad and Jamaica, and the use of a French-based patois in British-affiliated islands such as St. Lucia.

Peninsular-Creole Divisions

The loss of most of the greater Caribbean Basin stood as a highly visible sign of declining Spanish influence along the margins of the empire. Meanwhile, an internal schism building among the Spaniards of the mainland colonies ultimately played a major role in provoking the wars of independence in the early nineteenth century. At the heart of the problem was a distinction made by the extremely class-conscious Spaniards between peoples of European ancestry who had been born in Spain and those who had been born in the New World. The former were called **peninsulares** ("peninsulars," meaning those born on the Iberian peninsula) or *gachupines* (literally, "spur wearers"; yet another reference to the fact that only bluebloods could be horsemen and, therefore, wear spurs on their feet). The American-born Europeans were classified as *criollos* (**creoles**).

The *criollos* were discriminated against by the *peninsulares,* who reserved for themselves the highest civil and ecclesiastical offices. Being categorized a *criollo* was so detrimental to upward mobility during the colonial era that many Spanish women, on finding themselves pregnant, booked passage for Spain in order that their offspring might be born in the old country and thus have claim to peninsular standing. When the inevitable question arose among the squabbling Spaniards as to the status of an infant born at sea to peninsular parents, the courts ruled that the child would retain peninsular status, provided that the mother was en route to Spain when the birth occurred.

As long as peninsulars outnumbered creoles within the New World Spanish societies, the regulations went relatively unchallenged. With the passage of time, however, the creoles came to be far more numerous than the peninsulars and their frustrations became more pronounced. By the late 1700s, peninsulars constituted but a small minority in most Latin American cities and were virtually absent from the rural countryside. With the successful American and French revolutions providing inspiration that more free and egalitarian societies could be formed, the stage was set for the revolt of Spain's remaining American colonies.

INDEPENDENCE AND THE REPUBLICAN PERIOD

The wars of independence were triggered in part by political events in Europe, where in 1808 Napoleon Bonaparte forced King Ferdinand VII of Spain into exile and attempted to replace him with a puppet government led by the emperor's brother, Joseph. The restless creole leaders in America, some of whom (such as Venezuela's Francisco de Miranda) had been involved previously in unsuccessful revolts, sensed a historical turning point and demanded the right to organize *juntas,* or political councils, to govern themselves until Ferdinand was restored to the throne. This legal posturing was a ploy on the part of most of the revolutionaries, who had no intention of ever returning to the Spanish fold, regardless of what might happen to the royal family.

The restoration of Ferdinand to the throne in 1814 and his increasingly inept and absolutist rule thereafter fueled the revolutionary wars, whose leaders included Fathers Miguel Hidalgo y Costilla and José María Morelos in Mexico, Mariano Moreno and José de San Martín in Argentina, Bernardo O'Hig-

FIGURE 5.16

A mural depicting Father Hidalgo leading the uprising for Mexican independence. The painting is found inside Chapultepec Castle, Mexico City.

gins in Chile, and Francisco de Miranda, Simón Bolívar, and Antonio José de Sucre in Venezuela (Figures 5.16 and 5.17). What bound this diverse group of men together was an intense frustration with peninsular dominance in Spanish America, and a democratic idealism derived largely from the values of the American and French revolutions. Indeed, many of the Latin American liberators were personally acquainted with leading French and American political philosophers.

Most of the Latin American revolutionary leaders favored, in an intellectual and emotional sense, the extension of personal freedoms and opportunities to peoples of all economic and social ranks, as had occurred in the American and, to a lesser degree, the French revolutions (Harvey 2000; Robinson 1990; Humphreys and Lynch 1965). They soon came to realize, however, that the oppressed, illiterate Latin American masses were ill-prepared to assume civic responsibility, and that the elites, upon whom the success of the revolutions depended, were opposed to any form of power sharing. So while the Latin American revolutions appeared to be waged for the purpose of extending liberty to all, they consisted primarily of a violent creole-peninsular class conflict. The victory of the creoles resulted in a mere reshuffling of players at the top of Latin America's socioeconomic pyramid and did little to alter the old exploitative colonial attitudes and structures, which continued in most Latin American nations down through the nineteenth and into the twentieth century.

FIGURE 5.17

Simón Bolívar, the leader of the independence movements of the northern Andean nations.

This peculiar set of circumstances, in which a wealthy upper class fought for freedom and democracy for itself but not for others, led in Latin America to the development of a unique political environment, which on the surface might appear to include contradictory elements. On the one hand, the constitutions in force among the Latin American nations historically have contained eloquent guarantees of personal civil liberties. The governments established therein are generally patterned after the United States model, with power divided between the executive, legislative, and judicial branches, which function at local, state/provincial, and national levels.

On the other hand, the individual freedoms and constitutional procedures so beautifully expressed in these documents have been widely violated since independence was achieved over 180 years ago. Following various ill-fated attempts to form regional confederations, whose boundaries corresponded roughly with those of the old colonial viceroyalties, almost all of the Latin American nations lapsed into a long and difficult period of rule by military strongmen. Together with their followers, these *caudillos* looted and oppressed their subjects, thus perpetuating the underdevelopment initiated during the preceding colonial era.

There were a few exceptions to the pattern of violent despotism. Costa Rica began what has become one of the strongest traditions of peaceful, democratic rule in all of Latin America. Both Cuba and Puerto Rico chose to remain loyal to Spain during the revolutionary wars, and continued as remnants of the formerly vast Spanish colonial empire until taken from Spain by the United States as a consequence of the Spanish-American War of 1898.

Brazil, too, largely escaped the violent excesses of the early independence era, owing in part to a unique set of historical circumstances set in motion by Napoleon's invasion of Portugal in 1807 (Boxer 1961). The ensuing French occupation of Lisbon forced the Portuguese royal family to flee to Brazil, where a government-in-exile was established in Rio de Janeiro. With King John VI, or Dom João as he was known, residing in Rio, the hated colonial mercantile system was dismantled and Brazil was raised to a position equal to Portugal within the empire. Dom João moved the royal court back to Lisbon following the withdrawal of the French, but left his son, Dom Pedro, behind to govern Brazil. When the Portuguese government subsequently attempted to reinstate the former colonial limitations on Brazilian trade, Dom Pedro himself declared independence for Brazil in 1822.

Conservative–Liberal Schisms

No sooner did the nations of Spanish America gain their independence than a bitter ideological division arose between the conservative and liberal wings of the ruling creole classes. Generally speaking, the **conservatives** favored strong national governments with power centered in the capital cities and their immediate environs. The conservatives also represented the interests of the old ruling oligarchies, composed primarily of the landed aristocracy and high Catholic Church officials, most of whom distrusted democratic political practices that might bring the masses to power. In some nations, the conservative elements were so fearful of representative government that they favored the reestablishment of European monarchical rule. In short, the conservatives consisted primarily of the old power brokers who endeavored to prevent real, meaningful social and political change from occurring in the newly independent states.

The **liberal** factions stood opposite of the conservatives on virtually every issue. The political base of the liberals tended to concentrate in the distant, outlying regions of the various countries, where economic development lagged behind that of the national capitals. The inhabitants of these regions tended to favor a federal form of government that would result in greater power being vested in the states or provinces. They also tended to be

TABLE 5.1

Forms of *Caudillismo*

NAME	MEANING
golpe de estado	To remove or incapacitate the head of state through assassination or detention
cuartelazo	Army barracks revolt
imposición	Rigging an election to guarantee victory
candidato único	Potential opposition candidates intimidated into not running for office
continuismo	Peacefully maintaining the strongman in office beyond the initial expiration date of his legal term

Sources: Stokes 1952, 1953, and 1959.

more egalitarian in their political philosophies and more willing to allocate power to the common man. Because the leadership of the institutional Catholic Church historically had allied the Church with conservative interests, the liberals argued, on grounds of economic development, for the separation of church and state and for the ending of the Church's dominance of civil registries and education. The liberals thus favored the substantive social and economic change that had generally failed to come with the attainment of independence. The postindependence history of most Latin American nations can best be understood in the context of the continuing conservative-liberal schism that generally undergirds party and personal political rivalries.

Caudillismo

In addition to the ongoing conservative-liberal conflicts, a continuing thread throughout the republican era has been the presence of **caudillismo,** which can be defined as arbitrary rule by a strongman or chief—the so-called *jefe mandatario. Caudillismo* was formerly enforced mostly through violent means. Two of the most common were the outright assassination of one's opponent—a strategy sometimes referred to as *golpe de estado,* meaning to strike or take out the head of state—and an army revolt, or *cuartelazo* (Table 5.1). The latter, which requires a higher level of organizational skill than a simple assassination, involves a rapid and well-coordinated takeover by the leader of a key army barracks (*cuartel*) of vital government installations such as television and radio stations, airports, the mint, and the national palace itself.

Needless to say, violent forms of *caudillismo* no longer play well to American and European lawmakers whose economic aid is increasingly tied to the observance of human rights in the recipient nations. Most modern Latin American *caudillos* therefore prefer nonviolent means of perpetuating their power. Stokes (1952, 1953, 1959) has noted three common strategies,

all of which are becoming more and more difficult to carry out without detection. The first, called *imposición,* or imposition, involves rigging an election on behalf of the *caudillo,* who may even allow, for the sake of appearances, unsuspecting opposition candidates to campaign against him. Occasionally, such as in the Dominican Republic presidential election of 1994, the *caudillo* is shocked to learn he is actually losing in the vote count, and he is forced to come up with an embarrassing pretense for declaring the election null and void.

The second nonviolent strategy, which was in vogue as recently as the 1980s in Paraguay, is called *candidato único* and entails intimidating potential opposition candidates to such a degree that none dares to run against the *caudillo.* While this does not look good in the press, the *caudillo* can always claim that he is so popular that no one cares to run against him. He also knows that he will not be forced to cancel the election once it is under way.

The third nonviolent strategy, *continuismo,* consists of maintaining the strongman in power beyond the initial expiration of his term of office. This can readily be accomplished by rewriting the constitution, as occurred in Peru in 1993 and Venezuela in 1999. When rewriting the constitution is not possible or convenient, *continuismo* can also be achieved by the outgoing *caudillo* rigging the next presidential election in favor of his handpicked successor who governs as a figurehead until the strongman is eligible to be reelected. This latter strategy was successfully implemented in the late 1990s in Haiti.

One reason the military historically has intervened so often in the governance of many Latin American nations is that the officer corps has been imbued with the belief that it is the only segment of society capable of preserving or, if need be, reestablishing democratic practices threatened by *caudillismo* (Loveman 1999; Rodríguez 1994; Ramsey 1997). Time and again, the military has justified the removal of a *caudillo* from office by promising that the newly formed junta will govern only temporarily until honest elections can be held to select the next president. Unfortunately, one of the junta may consolidate his power and become himself the next *caudillo*.

POLITICAL MATURATION IN THE LATE TWENTIETH CENTURY

As social and economic change has accelerated among the Latin American nations in recent decades, *caudillismo* has begun to take a back seat to more open forms of political expression. In 1980, for instance, only thirteen Latin American nations were classified by the Inter-American Development Bank as possessing democratic governments at the national level and only three allowed municipal mayors to be popularly elected rather than appointed by political party officials. By 1997, those figures had risen to twenty-six and seventeen, respectively (Hausmann 1997). There is, therefore, reason to believe that the forces of pluralism transforming so many aspects of Latin American life are working to move the region as a whole toward ever-increasing levels of individual freedom (O'Loughlin et al. 1998; Domínguez 1998; Camp 1996).

Having noted the decline of some of the more visible expressions of *caudillismo*, we must also recognize that authoritarian rule continues to permeate many aspects of Latin American society. Labor unions, churches, neighborhood block organizations, and even charities are all commonly governed in authoritarian style by charismatic leaders who seek to punish those who oppose them and who distribute spoils to their supporters. Political authoritarianism at the national level has commonly expressed itself through the establishment of **"protected democracies."** A "protected democracy" is defined by Loveman (1994) as a government that outwardly appears to be free and democratic but whose policies and practices are inwardly limited by fear of violence from a group of military, political, or ecclesiastical elites. These guardians demand the right to function as guarantors of a long-term "national interest" or "common good" which they alone can define (Loveman and Davies 1997) (Figure 5.18).

In addition to authoritarianism, the peoples of Latin America continue to face other serious political challenges. Latin America has long been the object of foreign military intervention. The pattern began, of course, with European colonization. No sooner had the Latin American revolutionaries achieved their independence, however, than United States President James Monroe issued in 1823 a statement that came to be known as the **Monroe Doctrine.** The document has come to symbolize in many respects the love-hate relationship that

FIGURE 5.18

Soldiers maintaining a casual yet very visible presence on a Lima, Peru, street. The military continues to impose certain terms and limitations on the policies and actions of many of the democratically elected governments of the Latin American nations.

has characterized United States-Latin American relations for over 180 years. It contained a warning that the United States would not permit renewed European military involvement in the Western Hemisphere. While the leaders of the new Latin American republics generally welcomed United States protection during the period of national infancy, Latin Americans have been understandably sensitive to what they perceive as a continuing pattern of American intervention in their internal affairs (Table 5.2). These invasions, both overt and covert, historically have been rationalized by Americans as necessary evils in defense of freedom and democracy—a sort of latter-day American expression of manifest destiny (Smith 2000; Langley 2002; Johnson 1991). They have created, however, ever-increasing levels of anti-American feeling among Latin Americans, who as a whole have otherwise long admired America's democratic form of government and material development (Cottam 1994; Musicant 1991; Munro 1964).

The pattern of American military intervention has been most pronounced in Middle America, owing to the region's geographic proximity and strategic importance to the United States, as well as to the generally small populations, which can be managed more easily than those of the more populous countries of South America. From the 1960s through the 1980s, the former Soviet Union became yet another foreign military presence in the region, and Central America in particular became one of the most intensively militarized zones on earth. The tragic consequences for the common people included the loss of loved ones and livelihoods as well as the disruption of much of the social and cultural continuity that formerly provided stability and meaning to their lives.

In addition to actual or potential foreign military intervention, the political stability of the Latin American nations is further compromised by the continuing debt crisis, which has raised foreign economic intervention to new heights (chapter 13). Finally, the very foundation of civil order and the rule of law has been threatened in some countries in recent years by the illicit, international trafficking of narcotic drugs. However, we should not overlook the fact that for Latin America as a whole, there probably has never been a time when so many were so free and when hopes for an even better tomorrow were so justified.

TABLE 5.2

Overt United States Military Invasions/Occupations of Middle American Nations

COUNTRY INVADED	DATES
Mexico	1846–1848 1914 1917
Honduras	1911–1912 1924
El Salvador	1981
Nicaragua	1910–1933
Panama	1908–1918 1989–1990
Cuba	1898–1902 1906–1909 1961
Haiti	1915–1934 1994–1996
Dominican Republic	1916–1924 1965
Puerto Rico	1898
Grenada	1983–1985

SUMMARY

The Spanish and Portuguese settlers of Latin America and the Caribbean brought both Roman and Moorish values, institutions, and products to the New World. England, France, the Netherlands, and other non-Hispanic nations later took from Spain a number of territories in the greater Caribbean Basin. The Latin American and Caribbean peoples have made great progress in recent years toward more open and democratic forms of government. They nevertheless continue to be challenged by conservative-liberal schisms, by a tradition of authoritarian *caudillismo,* and by threats of internal and external military intervention.

KEY TERMS

Iberian Peninsula 90
latifundios 90
minifundios 90
absentee land ownership 90
patrón 91
noblesse oblige 91
primogeniture 91
Moors 91
la limpieza de la sangre 93
The Reconquest 94
Royal Patronage 94

conquistadores 94
Treaty of Tordesillas 95
Quetzalcoatl 98
Viracocha 100
capitanias 101
Supreme Council of the Indies 102
viceroyalties 102
House of Trade 104
mercantile system 104
Quinto Real 104

encomienda system 104
repartimiento system 105
Black Legend 105
peninsulares 106
creoles 107
conservatives 108
liberals 108
caudillismo 109
"protected democracies" 110
Monroe Doctrine 110

SUGGESTED READINGS

Boxer, Charles R. 1961. *Four Centuries of Portuguese Expansion, 1415–1825: A Succinct Survey.* Johannesburg: Witwatersrand University Press.

Butzer, Karl W. 1988. "Cattle and Sheep from Old to New Spain: Historical Antecedents." *Annals of the Association of American Geographers* 78:29–56.

Butzer, Karl W., et al. 1985. "Irrigation Agrosystems in Eastern Spain: Roman or Islamic Origins?" *Annals of the Association of American Geographers* 75:479–509.

Camp, Roderic Ai, ed. 1996. *Democracy in Latin America: Patterns and Cycles.* Wilmington, Del.: Scholarly Resources.

Chevalier, Francois. 1965. *Land and Society in Colonial Mexico.* Berkeley: University of California Press.

Clendinnen, Inga. 1987. *Maya and Spaniard in Yucatán, 1517–1570.* Cambridge: Cambridge University Press.

Cottam, Martha L. 1994. *Images and Intervention: U.S. Policies in Latin America.* Pittsburg: University of Pittsburg Press.

Crist, Raymond E. 1957. "Rice Culture in Spain." *The Scientific Monthly* 84:66–74.

de las Casas, Bartolomé. 1974. *In Defense of the Indians.* Trans. and ed. Stafford Poole. DeKalb: Northern Illinois University Press.

———. 1992. *The Devastation of the Indies: A Brief Account.* Trans. Herma Briffault. Baltimore: Johns Hopkins University Press.

Deagan, Kathleen A. 1992. "La Isabela, Foothold in the New World." *National Geographic Magazine* 181:40–53.

Díaz del Castillo, Bernal. 1989. *The Conquest of New Spain.* Trans. J. M. Cohen. New York: Viking Penguin.

Domínguez, Jorge I. 1998. *Democratic Politics in Latin America and the Caribbean.* Baltimore: Johns Hopkins University Press.

Duncan-Jones, R. 1974. *The Economy of the Roman Empire.* Cambridge: Cambridge University Press.

Dyson, Stephen L. 1992. *Community and Society in Roman Italy.* Baltimore: Johns Hopkins University Press.

Esquemeling, John. 1967. *The Buccaneers of America.* New York: Dover Publications.

Frayn, J. M. 1979. *Subsistence Farming in Roman Italy.* London: Centaur Press.

Gibson, Charles, ed. 1968. *The Spanish Tradition in America.* Columbia: University of South Carolina Press.

Glick, Thomas F. 1979. *Islamic and Christian Spain in the Early Middle Ages.* Princeton, N.J.: Princeton University Press.

Harvey, Robert. 2000. *Liberators: Latin America's Struggle for Independence 1810–1830.* Woodstock, N.Y.: Overlook Press.

Hausmann, Ricardo. 1997. "Municipal Money Managers." *The IDB* 24 (9–10):6–7.

Henige, David P. 1991. *In Search of Columbus: The Sources for the First Voyage.* Tucson: University of Arizona Press.

Humphreys, R. A., and John Lynch. 1965. *The Origins of the Latin American Revolutions 1808–1826.* New York: Knopf.

Johnson, John J. 1991. *A Hemisphere Apart: The Foundations of United States Policy towards Latin America.* Baltimore: Johns Hopkins University Press.

Lane-Poole, Stanley. 1967. *The Moors in Spain.* Beirut: Khayats.

Langley, Lester D. 2002. *The Banana Wars: United States Intervention in the Caribbean, 1898–1934.* Wilmington, Del.: Scholarly Resources.

Lovell, W. George. 1990. "Mayans, Missionaries, Evidence and Truths: The Polemics of Native Resettlement in Sixteenth-Century Guatemala." *Journal of Historical Geography* 16:277–294.

Loveman, Brian. 1999. *For La Patria: Politics and the Armed Forces in Latin America.* Wilmington, Del.: Scholarly Resources.

———. 1994. "'Protected Democracies' and Military Guardianship: Political Transitions in Latin America, 1978–1993." *Journal of Interamerican Studies and World Affairs* 36:105–189.

Loveman, Brian, and Thomas M. Davies, Jr., eds. 1997. *The Politics of Antipolitics: The Military in Latin America*, rev. and updated ed. Wilmington, Del.: Scholarly Resources.

Lynch, John. 1992. "The Institutional Framework of Colonial Spanish America." *Journal of Latin American Studies* 24 (Quincentenary Supplement):69–81.

Marden, Luis. 1986. "Tracking Columbus Across the Atlantic." *National Geographic Magazine* 170:572–600.

Munro, Dana G. 1964. *Intervention and Dollar Diplomacy in the Caribbean 1900–1921.* Princeton, N.J.: Princeton University Press.

Musicant, Ivan. 1991. *The Banana Wars: A History of United States Military Intervention in Latin America from the Spanish American War to the Invasion of Panama.* New York: McMillan.

O'Loughlin, John, et al. 1998. "The Diffusion of Democracy, 1946–1994." *Annals of the Association of American Geographers* 88:545–574.

Parsons, James J. 1962. "The Moorish Imprint on the Iberian Peninsula." *Geographical Review* 52:120–122.

Pérez-Mallaína, Pablo E. 1998. *Spain's Men of the Sea: Daily Life on the Indies Fleets in the Sixteenth Century.* Trans. Carla Rahn Phillips. Baltimore: Johns Hopkins University Press.

Prescott, William H. 1848. *History of the Conquest of Peru.* New York: Harper and Brothers.

Ramsey, Russell W. 1997. *Guardians of the Other Americas: Essays on the Military Forces of Latin America.* Lanham, Md.: University Press of America.

Richardson, Bonham C. 1992. *The Caribbean in the Wider World, 1492–1992.* Cambridge: Cambridge University Press.

Robinson, David J. 1990. "Liberty, Fragile Fraternity and Inequality in Early-Republican Spanish America: Assessing the Impact of French Revolutionary Ideals." *Journal of Historical Geography* 16:51–75.

Rodríguez, Linda A., ed. 1994. *Rank and Privilege: The Military and Society in Latin America*. Wilmington, Del.: Scholarly Resources.

Russell-Wood, A. J. R. 1998. *The Portuguese Empire*. Baltimore: Johns Hopkins University Press.

Sauer, Carl O. 1966. *The Early Spanish Main*. Berkeley: University of California Press.

Simpson, Lesley Byrd. 1966. *The Encomienda in New Spain: The Beginnings of Spanish Mexico*. 3rd ed. Berkeley: University of California Press.

Smith, Peter H. 2000. *Talons of the Eagle: Dynamics of U.S.–Latin American Relations*. 2nd ed. New York: Oxford University Press.

Stokes, William S. 1959. *Latin American Politics*. New York: Thomas Y. Crowell; especially chapter 13, "Violence," 299–334.

———. 1953. "National and Local Violence in Cuban Politics." *Southwestern Social Science Quarterly* 34 (2): 57–63.

———. 1952. "Violence as a Power Factor in Latin American Politics." *Western Political Quarterly* 5:445–469.

Woodward, Ralph Lee, Jr. 1999. *Central America: A Nation Divided*. 3rd ed. New York: Oxford University Press.

6

Political Change

Following are overviews of the political evolution, in the postindependence national period, of each of the independent Latin American mainland nations and of the island nations of the Greater Antilles. Included also are discussions of French Guiana and Puerto Rico, which are a part of France and the United States, respectively. These summaries are intended to increase our understanding of each country's political geography and its impact on economic and social development. They will also enable us to evaluate the significance of new developments in relation to their broader historical contexts.

MEXICO

Upon achieving independence in 1821, Mexico found itself physically devastated by the preceding decade of war, and bitterly divided between conservative and liberal Creole factions who fought among themselves for the chance to govern a nation consisting mostly of impoverished rural peasants. The first of the *caudillos* to gain power was a conservative *latifundista* named Augustín Iturbide. Iturbide admired greatly the authoritarian monarchies of Europe and, as Emperor Augustín I, attempted to expand his dominion into Central America. Iturbide's reign lasted only until 1823 and ushered in a chaotic half century in which Mexico averaged one ruler a year. The most prominent of these was a flamboyant, self-serving demagogue named Antonio López de Santa Anna, who, as Parkes (1970, 198) noted, "succeeded in becoming, for thirty years, the curse of his native country." One of Santa Anna's sorriest disgraces was losing Texas through an inept military campaign. The

loss led directly to the Mexican-American War of 1846 to 1848, which concluded with Mexico's ceding the northern half of its territory to the United States.

Santa Anna had somehow managed to represent both the liberals and conservatives in the course of coming to power. After being deposed on four separate occasions, he fled the country for the last time in 1855. There followed a brief period of liberal reform highlighted by the presidency of **Benito Juárez,** a pureblooded Indian from the southern state of Oaxaca, who governed with a moral rectitude that has rarely been approached in subsequent Mexican history (Figure 6.1). Juárez died in 1872, and in 1876 Porfirio Díaz began a corrupt and repressive dictatorial reign that was dominated by foreign economic interests.

In 1910, all of Mexico erupted into revolution. Out of the ferment and turmoil there emerged, in 1917, a new liberal and, in some respects, radical constitution that has governed Mexico to the present (Benjamin 2000; Brenner 1971). Among its principal provisions were widespread agrarian reform, strict restrictions on the economic holdings and political activities of churches, and broad labor rights.

Yet another outgrowth of the revolution was the emergence of a dominant political party that, since 1946, has been called the *Partido Revolucionario Institucional* **(PRI),** meaning the **Institutional Revolutionary Party** (Figure 6.2). PRI supporters were fond of calling Mexico a "one-party democracy," since the party comprised three major sectors representing business, labor, and the peasantry. In theory, the process of selecting candidates for office—whether at the local, state, or national level—followed democratic

FIGURE 6.1

This mural shows the nineteenth-century Mexican liberal reformer Benito Juárez. The painting is found in Chapultepec Castle, Mexico City.

FIGURE 6.2

The PRI state headquarters building, Campeche City.

procedures governing the political give-and-take between the three principal interest sectors. Critics of the system argued, however, that it was not fully open and that party leadership had frequently altered election results to avoid transferring power to one of the minority parties of the left or right (Camp 1999).

A historical turning point in the development of a more open multiparty democracy occurred in the 1988 national elections, when official results showed that the PRI candidate, Carlos Salinas de Gortari, won the six-year presidential term of office with a mere 51 percent of the vote. The remaining ballots were equally divided between the leftist Democratic Revolutionary Party (PRD) led by Cuauhtémoc Cárdenas Solórzano and the long-time conservative, right-wing National Action Party (PAN). The reform process experienced another milestone in 1999 with the decision of the PRI to aban-

don its time-revered tradition of the sitting president choosing his successor in favor of an open presidential primary election. This was followed, in July of 2000, by the election of the PAN candidate, Vicente Fox Quesada, as president of Mexico.

Mexico thus appears to have evolved into a true multiparty democracy. Great challenges, of course, remain, such as extending the newly won freedoms and opportunities to historically neglected sectors, including the rural peasantry. The small but much-publicized Zapatista rebellion of a group of highland Maya Indians in the isolated southern state of Chiapas is but one manifestation of the pressing need for continuing reform (Harvey 1998).

CENTRAL AMERICA

With the rise to power of a liberal government in Spain in 1820 and the granting of independence to Mexico the following year, the Central American regions, formerly tied to the Viceroyalty of New Spain by virtue of their belonging to the colonial Captaincy General of Guatemala, declared their independence. Chiapas, Guatemala, Honduras, El Salvador, Nicaragua, and Costa Rica all joined Iturbide's Mexican empire. Its collapse led in 1823 to the formation of a federation named the **United Provinces of Central America.** The federation faced almost insurmountable challenges from the beginning, the greatest being simply the isolation of the regional population centers. Severely restricted communication within the union magnified personal and regional rivalries and, together with deep conservative-liberal differences, led first to the loss of Chiapas to Mexico in 1824 and finally, in the late 1830s, to the dissolution of the federation itself (Figure 6.3). The remaining nations declared independence immediately thereafter, but have struggled ever since to overcome the disadvantages inherent in their small size and in their socially and culturally disunited populations (Walker and Armony 2000; Booth and Walker 1999).

Guatemala

Following the proclamation of Guatemalan independence in 1839, the country entered a period of conservative governance dominated by Rafael Carrera. Carrera was first elected president in 1844, designated "president for life" in 1854, and died in 1865. It was during the Carrera regime that Guatemala, in 1859, signed a treaty with Great Britain that Britain later used as legal justification for rejecting Guatemala's claim to British Honduras. This claim was weakened further with Guatemala's formal recognition in 1991 of Belizian independence. A liberal

FIGURE 6.3

A statue of Francisco Morazán in a San José, Costa Rica, park. Morazán was a champion of liberal causes and served as president of the United Provinces of Central America for eight years during the 1830s—the only period of political unity the region has ever known.

FIGURE 6.4

Guatemalan army forces on parade.

revolution, led in the 1870s and 1880s by Justo Rufino Barrios, broke the power of the Church and the landed oligarchy and promoted economic modernization and development, which by the turn of the century was funded to a significant degree by earnings from the export of coffee. Liberal governments continued to dominate down to the election of Jorge Ubico in 1931. Ubico imposed a ruthless dictatorship that lasted until 1944.

The post–World War II era has been a difficult one for Guatemalan democracy, as the nation became increasingly polarized between the extreme left and the extreme right (Jonas 2000; Carmack 1988; Lovell 2000)(Figure 6.4). In 1954, a United States Central Intelligence Agency–sponsored coup overthrew a democratically elected government which had enacted socioeconomic reforms. There then began, in 1960, a long and deeply divisive civil war that officially took an estimated 140,000 lives and left another 60,000 persons "missing." Marxist guerrillas were arrayed against the ultra-right-wing military, which threatened to stage a coup whenever it became too uncomfortable with the policies of the civilian government. Paramilitary death squads perpetuated the cycle of violence. Tens of thousands of peaceful Maya Indians were caught in the crossfire.

By the early 1990s, Guatemala was widely regarded as the worst violator of human rights in all of Latin America, and the once prosperous nation was facing its gravest economic crisis in history. The 1996 presidential campaign brought the election of Alvaro Arzú Irigoyen, a conservative businessman, and the peaceful resolution of the civil war. This was followed in 1999 by the victory of Arzu's rival, Alfonso Portillo, in the first peacetime presidential election in nearly forty years. The brutal murder in 1998 of Bishop José Juan Gerardi, the 75-year-old coordinator of the Guatemalan Archbishop's Office of Human Rights, and continuing high levels of civil violence are grim reminders, however, that the healing process is far from complete.

Belize

Belize occupies a portion of the southeastern Yucatán Peninsula, and for thousands of years was an integral part of the Maya civilization. Although officially falling under Spanish authority during the colonial period, the first permanent European settlements were established by seventeenth-century British buccaneers and later by English loggers and escaped slaves from the Caribbean Islands. It thus developed as a sparsely populated, predominantly English culture realm and was formally declared a British colony in 1862. Self-governance within the British Commonwealth was granted in 1964, and

FIGURE 6.5

A 2001 Guatemalan elementary school geography notebook with a cover map showing Belize as a province of Guatemala. Although Guatemala has officially recognized the independence of Belize, many of its citizens continue to feel that Belize was unjustly taken from them.

the official name was changed from British Honduras to Belize in 1973. Independence followed in 1981, at which time Great Britain promised to come to the country's military defense in the event of an outbreak of armed hostilities with Guatemala (Figure 6.5). Belize functions today as a stable parliamentary democracy and is widely viewed as more Caribbean than Central American in culture.

Honduras

The early political development of Honduras mirrored to a large degree that of its western neighbor, Guatemala. Following its declaration of independence in 1838, conservatives ruled until 1876, when a liberal reform era was introduced by Marco Aurelio Soto. There followed another period of conservative dominance in the later nineteenth century, during which time Honduras was subjected to frequent intervention by its Central American neighbors, particularly by more populous and prosperous Guatemala. The establishment of huge American-owned banana operations in the early 1900s brought to the northern Caribbean lowlands a degree of economic development, but also resulted in foreign fruit interests exercising an inordinate amount of political influence—a situation that critics maintain persists to some degree to the present day (Langley 2002).

Despite its problems, a democratic tradition has recently emerged in Honduras. After an extended period of military dominance, civilian government was reestablished in the early 1980s. Since that time, the country has held six consecutive free and open presidential elections. Much-needed judicial and penal reforms have been implemented, and civilian control over the military has been formalized. Widespread corruption and rising levels of violent crime, however, threaten the newly won gains.

El Salvador

El Salvador is a small but densely settled nation whose leaders historically have played prominent roles in Central American politics. The mid–nineteenth century was dominated by the usual conservative-liberal rivalries. Those rivalries came to be subsumed in the late 1800s by the impact of coffee cultivation, which united the nation's landed aristocracy into an extremely tightly knit oligarchy that has attempted to perpetuate itself in power to the present.

High population densities and the landed aristocracy's unwillingness to voluntarily promote a program of substantive agrarian reform have been the root causes of El Salvador's two most recent political crises. The first, the so-called "Soccer War" with Honduras in 1969, had its immediate provocation in a regional qualifying soccer match for the World Cup between the two countries, which inflamed passions to such a degree that the Salvadoran air force dropped a few bombs on Honduran airports and its army invaded Honduras at two points. Underlying the conflict, however, had been the migration in the previous decades of an estimated 300,000 landless Salvadoran peasants into the thinly populated western Honduran borderlands, caused by the Salvadorans' inability to obtain land to work in their own nation.

The second, and far more serious, crisis was an armed conflict between the right-wing Salvadoran military and Marxist rebels of the **Farabundo Martí National Liberation Front (FMLN),** who attempted to represent the interests of the growing rural landless class. The civil war, which broke out in 1980, took an estimated 75,000 lives, most of them civilian, and displaced approximately 1 million persons before its negotiated settlement in 1992. Remarkable progress has been achieved in recent years in building democratic political structures within the nation (Juhn 1998; Williams and Walter 1997; Brockett 1994). Electoral, judicial, and economic reforms have been implemented, and the FMLN is now one of many political parties functioning within an open, pluralist government. El Salvador is rapidly reassuming its historic leadership role within the region.

Nicaragua

Few people of Central America have as tragic a history as the long-suffering Nicaraguans, who from the early colonial period to the present have seldom experienced a time when war and conflict did not prevail. One source of continuing tension within the country has been the deep cultural differences between Mosquitia, the mixed Indian and black Caribbean coastal plains region that until 1860 functioned as an autonomous kingdom under British protection, and the Spanish-Indian realm that came to focus on the fertile western volcanic lowlands bordering the Pacific coast (Hale 1994).

In addition to Spanish-English antagonisms, Spanish society itself became bitterly divided during the colonial era between conservative and liberal elements. The former centered on the interior agricultural city of Granada and the latter on León, whose proximity to the Pacific Ocean and the world beyond led to its selection as the seat of the provincial government. The rivalry between the two factions became so intense following Nicaragua's declaration of independence in 1838 that the national capital was finally moved to the compromise site of Managua, then a small town situated roughly halfway between Granada and León.

Frustrated by their inability to defeat the conservatives, the liberals next staged a bizarre episode, hiring a band of American filibusters, led by an adventurer named William Walker, to engage the conservative armies on their behalf. Walker arrived with his men in 1855, and the next year managed to get himself "elected" president of the republic. Walker's rule quickly became so offensive to liberals and conservatives alike that both parties enlisted the aid of their Central American neighbors to drive him from the country in 1857.

There followed a period of conservative domination that ended in 1893 with the election of a young liberal named José Santos Zelaya. Zelaya succeeded in bringing Mosquitia under effective control for the first time in Nicaraguan history, but he also sowed the seeds of his own demise by attempting to provoke revolutions in neighboring countries. When two American citizens were killed in the 1909 revolution against Zelaya, the United States sent naval forces to patrol the Caribbean coastline of Nicaragua, and within three years had marines stationed in Managua, where they were to remain on a "peacekeeping" mission until 1933.

FIGURE 6.6

The Managua, Nicaragua, international airport is named for Augusto César Sandino.

The American occupation was resisted by the forces of General Augusto César Sandino, who was assassinated in 1934 by the U.S.–trained National Guard commanded by Anastacio Somoza García (Macaulay 1985). Somoza became president in 1937 and proceeded to establish an exceptionally corrupt and long-lived dynastic regime. By the time that Somoza's youngest son, Anastasio Somoza Debayle, governed as dictator in the 1970s, it was estimated that the extended family controlled fully half the wealth of the nation.

In 1979 a broad coalition of opposition groups drove Somoza into exile, leaving Nicaragua impoverished and deeply divided. A leftist group, which called themselves the **Sandinistas** in honor of General Sandino, established themselves in power (Figure 6.6). With Soviet assistance, they built up the military to unprecedented levels while attempting to discourage private enterprise. The United States, alarmed at what it perceived as the beginnings of the first Central American Marxist state, responded by supporting the *contra* counterrevolution, which cost the nation over 55,000 dead or wounded and untold economic losses. The counterrevolution failed to gain the support of the masses, partly because the *contra* leadership was composed largely of hated former *Somozista* commanders. In 1990, Violeta Barrios de Chamorro, the widow of a former newspaper editor who had been murdered in 1978 by presumed agents of Somoza, defeated *Sandinista* leader Daniel Ortega in a free election and took office for a six-year term. The following two elections were relatively open. The 2001 election resulted in another humiliating defeat of Ortega and the fragmented *Sandinistas*. The nation remains intensely divided, however, both politically and socially, and the placing of collective national good ahead of the pursuit of personal gain remains its greatest challenge.

Costa Rica

In contrast to the turbulence and political instability that have characterized most Central American nations, Costa Rica has been a model of relative calm, with only occasional short interludes of authoritarian rule interrupting long periods of peaceful, democratic government. The roots of Costa Rican democracy extend back into the colonial era, when the early Spanish settlers arrived in the temperate, volcanic uplands of the Meseta Central only to discover that the indigenous populations of the region were very scarce. Unable to live off the labor of the Indians as great *latifundistas*, as was generally the case elsewhere in Latin America, the Costa Rican Spaniards were forced to divide the land into smaller family farm-type units that they could work themselves. This, in turn, gave rise to a relatively egalitarian society that stressed democracy and its sustaining value, universal education, and that remained largely aloof from the never-ending military intrigues that undermined the political development of neighboring regions.

This is not to suggest that Costa Rica has escaped authoritarian influences altogether. The country was ruled from 1870 to 1882 by the dictator Gen. Tomás Guardia, and underwent a short-lived military coup in 1917. The father of modern Costa Rican democracy is widely considered to be José "Pepe" Figueres, who rose to national prominence in 1948 as the leader of a popular revolt (Lehoucq 1991). Elected president first in 1953 and later in 1970, Figueres's greatest achievement may have been the complete abolition of the army, which he replaced with a civilian-led national police force (Figure 6.7). Many of Figueres's successors have taken leadership roles in pursuing regional solutions to Central America's numerous military conflicts. One of those successors, Oscar Arias Sánchez, was awarded the 1987 Nobel Peace Prize for his contributions to the peaceful resolution of Nicaragua's *Sandinista-contra* war.

Panama

Panama's unique location, linking North and South America as well as the Atlantic and Pacific Oceans, has contributed since colonial times to making it a region where people, goods, and ideas all seemed to gather and mix in a manner so fluid that long-term values and traditions formed extremely slowly. Its strategic location also has encouraged intervention by foreign interests, which in the post–World War II era has been countered by a rising tide of nationalism.

Panama first achieved independence in 1821 following Bolívar's conquest of loyalist forces stationed in the isthmus. Shortly thereafter, however, the country requested union with Colombia, which governed the

FIGURE 6.7

The rural guard headquarters building in Heredia, Costa Rica. Costa Rica abolished its army in the midtwentieth century.

occasionally rebellious territory as a semi-autonomous region throughout the remainder of the nineteenth century. In 1903 the United States, frustrated at Colombia's failure to ratify a treaty that would have allowed the United States to build a canal linking the Atlantic and Pacific Oceans, deployed gunboats and troops to persuade the Colombian legislature to grant Panama independence yet a second time. Within weeks of Anglo American recognition of their independence, the jubilant Panamanians, who knew full well the source of their newfound freedom, signed the treaty that ceded in perpetuity to the United States the right to use, occupy, and control a strip of land now known as the **Canal Zone.** The following year, in yet another extraordinary abrogation of sovereign power, Panama adopted a constitution that specifically authorized American military intervention in the country, if such intervention was necessary to maintain order.

Having established near total control over Panamanian internal affairs, the United States proceeded to dig the canal, which was completed and put into operation in 1914. From that point until 1940, the country was ruled by a series of strongmen who, while not above profiting personally from their positions, were generally quick to do American bidding. The post–World War II era brought attitudinal changes in many Panamanians. While recognizing that much good had resulted from American economic involvement in the country, they came to resent U.S. ownership of Panama's heartland—the Canal Zone (Figure 6.8). This emerging nationalism was promoted, in part for personal political considerations, by several Panamanian *caudillos,* including the pro-Fascist Arnulfo Arias, who was elected president in 1940, 1950, and 1968—only to be ousted in coups that followed all three victories—and Colonel Omar Torrijos

FIGURE 6.8

The bilingual sign welcoming visitors to the Panama Canal is symptomatic of the extensive Anglo American influence in Panama.

Herrera, who reigned as dictator from 1968 until 1981. In 1977 Torrijos and American president Jimmy Carter agreed on a new pair of treaties that called for the canal to pass completely to Panamanian control on January 1, 2000, and for Panama to remain politically neutral thereafter.

Torrijos was followed as military dictator of Panama by General Manuel Antonio Noriega, who is widely believed to have been simultaneously employed for years by the United States Central Intelligence Agency and by Colombia's drug warlords. Noriega, having been indicted on drug trafficking charges by two U.S. grand juries, became increasingly anti-American in his behavior, and in the fall of 1989 made the tactical mistake of declaring war on the United States. The U.S., which had been anxious to have the general deposed, then invaded the country (Perez 2000; Millett 1993).

Since Noriega's ouster, there has been considerable progress toward the creation of stable democracy. The military was abolished in 1994 by constitutional amendment and replaced by a national police force, a coast guard, and an air service corps. Three successive democratic elections also have been held. However, the executive branch continues to dominate, and the political party system is weak and candidate—rather than issue—centered. In the most recent presidential elections, for instance, Mireya Moscoso, widow of former President Arias, defeated Martin Torrijos, the son of the late dictator.

THE GREATER ANTILLES

Cuba

The failure of Cuba and Puerto Rico to seek independence in the early nineteenth century dramatically increased Cuba's geopolitical importance. Spain, which had just lost its possessions on the mainland, determined to hold on to its remaining Caribbean colonies at all costs and imposed repressive military rule on the island. Meanwhile, American politicians from southern slaveholding states, locked in battle with representatives of the northern states for control of Congress, began to lobby for U.S. annexation of Cuba, ostensibly in order to bring liberty and freedom to the island. The Cubans themselves remained deeply divided between those who favored independence, those who wanted to seek union with the United States, and those who preferred to see Cuba continue as part of the Spanish empire.

The animosity between the revolutionary and royalist forces became so intense that a bitter, militarily inconclusive civil war erupted in 1868, lasting ten years and costing over 200,000 lives. Upon the conclusion of the war, Cuba was granted representation in the Spanish Parliament, or *Cortes*, and various reforms were enacted on the island that were intended to appease those who had fought for independence. The changes were too little and came too late, however, to stem the growing independence movement that coalesced around a young poet living in exile in the United States named **José Martí**. The invasion of the revolutionary forces in 1895 marked the start of yet another bloody civil war, one that took Martí's own life and that was still raging three years later when the U.S. battleship *Maine* exploded in Havana harbor. The United States' subsequent declaration of war on Spain brought about Spain's expulsion from its remaining Caribbean colonies, but it failed to free Cuba from foreign domination (Smith and Dávila-Cox 1999). This took the form first of direct American military rule and secondly, from 1902 to 1934, of limited self-government (Pérez 1991). During the latter period, American-Cuban relations were governed by the Platt Amendment, which granted to the United States the use of Cuban military bases, including the naval base that the U.S. still holds at Guantánamo, and the right to intervene militarily in Cuba for the "protection of life, property, and individual liberty."

In reality, the protection of property, especially American sugar and gambling interests, was of far greater concern to American politicians than was the preservation of individual Cuban liberties, which languished throughout the early 1900s under a series of corrupt governments chosen from both the liberal and conservative sectors. The abrogation of the Platt

FIGURE 6.9

The Moncada barracks in Santiago de Cuba, where Fidel Castro first attempted to overthrow the Batista regime. The building is now used as a school.

Amendment was accompanied by the ascension to power of an army sergeant named Fulgencio Batista, who, except during a democratic interlude from the mid-1940s to the early 1950s, controlled a country whose economy continued to be bound closely to that of the United States.

It was the political repression of the Batista regime, together with widening socioeconomic inequalities fed by American investments on the island, that inspired a band of youthful revolutionaries led by **Fidel Castro** to attack in 1953 the Moncada military barracks in Santiago de Cuba (Figure 6.9). The initial rebellion, staged on the 100th anniversary of Martí's birth, was put down. However, Castro and his followers later regrouped in Mexico and from there returned to Cuba in 1956. In January 1959, Castro assumed power in what to many appeared to be simply another Latin American coup. Within a few years, however, he had outlawed all political opposition and publicly declared himself a Marxist-Leninist. In April of 1961, Castro weathered the abortive U.S.-backed Bay of Pigs invasion of Matanzas province, and in October 1962 the American-Soviet missile crisis.

In 1965 all political parties other than the Communist were outlawed, and in 1968 private property was declared illegal. By that time, Cuba's economy was heavily subsidized by the Soviet Union, and its foreign policy mirrored that of the USSR and its allies, much to the consternation of the United States. Political and economic reforms within the Soviet Union and its former eastern European satellite nations in the late 1980s were greeted with disdain by Castro, who continued to enjoy a degree of support from many of Cuba's elderly and the agrarian poor who had benefited from improve-

ments in education and health care enacted by the regime. But by the early 1990s, the loss of Soviet economic subsidies had forced Castro to declare the existence of a "special period" that required the implementation of many of the reforms for which he had previously criticized the Russians (Susman 1998). These have included authorization for certain forms of private business and landownership, legalization of the holding of American dollars, and encouragement of foreign investment and tourism. These concessions to capitalism have resulted in growing numbers of highly educated professionals, such as doctors and engineers, leaving their peso-paid government jobs to work as dollar-paid restaurant workers and taxi drivers. Notwithstanding this limited economic restructuring, the government continues to resist political reform. Cuba remains a highly militarized state that criminalizes free speech and association and represses dissent (Moses 2000; Jatar-Hausman 1999). Meanwhile, the aging Castro continues to attribute the country's ills to American imperialism and its punishing economic embargo.

Jamaica

The English conquest of Jamaica in 1655 spared the island the horrors of the wars of independence that swept Spain's American colonies, but otherwise continued the pervasive underdevelopment common to the plantation-dominated peoples of the Caribbean. The political beginnings of modern Jamaica date to 1944, when Great Britain granted the islanders limited self-government. This was followed in 1958 by membership in the ill-fated Federation of the West Indies, which lasted until 1962 when the nation was given its full independence as a parliamentary democracy within the Commonwealth.

Jamaican foreign and economic policy has fluctuated widely since independence, further complicating attempts to achieve long-term development. During the 1970s, the dominant political force was the People's National Party (PNP) led by Michael Manley. Manley attempted to develop a modified socialist welfare state and to politically align Jamaica closer to Castro's Cuba—both to the detriment of the country's all-important tourist industry. Manley was followed as prime minister in the 1980s by Edward Seaga, whose Jamaica Labour Party (JLP) advocated a return to private enterprise and close ties with the United States. Needless to say, tourism increased, but widespread poverty and social unrest persisted. In 1990, an ideologically reformed Manley was again elected to office, this time on a platform of reinvigorating the private economic sector. Manley resigned from office in 1992 for reasons of health and was replaced by Percival J. Patterson of the PNP. Patterson was elected to a third consecutive term as prime minister in 1997.

Haiti

Haitian independence from France was achieved in the 1790s through a great slave uprising led initially by Pierre-Dominique Toussaint L'Ouverture. From the conclusion of that conflict to the present, Haiti has experienced consistently violent and repressive government. The cruelty and suffering have continued for so long that, to many, the nation has come to symbolize hatred and poverty—a horrifying example of the depths to which human society can descend when subjected to unchecked prejudice and personal ambition. What makes the situation all the more tragic is that prior to the uprising Saint-Domingue was one of the most prosperous regions in Latin America.

As was often the case in Latin America's other wars of independence, the early leaders of Haiti's rebellion were moderate visionaries who were determined to redress the evils of colonialism, including slavery, and whose character prevented them from participating in the tyrannical excesses of those who followed them in power. Toussaint L'Ouverture was willing for Haiti to become a self-governing part of the French empire if Napoleon would only abolish slavery in the colony. Napoleon's refusal to do so, and Toussaint L'Ouverture's subsequent kidnapping and death in France, resulted in Haiti's declaration of independence in 1804. There followed an attempt, under Haiti's first president-emperor, Jean-Jacques Dessalines, to eradicate every vestige of despised European influence, including the killing of all white people and the destruction of their material possessions. In the process, the economic infrastructure of the young country was devastated and the majority of the people reverted to a subsistence existence from which most of the population has yet to emerge.

From the fall of President Jean-Pierre Boyer in 1843 to the mid–twentieth century, Haiti alternated between dictatorial mulatto rule emanating from the capital city of Port-au-Prince and periods in which the national government, whose influence in the rural interior districts was never great, virtually ceased to function. United States marines occupied the country from 1915 to 1934.

In 1957, a black physician named Francois Duvalier was elected president. Duvalier, or "Papa Doc" as he came to be called, quickly established one of the most brutal dictatorships in modern history, one sustained by inciting the black masses to violence toward middle- and upper-class mulattoes (Heinl and Heinl 1996). Those whom Duvalier perceived as threats to himself were systematically hunted down and killed by his personal security force, called the Tontons Macoutes. After previously having declared, "I am the state," Duvalier had the constitution rewritten in 1964 to make him "President for Life," a title which he held until his death in 1971.

Papa Doc was succeeded by his son, Jean-Claude Duvalier, known as "Baby Doc," who fled the country in 1986 after having transferred much of the national treasury to personal overseas bank accounts. In December 1990, a leftist Catholic priest named **Jean-Bertrand Aristide** was chosen president in the first fully democratic voting in Haiti's history. Aristide was overthrown the following year by the Haitian military and took refuge in the United States, which restored him to power in 1994. Haiti's first peaceful transfer of power from one popularly elected leader to another occurred in 1996 with the inauguration of Rene Preval, Aristide's hand-picked successor. Preval's term was marked by international allegations of widespread corruption and electoral fraud and the consequent withholding of most multilateral and bilateral economic assistance. In 2000, Aristide won a second term in an election boycotted by the political opposition and has subsequently continued to oppose democratic initiatives in the environmentally degraded and poverty-stricken country.

Dominican Republic

In contrast to the almost continual turmoil of Haiti, its western neighbor, the political development of the Dominican Republic has proceeded along the fairly common Latin American pattern of *caudillismo* followed more recently by democracy. Independence was achieved initially in 1821 but was temporarily lost when Haiti, and later Spain, governed the country. The later nineteenth century and early twentieth was a time of generally short-lived dictatorships that brought modest economic development to the elites but little change in the lives of the poor masses. An internal power vacuum, together with growing American financial investments within the country, led to United States military occupation from 1916 to 1924 (Calder 1984). This was followed in 1930 by the beginning of the long, harsh dictatorship of **Rafael Trujillo,** who, notwithstanding his callous disregard for civil liberties, did much to modernize the economy.

Trujillo's assassination in 1961 led to a period of political instability that brought about a brief American occupation in 1965. The post-Trujillo era was dominated until 1996 by Joaquín Balaguer, a classic strongman who pillaged the national treasury in order to reward his supporters and who employed a variety of self-serving strategies to weaken his political opposition (Figure 6.10). Under Balaguer, the country acquired many symbols of modernity and prosperity while the masses continued to live, for the most part, under conditions of severe economic privation. In 1994 the 87-year-old, nearly blind Balaguer found himself trailing in the electoral count of an election he thought he was certain to win; panicked, he stopped the count and shortly thereafter announced that he

FIGURE 6.10

Joaquín Balaguer, who served part or all of seven terms as president of the Dominican Republic from the 1960s to the 1990s.

had won. Domestic and international outrage to the ploy was so strong that Balaguer agreed during subsequent negotiations to shorten his term to two years. In 1996 Leonel Fernández Reyna, a pro-business attorney who grew up in New York City, was chosen president in open, peaceful elections. Fernández Reyna was succeeded in 2000 by Hipólito Mejía, who was chosen president in another free election.

Puerto Rico

Puerto Rico, the second of the American colonies that remained loyal to Spain during the revolutionary wars, was, like Cuba, rewarded thereafter with liberalized trade privileges and representation in the *Cortes.* However, these concessions failed to prevent the growth of a liberal pro-independence party that had achieved significant concessions prior to the outbreak of the Spanish-American War. Although many Puerto Ricans resented the island's becoming a United States possession, opposition to the American occupation gradually diminished, owing largely to an impressive expansion of roads, schools, and public health services.

Unlike most other American territorial acquisitions that ultimately were granted either statehood or independence, Puerto Rico's political standing has evolved over the twentieth century in a somewhat haphazard fashion that may yet be modified still further. Following the conclusion of the Spanish-American War, the island was initially classified as a "territory." That status was changed in the Jones Act of 1917 to one of an "organized but unincorporated" part of the United States,

and in 1952 the island was officially designated an American **"commonwealth"** or, in the Spanish, *Estado Libre Asociado* (Free Associated State).

Under the provisions of the commonwealth agreement, Puerto Rico's relationship to the United States federal government is similar, but not identical, to that of the American states, with local home rule through popular election of a governor and members of a senate and a house of representatives. Puerto Ricans are considered to be full American citizens, eligible for federal employment, military service, and welfare assistance. They are free to travel to and from, and to seek employment on, the United States mainland. Virtually the only privileges withheld from them are those of voting in American presidential elections and of having voting representatives in the United States Congress. Those limitations are tempered, however, by their having the opportunity to vote in national primary elections, representation by an elected (but nonvoting) resident commissioner in the U.S. Congress, and exemption from all federal income taxes, except taxes on income earned through employment with the federal government or while residing on the mainland. The islanders also benefit collectively from approximately U.S. $10 billion in annual federal spending.

Not surprisingly, many Puerto Ricans are relatively content with the current commonwealth status, while others are seeking either full statehood or full independence. In a 1993 plebiscite, 48.4 percent of those voting favored continued commonwealth status, 46.2 percent statehood, and 4.4 percent independence. In 1998, yet another nonbinding referendum resulted in 50.2 percent of voters favoring the existing commonwealth arrangement and 46.5 percent statehood. One additional disadvantage of statehood would be the probable loss of Spanish as the official language and the resultant dilution of the island's predominantly Hispanic culture.

The pro-commonwealth position is represented today by the Popular Democratic Party (PPD), which was founded in 1938 by **Luis Muñoz Marín,** the pro-statehood by the New Progressive Party (PNP); and the pro-independence by a variety of smaller parties, including the Socialist Party (PSP) (Figure 6.11). Because the PPD has traditionally been allied with the Democratic Party of the U.S. mainland, recent Republican presidents have expressed support for Puerto Rican statehood. The American Congress has generally tried to assume a neutral position.

ANDEAN SOUTH AMERICA

After Simón Bolívar's final defeat of the Spaniards at Carabobo in 1821, Creole leaders from throughout the northern Andes gathered at Cúcuta and formed a loose federation that came to be known as *La Gran Colombia.* The new nation comprised much of present-day

FIGURE 6.11

Luis Muñoz Marín, the father of modern Puerto Rico.

Venezuela, Colombia, Panama, and Ecuador and was administered from Bogotá. It was largely the creation of Bolívar, who dreamed of uniting all of Spain's former New World colonies into a voluntary association of free and enlightened states.

Bolívar's idealism was far ahead of his time. La Gran Colombia never functioned effectively, owing to the isolation and separatist sentiments of the regional population centers, to the personal ambitions of its leadership, and to the extended absences of the great Liberator himself, who immediately turned his attention to freeing Peru and Bolivia from the Spanish grip. In 1828 Bolívar resigned as president amidst an outbreak of regional rebellions, and in December of 1830 he passed away of tuberculosis. Bolívar had played a prominent role in the liberation of five of South America's modern nations, yet died saddened by the collapse of his dream of hemispheric unity—an ideal that has largely eluded Latin America to the present day.

Venezuela

Venezuela's secession from La Gran Colombia was led by José Antonio Páez, a *llanero* from the Orinoco plains, who broke with Bolívar and ruled either openly or behind the scenes from 1830 until his exile some eighteen years later. Although Páez derived his support from the conservative oligarchy and was reluctant to relinquish control of the country, he proved himself a capable administrator whose moderate policies promoted regional reconciliation and the economic development of the fledgling republic.

With the fall of Páez and the conservatives, Venezuela was dominated throughout the remainder of the nineteenth century by a series of dictators whose avowed liberalism was masked by their strong ties to the military. Chief among these was General Antonio Guzmán Blanco, who ruled directly or indirectly from 1873 until his exile to his beloved Paris in 1888. Guzmán Blanco reduced still further the power of the already weak Venezuelan Catholic Church, promoted education, and attempted to build up Caracas as a showcase of his greatness. Thus, despite Guzmán Blanco's legendary capacity for embezzlement, Venezuela advanced under his reign.

The same was true, in many respects, of the rule of Juan Vicente Gómez, who from 1908 to 1935 crushed his opponents with a cruelty unmatched in Venezuelan history and treated the national treasury as if it were his personal spending account. Ironically, Gómez also laid the foundations of modern Venezuela through his encouragement of foreign investment, particularly in the petroleum sector, which financed the country's development throughout the twentieth century. Even the progressive democratic tradition of the post-Gómez era may be traced in part to the revulsion felt by most Venezuelans to the tyrannical excesses of the illiterate *caudillo*.

Gómez was succeeded as president by Eleazar López Contreras, who used the proceeds of the expanding petroleum industry to fund much-needed schools, highways, and hospitals—a policy known as *sembrar el petróleo* (sow the oil). Welcome as the ensuing economic growth was, López Contreras made his greatest contribution in the political sphere, allowing the return of numerous exiles and voluntarily relinquishing power at the completion of his term of office (Figure 6.12).

Among the returning exiles was Rómulo Betancourt, whose Democratic Action (AD) party was placed in power by the military in 1945. Betancourt and his successor, Rómulo Gallegos, implemented widespread social and agrarian reform, which triggered a conservative military coup and a decade of corrupt dictatorship by Marcos Pérez Jiménez. Following Pérez Jiménez' ouster in 1958, Venezuela established a democratic tradition, with presidents from the elite-dominated Democratic Action and Christian Democratic (COPEI— Committee for Independent Political Electoral Organization) parties guiding the country on a slightly left-of-center course.

Eleazar López Contreras who, as president of Venezuela from 1935 to 1940, did much to promote the development of democracy by allowing the return of political exiles and by relinquishing power at the completion of his term in office.

Frustrated by what he perceived to be mounting corruption and declining income levels for the middle and lower socioeconomic classes of the oil-rich country, a young army colonel named Hugo Chávez tried twice in 1992 to organize military coups to overthrow President Carlos Andrés Pérez. Chávez failed in both attempts and was imprisoned for two years, but Pérez was impeached on corruption charges. Chávez then determined to seek the presidency on a platform of far-reaching structural reform. Following his election in 1998, Chávez directed the formation of a constitutional assembly, which wrote a new Constitution in 1999. The Constitution enabled Chávez, as president, to disband the opposition-dominated Congress and replace it with a unicameral legislature called the National Assembly that was controlled by Chávez's political party. Chávez also gained control over the judicial branch of government. In addition, the new Constitution extends the president's term of office from five years to six and permits the president to seek immediate reelection. Venezuela's upper and middle classes have reacted with alarm to the constitutional changes and to Chávez's leftist political rhetoric, and in April 2002 made an unsuccessful attempt to oust him through a military coup. Many members of the lower classes, however, view him as their best hope for justice and opportunity.

Colombia

Colombia, or New Granada as it was officially known until 1863, has long struggled, with only limited success, to achieve the political unity and stability necessary to develop fully its rich natural resource base. Liberal and conservative parties had formed as early as the 1830s, and the first of the nation's periodic civil wars was fought from 1840 to 1842. Both liberals and conservatives initially favored a weak central government with power concentrated in the states, whose isolation perpetuated the strong regional economic and cultural differences that had evolved during the colonial era. However, the failure of federalism to eradicate the endemic poverty and violence finally led to the passage of a Constitution of 1886, which created a strong central government headquartered at Bogotá. The new constitution laid the foundation of the modern Colombian state but failed to prevent a second liberal-conservative civil war. This "War of a Thousand Days" raged from 1899 until 1901 and left the country virtually powerless to oppose the U.S.-supported secession of Panama. So unstable was the national government throughout this early republican period that the first peaceful transfer of presidential power from one political party to another did not occur until 1930.

In the midst of the never-ending political turbulence, the Colombian population continued to grow, and millions upon millions of rural peasants found themselves reduced to a near-hopeless existence, struggling to sustain life on tiny subsistence plots of land owned but scarcely used by the urban-based aristocracy. In an effort to assist these families, the Colombian Congress in 1936 passed Law 200. Law 200 promised title to the rural *colonos*—most of whom were either squatting on or renting lands that would never be used by the *latifundistas*—of properties that they could show they had brought under "economic exploitation." Contrary to the legislature's intent however, the law had the effect of spurring the previously indifferent absentee landlords to oust the peasants before they could improve the land. This, in turn, led to the dark years of **La Violencia,** an era of almost indiscriminate rural violence, which produced the assassination in 1948 of Jorge Eliécer Gaitán, the leader of the Liberal Party. Gaitán's death sparked a riot in Bogotá now referred to as the *Bogotazo*, and Colombia was plunged for a third time into bloody civil war. By 1957 the war had resulted in a quarter of a million deaths and over 750,000 refugees—mostly rural folk fleeing the lawlessness in the countryside (Sanchez and Meertens 2001; Bushnell 1993). Those peasants who remained in the villages came to form a vast discontented, landless rural proletariat whose children, *los* *hijos de la violencia,* have formed the nuclei of the numerous leftist guerrilla groups that have persisted to the present (Bergquist, Peñaranda, and Sánchez 2001; *Violence in Colombia* 2000).

La Violencia was brought to an end by the National Front Constitutional Reform of 1957, in which liberals and conservatives agreed to form a coalition government wherein the presidency alternated and governorships and mayoral offices were divided in proportion to the number of votes obtained by each party. This arrangement was abandoned by mutual consent in 1974, and the late twentieth century was dominated by the Liberal Party.

Regrettably, the end of *La Violencia* as a historical period has not been accompanied by a decline in violence as a feature of the Colombian political landscape. Frustrated by their exclusion from meaningful political representation during the National Front period, numerous nontraditional groups have turned to guerrilla warfare. Thus, the killing and maiming that were formerly perpetrated by liberals and conservatives have simply been replaced by crimes committed by leftist and rightist guerrilla and paramilitary groups, by drug traffickers, and by the military itself. By 2000, Colombia's homicide rate of 57 per 100,000 population was the highest in the world and kidnappings were averaging over ten per day ("Drugs, War, and Democracy" 2001). Guerrilla groups were estimated to control nearly half the national territory, and the country had descended into a state of uncontrolled violence that threatened the very survival of constitutional government. Political independent Alvaro Uribe won election as president in 2002, promising to wage a relentless war against both the guerrillas and paramilitary death squads. (Case Study 6.1).

Ecuador

Ecuador, like Colombia and Venezuela, achieved political independence with the collapse of Bolívar's Gran Colombia. The new nation immediately found itself deeply divided between the ultraconservative *sierra,* or Andean highland region dominated by the residents of Quito, and the liberal Pacific coastal settlements led by the citizens of Guayaquil. Looming over the conservative-liberal divisions was an ever-present military-Church alliance, which produced a lengthy string of short-lived and undistinguished dictatorial regimes down through the nineteenth century.

An era of liberal rule from 1895 to 1925 resulted in a lessening of Church influence in secular affairs and in considerable economic development. In 1941 Peru declared war on Ecuador, and in a Protocol signed the following year at Rio de Janeiro, succeeded in annexing a large sector of Amazonia previously controlled by Ecuador (Figure 6.13).

The post–World War II era has been characterized by continued political instability and high levels of military influence, but with little of the intense regionalism and violence that have marred Colombian politics (Isaacs 1993). One individual, José Maria Velasco Ibarra, managed to assume the presidency five times between 1934 and 1972 without ever completing a full term in office.

One of the more bizarre Latin American political episodes in recent years was the election in 1996 of Abdalá Bucaram, a fiery populist from Guayaquil whose screaming, singing, and dancing at campaign stops earned him the nickname "El Loco," the crazy one. Bucaram lasted only a few months before he was removed from office by Congress on grounds of "mental incapacity." Following the brief and chaotic reign of Bucaram's interim replacement, the Harvard-trained Jamil Mahuad, mayor of Quito, was elected president in 1998.

One of Mahuad's most enduring political accomplishments was the conclusion in October 1998 of a peace accord with Peru that settled a border dispute that had flared off and on since the Rio de Janeiro Boundary Protocol of 1942. Mahuad was not as successful, however, in reversing Ecuador's long economic downturn and in January 2000 was ousted by Indian protesters and junior army officers who promptly announced the formation of a three-man junta. Under intense pressure from the United States and other foreign nations, the junta peacefully relinquished power the next day to Vice President Gustavo Naboa, who was immediately ratified as president by congress.

Peru

Peru's position as the wealthiest and most powerful of Spain's South American colonies contributed to the growth of a strong royalist faction that was able to repress several Creole revolts in the early revolutionary period. Independence was achieved through the assistance of foreign forces, initially those of the Argentine general José de San Martín, whose Army of Liberation arrived from Valparaíso, Chile, in 1821 (Figure 6.14). The following year at a historic meeting in Guayaquil, San Martín turned over command of his Peruvian forces to Simón Bolívar, whose assistant, Antonio José de Sucre, defeated the remaining royalist forces both in Lower Peru and in Upper Peru, the latter being renamed by Sucre the Republic of Bolívar (later Bolivia).

Bolívar himself had no desire to remain permanently in Peru and transferred his authority to a governing council led by a liberal scientist named Hipólito Unánue. Unfortunately, Unánue's democratic ideals were not shared by a number of military *caudillos,* who plunged the country into virtual anarchy, notwithstanding a period from 1836 to 1839 when Peru and Bolivia joined in a confederation. In the era

Case Study 6.1: Colombia's Violent Drug War

May 1986   Virgilio Barco, candidate of the Liberal Party, is elected president of Colombia.

August 1989   Luís Carlos Galán, Liberal Party senator and the leading candidate to succeed Mr. Barco as president, is assassinated on orders of the Medellín drug cartel. President Barco responds by declaring war on the drug cartels and authorizes the extradition of drug suspects to the United States. César Gaviria replaces Galán as Liberal Party candidate for president.

November 1989   Colombia's judiciary system closes down as judges refuse to hold court unless the government increases their protection.

December 1989   Medellín drug lord José Gonzalo Rodrígues Gacha is killed in a bloody battle with national security forces.

May 1990   César Gaviria is elected successor to President Barco. While stating publicly his intention to continue the deadly drug war, Gaviria hints of his desire for a negotiated settlement.

Fall 1990   The Medellín cartel kidnaps seven journalists, including Diana Turbay, the daughter of a former president, and Francisco Santos, news editor of *El Tiempo*, Colombia's most prestigious newspaper. President Gaviria softens his official position toward the cartel by promising not to extradite to the United States those traffickers who turn themselves in and confess their crimes, but rather to try them in Colombia.

April 1991   Medellín cartel leader Pablo Escobar and several lieutenants "surrender" to Colombian government officials on condition that they not be extradited to the United States and that they be imprisoned at a luxurious country ranch.

July 1992   Mr. Escobar and a number of associates "escape" the luxury prison ranch where they had been held.

December 1993   Pablo Escobar is killed by Colombian security forces, and the Cali cartel, allegedly led by brothers Gilberto and Miguel Rodríguez Orejuela, increases its influence.

Fall 1994   Colombian police find two miniature "narcosubmarines," which they suspect have been used to smuggle cocaine from port areas to ships waiting offshore.

Fall 1994   News accounts report that Colombia's narcotraffickers have gained control of much of the country's best farm and grazing lands, becoming, in the process, the nation's newest class of *latifundistas*.

Summer 1995   Gilberto Rodríguez Orejuela and his brother, Miguel, are captured by Colombian police, who also obtain a list containing the names of 2,800 persons allegedly on the payroll of the Cali cartel. Those named on the list included members of Congress, governors, and judges as well as high-ranking military officers, prominent journalists, and athletes.

February 1996   President Ernesto Samper is formally charged before Congress by Colombia's chief prosecutor with having accepted millions of dollars from cocaine traffickers for his 1994 presidential campaign.

May 1996   Attorney General Orlando Vásquez surrenders to authorities to face charges he took money from cocaine traffickers.

November 1997   The Colombian government seizes more than 300 properties belonging to slain drug boss José Santacruz Londono and his allies, including 43 ranches, 68 apartments, 103 parking garages, and assorted real estate agencies and construction companies. The properties were worth over U.S. $150 million.

Summer 1998   Media accounts report that approximately 3,000 of Colombia's 15,000 left-wing guerrillas are active in protecting the drug trade, which has come to be dominated by smaller, more technologically sophisticated and less flamboyant trafficking organizations headquartered mostly in Medellín. Another important source of guerrilla revenues has become the kidnapping of civilians and foreign visitors and holding them for ransom.

January 2000   U.S. President Bill Clinton announces a $1.3 billion emergency aid package intended to strengthen the Colombian military war on drugs. Many commentators express concern that the program could lead to increased U.S. military involvement in the Colombian civil wars.

February 2002   The Colombian military launches an offensive to retake a Switzerland-sized zone in Caquetá and Meta departments that had been ceded as a peace gesture by the government three years before to the Revolutionary Armed Forces of Colombia (FARC) guerrilla group.

FIGURE 6.13

South American boundary changes in the republican period.

The continuing failure of Peru's civilian and military aristocracy to improve the living conditions of the highland Indians led in the 1920s to the emergence of two radical political movements. One, known as **APRA (American Popular Revolutionary Alliance),** was founded in 1924 by Víctor Raúl Haya de la Torre, who drew much of his inspiration from revolutionary Mexico, where he was living in exile. The party platform called for a unified Latin America dedicated to stemming foreign imperialism, gaining Latin American ownership and control of the Panama Canal, and nationalizing land and industry on behalf of the poor. The movement immediately gained widespread popularity among the Peruvian intelligentsia and lower classes and was defrauded of several election victories in the 1930s. More recently, it has provided a philosophical rationale for a number of strongly nationalistic regimes—both military and civilian—that have discouraged free-market economies.

The second radical movement, Marxism, attracted little popular support during the twentieth century but became highly visible in the 1980s through the terrorist activities of a Maoist wing known as **Shining Path,** or *Sendero Luminoso* (Barton 1997; Starn 1992; Palmer 1992). Operating mostly out of the southern and central *sierra* regions, the guerrilla group was responsible for over 25,000 deaths, hundreds of them local and regional government officials, and for over $20 billion dollars' worth of property damage.

A much smaller leftist guerilla group, known as the Tupac Amaru Revolutionary Movement (MRTA), also prospered to some degree during the 1980s but steadily lost influence thereafter. In a desperate attempt to force the national government to release about 300 of their imprisoned associates, a group of 14 MRTA rebels captured 340 diplomats and prominent Peruvians attending a social event at the Lima residence of the Japanese ambassador in December of 1996. The daring recapture of the compound by Peruvian military commandos the following April and the resultant deaths of the guerrillas placed the movement on the verge of extinction.

In the 1990 presidential election, Alberto Fujimori, an agricultural engineer and son of Japanese immigrants, won an upset victory over the celebrated novelist Mario Vargas Llosa. Frustrated at the slow pace of economic and political reforms, Fujimori staged what has come to be known as an *autogolpe,* or "self-coup," in April 1992. Assuming, with the support of the powerful military, all civil political authority, Fujimori dissolved Congress and oversaw the adoption in 1993 of a new constitution that made legal his reelection in 1995.

By the late 1990s, Fujimori enjoyed high voter approval ratings, having been widely credited with defeating both terrorism and inflation. Buoyed by a controversial supreme court ruling, the strongman sought

FIGURE 6.14

General José de San Martín, who assisted in the liberation of Argentina, Chile, and Peru.

of "Restoration" that followed, Peru was ruled by a series of progressive leaders, including Ramón Castilla, under whom the country's economic well-being became increasingly dependent on the export both of the rich coastal deposits of sea bird dung known as *guano* and of nitrate reserves mined in the northern Atacama region.

The riches of the desert fertilizers soon attracted the attention of Spain, which in 1865 attacked by sea a number of positions from Valparaíso to the Chincha Islands north of Lima. This led to a brief alliance between Chile, Peru, and Bolivia, each of whom at the time controlled portions of the Pacific coastline then under siege. Victory over Spain was followed by the election of the first civilian president in Peruvian history, Manuel Pardo, who governed democratically from 1872 until 1876.

Conflict over the Atacama nitrate fields led, in 1879, to the **War of the Pacific,** which resulted in the humiliating defeat of Bolivia and Peru by Chile. In the 1883 Treaty of Ancón, Chile gained Bolivia's Pacific territories as well as the southern Peruvian provinces of Tarapacá and Arica (see Figure 6.13). Tarapacá was returned to Peru in 1929, by which time the country was governed by a civilian administration headed by Augusto B. Leguía.

FIGURE 6.15

This poster showing Alejandro Toledo standing among the Inca ruins of Machu Picchu dressed in the presidential sash and staff symbolizes Peruvian pride in his Amerindian ancestry.

and won a tainted reelection in May of 2000. By November, however, Fujimori had resigned in disgrace, having been dismissed on grounds of "moral incapacity" by Congress after revelations that his feared intelligence chief had bribed an opposition legislator. In 2001, Alejandro Toledo, a former World Bank economist, became the first Amerindian president of modern Peru (Figure 6.15).

Bolivia

Bolivia, like Peru, long was divided between a small, elite European class, which monopolized the wealth of the nation's mining and agricultural sectors, and the masses of highland Indians, who since the European conquest had lived as if in a world completely separate from but subservient to that of the *blancos*, or whites. While the Indians toiled and suffered in silence, fully aware of the awful consequences of rebellion, the aristocracy engaged in an incredible number of coups and countercoups, averaging in the process more than one government per year. Meanwhile, the political instability did much to weaken Bolivia's military and economic position and led to the loss of three valuable peripheral

regions: the northern Atacama to Chile through the War of the Pacific, the territory of Acre to Brazil in the early 1900s, and over 233,000 square kilometers (90,000 square miles) of lowland Chaco to Paraguay in the 1930s (see Figure 6.13).

The **Chaco War** marked a historic turning point in Bolivia's social development, as the white and *mestizo* officers found themselves dependent for their very lives upon the despised *indios*. For their part, the previously docile Indians were awakened both to a sense of their individual worth as human beings and to their collective capacity to influence the course of the nation.

In the years following the war, increasing numbers of young, nationalistic Bolivians turned to socialist and Marxist thought in their quests to redress the evils of the past. These stirrings were met by a series of liberal reform laws whose effect was negated by the power of the established conservative aristocracy. A new political party called the *Movimiento Nacionalista Revolucionario* (**MNR, or Nationalist Revolutionary Movement**) was formed in 1940 by a group of intellectuals headed by Víctor Paz Estenssoro. Paz Estenssoro fled to Argentina in 1946, where he became a symbol of the budding Bolivian revolution. Finally, in 1952, the country erupted into civil war and Paz Estenssoro returned from exile to lead the new revolutionary government. In contrast to the countless coups of the past, Bolivian society was fundamentally and irreversibly altered. The government nationalized the holdings of the Patiño, Aramayo, and Hochschild mining corporations, each of which had annual budgets larger than those of the national government and which had exploited the labor of the Indians for centuries. Nationalized too were the huge rural *latifundia*. Laws calling for universal suffrage and education and for expanded labor rights were also passed in an outpouring of social legislation that continued until the ouster of the MNR by General René Barrientos in 1964. It was in 1967, while Barrientos served as president, that Ernesto "Che" Guevara, the noted Cuban revolutionary of Argentine birth, was killed in the eastern Andean slopes of Santa Cruz province.

The intervening years have brought halting progress in the efforts of Bolivians to establish a stable democracy. The forced tranquility of the conservative military dictatorship of General Hugo Banzer from 1971 to 1978 was followed by a four-year period when the country experienced nine different governments. Bolivia has enjoyed democratic rule since the aged Paz Estenssoro was elected to a second term in 1982. He was reelected to a third term in 1985 (Figure 6.16). Gonzalo Sánchez de Lozada, head of the MNR, was chosen president in 1993 and succeeded in implementing numerous economic and political reforms. He was followed in 1997 by the former dictator Banzer, who campaigned as a reformed proponent of democracy and socially equi-

FIGURE 6.16

Victor Paz Estenssoro founded Bolivia's Nationalist Revolutionary Movement and served as president of the nation on three occasions from the 1950s to the 1980s.

table free-market economic policies. Sánchez de Lozada was elected to his second, non-consecutive, term in 2002. As with Colombia and Peru, Bolivia's leaders must address not only the traditional social and technological barriers to development but also the economic and political consequences of dependence on the export of narcotic drugs.

SOUTHERN SOUTH AMERICA

Chile

Chile today has one of the most elongated shapes of any nation on earth. During the colonial era, however, the region of Spanish settlement referred to as Chile was small and compact, extending from the River Bío-Bío south of Concepción to Santiago and Valparaíso on the north. Here, the Central Valley was divided into huge haciendas whose owners lived in the style of classic Spanish aristocrats, supported by masses of tenant laborers known as *inquilinos*.

Creole revolutionaries led by Juan Martínez de Rozas and Bernardo O'Higgins first declared independence in 1810 but were unable to defeat the royalist forces. O'Higgins, whose Irish father had served Spain as governor of Chile and viceroy of Peru, fled to Mendoza, Argentina, where he joined the army of José de San Martín, who then led his troops across the Andes, driving the Spaniards from Chile in 1818 before moving on to liberate Peru.

The English-educated O'Higgins ruled Chile as supreme director until 1823, imposing a number of social and economic reforms that alienated the Church and landed aristocracy but also set a standard of enlightened government that would become the hallmark of Chilean politics for a century and a half. Following his ouster, O'Higgins fled to Peru, and Chile entered a period of conservative dominance that lasted until 1861. The conservatives reinforced the Chilean tradition of a strong central government and defeated the Peruvian-Bolivian Confederation in a war that lasted from 1836 until 1839. The country's southern frontiers were extended to the Gulf of Ancud through the promotion of German settlement in response to continued Araucanian Indian resistance. Under a series of liberal administrations that governed Chile from 1861 until 1891, the strong sense of national pride that flowed from the conservative military and immigration successes continued to grow. The Chilean victory in the War of the Pacific brought additional territorial gains and considerable wealth from the export of Atacama nitrates. This wealth, together with the sale of agricultural produce from the Central Valley, financed the establishment of a modern industrial sector. Economic prosperity was accompanied by additional social reforms, including a further reduction in the influence of the Church, and Chile came to be viewed both at home and abroad as a model of stable and progressive government.

The early twentieth century was characterized by an expansion of parliamentary power relative to that of the chief executive, and also by the beginnings of a significant labor movement that gained widespread support among the nitrate and copper miners of the north and the urban industrial workforce. Communist, socialist, and other radical leftist parties were established and grew rapidly, fed by rising unemployment stemming from reduced foreign demand for nitrates, the worldwide depression, and the failure of the landed aristocracy to enact meaningful agrarian reform. In 1936, the leftist parties united in a Popular Front that governed Chile as the Radical Party from 1938 to 1952 and attempted to pattern the country's economic and social programs after those of the welfare states of western Europe.

With the defeat of the Radicals in the 1952 election, even greater pressures for change emerged from the far left. In the 1958 presidential election, **Salvador Allende,** the candidate of the Marxist party, lost by less

Salvador Allende became Latin America's first freely elected communist head of state with his selection as president of Chile in 1970.

General Augusto Pinochet was the right-wing military dictator of Chile from 1973 to 1990.

than three percentage points. Allende placed second again in the 1964 election and then, in 1970, stunned the world by becoming Latin America's first freely elected communist head of state (Figure 6.17).

Allende immediately began to restructure Chilean society. The huge American-owned copper mines in the northern Andes were nationalized in 1971, further incurring the wrath of the United States. Within months, large segments of the banking, transportation, and industrial sectors had also passed into government control and agrarian reform was under way. Inflation and unemployment soared, and homemakers, businessmen, and factory workers took to the streets in a never-ending series of demonstrations and counter-demonstrations. Allende, who had won less than 37 percent of the popular vote, had moved too far too fast and was losing control of even his own supporters (Carr and Ellner 1993).

In September 1973, the armed forces attacked the national palace, killing Allende in the process. South America's strongest democratic tradition was replaced by a military dictatorship headed by General Augusto

Pinochet (Figure 6.18). Pinochet proceeded to establish a ruthless police state that was responsible for the disappearance and killing of thousands of alleged subversives. He also undid much of the social legislation of the previous half century and moved the nation into a free-market economy that resulted in large economic gains beginning in the 1980s.

Since Pinochet's resignation as president in 1990, Chilean presidential politics have been dominated by a coalition of center-left parties known as the **Concertación por la Democracia** (Accord for Democracy). The priorities of the *Concertación* have included the maintenance of Chile's strong economic growth and the completion of the institutional transition to full democracy (Mares and Aravena 2001). The latter was greatly facilitated by Pinochet's retirement as army commander in 1998.

Paraguay

Paraguay has been ruled through most of its history by long-lived *caudillos*, whose actions have made the nation one of the most socially and economically underdeveloped countries in the Western Hemisphere. The first of the dictators was José Gaspar Rodríguez de Francia, who from 1814 until his death in 1840 did his utmost to physically and intellectually isolate Paraguay from the outside world. Francia was followed by Carlos Antonio López and his son, Francisco Solano López, who ruled until 1870. During the reign of the latter, Paraguay, angered at the Argentine closing of the Rió de la Plata, unwisely declared war in 1865 not only on Argentina but on Uruguay and Brazil as well. In the ensuing **War of the Triple Alliance,** Paraguay suffered immense economic and emotional devastation, losing over one-half of its population and almost all of its adult males. The chaotic period that followed brought, in 1887, the formation of two political parties, the conservative Colorados and the Liberals.

From 1932 to 1935 the nation fought the Chaco War, which resulted in Bolivia's defeat and the annexation of a large but sparsely settled area (see Figure 6.13). The end of the war brought Colonel Rafael Franco to power. Franco attempted to modernize the impoverished country through a program of agrarian reform and labor legislation. Both moves alienated the traditional elites, who promptly deposed Franco. However, they failed to eliminate altogether his supporters, who under the name *Febreristas* have continued to the present time as a muted political opposition group.

In 1954, General Alfredo Stroessner of the Colorado Party assumed power and, operating under a perpetual state of siege, was reelected without meaningful opposition every five years thereafter, until he was deposed in 1989 by General Andrés Rodríguez. Rodríguez, whose daughter was married to Stroessner's son, had previously gained control of much of the wealth of the country and did little to alter its reputation as a center of South American contraband trade, drug trafficking, and money laundering (Nickson 1996).

One of Rodríguez's positive accomplishments, however, was the passage in 1992 of a new constitution that greatly strengthened governmental checks and balances and protections of human rights. Paraguay's tender democracy has since survived an unsuccessful military coup in 1996 against Rodríguez's successor, Juan Carlos Wasmosy, and the murder in 1999 of Vice President Luís Argana. The latter led the Chamber of Deputies to impeach then-President Raul Cubas who, with former army general Lino Oviedo, subsequently fled to Brazil to escape prosecution for Argana's murder.

Argentina

Argentina's modern political development can be traced to 1776, when the Spanish government, concerned with the threat of Portuguese expansion southward from Brazil along the Atlantic coast, established the Viceroyalty of La Plata, with administrative headquarters at Buenos Aires at the mouth of the River Plate. In the ensuing four decades, the growth and influence of Buenos Aires and the Pampean region it served increased steadily, much to the consternation of the remainder of the viceroyalty, which in addition to Paraguay included portions of Upper Peru or Bolivia, Uruguay, and the interior Andean areas centered on Salta, Tucumán, and Córdoba.

In 1816 the region declared its independence from Spain as the United Provinces of the Río de la Plata. A power struggle immediately arose between the more outward-looking leadership of Buenos Aires and that of the inward-looking interior provinces, which historically had been oriented toward the Andean mining centers of Upper Peru. By the late 1820s, Paraguay, Bolivia, and Uruguay had each become independent, and the competition between Buenos Aires and the Andean regions was framed in the context of a bitter rivalry between two groups: federalists, who favored a weak national government with power concentrated in the provinces, and centralists or unitarianists (*Unitarios*), who sought to build a strong central government.

The federalist-centralist schism was subsumed into a larger conservative-liberal conflict. The conservative federalists favored the continued dominance of the classic alliance of the Church and landed aristocracy, while the liberal centralists favored expanded trade and a secular state patterned after the United States and Great Britain. The initial success of the liberals and their enlightened leader, Bernardino Rivadavia, soon gave way to an extended period of conservative dominance under the leadership of **Juan Manuel de Rosas,** who enjoyed broad support among the *gaucho caudillos* of the Pampa (Lynch 2001).

Rosas' forced exile to Great Britain in 1852 ushered in a golden age of liberalism, whose leaders, including Bartolomé Mitre, Domingo Faustino Sarmiento, and Nicolás Avellaneda, promoted foreign immigration and investment, public secular education, and freedom of the press and religion. The influx of foreigners continued into the late 1800s, which saw the opening of the Pampa to farming and increased investment by and trade with Great Britain. Two new political parties were also formed in the 1890s. The first, the Socialist Party, attracted moderate support among the emerging urban labor force, while the second, the Radical Civic Union, grew quickly as a reform movement advocating universal male suffrage and an end to public corruption. The election of Hipólito Yrigoyen in 1916 marked the beginning of a fourteen-year Radical era that brought unprecedented economic and social development.

In the succeeding decades Argentina fell from one of Latin America's most developed and enlightened countries to one of its most financially and politically troubled—a tragic illustration of the consequences of unchecked demagoguery and misguided economic policies. The collapse began with the Great Depression of the 1930s, which seriously damaged Argentina's export-oriented economy and fostered a longing for the grandeur of the past. Into this void stepped a group of military officers, who in 1943 overthrew the conservative government. Included among the rebels was Colonel **Juan Domingo Perón,** who after his election as president in 1946 imposed extremely high taxes on privately owned businesses and the *estancias* of the Pampa. Perón then invested the tax revenues in government-subsidized industries that provided jobs to thousands of rural-to-urban migrants known collectively as the **descamisados,** or "shirtless ones." Perón thus succeeded in creating, in classic demagogic fashion, a huge urban proletariat that looked to him for support and direction. In this he was greatly aided by his wife, María Eva Duarte, who, as "Evita," became virtually a demigod to the masses, whom she courted through her Eva Perón Foundation.

Perón was reelected in 1952, the same year Evita died of cancer. Three years later, with the output of the overtaxed *estancias* and government-operated industrial sector both in rapid decline, the military staged a coup and Perón fled to Europe. There followed a parade of short-lived military and civilian governments, none of which was capable of restoring economic discipline to the nation. As conditions continued to worsen, the exiled Perón became almost a cult figure to many who convinced themselves that prosperity could be restored through his return. The climactic return occurred at last in June of 1973, but a year later Perón died. An economically reeling Argentina found itself ruled by Perón's widow, María Estela (Isabel), whose sole qualification to govern was that she had been married briefly to Juan.

With Marxist terrorism on the rise and the treasury depleted, Isabel was removed from office by the military in 1976. Years of harsh authoritarian rule and rightist death squads followed, which resulted not only in the near elimination of the communist guerrillas but also the kidnapping, torturing, and/or killing of some 30,000 Argentine citizens during a period of state terrorism that has come to be known as the **Dirty War** (Feitlowitz 1998; Bouvard 1994). Most of the victims were guilty of nothing other than disagreeing with the nation's military rulers. Meanwhile, the downward economic spiral continued. In April 1982, Argentina invaded the Falkland, or Malvinas, Islands and declared war on Great Britain. The action's ostensible purpose was to recover the islands, which, notwithstanding Argentine claims of sovereignty, had been administered by Great Britain continuously since 1833. However, the nation's military rulers clearly hoped the conflict would divert attention from the deteriorating domestic scene (Freedman 1988; Escudé 1988; Caviedes 1984). Argentina's humiliating defeat led to a return of civilian rule in 1983. In 1989, Carlos Saul Menem of the Peronist Party took the presidential oath of office, marking the first time in Argentine history that one freely elected president had succeeded another. Menem did much in the early 1990s to return the financially ravaged country to a free-market economy.

By the end of Menem's second term, however, Argentina's economy was again in sharp decline, with unemployment, corruption, and crime at unacceptable levels. Menem was succeeded as president in 1999 by Fernando De la Rua of the Rival Alliance for Work, Justice, and Education. Regrettably, De la Rua proved to be no more effective than Menem in instilling fiscal discipline upon the Argentine people, and his resignation in 2001 plunged the country into a grave economic and political crisis (see Case Study 13.2).

Uruguay

Uruguay, or *La Banda Oriental,* as the land east of the Uruguay River was widely known, long was a buffer zone between Brazilian settlement to the north and Spanish colonies along the Río de la Plata. It first declared its independence in 1810 under the leadership of José Gervasio Artigas, whose *gaucho* forces were repulsed in an attack on the Spanish garrison at Montevideo. Following additional periods of foreign occupation and British mediation, Argentina and Brazil finally agreed to relinquish their claims to the area and independence was truly realized in 1828.

No sooner had freedom come, however, than the nation lapsed into a civil war between the followers of two rival *caudillos,* Fructuoso Rivera and Manuel Oribe. Because Rivera's forces incorporated red-colored ribbons into their clothing and Oribe's followers wore white, the factions became known as the *Colorados* (reds) and *Blancos* (whites). Although the conflict focused initially on personalities, it became in time a liberal (*Colorado*)-conservative (*Blanco*) division that has continued to the present.

Following the conclusion of the civil war in the early 1850s, Uruguay was dominated for half a century by *gaucho caudillos,* who from 1865 onward were all *Colorados.* By the later decades of the nineteenth century, both the *Colorado* and *Blanco* organizations were maturing into traditional political parties as the nation as a whole evolved into a leading exporter of meat and wool. Numerous social reforms, including free public education, were also enacted. A significant milestone on the road to democracy was reached in 1890 when the outgoing president, General Máximo Tajes, supported a civilian successor, Julio Herrera y Obes, thus ending military control over the country.

In 1903 yet another *Colorado* candidate, José Batlle y Ordóñez, was chosen president (Figure 6.19). After putting down a revolt led by the *Blancos,* who by then had changed their party name to Nationalist, Batlle set about trying to transform Uruguay into a South American Switzerland—the country he most admired. His reforms, some not fully implemented until after the completion of his second term as president in 1915, involved major political and socioeco-

José Batlle y Ordóñez attempted to mold Uruguay into Latin America's first welfare state in the early 1900s.

nomic restructuring. Politically, Batlle not only succeeded in establishing peace between the rival political parties but argued as well for the creation of a plural executive system. This was implemented in 1919 through the formation of a nine-member National Council for Administration that shared executive power with the president. Economically, Batlle did much to convert Uruguay into Latin America's first welfare state, nationalizing numerous enterprises and passing liberal social and labor legislation.

Uruguay's socialist economy continued long after Batlle's death, bringing the nation international acclaim as a progressive state but gradually sapping its financial strength as federal outlays increased and private investment declined. The country's lingering economic malaise finally led in 1958 to a conservative Nationalist victory, ending ninety-three consecutive years of *Colorado* rule. Rather than implementing fiscal reforms, however, the Nationalists continued in deficit spending, hoping thereby to gain an even greater degree of popular support. Batlle's concept of a plural executive form of government had been reimplemented in 1951 but proved again to be unworkable.

The 1960s were marked by deepening inflation and the emergence of a violent Marxist guerrilla group named the Movement of National Liberation, or *Tupamaros*.

By 1973, the violence and economic disintegration had reached such levels that President Juan M. Bordaberry of the *Colorado* Party agreed to military control of his administration. There followed a dark period in which Congress was abolished and civil liberties were reduced. Bordaberry himself was removed by the military in 1976. Civilian rule was reestablished in 1985, although the military has continued to exert a strong political influence.

The post-military era has been characterized by the emergence of a leftist coalition known as the Broad Front, which won 40 percent of congressional seats in the 1999 election. In response, the *Colorado* and Nationalist parties have formed their own legislative coalition and have increasingly supported a single presidential candidate in order to maintain control of the executive branch. Jorge Batlle, the current president, and his two immediate predecessors have steered the country on a centrist course while emphasizing the implementation of neoliberal economic reforms.

Brazil

We have noted previously that the Napoleonic wars of early nineteenth century Europe led to the establishment by Portugal's royal family of a government-in-exile in Rio de Janeiro and to Dom Pedro's declaration of Brazilian independence in 1822. With the sovereignty of the new nation assured through the protection of the British navy, Pedro moved to legitimize his reign by drafting in 1824 a highly authoritarian constitution. The constitution gave the emperor the authority to appoint both federal and state officials and also the right to resolve conflict between different government branches through the use of a royal "moderating power." However, Pedro's virtually unlimited constitutional authority failed to prevent a steady erosion of public support, which was linked to his arbitrary rule, to his forcing into exile his most competent advisor, José Bonifacio de Andrada e Silva, and to a number of foreign policy reversals, including the loss of Uruguay. Few of the dominant Brazilian aristocracy were disappointed, then, when Dom Pedro returned to Portugal in 1831, leaving his 6-year-old son to rule in his stead.

There followed nine years of liberal-conservative jousting that ended with Dom Pedro II being declared of age in 1840. In contrast to his sire, **Dom Pedro II** matured into a kind and scholarly father-figure type of ruler who felt a sincere love for Brazil and its people, and who generally ruled wisely and effectively during his nearly five decades in office (Skidmore 1999; Prado 1967) (Figure 6.20). By creating in 1847 the post of prime minister and by carefully using his "moderating

Emperor Dom Pedro II governed Brazil in a generally progressive and paternalistic fashion from 1840 to 1889.

Getúlio Vargas functioned as the authoritarian ruler of Brazil from 1930 to 1945 and from 1950 to 1954.

power" to balance conflicting liberal-conservative interests within the traditional oligarchical power groups, Pedro II was able to establish a degree of freedom and order that Brazil has yet to duplicate. Foreign capital and immigrants, attracted by the country's stability and economic potential, arrived in ever-increasing numbers; slavery was gradually eliminated; the military was barred from political activity; and the Church was reigned in. Artistic and other forms of individual expression were encouraged.

Yet as time passed, the emperor gradually lost influence. Many of the landed aristocracy resented his efforts to abolish slavery, his daughter and son-in-law were most unpopular, and his handling of Brazil's role in the War of the Triple Alliance was criticized by an increasingly restive officer corps. Afflicted with diabetes, Pedro II finally abdicated in 1889. Brazil entered a republican period that brought a new constitution, drafted in 1891 by the liberal statesman Ruy Barbosa, and the transfer of economic and political

power from the national government back to local *caudillos,* known as *coroneis.*

By the 1920s, three distinct political movements were under way. The first was the formation of a group of young military technocrats known as the *tenentes,* or lieutenants, who became convinced that they alone possessed the collective scientific knowledge, managerial skills, and commitment to individual freedoms necessary to assure the future development of the nation. The second movement was communism, which came to be identified with the National Liberation Alliance (ANL). The third was fascism, which was eventually represented by the Brazilian Integralist Action (AIB) and attracted considerable support among the German immigrant groups of the south.

In the midst of the growing political polarization and the economic aftershocks of the Great Depression, Getúlio Vargas assumed power in 1930. Over a fifteen-year period of demagogic rule, Vargas restructured Brazilian society into what he called the New State, or **Estado Nôvo** (Figure 6.21). Economically, the *Estado Nôvo* was a hybrid blending of the *tenente* faith in technology with socialism, and entailed the use of federal resources to promote industrial and agricultural modernization. Politically, the New State was fundamentally the old *caudillismo* at a national level, with Vargas offering himself as the champion of the emerging urban labor force and the embodiment of Brazilian nationalism, which he promoted through a renewed emphasis on sports and the arts. After being forced by the army to resign in 1945, Vargas was elected for yet another term in 1950 but accomplished little other than nationalizing foreign oil holdings. He committed suicide in 1954.

His successor, Juscelino Kubitschek, continued Vargas's program of grandiose, state-sponsored industrialization, particularly in the mining and energy sectors, but did so only through massive borrowing that enlarged the foreign debt and fueled the chronic inflation of the modern era. One of Kubitschek's legacies was the building of the new capital city of Brasília, in fulfillment of the centuries-old dream of developing the sparsely settled interior of the country.

Kubitschek was followed by Jañio Quadros, a reformer who mysteriously resigned after only seven months in office, and by João Goulart, an old left-wing Vargas-era politician whose years of service as vice president failed to prepare him to be an effective president. Within a few years of Goulart's assuming office, Brazil's economy was on the verge of collapse, and the middle and upper classes were convinced that a leftist dictatorship was in the offing. It came as no surprise, then, when the military removed Goulart in 1964, suspended constitutional guarantees of civil liberties, and began to hunt down and torture leftist "subversives," some 35,000 of whom were killed or disappeared in the following two decades (Skidmore 1988; Stepan 1988; Archdiocese of São Paulo 1986). The political repression was gradually relaxed in the late 1970s and early 1980s as the economy continued to struggle under the weight of triple-digit inflation and growing income gaps between the wealthy and the poor, between urban and rural dwellers, and between the south and the north.

The transition back to democracy began in 1985 with the holding of open municipal elections and the selection of a transitional civilian president. A new constitution was approved three years later, and was followed in 1989 by an open presidential election in which the populist center-right Fernando Collor de Mello defeated the socialist-worker candidate, Luiz Inacio Lula da Silva (Weyland 1993). Collor campaigned on a platform of overcoming Brazil's hyperinflation by privatizing state-owned corporations and by balancing the budget through reduced government spending. He resigned from office in 1992 after the lower house of Congress impeached him on corruption charges. Former finance minister Fernando Henrique Cardoso became president in 1995 and was reelected in 1998. Da

Silva's election as president in 2002 was viewed as a sign of growing voter discontent with Brazil's continuing economic stagnation and social inequalities.

THE GUIANAS

The so-called "Wild Coast" of South America's northern Guiana country was colonized intermittently during the sixteenth and seventeenth centuries by Spain, Portugal, the Netherlands, England, and France, all with an eye to establishing Caribbean-like plantation economies along the coastal plains. By the late 1700s, the Dutch had emerged as the dominant force throughout the region with the exception of Cayenne on the east, which was controlled during most of the colonial era by the French.

Guyana

The Dutch colonies of Demerara, Essequibo, and Berbice were ceded in 1814 to Great Britain, which united them as British Guiana in 1831. Just three years later, slavery was abolished throughout the British empire and the blacks began abandoning the Guiana plantations in great numbers, wanting at all costs to avoid agricultural labor, which in their minds carried a stigma of inferiority by virtue of its association with slavery. As the blacks migrated to Georgetown, the capital city, and to other newly established settlements, the European planter class replaced them with **indentured laborers,** mostly from India. As time passed, the East Indians, as they were known, came to control not only the agricultural sector but the urban commercial and technical professions as well. After the cessation of indentured labor in 1917, black feelings of economic and political insecurity were compounded by the higher East Indian birth rates, which by the mid-twentieth century were resulting in an ever-increasing numerical advantage for the Indo-Guyanese.

The nation's deep racial divisions led, in the 1950s, to the formation of two political parties. The first, the People's Progressive Party (PPP), was led by an East Indian dentist named Cheddi Jagan, an acknowledged communist. The second was a black African party known as the Peoples National Congress (PNC), whose leader, Forbes Burnham, described himself as a socialist (Figure 6.22). Independence within the British Commonwealth was achieved in 1966, but political fraud and racial violence have continued in the "Cooperative Republic" through a postindependence era dominated by Burnham, Jagan, and Burnham's PNC successor, Desmond Hoyte. Chicago-born Janet Jagan was elected president in 1997 following her husband's death but resigned in 1999 for reasons of health. The country has since been governed by Bharrat Jagdeo, an Indo-Guyanese.

FIGURE 6.22

Forbes Burnham, founder of Guyana's Peoples National Congress Party, was chosen the first president of the newly independent country in 1966. He ruled Guyana for twenty years, until his death in 1985.

Guyana has also had a long-standing dispute with Venezuela over the ownership of the land between the west bank of the Essequibo River and the present shared boundary, which was fixed in 1899. Guyanese desires to strengthen their claim to the isolated region, which amounts to slightly more than half of the national territory, contributed to the government's authorizing the Reverend Jim Jones's American People's Temple group to settle an undeveloped portion of the rain forest. The mass murder-suicides of the group resulted in the deaths of nearly 1,000 men, women, and children in November of 1978.

Suriname

The land that today makes up the nation of Suriname was colonized in the mid–seventeenth century by the English, who traded it to the Dutch in 1667 in exchange for New Netherlands (New York). As in Guyana, Suriname's colonial economy centered on coastal sugar and cotton plantations, which were manned through the importation of hundreds of thousands of black slaves from Africa. Some of these were able to escape into the heavily forested interior, where their descendants, called Bush Negroes or Maroons, continue a largely subsistence existence (chapter 7). The abolition of slavery in 1863 led, as in Guyana, to a black flight to the towns and mining centers, and to the importation of mostly East Indian and Javanese indentured laborers, who eventually came to own most of the agricultural land and to dominate the urban professional classes.

Internal self-government was achieved in 1954 when the Dutch declared Suriname to be equal to the Netherlands and the Netherlands Antilles within the Kingdom of the Netherlands. Many at this time expressed the belief that Suriname would mature into a prosperous and peaceful state, a model of polyethnic democracy for other developing nations. Others, especially Surinamers of Asian descent, feared that the racial and economic divisions were far more serious than was officially acknowledged. These fears led approximately 40 percent of the population, including much of the educated and skilled workforce, to emigrate to the Netherlands prior to the granting of independence in 1975.

In February of 1980, Colonel Desi Bouterse led a Creole military coup that overthrew the civilian government. The action was welcomed by much of the population, which perceived the outgoing administration to be incapable of managing the economy effectively. The initial euphoria quickly vanished, however, as Bouterse, governing through a nine-member National Military Council (NMC), began to systematically assassinate political opponents and to implement a reign of terror. The Bush Negroes, under the leadership of Ronnie Brunswijk, responded to Bouterse's attempts to forcibly resettle them by starting in 1986 a guerrilla movement called the Jungle Commando. The movement crippled bauxite mining operations in the northeastern portion of the country and led to 10,000 Maroons taking refuge across the Maroni River in French Guiana. A peace agreement was signed between the rebel groups and the government in 1992.

Pressured by his political and economic problems, and by long-standing territorial disputes with Guyana and French Guiana, Bouterse allowed free elections in 1987. He staged yet another coup in 1990 and then permitted elections again in 1991 and 1996. Suriname continues today to be a deeply divided nation whose fragile democracy is threatened by ethnic intolerance and a weak economy.

French Guiana

French Guiana, or Guyane Française, has belonged to France almost continuously since 1667. Its eighteenth- and nineteenth-century agricultural development paralleled that of Guyana and Suriname, with indentured Asians replacing freed blacks as the principal source of labor on the coastal sugar plantations. The development of the plantations was hindered by periodic outbreaks of typhus, malaria, and yellow fever that cost thousands of lives. Owing in part to the region's reputation as one of the least healthy places on earth, it attracted few settlers, and the French resorted in the mid–nineteenth century to using it for various penal colonies. The most infamous of these, **Devil's Island,** was closed in 1945.

The people of French Guiana have held full French citizenship since 1848 and have had representation in the French Parliament since 1877. French Guiana was declared an overseas department of France in 1946, since which time it has experienced modest population growth through the settlement of Hmong refugees from Laos and French continentals. Although the latter are a numerical minority to the black Creoles, they have formed an ultraconservative National Front movement that is strongly opposed to independence.

Summary

Although sharing many challenges arising from the colonial past, the nations of Latin America and the Caribbean have followed distinct political paths. The political geography of these countries has been shaped by their leaders' values and behaviors and also by the interactions of competing interest groups, many of which are centered on specific geographical regions. Understanding the political development of the Latin American and Caribbean nations allows us to see more fully the national and international significance of events presently occurring within them. Many of these events have some basis in the social and racial tensions of the region, which we will address in chapter 7.

Key Terms

Benito Juárez 114
PRI (Institutional Revolutionary Party) 114
United Provinces of Central America 115
FMLN (Farabundo Martí National Liberation Front) 117
Sandinistas 118
Canal Zone 119
José Martí 120
Fidel Castro 121
Jean-Bertrand Aristide 122

Rafael Trujillo 122
commonwealth 123
Luis Muñoz Marín 123
La Gran Colombia 123
La Violencia 125
War of the Pacific 129
APRA (American Popular Revolutionary Alliance) 129
Shining Path (*Sendero Luminoso*) 129
Chaco War 130
MNR (Nationalist Revolutionary Movement) 130

Salvador Allende 131
Concertación por la Democracia 132
War of the Triple Alliance 133
Juan Manuel de Rosas 133
Juan Domingo Perón 134
descamisados 134
Dirty War 134
Dom Pedro II 135
Estado Nôvo 136
indentured laborers 137
Devil's Island 139

Suggested Readings

Archdiocese of Sao Paulo. 1986. *Torture in Brazil*, trans. Joan Dassin. New York: Vintage Books.

Barton, Jonathan R. 1997. *A Political Geography of Latin America*. London: Routledge.

Benjamin, Thomas. 2000. *La Revolución: Mexico's Great Revolution as Memory, Myth, and History*. Austin: University of Texas Press.

Bergquist, Charles, Ricardo Peñaranda, and Gonzalo G. Sánchez, eds. 2001. *Violence in Colombia: 1990–2000: Waging War and Negotiating Peace*. Wilmington, Del.: Scholarly Resources.

Booth, John A., and Thomas W. Walker. 1999. *Understanding Central America*. 3rd ed. Boulder, Colo.: Westview Press.

Bouvard, Marguerite Guzmán. 1994. *Revolutionizing Motherhood: The Mothers of the Plaza de Mayo*. Wilmington, Del.: Scholarly Books.

Brenner, Anita. 1971. *The Wind that Swept Mexico*. Austin: University of Texas Press.

Brockett, Charles D. 1994. "El Salvador: The Long Journey from Violence to Reconciliation." *Latin American Research Review* 29 (3): 174–187.

Bushnell, David. 1993. *The Making of Modern Colombia: A Nation Inspite of Itself.* Berkeley: University of California Press.

Calder, Bruce J. 1984. *The Impact of Intervention: The Dominican Republic During the U.S. Occupation of 1916–1924.* Austin: University of Texas Press.

Camp, Roderic Ai. 1999. *Politics in Mexico.* 3rd ed. New York: Oxford University Press.

Carmack, Robert M., ed. 1988. *Harvest of Violence: The Maya Indians and the Guatemalan Crisis.* Norman: University of Oklahoma Press.

Carr, B., and S. Ellner, eds. 1993. *The Latin American Left from the Fall of Allende to Perestroika.* Boulder, Colo.: Westview Press.

Caviedes, César. 1984. *The Southern Cone: Realities of the Authoritarian State in South America.* Savage, Md.: Roman & Littlefield.

"Drugs, War, and Democracy." *The Economist,* April 26, 2001. http://economist.com/surve.

Escudé, Carlos. 1988. "Argentine Territorial Nationalism." *Journal of Latin American Studies* 20 (1): 139–165.

Feitlowitz, Marguerite. 1998. *A Lexicon of Terror: Argentina and the Legacies of Torture.* Oxford: Oxford University Press.

Freedman, Lawrence. 1988. *Britain and the Falklands War.* Oxford: Blackwell Scientific Publications.

Hale, Charles R. 1994. *Resistance and Contradiction: Miskitu Indians and the Nicaraguan State, 1894-1987.* Stanford: Stanford University Press.

Harvey, Neil. 1998. *The Chiapas Rebellion: The Struggle for Land and Democracy.* Durham, N.C.: Duke University Press.

Heinl, Robert Debs, and Nancy Gordon Heinl. 1996. *Written in Blood: The Story of the Haitian People 1492–1995.* Lanham, Md.: University Press of America.

Isaacs, A. 1993. *Military Rule and Transition in Ecuador, 1972–92.* Pittsburgh: University of Pittsburgh Press.

Jatar-Hausmar, Ana Julia. 1999. *The Cuban Way: Capitalism, Communism, and Confrontation.* West Waterford, Conn.: Kumarian Press.

Jonas, Susanne. 2000. *Of Centaurs and Doves: Guatemala's Peace Process.* Boulder, Colo.: Westview Press.

Juhn, Tricia. 1998. *Negotiating Peace in El Salvador: Civil-Military Relations and the Conspiracy to End the War.* New York: St. Martin's Press.

Langley, Lester D. 2002. *The Banana Wars: United States Intervention in the Caribbean, 1898–1934.* Wilmington, Del.: Scholarly Resources.

Lehoucq, Fabrice Edouard. 1991. "Class Conflict, Political Crisis and the Breakdown of Democratic Practices in Costa Rica: Reassessing the Origins of the 1948 Civil War." *Journal of Latin American Studies* 23:37–60.

Lovell, W. George. 2000. *A Beauty That Hurts: Life and Death in Guatemala.* Austin: University of Texas Press.

Lynch, John. 2001. *Argentine Caudillo: Juan Manuel de Rosas.* Wilmington, Del.: Scholarly Resources.

Macaulay, Neill. 1985. *The Sandino Affair.* Durham, N.C.: Duke University Press.

Mares, David R., and Francisco Rojas Aravena. 2001. *The United States and Chile: Coming in From the Cold.* New York: Routledge.

Millett, Richard L., ed. 1993. Special issue on "The Future of Panama and the Canal." *Journal of Interamerican Studies and World Affairs* 35 (3).

Moses, Catherine. 2000. *Real Life in Cuba.* Wilmington, Del.: Scholarly Resources.

Nickson, R. Andrew. 1996. "Democratization and Institutionalized Corruption in Paraguay." In *Political Corruption in Europe and Latin America,* eds. Walter Little and Eduardo Posada-Carbó, 237–266. New York: St. Martin's Press.

Palmer, David Scott. 1992. "Peru, the Drug Business, and Shining Path: Between Scylla and Charybdis?" *Journal of Interamerican Studies and World Affairs* 34 (3):65–88.

Parkes, Henry Bamford. 1970. *A History of Mexico.* Boston: Houghton Mifflin.

Pérez, Louis A., Jr. 1991. *Cuba Under the Platt Amendment, 1902–1934.* Pittsburg: University of Pittsburg Press.

Perez, Orlando J., ed. 2000. *Post-Invasion Panama: The Challenges of Democratization in the New World Order.* Lanham, Md.: Lexington Books.

Prado, C., Jr. 1967. *The Colonial Background of Modern Brazil.* Berkeley: University of California Press.

Sanchez, Gonzalo, and Donny Meertens. 2001. *Bandits, Peasants, and Politics: The Case of "La Violencia" in Colombia,* trans. Alan Hynds. Austin: University of Texas Press.

Skidmore, Thomas E. 1999. *Brazil: Five Centuries of Change.* New York: Oxford University Press.

———. 1988. *The Politics of Military Rule in Brazil, 1964–85.* New York: Oxford University Press.

———, and Peter H. Smith. 2001. *Modern Latin America,* 5th ed. New York: Oxford University Press.

Smith, Angel, and Emma Dávila-Cox, eds. 1999. *The Crisis of 1898: Colonial Redistribution and Nationalist Mobilization.* New York: St Martin's Press.

Starn, Orin. 1992. "New Literature on Peru's Sendero Luminoso." *Latin American Research Review* 27 (2):212–226.

Stepan, A. 1988. *Rethinking Military Politics: Brazil and the Southern Cone.* Princeton, N.J.: Princeton University Press.

Susman, Paul. 1998. "Cuban Socialism in Crisis: A Neoliberal Solution?" In *Globalization and Neoliberalism: The Caribbean Context,* ed. Thomas Klak, 179–208. Lanham, Md.: Rowman and Littlefield.

Violence in Colombia: Building Sustainable Peace and Social Capital. 2000. Washington, D.C.: World Bank.

Walker, Thomas W., and Ariel C. Armony, eds. 2000. *Repression, Resistance, and Democratic Transition in Central America.* Wilmington, Del.: Scholarly Resources.

Weyland, Kurt. 1993. "The Rise and Fall of President Collor and Its Impact on Brazilian Democracy." *Journal of Interamerican Studies and World Affairs* 35 (1): 1–37.

Wiarda, Howard J., and Harvey L. Kline, eds. 2000. *Latin American Politics and Development.* 5th ed. Boulder, Colo.: Westview Press.

Williams, Philip J. and Knut Walter. 1997. *Militarization and Demilitarization in El Salvador's Transition to Democracy.* Pittsburgh, Pa: University of Pittsburgh Press.

Electronic Source:

U.S. Department of State, Bureau of Public Affairs Electronic Information Office: http://www.state.gov/r/pa/bgn/

Race, Ethnicity, and Social Class

Knowing how race, ethnicity, and social class are interrelated in Latin America and the Caribbean is essential to understanding the cultural geography of the region. This chapter presents an overview of the principal pre-Conquest Indian peoples, the collapse of their populations following the arrival of the Europeans, and the racial mixing that then occurred. We also analyze the migrations to the New World of blacks and East Asians, and perceptions of race and ethnicity today. We conclude by studying the development of social races in Latin America and the Caribbean and the manner in which they presently define class relationships.

NATIVE POPULATION LEVELS AT THE TIME OF THE CONQUEST

When the Spanish conquistadores and clerics arrived in the New World, they were consistently impressed with both the number of native peoples they encountered in the Americas and the advanced levels of their civilizations, many of which we now recognize to be equal or superior to those of sixteenth-century Europe. One of the most intensely debated and controversial topics of Latin American studies in recent decades has been the population levels of native Americans at the time of the European Conquest. The issue is not merely an abstract statistical matter. It carries widespread implications with respect to the sophistication of indigenous cropping systems and natural resource utilization and their applicability to current development efforts. The findings can also show us the consequences of prolonged and intensive contact between two culture groups that previously were isolated from each other.

Most native American peoples failed to keep census types of data, and the records of those that did were largely destroyed—often in the name of protecting Christianity from the potential influences of so-called pagan documents. So researchers have been forced to estimate the population of Latin America on the eve of the Spanish Conquest through extrapolation of circumstantial evidence. This includes archaeological remains, our understanding of the social complexity of pre-Columbian Indian societies, and the food production capacity of various groups that occupied distinct natural environments. The conclusions drawn from these different kinds of evidence have proven to be extremely inconsistent. This inconsistency is influenced not only by the disciplinary methods employed by the scholars studying the topic but also to some degree by their cultural backgrounds. For example, Latin Americanist scholars who are sympathetic to the region's Iberian heritage have tended to arrive at relatively low pre-Conquest population levels. If correct, these estimates would suggest that the Spanish and Portuguese conquests brought about less destruction of native American cultures than what is widely suggested by scholars whose research has stressed the positive qualities of indigenous groups and their contributions to Latin American development. Studies representative of low, medium, and high pre-Conquest population estimates are summarized in Table 7.1 and show populations ranging from approximately 12 million by Rosenblat to over 100 million by Dobyns.

TABLE 7.1

Estimates of the Aboriginal Population of Latin America and the Caribbean on the Eve of the European Conquest (in thousands)

REGION	ROSENBLAT	DENEVAN	DOBYNS
Mexico	4,500	21,400	30,000–37,500
Central America	800	5,650	10,800–13,500
Caribbean	300	5,850	443–554
Andes	4,750	11,500	30,000–37,500
Lowland South America	2,030	8,500	9,000–11,250
Total	12,380	53,900	80,243–100,304

Sources: Rosenblat 1954; Denevan 1992a; Dobyns 1966.

FIGURE 7.1

These raised fields, or *sukacollos,* of the Bolivian Altiplano near Lake Titicaca are a modern form of an ancient and highly productive pre-Conquest indigenous cropping system.

We will never know for certain the exact number of native Americans living in what was to become Latin America and the Caribbean at the time of European conquests. However, a growing body of evidence suggests that the agricultural technologies employed by the Indians were sufficiently advanced to support dense population levels throughout much of the region. One such technology appears to have been the erection of massive hydrologic and land modification systems, which, in addition to irrigating hundreds of thousands of hectares of arid lands, included the construction of equally extensive areas of raised fields in poorly drained lands that today lie largely abandoned (Figure 7.1). Other evidence of the capacity of the indigenous Americans to support dense populations is their domestication of food crops suited to cultivation within every natural environment found within the re-

gion and their masterful integration of these genetic resources into highly productive cropping systems (Patiño 1963) (Table 7.2).

These findings do not document in a statistical sense that the native populations were as high as some observers believe. They do suggest, however, that de las Casas and other early Spanish chroniclers, who wrote of major population clusters, may not have exaggerated, as some have assumed. Regardless of the precise numbers, it can be concluded that the native population of Latin America and the Caribbean was likely relatively high when compared both to the population of the remainder of the world at the time of the Conquest and to the population of the region for the ensuing four hundred years.

NATIVE AMERICAN CIVILIZATIONS ON THE EVE OF THE CONQUEST

However many indigenous peoples occupied the Americas prior to the Spanish and Portuguese conquests, it is necessary for us to study how the various Indian groups lived and what they accomplished. This will enable us to appreciate the massive changes and dislocations that accompanied the European invasions and that formed the basis of race, ethnicity, and social class in modern Latin America and the Caribbean.

At the time of Columbus' discovery of the New World, there may have been as many as 5,000 distinct native American groups or, as they came to be called, Indian groups, occupying South America, with hundreds of others residing in Central America, the Caribbean, and Mexico (Mason 1950). The primary basis of differentiation between these peoples was language or dialect, but they also varied greatly in levels of social and political organization, customs, and physical and economic adaptations to the distinct natural habitats they occupied (Payne 1990; Salzano and Callegari-

TABLE 7.2

Selected Crops Domesticated by Native American Indians Prior to 1492

COMMON NAME	BOTANICAL NAME	COMMON NAME	BOTANICAL NAME
Highlands		**Lowlands (continued)**	
quinoa	*Chenopodium quinoa*	cherimoya	*Annona cherimola*
kaniwa	*Chenopodium pallidicaule*	cashew	*Anacardium occidentale*
kiwicha/amaranth	*Amaranthus caudatus*	cacao	*Theobroma cacao*
potato	*Solanum spp.*	vanilla	*Vanilla fragrans*
oca	*Oxalis tuberosa*	papaya	*Carica papaya*
ulluco	*Ullucus tuberosus*		
mashua/añu	*Tropaeolum tuberosum*	**Both Highlands and Lowlands**	
arracacha	*Arracacia xanthorrhiza*	maize	*Zea mays*
narranjilla	*Solanum quitoense*	common bean	*Phaseolus spp.*
		squash	*Cucurbita spp.*
Lowlands		chile pepper	*Capsicum spp.*
manioc/cassava/yuca	*Manihot esculenta*	tomato	*Lycopersicon esculentum*
sweet potato	*Ipomoea batatas*	chilacayote	*Cucurbita ficifolia*
yampee	*Dioscorea spp.*	chayote	*Sechium edule*
yautia/ocumo/tania	*Xanthosoma spp.*	yacon	*Polymnia sonchifolia*
arrowroot	*Maranta arundinacea*	achira	*Canna edulis*
jicama	*Pachyrhizus erosus*	pepino/pear melon	*Solanum muricatum*
peanut	*Arachis hypogaea*	tamarillo/tree tomato	*Cyphomandra betacea*
pineapple	*Ananas comosus*	avocado	*Persea americana*
pejibaye palm	*Guilielma gasipaes*		
peach palm	*Bactris gasipaes*	**Nonfood Crops**	
Brazil nut	*Bertholletia excelsa*	agave	*Agavaceae spp.*
Babassu palm	*Orbignya martiana*	silk cotton	*Ceiba pentandra*
guava	*Psidium guajava*	cotton	*Gossypium spp.*
sapodilla	*Achras sapota*	tobacco	*Nicotiana tabacum*
soursop	*Anona muricata*	coca	*Erythroxylon coca*
passion fruit	*Passiflora edulis*	quinine	*Cinchona calisaya*
calabash gourd	*Crescentia cujete*	Pará rubber	*Hevea brasiliensis*
uvilla	*Pourouma cecropiaefolia*		

Jacques 1988). In brief, the indigenous Americans were nowhere close to being a monolithic body and, in reality, differed almost as much from one another as they did from their post-Conquest European masters.

Having noted the differences between the Indian groups, we must also recognize that they did not live in total isolation from one another and that there was considerable trade and contact among them, both by land and by sea. This association was facilitated by numerous linguistic "pan-Americanisms," which to many scholars of American Indian languages suggests a common ancestry for virtually all of the indigenous New World peoples (Kaufman 1990).

Despite the complexity of the Indian cultures, it is possible to organize all of the individual tribes and clans into two broad groupings at the time of the Conquest: those occupying the mountainous, western regions, and those residing on the lowlands of central

and eastern Latin America and the Caribbean. These groupings correspond generally to the use of the highland and lowland crops summarized in Table 7.2.

Western Highland Peoples

Latin America's western highlands included four advanced civilizations and a number of culturally less developed regions at the time of the Conquest (Figure 7.2). The arid deserts of northern Mexico and the southwestern United States were occupied by various seminomadic hunting and gathering tribes of low material and social achievement who came to be known collectively as the **Chichimecs.** From time to time one or more of these tribes, driven possibly by a need to obtain a more secure food supply, would migrate southward into the high volcanic valleys of the Mesa Central. From the ninth through the twelfth centuries

FIGURE 7.2

Pre-Columbian Indian populations on the eve of the European Conquest.

that region was dominated by the Toltec peoples, whose culture's hero, Quetzalcoatl, was represented as a feathered serpent.

THE AZTECS

One of the tribes that migrated into the Valley of Mexico during the thirteenth century was the **Aztec** or, as they called themselves, the Mexica. The Mexica were received by the far more advanced tribes of the Valley not as conquerors but as outcasts and were forced to wander about the lakes and marshes in search of a place where they could settle. They could do no better than an island in Lake Texcoco where their priests found an eagle perched on a prickly pear cactus in fulfillment of a prophecy attributed to their patron god, Huitzilopochtli. There, in 1325, they founded a city named Tenochtitlán and subsisted on fish, water fowl, and small-scale farming practiced on raised platforms, called *chinampas*, that were built out of the shallow lake bed.

During the first century of their residence in the Valley, the Mexica, or Aztec, steadily expanded their sphere of influence, largely adopting the preexisting Toltec culture rather than developing a new one of their own. By 1430, the power of Tenochtitlán had grown to such a degree that the Aztecs had joined with the rulers of the neighboring cities of Texcoco and Tlacopan to form the Triple Alliance, which was to gain control of most of central Mexico in the ensuing ninety years (Figure 7.3).

The empire built by the Aztecs and their allies came to consist of over 400 settlements organized into thirty-eight provinces, extending from the independent Tarascan kingdom to the northwest to the lowlands of the gulf coast bordering the Maya realm (P. Carrasco 1999). The principal object of their conquests was to expand the quantity and diversity of tribute sent to support the Aztec nobility residing in Tenochtitlán. Each province was assessed tribute annually, with the list of commodities including food items such as maize, beans, dried chile, honey, and salt, various articles of clothing and ceremonial and warring garb sewn from cotton or maguey, wood and furnishings, incense, feathers, gold, turquoise, and other precious metals and gems.

One of the most grisly and repulsive of the "products" consumed in large quantities by the Aztecs was human ritual sacrificial victims, who by most accounts numbered 20,000 to 50,000 per year. For reasons that we do not fully understand, the Aztecs were convinced that Huitzilopochtli required the offering of a continual supply of human hearts and blood in exchange for maintaining order in the universe. Even more gruesome was their custom of cannibalizing some of the victims (D. Carrasco 1999; Harner 1977).

FIGURE 7.3

The Aztec was but the latest of a series of Indian empires that dominated central Mexico prior to the European Conquest. One of the largest of the earlier urban centers was Teotihuacán, which supported a population of between 85,000 and 200,000 in the fifth through seventh centuries A.D. Shown is the Pyramid of the Moon, one of the largest pre-Conquest structures of the New World.

The large-scale offering and eating of human sacrificial victims was but one expression of the extremely fatalistic and pessimistic world view of the Aztecs, who were seemingly preoccupied with war, death, and the need to appease the gods in a never-ending attempt to preserve the cosmic order. To the Aztecs, the chaos and unpredictability of this world were but reflections of the confused and turbulent heavens, where gods conquered one another and, in so doing, assumed each other's traits. As the empire grew, the gods of the newly subjected peoples were simply incorporated as lesser deities within the troubled Aztec pantheon.

In touching on some of the unstable dimensions of the Aztec mindset, we must not lose sight of their accomplishments. One of the most significant was urban planning and development. At the time of Cortés' arrival, no city in Spain and few in all of Europe could begin to approach the size and grandeur of Tenochtitlán. Together with its twin city of Tlatelolco, to which it was connected by three artificial causeways, Tenochtitlán is estimated to have had a surface area of twelve to twenty square kilometers (five to eight square miles) and a population of between 200,000 and 250,000 (Figure 7.4).

Not only was the city among the largest in the world, it was also among the most wealthy and orderly. At its heart was a magnificent plaza where the four principal avenues joined. Rising some sixty meters (200 feet) above the plaza was the great pyramid capped by two temples, one dedicated to Huitzilopochtli and the other

FIGURE 7.4

Mexico City was built literally on the ruins of the Aztec capital of Tenochtitlán, many of which, such as these close to Mexico City's main plaza, have been uncovered during excavation for the city's subway system and are now being restored.

FIGURE 7.5

The Mayan center of Altun Ha was built from the Early Classic to the Late Classic Period. The complex is situated within the swampy coastal plain of Belize that today supports only minimal agricultural development and low population densities.

to Tlaloc, the God of Rains and Harvests. Surrounding the plaza were the palaces of the nobility and, beyond those, ball courts, beautiful gardens, a zoo and aviary, and thriving markets stocked with produce from throughout Mesoamerica. Finally, the outlying zones were occupied by the residences and farms of the common people.

In addition to city planning, the Aztecs also excelled in the arts of leather and woodworking, weaving, pottery making, and metallurgy—the latter having spread from the Andean civilizations to central Mexico in the eleventh century A.D. They and their subject peoples were also accomplished agriculturalists. Access to the land was controlled by small communal units called *calpulli*, which also controlled the payment of taxes in the form of labor on public works and the lands of the aristocracy. Medical treatments centered on herbal remedies, libraries were filled with books, manuscripts, and codices written in ideographic script, and music, poetry, and the arts flourished (D. Carrasco 1998; Davies 1980; Berdan 1982).

THE MAYA

Unlike the short-lived Aztec Empire, which flourished for less than a century before its fall to the Spaniards, the **Maya** civilization is of great antiquity, having formed by 3000 B.C. in the highlands of northern Central America and then diffused northward onto the lowlands of the Yucatán peninsula. Life during the Formative Period centered on subsistence agricultural production and religious ritual, both of which were influenced strongly by a concern for understanding time and its perceived effects on human events. Little

change or development appears to have occurred until the seventh century B.C., when, for reasons that we do not fully understand, there suddenly appeared in the Petén region of northern Guatemala large cities containing pyramids and stone buildings that likely required the coordinated labor of thousands of commoners working under the direction of a highly structured bureaucratic hierarchy.

By the fourth century B.C., the Maya had entered a golden era referred to as the Classic Period. In the process they expanded into the Copán River Valley of northern Honduras and as far north as the Mexican gulf coastal plains of Tabasco, where it is possible that they were influenced by the ancient Olmec civilization. Towns and cities with populations ranging as high as 40,000 to 45,000 persons aligned themselves into some fifty independent states that participated in a complex regional network of economic exchange and shifting military alliances. Monumental pyramids and associated structures rose above the forest floor, and extraordinary advances were achieved in mathematics, astronomy, and writing (Figure 7.5). Population densities as high as 120 to 190 per square kilometer (300 to 500 per square mile) supported a total population numbering in the millions within a culture realm that encompassed over 260,000 square kilometers (100,000 square miles).

Then, in the ninth century A.D., the civilization began to decline as suddenly as it had risen over a thousand years before. During the next several hundred years the great cities were abandoned, societal organization reverted to the level of the hamlet, and population densities fell to as low as 5 persons per square kilometer (13 per square mile). In the tenth century A.D., much of northern Yucatán was conquered by Toltec peoples from central Mexico, who established their capital at Chichén-Itzá. Despite the use of much violence, there was a modest resurgence of Mayan culture from the thirteenth through the fifteenth centuries, before the area once again declined in the decades preceding the Spanish Conquest.

The sudden rise and fall of the Maya civilization has perplexed generations of scholars, who have offered a wide range of explanations. Many have speculated that the fall of the Maya was linked to the destruction of their natural environment through hurricanes, earthquakes, or long-term climatic change. Others have argued that the decline was caused by epidemic disease or by population growth that could have led to widespread deforestation and to the exhaustion of the region's thin lateritic soils. Critics have countered the overpopulation thesis by noting that recent research has revealed extensive regions of fertile muck soils that were often worked into terraced or raised fields called *pet kotoob*. Together with the intervening canals and water bodies, the *pet kotoob* were capable of supporting dense populations. We know of few, if any, civilizations whose decline can be traced solely to physical causes. So others have hypothesized that the abrupt demise of the Maya must be attributable to sociocultural factors such as long-term warfare, or to other circumstances that could have led to a loss of social discipline and control and, eventually, to an abandonment of the cities and a return to the forest. In reality, no single explanation is capable of accounting for all that we know about the Maya. The causes of their sudden rise and fall may remain forever a mystery to us.

We do know, however, that the Maya were in many ways the most culturally advanced of all the pre-Columbian Indian peoples. They developed a sophisticated form of hieroglyphic writing through which they recorded, both on stone monuments called stelae and on books made from folded tree bark, the most significant events of their lives (Figure 7.6). Although the Spanish priests burned thousands of the bark books during the early colonial period, three texts known as the Dresden, Paris, and Madrid codices, and fragments of a fourth, miraculously survived.

Driven by a desire to correlate the occurrence of earthly and heavenly actions, the Maya developed remarkable skills in astronomy and mathematics (Aveni 2001; Hassig 2001) (Figure 7.7). Their practice of simultaneously using two permutating annual calendars to reckon time was adopted by the Aztecs and numerous other Middle American peoples. The first was a 260-day ritual calendar that consisted of 13 months of 20 days each (Malmström 1997). It was, and in the highlands of Guatemala continues to be, used mostly for purposes of divination, such as knowing when to plant or harvest crops, inaugurate rulers, marry, or attempt to prevent or cure certain diseases. A solar calen-

FIGURE 7.6

This stone stele is a part of the magnificent Mayan ruins of Calakmul which recently have been excavated deep in the lowland rain forests of the Mexican state of Campeche near the Guatemalan Petén.

FIGURE 7.7

The ruins of the Observatory at Chichén-Itzá indicate the accomplishments of the Maya and other pre-Columbian Indian civilizations in the areas of astronomy, mathematics, and the reckoning of time.

dar of 365 days was also used, consisting of 18 months of 20 days each, with an additional greatly dreaded five-day unlucky period during which important events were not scheduled. The ritual and solar calendars joined every fifty-two solar and seventy-three sacred years, at which time the earth was believed to be subject to destruction and renewal. Time was further reckoned through a Long Count, which began at 3114 B.C. So precise were the ancient Maya astronomical calculations that their lunar month differed from our present mathematical calculations by just 33.7 seconds, and the revolution of Venus was reckoned to an error of only 1 hour 55 minutes every 584 days (Sharer 1994). Other

FIGURE 7.8

The Chibcha linguistic realm of the Colombian Andes emerged from a number of highland civilizations of great antiquity. One of these, the San Agustín culture, flourished from the first to the eighth centuries A.D. and left over 500 monumental statues, many of which are now found in an archaeological park situated near the headwaters of the Magdalena River.

features of Maya life included a ferocious pan-Caribbean game whose object was the placement, without the use of hands, of a hard rubber ball through a stone ring or collar positioned high above the ground; the use of the corbelled vault in architecture; and the maintenance of a vast sea-based trading network that extended from Veracruz to Honduras (Scarborough and Wilcox 1991).

THE CHIBCHA

The third of the advanced Indian civilizations of the western highlands was the Chibchan-speaking peoples of Colombia (Figure 7.8). Their level of social organization was in many ways a notch below that of the great Inca empire to the south but yet clearly above that of the forest dwellers of lowland South America. At the time of the Spanish Conquest, the **Chibcha** were concentrated in small agricultural villages scattered throughout the basin of Cundinamarca and the neighboring valleys of the Eastern Cordillera. Large cities were lacking and political governance was based pri-

marily on chiefdoms, although two loosely organized states had been formed and were beginning to expand and exact tribute from conquered peoples.

Unlike the Inca, who possessed massive stone structures and large cities, Chibcha villages consisted of small, family-sized dwellings constructed of wooden poles lined with clay plaster. The majority of the people were farmers or artisans, with many of the latter producing exquisite gold work. Although cultivating the same Andean crops as the Inca, Chibcha agriculture was characterized by an absence of irrigation and by limited use of terracing and fertilizers. By the early sixteenth century, Chibchan-speaking peoples were found throughout the Andes of northern Ecuador and Colombia, as well as in much of eastern Central America, where they included the Cuna of Panama, the Guetar, Cabécar, and Bribri of the Talamanca-linguistic family of Costa Rica and Panama, and possibly the Miskito of eastern Nicaragua and Honduras. Interestingly, the Chibchan expansion into western Central America was blocked by Aztec or Nahuatl-speaking tribes, including the Pipil and Lenca of El Salvador and Honduras and the Chorotega and Nicarao of Nicaragua, who had fled central Mexico and migrated southward along the Pacific coastal lowlands.

THE INCA

The emergence of the Inca Empire, like that of the Aztecs thousands of kilometers to the north, can be traced to the migration in the thirteenth century of a small, previously inconsequential tribe to a new homeland. There they rapidly built a militaristic state, which crumbled with stunning quickness three centuries later in the face of the Spanish onslaught. The true origins of the **Inca** are shrouded in myth, but they themselves maintained that their first ruler, Manco Copac, and his sister/wife, Mama Ocllohuanco, had been born of the sun god, Inti, on an island situated within Lake Titicaca on the high Altiplano. Prompted, so they said, by Inti's instructions to them to teach and govern all other peoples, the tribe traveled northwestward until they came to the well-watered and densely populated Cuzco Valley. There they established a new home and began to expand by incorporating the previous occupants, who were given the title of Incas-by-privilege (Figure 7.9).

In the Quechua language of the Inca, Cuzco means "the navel," or the center. Cuzco emerged as the sacred city of the Inca world, with the presiding priests and nobles residing within the inner district and the commoners within the periphery. Although Cuzco never rivaled Tenochtitlán in size or grandeur, massive stonework had begun to appear by the reign of the sixth ruler, Inca Roca. By the fifteenth century, the empire, named Tawantinsuyu, was expanding rapidly up and down the Andean highlands and the dry Pacific

FIGURE 7.9

The Cuzco Valley of southern Peru served as the administrative center of the vast Inca Empire.

coastal lowlands to the west (Bauer 1992; Davies 1997). By around 1430 they had subdued the Aymara-speaking peoples south of Lake Titicaca, and followed that with the conquest of northern Peru and Ecuador by 1495, designating Quito and Tumebamba (Cuenca) as the military and administrative centers of the northern provinces. Northern and central Chile south to the Maule River were annexed by 1525, giving the empire a total length on the eve of the Spanish Conquest of approximately 4,000 kilometers (2,500 miles). Of course, we will never know what the long-term fate of the Inca state would have been. However, shortly before the arrival of the Spaniards, civil war had broken out between the half-brothers Inca Atahualpa, who represented the northern territories, and Inca Huascar, who was supported by the southern provinces, suggesting that the empire was possibly overextended and may have eventually disintegrated from within.

At its height, however, Tawantinsuyu was one of the most efficiently administered kingdoms in the history of the world. The ruling Inca class attempted to exercise complete control over the life of every subject. The entire state was broken down into groups of 10, 50, 100, 500, 1,000, 5,000, 10,000, 20,000 and, finally, 40,000 families, each strictly governed by a leader who reported on all matters to his superior. All administrative officials down to rulers of 100 belonged to a privileged caste known as the *curacas*. When the Inca armies conquered a new region, the local elites and certain of their descendants were taken to Cuzco, where they were treated as royalty, reeducated, and adopted before being returned to their home areas and reinstalled as midlevel administrators. If they remained loyal, they and their subjects were rewarded with additional lands, wives, and other benefits. If they rebelled, however, they were shipped off to distant, isolated regions and were replaced by loyal Quechua-speaking colonists known as *mitimaes*.

Every worker in the empire was assigned a trade or profession to which he or she was bound for the remainder of his or her life. The vast majority of the population resided in dispersed hamlets and villages, from which they worked small and steeply sloped or terraced fields of highland tubers and grains and tended herds of llamas and alpacas (Case Study 7.1). Cooperative, reciprocal labor, called a *minga*, was commonly used on the more difficult tasks, such as harvesting crops or raising a house.

The basic unit of Inca society was the *ayllu*, which in theory but not always in practice was an endogamous social unit comprising a number of extended families whose ancestry could be traced to a common male progenitor. Although the head Inca claimed ultimate ownership of all land and resources by virtue of his divine lineage through the sun god, de facto rights to the lands resided permanently with the *ayllus*, whose leaders reallocated each year, on the basis of need, the fields under their domain to the families belonging to their *ayllu*. Water rights and crop selection were also governed by the *ayllu*, which thus functioned as a mechanism for the equitable distribution of resources. The intensive use of the land for the production of human food crops and the practice of crop rotation both contributed to the maintenance of soil fertility and to the sustaining of relatively high yields over time. One of the most impressive aspects of Inca life is that there appears to have been no hunger. Even among the commoners there was food enough for all—and this despite the fact that much of the land and its produce was reserved for the sun god and his priestly representatives and for the bureaucracy that administered the affairs of state.

Labor to work the land belonging to the sun god and to the secular officials of the empire was conscripted through a form of nonmonetary taxation known as the *mita*. Each *ayllu* was required to provide *mita* laborers not only to work the lands of the high officials but to perform military service, work in the mines, and construct public roads, buildings, and irrigation networks. The latter rank among the most sophisticated in the world, conducting water for miles along the sides of mountains and, in places, through underground stone channels to distant farms and fields.

The Inca had other impressive technical achievements: the construction of tiny suspension bridges sixty meters (200 feet) or more in length across deep chasms and rivers situated hundreds of meters below; over 29,000 kilometers (18,000 miles) of paved roads centering on a highway built along the coast and another winding its way through the highlands, both of which were provided with well-stocked rest houses called *tambos;* a human runner-based postal service called the

Case Study 7.1: Native Animals of the Andes

When the Spaniards conquered the Inca Empire, they discovered that the Andean world, like the remainder of Latin America, was devoid of the large grazing animals common to the Old World, such as sheep, goats, pigs, donkeys, horses, and cattle. Instead they encountered dogs, deer, and rabbits, and four New World members of the camel (*Camelidae*) family that have survived to the present day.

The largest of these is the llama (*Lama glama*), which was used by the Inca as their only domesticated beast of burden, as a sacrificial animal, and as a source of hides, meat, and dung. Although llamas commonly achieve a shoulder height of one and a half meters (four and a half feet) or more, as pack animals they are relatively small, weighing less than 140 kilograms (300 pounds) each (see Figure 3.11). However, their limited size is compensated for by their ability to transport small loads across some of the highest and most treacherous terrain on earth, to withstand extended periods of thirst, and to survive off the natural highland pastures.

Only male llamas from three to twelve years of age are used as pack animals—the females being sheared for a rather low-grade wool and saved for breeding purposes. A healthy male llama can carry roughly 35–45 kilograms (75–100 pounds) of cargo for a three-week trip, averaging 16–24 kilometers (10–15 miles) per day. When overloaded or pushed to walk too far, however, the llama will simply sit down and quit, hissing and spitting and refusing to budge until its problem is resolved. Although supplanted somewhat by mules and trucks in Andean mining camps, it continues to thrive in the agrarian sector, where no other animal nor machine can match its superb combination of surefootedness, frugality, and versatility.

In addition to the llama, the arid Andes are also home to three smaller New World camel species: the alpaca (*Lama pacos*), the vicuña (*Lama vicugna*), and the guanaco (*Lama guanicoe*). The alpaca, like the llama, has been domesticated for thousands of years, but it is not used to transport cargo. It is quite restricted in range, being found today mostly on those portions of the Altiplano bordering the shores of Lake Titicaca, with secondary concentrations on the high, windswept *punas* of southern Bolivia and northern Argentina. Its flesh is edible, but it is raised primarily for its long, highly prized hair or wool, which is extraordinarily light in weight, strong, and warm and which commands a premium price in the world's textile markets. Neither the vicuña nor the guanaco has been domesticated, but the native Indian peoples have long harvested the high-grade wool of the vicuña by driving the small wild herds into pens where the animals are sheared and released. The guanaco is valued most for its meat. Both the vicuña and the guanaco have been overhunted, with the latter surviving today only in Tierra del Fuego and the neighboring southernmost islands of the continent.

A final animal found throughout the Andes is the domesticated guinea pig (*Cavia aperea porcellus*), which evolved originally in the mountains of central and southern Chile. By the time of the arrival of the Spaniards, the animal had become a favorite meat source throughout the highlands of the continent, being allowed free run of the homes and yards of the Indian peoples. Guinea pigs frequently host fleas and, in modern times, have been implicated in the spread of typhus and bubonic plague. They remain to this day, however, a leading source of animal protein for the poor, owing to their flavorful, fatty meat, their ability to forage, and their capacity to multiply rapidly.

chasqui, which could relay information at approximately 240 kilometers (150 miles) per day and which was four times faster than the Spanish horseback-carried mails of the seventeenth century; and rectangular and polygonal mortarless masonry, which entailed rubbing huge stones weighing up to 54,000 kilograms (60 tons) against one another until they fit so tightly that, to this day, a knife blade cannot penetrate many of the joints (Figures 7.10 and 7.11).

While the Inca were unsurpassed among the pre-Columbian Americans in the areas of social control and engineering, they placed less stress on handicraft production and intellectual pursuits than the Aztecs and Maya or even the Peruvian civilizations preceding them. Ceramics, textile production, and weaving, for example, were all less developed than under the previous Nazca, Moche, and Chimú coastal desert cultures. Nor did a written language exist, although laws and numerical accounts were recorded on *quipus*, collections of knotted strings attached to a larger cord or rope. The Inca, like the Aztecs and Maya, also failed to develop large wheeled vehicles and plows—neither of which had a function among a people lacking large beasts of burden.

FIGURE 7.10

A flock of llamas traveling a road in the high Andes of Bolivia.

FIGURE 7.11

Massive polygonal, mortarless Inca stonework.

THE SOUTHERN ANDEAN INDIANS

The Andean regions immediately to the south of the Inca Empire were occupied by a number of politically disunited tribes belonging to the **Araucanian** linguistic family. The words *Auca, Aucano,* and *Aucanian* all mean "rebel or enemy" in the Quechua language—a designation justly earned by Inca armies' repeated failures to conquer these fiercely independent peoples. The Spanish initially fared little better than the Inca in their attempts to subdue the Araucanians, who continued to control most of the territory south of the Bío-Bío River through the seventeenth and into the eighteenth centuries (Padden 2000). Chile's "Indian Wars" did not end until the 1880s, when some of the Araucanian lands were given reservation status.

In the meantime, the Araucanians had adopted the European horse. With considerable encouragement from the Chilean Spaniards, by the 1600s they were crossing the mountains and raiding the animals of the Argentine settlements of the eastern Andean foothills and western Pampa. By 1709, the raids were approaching the very gates of Buenos Aires. Not until the 1870s and 1880s were the Argentines able to defeat the Mapuche (as the Araucanians were called) and begin to open the Pampa to European development.

Araucanian material culture reflected the natural resources common to the home regions of the various tribes. Both the northern Araucanians, who called themselves the Picunche, and the Middle Araucanians, known as the Mapuche, herded llamas and practiced irrigated agriculture centering on the production of Andean grains and tubers, but they lacked terracing and other advanced farming techniques of the Inca. The southern coastal tribes called themselves Huilliche and practiced slash and burn farming within the dense forests. The peoples of the higher elevations to the east, called the Pehuenche, hunted guanaco and gathered Araucaria seeds, roots, and berries. The Huilliche were also expert fishermen and relied much on canoes for water transportation. Land resources were held communally throughout the Araucanian region, with settlements consisting mostly of tiny clusters of three to eight single-family dwellings.

Although the Araucanians' level of social and political organization was far below that of the Incas to their north, it was considerably more advanced than that of the **Chono, Alacaluf,** and **Yahgan** peoples, who occupied the cold, wind-battered islands from the Gulf of Corcovado south to the southern shores of Tierra del Fuego. Distinct in a linguistic sense, these groups were among the most technologically primitive peoples of pre-Columbian America. Their "homes" were, at best, large sticks bent across one another to form a crude frame upon which tree boughs or animal skins were cast to shelter the occupants from the driving rains and snows. Systematic agriculture was unknown, and their diets consisted mostly of shellfish, sea mammals, and marine birds and eggs. Population densities were very low, with the nuclear family forming the basic social unit. Dogs were the only domesticated animals, and travel was accomplished almost entirely by water. Interestingly, although the Chono, Alacaluf, and Yahgan went barefoot and wore almost no clothing to protect themselves from the frigid elements, they appear to have been surprisingly healthy and largely free of disease.

Eastern Lowland Peoples

In contrast to the western highlands, which supported relatively dense and, frequently, highly advanced pre-Columbian Indian populations, Latin America's eastern lowlands were characterized by low population densities and by more modest levels of indigenous social organization and material achievement. The various groups differed, nevertheless, in their adaptations to the distinct natural environments that they occupied (see Figure 7.2).

THE SOUTHERN HUNTERS

The low, rolling prairies of eastern and northern Tierra del Fuego, the arid tablelands of Patagonia, the plains of the Pampa, and the hill country of Uruguay were all occupied, in pre-Columbian times, by nomadic hunting and gathering peoples who organized themselves into small bands capable of pursuing herds of guanaco, rhea (*Rhea darwini* and *R. americana*), and other small game. The southernmost Indians, the **Ona,** controlled the lowlands of Tierra del Fuego, sharing the island with the sea-oriented Yahgan of the western highlands. To the north of the Ona, in Patagonia, were the **Tehuelche,** whose lives were altered to such a degree by the Mapuche-led Araucanian migrations of the late 1600s and early 1700s that eventually they were largely assimilated into the warlike, horse-oriented culture of the latter. The Pampa was home to the **Puelche** and Uruguay home to the **Charrúa** peoples. Like the Tehuelche and Ona to the south, the Puelche and Charrúa engaged frequently in inter-band warfare.

THE TROPICAL FOREST DWELLERS

In the pre-Conquest period, the tropical rain forest and savanna regions of South America and the Caribbean were home to literally thousands of Indian tribes and peoples. While neither organized into great political empires nor possessing advanced architectural or metallurgical skills, the forest dwellers had come to understand, in many ways better than we do now, the natural resources and limitations of the wet, lowland tropics, and they had developed sustainable utilization strategies that provided a rich and varied diet.

TABLE 7.3

Food Animals of Lowland Latin America

COMMON NAME	ZOOLOGICAL NAME
Collared peccary	*Tayassu tajacu*
White-lipped peccary	*Tayassu pecari*
Tapir	*Tapirus terrestris*
Howler monkey	*Alouatta seniculus*
Spider monkey	*Ateles geofroyi*
Woolly monkey	*Lagothrix lagothricha*
Capybara (rodent)	*Hydrochoerus hydrochaeris*
Agouti (rodent)	*Dasyprocta spp.*
Armadillo	*Dasypus novecinctus*
Opossum	*Didelphis spp.*
Raccoon	*Procyon lotor*
Coati	*Nasua spp.*
Giant river turtle	*Podocnemis expansa*
Giant otter	*Ptenoura brasiliensis*
Deer	*Cervidae spp.*
Anaconda (snake)	*Eunectes murinus*
Iguana (lizard)	*Iguana iguana*
Manatee (aquatic mammal)	*Trichechus spp.*
River dolphin (aquatic mammal)	*Inia geoffrensis*
Cayman (aquatic reptile)	*Paleosuchus spp.*
Pirarucu (giant fish)	*Arapaima gigas*
Muscovy duck	*Cairina moschata*
Curassow (bird)	*Mitu salvini*

A few of the groups, such as the **Gê** of interior eastern and central Brazil and the **Ciboney** of western Cuba and Hispaniola, were unsophisticated hunters and gatherers, possessing at best only incipient agriculture. However, most of the forest people practiced shifting cultivation that focused on the production of manioc or cassava and other lowland tubers, fruits, and vegetables. In addition, a great array of nuts, medicinal substances, and other products were harvested from the trees of the forest. Animal flesh was secured by hunting a wide variety of small game with lances, bows and arrows, pits, traps, and curare (*Strychnos toxifera*)-tipped darts shot from blowguns. Fish were speared, trapped, and stunned through the use of natural stupifiers or piscicides. Insects and mollusks also contributed substantially to the diet of many groups (Table 7.3).

Settlement was concentrated along the inland waterways and coastlines, and both short- and long-distance travel were accomplished by canoe. The largest of the boats may have been associated with the Caribs, whose dugouts averaged eighteen to twenty-four meters (sixty to eighty feet) in length and could readily transport forty to fifty persons at speeds exceeding six knots, prompting

FIGURE 7.12

Shipibo Indians of western Amazonia.

Columbus to report that they could easily outrun his ships at sea (Benson 1977). Some of the tribes preferred to construct individual residences for each nuclear family, while others resided in villages comprising four to eight large communal structures, each up to 150 meters (500 feet) long and capable of housing 200 or more occupants (Figure 7.12). Furnishings were simple, with hammocks used for sleeping and large woven baskets for storage.

Except in the Caribbean islands, warfare, female infanticide, and prolonged periods of sexual continence all contributed to low population densities. When the carrying capacity of the land was exceeded, the village generally divided, one of the groups migrating elsewhere. Yet another form of population control was ritual cannibalism, which was especially common among the Tupinambá and the Caribs. Both groups relied on the eating of captured enemy prisoners as a limited, specialized form of psychological intimidation and warfare.

The lowland forest peoples consisted of four great linguistic families (see Figure 7.2). The **Guaraní**-speaking peoples of the Paraguay area mixed with the Tupí of southern and eastern Brazil and the southern bank of the lower Amazon River to form a linguistic group known as the Tupinambá. The Tupinambá were the first Indian peoples encountered by the Portuguese missionaries, who reduced their language to a European alphabet and called it *Lingua Geral,* meaning the general or common language. *Lingua Geral* continued as the most widely spoken language of Brazil into the mid-eighteenth century and greatly enriched Brazilian Portuguese. Gê, the second major linguistic family of southern Brazil, was spoken mostly in the interior uplands.

Northern South America was dominated by the Cariban and Arawakan linguistic families, with the latter also extending southward along the eastern flank of the Andes into western Amazonia. The **Arawaks** overspread the Caribbean Islands between 300 B.C. and A.D. 1000, following which a separate branch, called the **Island Caribs,** emerged in the Lesser Antilles (Elbow 1992; Davis and Goodwin 1990; Rouse 1992; 1986). At the time of the European Conquest, the language of the Island Caribs consisted of an Arawakan grammar base with heavy infusions of northern South American Carib vocabulary.

Common Indian Traits

As we conclude our overview of the characteristics of the Latin American and Caribbean Indians on the eve of the Spanish and Portuguese Conquests, it is apparent that while they differed markedly among themselves, they also shared many traits. These included the concept of communal, rather than private, ownership or use of the land; reciprocal labor exchanges; residence for the most part in small, agrarian settlements; an emphasis on the production or gathering of human food crops rather than on animal grazing; adequate food supplies; surprisingly high population densities; and the absence—as best as we can determine—of major infectious disease.

This is not to suggest that the Indians lived idyllic and carefree lives. To the contrary, warfare among many groups was almost continuous, slavery was a nearly universal practice, and human sacrifice and cannibalism were widespread. Rigid social stratification was characteristic of the higher civilizations, and the lives of people everywhere were enshrouded by uncertainty and fatalism. Taken collectively, however, these problems, as serious as they were, pale in comparison to those soon introduced by the Spaniards and Portuguese.

Impacts of European Conquest and Settlement

As the Europeans arrived in the New World and set about establishing control over the Indians, they initiated a chain of events that, within less than 150 years, produced what may have been the most massive population decline in the history of all humankind. Precise estimates of the numbers of Indians who died as a direct or as an indirect result of the Spanish and Portuguese conquests are as difficult to develop as are estimates of the original total indigenous populations. Yet one respected eyewitness, Fray Bartolomé de las Casas, reported that over 40 million Indians had died by 1560 in the Spanish-controlled areas of Latin America alone. This excludes, of course, the losses experienced in Brazil and those that occurred subsequent to the date of his writing. More important than the exact mortality counts, however, is the realization that likely 90 to 95 percent of the Indians

---

TABLE 7.4

Post-Conquest Population Loss on Mainland Latin America

COUNTRY/REGION	Estimated Population (in 1000s)		SOURCE
	PRE-CONQUEST	POST-CONQUEST (YEAR)	
Central Mexico	10,000–25,000	1,000 (1595)	Borah and Cook 1963; Berdan 1982; Whitmore 1992
Chiapas	275	53 (1800)	Lovell and Lutz 1995
Guatemala	2,000	128 (1625)	Lovell and Lutz 1997
El Salvador	700–800	70 (1570)	Lovell and Lutz 1995
Honduras	800	48 (1690)	Newson 1986
Nicaragua	826	61 (1690)	Newson 1987
Costa Rica	400	1 (1682)	Lovell and Lutz 1995
Ecuador	1,508	217 (1690)	Newson 1995
Peru	9,000	620 (1620)	Denevan 1992b

---

who had come into contact with the Europeans had died by the year 1650. The greatest numbers of deaths among the Indians occurred in the densely settled highlands of Mexico and Peru, but the highest death rates occurred in the eastern lowland regions, some of which were almost totally depopulated within the first century and a half of the Conquest.

Recent studies of population loss within individual regions illustrate the calamitous declines that likely characterized all of Latin America and the Caribbean. Within the Greater Antilles, Hispaniola's indigenous population, which had probably numbered between 3 and 4 million prior to the Conquest, had been reduced to approximately 60,000 within two decades of the Spanish arrival. Similarly, Puerto Rico's 600,000 Indians had largely disappeared within a decade of Ponce de Leon's arrival. Jamaica, whose Arawak densities were close to those of Hispaniola's, found itself with only 74 Indians by 1611 (Watts 1987).

Indian mortality rates were also catastrophic on the mainland, with central Mexico's pre-Conquest population of between 10 and 25 million falling to approximately 1 million by 1595 and equally precipitous declines being reported for Central America and the Andes (Table 7.4). Even those Indian groups that were initially spared contact with the Europeans eventually experienced the same kinds of losses when their isolation ended. Denevan (1992a), for example, reported that in Amazonia the number of Tapirape Indians fell from 1,000 in 1890 to 147 in 1930. Crosby (1972) noted further that the number of Yahgan and Ona peoples of Tierra del Fuego had dropped from between 7,000 and 9,000 in 1871 to only 150 by 1947. So it was that the very physical survival of the Indians came to be threatened in much of Latin America.

Causes of Indian Depopulation

We have noted previously that as reports of the rapid decline of the Indian population filtered out of early colonial Latin America, the rival Protestant nations of northern Europe propagated what came to be known as the Black Legend, in which they hypocritically accused the Spaniards of intentionally killing as many natives as possible. No doubt individual *encomenderos* were indeed guilty of committing heinous crimes against the Indians "entrusted" to their care. However, it would be a gross distortion to suggest that the Spaniards as a whole were systematically endeavoring to destroy the very people that they hoped to Christianize and be supported by. In order to appreciate fully the reasons for the Indian death rates, then, it is necessary to probe deeper and to analyze the inadvertent consequences of the Conquest as they related to Indians.

The first reason was the basic incompatibility of Iberian landuse patterns with those that had sustained the Indians so effectively prior to the Conquest. The landuse changes were driven by several factors. The first was the Spanish *caballero* ethic, which resulted in the Europeans taking much of the best farmland of the Indians and converting it from human food crop production to pasture for the grazing of horses, cattle, and other large Old World animals. Grazing contributed far less to local food supplies than had the Indian farms (Figure 7.13). In this regard, one of the most tragic developments of the early colonial era was the inverse relationship between Indian and large Old World animal populations (Figure 7.14).

A related inadvertent cause of Indian deaths was the Spanish and Portuguese conversion of additional prime farmland to the cultivation of introduced Old World plantation crops, the great majority of which

FIGURE 7.13

The European conversion of intensively worked farmlands producing human food crops to pasture for the extensive grazing of large animals was one of the leading reasons for the food shortages and native population declines of the early colonial era. This scene is from Guanacaste province, northwestern Costa Rica.

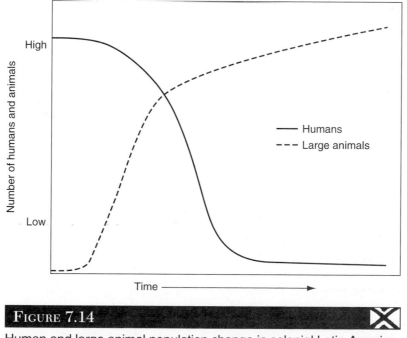

FIGURE 7.14

Human and large animal population change in colonial Latin America.

went to feed European peoples overseas rather than to supplying foodstuffs for local consumption. The result of both of these practices was a significant reduction in local food supplies. This, together with the removal of millions of Indian farmers from their homelands to be sold as slaves in the mines and cities and as human porters in such places as Panama, led to widespread food shortages and inevitably thereafter to malnutrition and death.

As serious as were the famines and labor dislocations, the principal unplanned cause of death among the Indians was the introduction of Old World diseases to which the natives had little or no natural immunity (Lovell 1992; Cook and Lovell 1991; Alchon 1991). Among the long list of communicable diseases brought by the Spanish and Portuguese were bubonic plague, measles, influenza, diphtheria, whooping cough, typhus, chicken pox, and tuberculosis. Malaria and yellow fever were introduced shortly thereafter through the Negro slave trade.

The greatest killer of all, however, was smallpox, which commonly took from one-third to one-half of the Indians of a given area within a few years of the arrival of the Europeans. We can also assume that some of the Indian deaths were caused, at least indirectly, by a progressive loss of a will to live under the traumatic changes to which they were subjected. It is further possible that the cataclysmic conditions resulted in lower than normal Indian birth rates, owing to a reluctance to bring children into such tumultuous circumstances and to the difficulty in some cases of finding a suitable mate.

Aftermath of the Conquest

Even as the Indian populations were dropping, the Spanish were rounding up those who had not fled and compelling them to relocate into new, nucleated settlements, variously called *resguardos, reducciones,* and *congregaciones,* from which the natives could be more easily controlled. Rather than dealing directly with the Indians themselves, which would have forced the Europeans to live without the amenities of urban life, the Iberians appointed the traditional local Indian governing class, such as the Inca *curacas* and Aztec *tlatoani,* as their representatives. These were responsible for the collection of tribute and the administration of forced labor, which in addition to being called *mita* came to be known by such euphemisms as "*accion cívica*" (civic service) and "*república*" (service to the republic).

From an Indian point of view, one of the problems with this system was that the *curacas* and other intermediate administrators were paid so little that they resorted to extorting and otherwise exalting themselves above their fellow Indians, such as by riding horses, carrying firearms, dressing in Spanish clothing, and eating and drinking from gold and silver vessels. The abuses of the *curacas* eventually became so severe that the Crown appointed men called *corregidores,* or correctors, who were charged with undoing the evils of the *curaca* system. Unfortunately for the Indians, however, the *corregidores* proved to be as corrupt as their predecessors. One particularly insidious practice instituted during the *corregidor* period was called the *reparto de efectos,* meaning the distribution of goods. Under this arrangement, the Indians were forced to purchase from the *corregidor,* at prices set by him, European products for which the natives had absolutely no use, such as silk stockings (presumably to be worn with one's sandals), razors (for a people who had little facial hair), snuffboxes, and, even though they were illiterate, pens and paper. Those Indians who could not pay for these "necessities" were informed that their cost was being added to their tribute debts.

Meanwhile, the Spaniards were petitioning the Crown for permission to acquire the "vacant" Indian lands now left unworked by the hapless natives. Over time, even those lands that the Indians were able to retain were taken from the *ayllus* and other traditional communal entities and assigned by misguided urban bureaucrats to individual native farmers. Because the Indians lacked experience in working within a capitalist system and had no control over the prices their produce would bring, they gradually fell into debt, lost their farms, and became serfs on the lands that had for countless generations supported their ancestors. Ultimately, the Indians, who prior to the Conquest had rarely wanted for food and who had built some of the most socially complex and intellectually advanced civilizations in the world, were reduced to little more than a nebulous, poverty-stricken rural underclass. Stripped of their dignity, they were compelled when encountering their European masters to silently bow and otherwise pretend to honor the very people who viewed and treated them almost as animals.

BLACKS IN THE NEW WORLD

Our study of race and social class has focused to this point on Europeans and Indians as components of colonial Latin America's ethnic mosaic. Before proceeding to assess current conditions, it is necessary to return to the early colonial era to trace the development of a third racial and ethnic group—the blacks, or Negroes, imported forcibly from Africa as slaves.

Antecedents of Latin American Black Slavery

To appreciate why the Spaniards and Portuguese resorted to the enslavement of blacks in the New World, it is helpful to recognize that the practice of enslaving peoples of other racial or ethnic or political backgrounds, as repulsive as it is to us today, is of great antiquity and was common to most if not all of the high civilizations of the Old World. For instance, slaves were widely used in the Greek civilization as urban artisans, and constituted an estimated 30 to 40 percent of the population of the Italian peninsula at the height of the Roman Empire, when the growth of a regional Mediterranean market economy created severe shortages of free laborers (Klein 1986). When the Spaniards and Portuguese arrived in the New World, they immediately set about enslaving the Indians living on the islands of the Bahamas and the Lesser Antilles, rationalizing their behavior by persuading the Crown to classify these lands as "useless" since they lacked gold deposits. Las Casas estimated that more than 2 million Indians from these regions were imported as slaves into Hispaniola's mining and plantation districts between 1509 and 1519 (Watts 1987). Radell (1992) calculated that approximately 500,000 Indians were removed from Nicaragua as slaves between 1527 and 1548 for sale in Panama and Peru.

At the same time that great numbers of Indians were being enslaved in the New World, the Portuguese were already actively engaged in the trafficking of black African slaves to Europe and to the sugar-producing Madeira, Cape Verde, and Azore islands of the eastern Atlantic. The terrible loss of Indian life in the Americas coincided with Spanish legislation intended to protect the surviving natives. So the Europeans turned instinctively to Africa as the source of replacement labor, the demand for which increased dramatically following the adoption of sugar cane as the predominant plantation crop of the Latin American and Caribbean lowlands.

The African Slave Trade

As part of its mercantile system, Spain initially insisted on transporting its own slaves to the New World in its own ships. It soon found, however, that the Portuguese control of most of the west African coastline greatly hindered its efforts. Spain then turned to an *asiento,* or licensing, system in which it contracted with British, Dutch, French, or Portuguese carriers to deliver a specified quantity of slaves to a designated port over a given period of time. By the eighteenth century, slaves had become the leading export from Africa to the Americas, with each of the European nations, Spain excepted, focused on supplying its own New World colonies (Davis 1995).

It is important to recognize that the blacks brought as slaves to the Americas were as diverse in their own cultures as the Indians they were replacing. The initial source region centered on Senegal and Gambia and the west coast of northern Africa. As time passed, the principal purchasing region shifted first to the central

FIGURE 7.15

A black mother and her children residing in the interior hill country of northeastern Jamaica.

African coast bordering the Gulf of Guinea and later to the Congo and Angola in the southwestern part of the continent. By the early 1800s, slaves were being secured even from Mozambique and other points along the Indian Ocean.

Regardless of where the slaves were obtained, their sufferings were beyond description. After being captured and torn forever from their loved ones, they were driven to the port cities where they were branded and baptized. Approximately half of them died as they awaited the terrible "middle passage" across the Atlantic in ships called "slavers" or "coffin ships" (*tumbeiros*) (Russell-Wood 1998). Living conditions on the ships were horrendous, with extreme overcrowding, trauma, appalling sanitation, disease, and contaminated food and water all contributing to death rates that ranged from 5 to 55 percent. While we will never know the exact number of Africans who died while en route from their homelands to the New World, they surely numbered in the millions. Their destinations were primarily the lowland plantation zones of Middle America and Northeast Brazil (Figure 7.15).

Living Conditions in the New World

Although slavery under any circumstances was an inherently immoral practice, the Africans brought to the New World were subjected to legal codes that varied according to the dominant European culture into which they were placed. Blacks who were sold to Portuguese or Spanish owners, for instance, were guaranteed certain rights and protections by virtue of Roman legal precedents. These included: (1) the right to life, meaning that they could not be legally killed by their masters;

TABLE 7.5

Growth of Free Black Populations in Latin America

COUNTRY/REGION	YEAR	WHITES	BLACK SLAVES (IN 1000s)	BLACK FREEDMEN	FREEDMEN % TOTAL BLACK POPULATION
British West Indies					
Barbados	1645	18.1	5.7	0	0.0
	1690	20.0	60.0	0	0.0
	1768	16.1	66.4	0.5	0.8
	1833	12.8	80.9	6.6	7.5
Jamaica	1698	7.4	40.0	0	0.0
	1768	17.9	176.9	3.5	1.9
	1834	15.0	310.0	35.0	10.2
French West Indies					
Martinique	1664	2.7	2.7	0	0.0
	1751	12.1	65.9	1.4	2.1
	1826	9.9	81.1	10.8	11.8
St. Domingue	1681	4.3	2.3	0	0.0
	1754	14.3	172.2	4.7	2.7
	1791	30.4	480.0	24.0	4.8
Hispanic Latin America					
Panama	1778	29.5	3.5	33.0	90.4
Mexico	1810	NA	10.0	70.0	87.5
Brazil	1872	3,800.0	1,500.0	4,200.0	73.7
Puerto Rico	1775	29.3	7.5	34.5	82.1
	1860	300.4	41.7	241.0	85.3
Cuba	1792	153.6	64.6	54.2	45.6
	1877	1,032.4	211.3	265.6	55.7

Sources: Schmidt-Nowara 1999; Watts 1987; Klein 1986; Rogozinski 1999.

(2) the protection of black women and children from abuse by their masters; (3) the right to own personal property; (4) the right to enter into personal contracts; and (5) the right to purchase their personal freedom. These privileges greatly exceeded those available to slaves living within the English- and French-controlled portions of the Caribbean, where, for example, the harsh French Code Noir denied to slaves the right to possess personal property, the right of personal protection, and even the right to marry without the approval of the master.

The right of slaves to purchase their own freedom in Iberian America was most significant and was strengthened by the declaration of the Roman Catholic Church that the blacks did indeed possess immortal souls and therefore were entitled to receive the Church sacraments. This position, in turn, forced the Portuguese and Spanish slave owners to give their slaves Sundays and holidays off, which many of the blacks used to tend small personal plots of land, the proceeds of which were saved for their eventual purchase of freedom.

A slave could also obtain his or her freedom by having it granted freely and without restriction by the slave owner, a practice called **manumission.** Those so favored tended to be females, American-born creole blacks rather than African-born *bozales,* young children, and persons of mixed racial ancestry called *pardos* rather than pure Negroes. These individuals were often, but not always, the black mistresses, children, and other relatives of the master (Nishida 1993). Because slave self-purchase and manumission were common practices within the Portuguese and Spanish spheres, Hispanic Latin America's black population came over time to be dominated by freedmen. By the end of the colonial era, for example, free blacks outnumbered slaves in Mexico and Panama by a ratio of approximately nine to one. In Brazil, which had by far the largest black population of any political entity in Latin America, there were three times more free blacks than slaves in 1872, and indeed free blacks outnumbered whites in the country. These figures stand in stark contrast to those of the Caribbean colonies of Great Britain and France, where freedmen generally made up only 5 to 10 percent of the black population (Table 7.5).

Case Study 7.2: Maroon Societies in Latin America

From time to time during the colonial era, relatively large numbers of escaped slaves, called Maroons in English-speaking areas and *cimarrones* in the Hispanic regions, were able to band together and establish villages and towns whose freedom was eventually recognized by the European rulers in exchange for the Maroons agreeing to support the policies of the government. One of the largest of the Maroon movements began in the backlands of the northeastern Brazilian Captaincy of Pernambuco during the period of Dutch control in the early 1600s. Here a number of small, fortified agricultural settlements, locally called *quilombos*, banded together to form the independent Palmares Republic, which at its zenith in 1690 contained a population of over 20,000, an army, and a king who collected taxes from his subjects.

In the Guiana country of northern South America, thousands of slaves were able to escape into the densely forested interior during the Dutch and British wars of the seventeenth and eighteenth centuries. By the early 1700s, over 6,000 so-called Bush Negroes were residing in areas claimed by the Dutch. Recognizing the futility of their efforts to recapture the former slaves, the Dutch finally granted them freedom and the right to self-governance in exchange for the Bush Negroes promising to return any future escaped slaves who might attempt to join their communities. Previously, in Jamaica, the English had offered freedom to any slaves who would join them in driving out the Spanish. To the dismay and surprise of the English, however, many of the Negroes refused to aid the British and fled to the interior mountains, where they joined existing Maroon communities that became a symbol of black determination to resist white domination.

Yet another major center of Maroon activity developed in the interior escarpment country of central Veracruz state, Mexico. The most prominent of the *cimarrón* bands was founded by a man named Yanga, who established a *palenque*, or settlement, about 1580 in the region bordering the Veracruz-Mexico City highway. Yanga and his men were so successful in capturing Indian and black women that the *cimarrones* eventually found it necessary to settle down with their wives and children and abandon their raiding in favor of farming and commerce. The Spanish government then struck a deal with the village leaders, granting the community the status of a legal *pueblo* officially named San Lorenzo de los Negros, while establishing the nearby city of Córdoba as a residence for the elitist European families that eventually governed the region.

Forms of Slave Resistance

Despite the much-publicized "paternalistic" nature of Portuguese and Spanish slavery, those who were not freed were still in bondage and protested their servitude in a variety of ways. The least obvious but most widespread was passive resistance, which included behavior such as feigning illness or laziness and sabotaging crops and property. In the British and French Caribbean plantation regions, especially, resistance also included the oral expression through poetry and song of black hatred of whites.

Violence was a second form of slave protest. While not as common as passive resistance, rebellions did occur from time to time, generally as acts of last resort. Most of these had little lasting impact on the region as a whole. The obvious exception is the slave uprising on Haiti, or Saint Domingue, which inspired a number of other rebellions in the greater Caribbean. Violence could also be individual, as in a physical assault by a slave on his or her master or overseer. It also took the form of suicide among the black populations.

A third option available to many slaves was to attempt to escape, an undertaking that was called marronage. Most of the escapes were individual acts of short duration, with the slave being caught while hiding near the plantation or mine. On occasion, however, relatively large numbers of slaves were able to flee during periods of political turmoil into physically isolated backlands, where they successfully resisted all European attempts to recapture them (Price 1996). Mass escape was most feasible on the mainland or on the larger Caribbean islands, and powerful **Maroon** societies were established during the colonial period in the interior forested regions of Northeast Brazil, British and Dutch Guyana, Colombia, Mexico, and Jamaica (Case Study 7.2). The Lesser Antilles offered far less physical space for escape and defense, and many of the slaves who fled these islands sought refuge along the sparsely settled Caribbean lowlands of Central America, where they subsequently intermarried with local Indian peoples.

Abolition

With the granting of freedom to large numbers of slaves, the escapes and premature deaths of many others, and a low black birth rate brought about by there being only half as many females as males brought out of Africa, the slave populations seldom reproduced

TABLE 7.6

Estimated Slave Imports into Latin America and the Caribbean

REGION	TOTAL
Brazil	3,646,800
Saint Domingue	864,300
Cuba	780,200
Jamaica	747,500
Dutch Caribbean and Guiana	550,000
Barbados	387,000
Martinique	365,800
Guadeloupe	290,800
Mexico	200,000
Colombia, Panama, Ecuador	200,000
Venezuela	121,000
La Plata and Bolivia	100,000
Peru	95,000
Puerto Rico	77,000
Grenada	67,000

Sources: Compiled from Curtin 1969; Russell-Wood 1998; Schmidt-Nowara 1999; Rogozinski 1999.

themselves. This led to the continuing importation of new slaves during the colonial era. All told, some 9 million Africans were taken as slaves to Latin America, with the sugar-dominated Brazilian and Caribbean lowlands receiving the great majority (Table 7.6). With respect to the European spheres of influence, Portuguese Latin America took approximately two-fifths of the slaves, the remainder being almost evenly divided between the French, British, and Spanish realms.

However, pressures were building in Europe that would eventually lead to slavery's collapse. Two of the most powerful intellectual movements of the eighteenth century, the French Enlightenment and British millenarian Protestantism, argued that slavery was both irrational and immoral. Economic theorists soon added their opposition on grounds that slavery was not a viable economic institution in light of the cyclic price fluctuations for the crops generated by the Latin American and Caribbean plantations. Slowly but surely then, abolitionist movements gained momentum, which led in the 1830s to the banning of all forms of slavery in the British and French empires. This was followed by abolitionist legislation in most of the independent Latin American nations in the mid-1800s and finally in Brazil in 1888. Freed at last, large numbers of the former slaves fled the agrarian countryside, which had come to symbolize in their minds all the sufferings and wrongs that they and their ancestors had for so long endured, and moved to the towns and cities where they became laborers, craftsmen, and merchants.

EAST ASIANS IN THE CARIBBEAN

With the actual or impending loss of their slaves, the Latin American planters turned to the importation of indentured workers as yet another form of inexpensive agricultural labor. The initial source regions were East and South Asia. Substantial numbers of Chinese "coolies," as all the Asians came to be called, were introduced into the sugar zones of coastal Peru and Cuba, and many Javanese were brought by the Dutch to Suriname. The British began transporting Indian and other South Asian laborers to Guyana, Trinidad, Jamaica, and other Caribbean destinations in the late 1830s and early 1840s and continued into the early twentieth century. By then, approximately 800,000 Asians had been brought to the New World as contract laborers.

Asians were willing to come to Latin America for various reasons: the desperate poverty of their native lands, unhappy family circumstances, the confinement of the Hindu caste system, and the perceived opportunities of beginning a new life in a faraway place. Male immigrants outnumbered females by a ratio of approximately 3.5 to 1.0 (Lai 1993). Reasons for the severe gender imbalance included the reluctance of Indian men to allow their wives and daughters to migrate to the New World where they would be subjected to the uncertainties of indentured life, and the Indian custom of child betrothal and marriage at puberty, which greatly reduced the number of potential single female immigrants. In regard to this, Dabydeen and Samaroo (1987) have noted that in Bengal in the 1880s, over 93 percent of all Hindus were listed as married before reaching the age of 14.

Although the specific contractual arrangements varied over time and between culture spheres, most of the Asian immigrants were given free passage to America in exchange for a commitment to work for five years (three years or less for women) on a plantation or other agricultural enterprise. Wages were miserable and based on piecework in order to force the workers to put in the longest possible hours. Housing consisted mostly of former slave quarters, and the absence of adequate health care and sanitation resulted in repeated outbreaks of hookworm and infectious diseases. As a consequence of the shortage of Asian women, the unwillingness of most Asian men to enter into sexual unions with black or American Indian women, and, in the case of the British colonies, European-placed obstacles to Hindu and Muslim marriages, Asian family ties for the first few generations were generally weak. Spouse betrayal and polyandry, which is the practice of women having more than one husband or male companion, were common, and led to widespread physical abuse of women by jealous lovers.

FIGURE 7.16

Following their release from indenture, many East Asians became successful business owners. This establishment is in Panama City, Panama.

Upon completing their terms of service, the Asians could elect to reenlist, remain in their adopted countries as independent farmers or businessmen, or accept an offer of free passage back to the lands of their birth. Generally, fewer than a third chose the latter course, which in Hindu India required the returning person to submit to a purification ceremony and to spend much of his or her hard-earned savings on a community feast. Most instead opted to remain in their new homelands, where they became highly productive smallholder farmers, urban shopkeepers, traders, restaurateurs, and other petty capitalists, and in later generations white-collar professionals (Rustomji-Kerns 1999; Clarke 1986) (Figure 7.16). Intense and bitter hatreds often developed between them and the black Creole populations, who came to view the "East Indians" as clannish, materialistic, and a threat to Afro-American political dominance.

A smaller, yet historically significant, source of indentured laborers in the greater Caribbean was Portuguese peasants from the Madeira, Cape Verde, and Azore Islands. Indentured labor was also largely responsible for opening the coffee frontier of southern Brazil and the wheat and cattle grazing regions of the Argentine Pampa. In these latter instances, the great majority of the immigrants were Italians, whose arrival simply reinforced the predominantly Caucasian racial composition of those regions.

RACIAL MIXING

Having traced the origins and characteristics of the principal races of Latin America and the Caribbean, we will now turn to an analysis of racial mixing, or **miscegenation,** and its consequences throughout the re-

gion. We have noted previously the shortage of European women in the early colonial period and the willingness of the Spanish and Portuguese men to mix genetically with women of darker skin color. With the exception of many of the Catholic clergy, who remained true to their vows of celibacy, most of the European men who settled Latin America entered into one, and often far more than one, union with Indian and/or black women.

The result was the growth, from the beginnings of the colonial era, of mixed races, or *castas*. These went by many and varied names. Persons of mixed European and American Indian ancestry were called **mestizos** in Spanish Latin America and **mamelucos** or *caboclos* in the Portuguese sphere. The offspring of white and black unions were called **mulattos** or coloreds, and those of black and brown parentage, **zambos** or *pardos*.

Because race correlated almost perfectly with socioeconomic class standing in the colonial era, the elitist Latin Americans soon developed an intense fixation with its physical expressions. The greatest prestige within the social pigmentocracy was accorded those persons with European physical features including light-colored skin, straight hair, and thin nose and lips. The further one's body appearance deviated from this ideal, the lower one's socioeconomic standing was perceived to be by others.

Because of the almost limitless number of genetic crossings that occurred as *mestizos, mulattos,* and *zambos* mixed with one another and with Europeans, blacks, and Indians, an incredible number of additional racial subgroups evolved in what was fast becoming the most racially heterogeneous of all the world's major culture regions. The capacity of the priests to maintain consistency in assigning young children their racial classification was often severely tested. In one instance reported by Carroll (1991), a well-intentioned seventeenth-century Mexican cleric designated twin sisters as belonging to different races—the one being listed in the parish records as a *parda* (mixed black and Indian parentage) and the other as an Indian.

The Evolution of Social Races

The obvious problem with having so many racial subgroups was that their names and categories eventually came to mean different things to different people. This led to the construction of social or cultural races in which one's racial standing was determined more by one's lifestyle than by one's physical features.

Under this arrangement, occupation, language, mode of dress, diet, and religion became the principal determinants of race. In Peru, for example, Indians, or *naturales* as they were derisively called by the Spaniards,

FIGURE 7.17

Indian mother and daughter of the Ecuadorian highlands. Note the mother's use of a hat, her long braids, and other native attire. The donkey, or *burro,* was a European introduction that is now in wide use among rural Latin American peoples.

came to include anyone who was a small-scale farmer and who spoke an Indian language or dialect. Furthermore, Indians dressed differently. Both men and women wore hats, the latter over long braided hair. Indians either went barefoot or wore sandals, the women used shawls, and the men wore vests, jackets, and knee-length trousers patterned after Spanish colonial dress (Figure 7.17). Indian diets consisted largely of potatoes and other native tubers and grains (Figures 7.18 and 7.19). Even the Indian form of folk Catholicism varied significantly from the formal European type (chapter 9).

Europeans, on the other hand, arrogantly referred to themselves as *gente decente* (decent people), *gente civilizada* (civilized people), or *gente de razón* (people capable of reasoning or thinking)—as if to say that no other groups could possess those traits. Europeans, of necessity, resided in urban areas, would never allow themselves to take anything but a white-collar job, and spoke only Spanish. Their diets were based on wheat bread and animal products, and their involvement with the Church was carefully calculated to enhance their socioeconomic standing (Figure 7.20).

Occupying an intermediate position in the caste-like Peruvian social structure were the ever-growing number of *mestizos.* Biologically, a *mestizo* was an individual of mixed Spanish and Indian ancestry, but culturally the term came to refer to a middle-class person who lived in a town or city, spoke Spanish, knew how to read and write, often acted as an economic intermediary, and purposely allowed one fingernail to grow long as the sign that he or she was avoiding the social degradation of manual labor. Although attempting publicly to appear to live as Europeans, poverty forced

FIGURE 7.18

Maize and maize tortillas historically have been Indian foods in Latin America.

FIGURE 7.19

Quinoa is a highland Andean herb, which tolerates well extremes of cold and drought. Its seeds have long been consumed by Indians of the region.

FIGURE 7.20

Wheat was introduced into Latin America by the Spaniards and historically has been associated with upper-class European diets in the highlands of Latin America.

FIGURE 7.21

These young Maya girls of the western Guatemalan highlands are dressed in *traje,* or traditional Indian clothing. Their younger brother's shirt is *vestido,* or Western-style clothing.

the *mestizos* to rely mostly on an Indian diet and to live in small apartments or houses in the poorest sections, or *barrios,* of town. Generally speaking, their involvement with institutional Catholicism was minimal.

Significantly, Latin American use of social races has made it possible for individuals to change their racial classification by altering their lifestyles. A person living in the Guatemalan highlands, for instance, can be identified as a Maya Indian by wearing traditional clothing known as *traje,* or as a *Ladino* (*mestizo*) by dressing in Western-style clothing called *vestido* (Hendrickson 1995; Schevill, Berlo, and Dwyer 1996) (Figure 7.21). In other parts of Latin America an Indian can begin the process of becoming a *mestizo* simply by putting on a pair of Western shoes, throwing away his hat or woolen cap, and taking up residence in the city. There he will communicate in Spanish and adopt, as much as possible, Western ways (Figures 7.22 and 7.23). In modern Peru, persons who are endeavoring to make the Indian-to-*mestizo* transition are called *cholos.* In other parts of Latin America, one moves directly from Indian to *mestizo* standing. In either case, the *cholos* or lower-class *mestizos* are employed as manual laborers, be it as street sweepers, market vendors, factory workers, or domestic servants.

RACIAL WHITENING

Although it is possible for an individual to move down the racial ladder through the voluntary or involuntary adoption of a lower-class lifestyle, historically the vast

FIGURE 7.22

An Indian woman of the highlands of central Mexico.

majority of changes in racial classification, such as from Indian to *mestizo,* have been upward. As a result, the population has become more and more European-like, with corresponding declines in the numbers of Indians and blacks. This process of upward mobility is widely referred to as "whitening," or "bleaching," and has had both physical and cultural dimensions.

FIGURE 7.23

The *mestizo* son of the Indian woman pictured in Figure 7.22.

Latin America's physical whitening began with the catastrophic losses of Indian life in the early colonial period and continued with the failure of the enslaved black race to reproduce itself. One factor in the whitening, the shortage of females, was discussed previously. Gudmundson (1986) has noted that in Argentina and Costa Rica, black population declines were also attributable to black women marrying at a later age than their white counterparts, to lower black birth rates, to higher infant and adult death rates among blacks, and to racial miscegenation (Table 7.7). In sum, blacks were less likely to marry and bear children, more likely to lose the children they did give birth to, and more likely to marry someone of another race, thus lightening their skin color from one generation to the next. Another factor that contributed to the virtual disappearance of blacks in Argentina was the deaths of large numbers of black men while serving in military campaigns (Andrews 1995). The result was that the black percentage of the total population of Buenos Aires dropped from 25–33 percent in the early 1800s to only 1.8 percent in 1887. Likewise, blacks in Costa Rica fell from 10–20 percent of the total population at the start of the nineteenth century to almost none a hundred years later. They then increased again in the early twentieth century as West Indian blacks were imported as railroad laborers and workers on the Caribbean banana plantations. Similar declines of black populations occurred in many other Latin American regions.

The physical whitening of Latin America further accelerated in the late nineteenth and early twentieth centuries with the introduction of a philosophy developed originally by the English philosopher Herbert Spencer and known as **Social Darwinism.** Advocates of Social Darwinism argued that human societies could be compared to competing natural species, only the fittest of which would ultimately survive. Social Darwinism was enthusiastically embraced by the Latin American elites who interpreted the term "societies" to mean "races," and who convinced themselves that the only way that Latin America could ever develop economically would be through the growth of the white race. This led to systematic efforts in many nations, including Argentina, Uruguay, Brazil, Venezuela, Costa Rica, Mexico, Cuba, and the Dominican Republic, to "improve the race" by both limiting the number of blacks admitted into the country and encouraging white indentured and nonindentured immigration.

In addition to an ongoing physical whitening, social whitening or bleaching continues to transform the Latin American racial landscape. "Money whitens" is a common phrase throughout the region, and it is equally true of high political power. It was said, in this context, that Porfirio Diaz, the dark-skinned dictator of late nineteenth-century Mexico, "grew whiter with age." Wagley (1971) observed that in a small Amazonian community, the most important woman in town was biologically a *mulatta*, but the townspeople consistently referred to her as "white"; conversely, the town drunk, who was physically white, was never recognized as such.

Generally speaking, social whitening occurs in small rather than large increments. It is almost impossible for a person who is physically black skinned to become a white, but it is relatively easy for a black to become a *mulatto* or for a fair-skinned *mulatto* to become a white. Yet another common vehicle for social whitening is "marrying up," meaning being accepted into the social circles of a lighter-skinned spouse. Taken collectively, physical and social whitening have progressed to the point where many Latin American nations now claim, in contrast to their past, to have almost no Indians or blacks among their populations. Whether these statements are factual or not in a biological sense, they are symptomatic of a widespread rejection of non-European cultures. It is this rejection that forms the basis of present-day racism in the region.

RACISM AND ITS RAMIFICATIONS

The widespread physical mixing of the region's peoples and the opportunities for upward mobility afforded by the use of racial groupings based upon culture and lifestyle have combined to create in Latin America and the Caribbean a very different set of racial circum-

TABLE 7.7

Black and White Demographic Patterns in Nineteenth-Century Buenos Aires, Argentina

CATEGORY	BLACK	WHITE
Average age of first marriage of females	29	21–22
Average number of children under 5 years of age per 1,000 women aged 15–44	183	365
Average infant mortality rate of children less than 1 year of age per 1,000 live births	350	284

Source: Gudmundson 1986.

stances than what evolved in the United States. Whereas discriminatory laws and practices based on a person's physical appearance long existed in the United States, Latin Americans historically have rejected such behavior. In so doing, they have often been quick to praise themselves for what they have viewed as their own progressive state of race relations, and they have been equally enthusiastic in criticizing race relations in the United States, where even a small degree of non-white ancestry is sufficient to have one classified as a racial minority. Without minimizing the seriousness of the United States' problems, we should in fairness, when discussing race relations in the Americas, ask ourselves if prejudice grounded in social race is any less evil, or its effects any less hurtful to its victims, than prejudice based on physical race. The answer of course is no. Unfortunately, Latin America today is afflicted every bit as much by social racial prejudice as the United States has been by physical racial prejudice.

Evidence of this is everywhere. Perhaps the most fundamental and most obvious—even to the casual visitor to the region—is the fact that the darker one's skin color, the lower one's social class and the poorer one's economic circumstances are likely to be (Telles 1992; Wright 1990). Throughout the black Caribbean realm, for instance, the upper class is still dominated by whites and lighter-skinned *mulattos*. The same pattern holds true for the predominantly *mestizo* and Indian countries of the mainland, where the old European families continue to exploit the darker-skinned masses.

Those who understand the use of language and greetings in Spanish-speaking Latin America are aware that racism is widely perpetrated in conversations through the use of either formal *usted* or familiar *tú* verb and pronoun forms. Normally the formal *usted* is used to address persons to whom one is not well known, and *tú* is used among family and close associates. An additional context occurs, however, in the area of race and social class relations, where by using the *tú* form, a person can communicate in a very brutal yet subtle way his or her self-ascribed superiority to the person judged to be inferior. Thus, Europeans and *mestizos* use *tú* when speaking to Indians and other "inferior" racially mixed groups, while the latter are expected to respond with *usted* as an outward sign of deference. Similarly, an upper-class European is generally reluctant to bestow an embrace (*abrazo*) or formal kiss on the cheek of a social inferior, and if a European offers to shake the hand of an Indian, the latter is expected to hold his own out limply as the superior grasps and releases it.

Numerous other examples of racism could be given. Perhaps the most tragic is the sense of inferiority and even self-contempt that evolved among the blacks and Indians. As Lowenthal (1972, 250) noted with reference to the Caribbean: "when blackness meant slavery and whiteness meant mastery, it was little wonder that blacks wished to be white." In other words, one of the most insidious historical consequences of black and Indian slavery has been black and Indian repudiation of their own culture and, in a very real sense, repudiation of themselves in an attempt to survive and perhaps even advance a little in a white-dominated world.

This repudiation of darkness expresses itself in many incongruous ways. Television programs and motion pictures are filled with light-skinned actors cast in upper-class settings totally removed from the everyday life of the common people. Great amounts of money continue to be spent by rich and poor alike on lotions and powders designed to lighten one's natural skin coloring. Hair straightening is widely practiced in black areas, while upper-class Europeans warn their children not to stay out too long in the sun and envy those who bear blond-haired offspring. Light-skinned beauty pageant contestants fare better than do their dark-skinned competitors. European-style business suits are worn by the socially ambitious regardless of the physical discomfort. In short, everything European—from mannerisms to sports to educational and political philosophies—has been emulated, and that which is Indian or African deprecated.

FIGURE 7.24

Mexico's National Museum of Anthropology is one of many expressions of *indigenismo* in Latin America.

Inevitably, the collective identity crisis has sparked nativist reactions. The two most influential of these have been *indigenismo* and *négritude,* which have responded to the historic abuses of Latin America's dark-skinned peoples by arguing that the Indian and black races are, in reality, superior to the European. By glorifying and, in some respects, romanticizing the Indian past, ***indigenismo*** has been responsible for the return of some farmlands to Indian communities, called *ejidos* in Mexico and *comunidades indígenas* or *comunidades campesinas* in Peru. It has also provided the rationale for an ongoing movement to preserve Indians and Indian ways through such diverse mechanisms as teaching Indian history in schools, preserving modified indigenous cultural practices, and establishing Indian reservations, homelands, and museums (De la Cadena 2000; Fischer and Brown 1996; Field 1994) (Figures 7.24 and 7.25) (Case Study 7.3). While achieving some success, the movement has been hampered by the fact that it has been and continues to be led and defined by diverse non-Indian interests. It is ironic, for instance, that much of the teaching of Latin American youth about the greatness of their Indian heritage takes place in the Spanish and Portuguese languages, thereby contributing to the long-term disappearance of the cultures the instruction is allegedly intended to strengthen.

Négritude, like *indigenismo,* originated as a romanticized reinterpretation of the past by upper-class intellectuals, specifically Europeanized Caribbean coloreds or *mulattos* who, with few exceptions, had no intention whatsoever of living like those they were glorifying. It has consequently taken a number of forms, including a literary movement extolling the richness of black culture and its closeness to nature, a

FIGURE 7.25

This stone monument to past, present, and future Indian greatness is located in the main plaza of Otavalo, Ecuador.

variety of Jamaican back-to-Africa movements including Rastafarianism and Reggae music, and the radical black power movements of the 1970s. Its failure to attract large numbers of followers can be attributed to the fact that, by extolling all things African and attacking European values, it has placed itself in open opposition to the majority of the Caribbean peoples, who do not see themselves as blacks in a cultural sense.

In summarizing race relationships in Latin America, it is important to point out that, despite the problems, racial prejudice has been reduced over the past several decades. Much of the progress has occurred as a natural outgrowth of the social and political restructuring that has come to the region. It is to be hoped that racism and prejudice will decline even further as the restructuring process continues.

Case Study 7.3: Indian Homelands and the Mesoamerican Biological Corridor

As tropical rain forests and other natural habitats continue to disappear at alarming rates in much of Latin America, the indigenous peoples who have resided within these physical environments are increasingly threatened with cultural extinction. Nowhere is this process more evident than in Central America, where it is estimated that only 45 traditional Indian groups survive, the most numerous of which are branches of the Mayan linguistic family residing in Guatemala. Until the mid to late twentieth century, the Central American elites as a group showed little concern for the plight of the lower-class Indians residing in distant wilderness areas, and indeed tended to view the loss of Indian peoples as an expression of economic and cultural modernization. This long-standing disregard by government leaders for the welfare of the Indian masses began to change, however, with the emerging recognition that the region's remaining rain forests would soon disappear if uncontrolled clearing was not halted. The cultural survival of the Indians thus came to be inseparably tied to the physical survival of their biologically rich tropical homelands.

By 1992, the Central American nations had designated 162 protected areas and 71 limited use areas whose collective holdings totaled 16 percent of the region's surface area. These units functioned under a variety of classifications, including biosphere reserves, marine and coastal reserves, Mayan archeological sites, national parks, and indigenous reserves. Some of the more prominent included the Maya Biosphere Reserve of northern Guatemala and neighboring portions of southeastern Mexico and Belize, the Copán Ruins of northern Honduras, the Miskito Cays Marine and Coastal Reserve of northeastern Nicaragua, Tortuguero and Guanacaste national parks and the Cocles-Kekoldi Indigenous Reserve of Costa Rica, and the Darién National Park of eastern Panama.

While the designation of protected and limited use areas is an essential first step in conserving the cultural and biological diversity of Central America, numerous obstacles limit the effectiveness of the well-intentioned legislation. Chief among these are the pervasive poverty of the region and its rapid population growth, which together have led to accelerated *mestizo* colonization and deforestation within many of the restricted use areas. The absence of management plans and poorly defined boundaries are other obstacles to the sustainable development of the protected zones, as are graft and bureaucratic ineptitude, which have led to rampant illegal commercial logging. National governments have also continued in many areas with plans to develop the mineral resources of the reserves. Somewhat surprisingly, the leaders of the indigenous federations have generally opposed attempts to create cultural reserves on grounds that they would further isolate and marginalize their occupants while offering little or no local autonomy to the Indians they are supposed to benefit. In this regard, Panama's *comarca* homeland concept may prove to be an acceptable alternative development model. It consists of granting the native peoples broad authority in regulating local cultural, political, and economic matters while requiring them to work with national authorities in the establishment of basic resource utilization policies.

While the long-term cultural survival of the indigenous peoples is uncertain, a consensus is emerging among conservationists that the best hope of slowing the loss of the region's immense biological diversity would be to create a Mesoamerican ecological corridor connecting the separated protected areas that have hitherto existed as biological and cultural islands. The environmental land bridge would permit the unimpeded movement of endangered plant and animal species, thus increasing their chances for survival. Originally called *El Paseo Pantera* (the path of the panther), the Mesoamerican Biological Corridor is now receiving funding from a host of national and international agencies. Although enormous obstacles remain, the *Paseo Pantera* has the potential to rank with the Inca Trail of the South American Andes and the Appalachian Trail of the United States and Canada as examples of the successful blending of cultural and biological conservation with sustainable economic development.

Sources: Lovejoy 2000; Herlihy 1997 and 1989; Illueca 1997; Sundberg 1998.

CURRENT RACIAL PATTERNS

We now summarize the current racial patterns of Latin America and the Caribbean (Figure 7.26). In doing so, we will use a minimum of statistics, since ever-evolving social perceptions of racial types mean that most quantitative country profiles are very subjective and of little comparative value. One example of how imprecise racial categories are in Latin America emerged from a study conducted by Harris (1964) of racial perceptions among the people of a small fishing village of northeast Brazil. In the study, a sample of adult townspeople was shown a set of nine portrait drawings, each differing in hair shade, hair texture, nasal and lip width, and skin tone.

Figure 7.26

Generalized racial patterns of Latin America.

The 100 participants responded with 40 different named racial types to describe the nine people shown in the drawings. One drawing elicited 19 racial categories! In another part of the study, one living person was assigned to 13 distinct racial subgroups by those who knew him. Harris further observed that the racial category to which a person assigns another person may change over time according to his or her changing feelings about that person. Given such inconsistency of racial perception within individuals and communities, one can appreciate how unreliable national statistics can be.

Mexico and Central America

While statistical analyses differ widely in their estimates of the percentages of the Mexican population that belong to the various racial categories, there is universal agreement that Mexico is an overwhelmingly *mestizo* nation with a moderate black admixture in the central gulf coastal lowlands of Veracruz state. The main point of controversy regards the racial classification of the millions of peasant farmers who reside in rural villages in the central and southern portions of the country. Most of these, as rural smallholder farmers, are Indians in an economic sense, but they are becoming more and more *mestizo*-like in their culture. Only a few unacculturated Indian groups have survived, mostly in the Western Sierra Madre, the Yucatán lowlands, and the highlands of Oaxaca and Chiapas.

Guatemala is divided racially between the peoples of the Western Highlands, who consist of more than twenty linguistically different Maya Indian groups, and the *Ladinos* of the larger cities and the Pacific and Caribbean lowlands. The term *Ladino* is derived from *Latino*, meaning belonging to the Latin language or Hispanic culture, and includes both whites and *mestizos* living European lifestyles. The neighboring Belizean population has historically been largely black and *mulatto*, although sizeable numbers of Maya Indian and *mestizo* immigrants from neighboring Guatemala and Mexico have diversified its ethnic composition in recent decades.

El Salvador, Honduras, Nicaragua, and Panama are all predominantly *mestizo* countries, with significant black infusions along the Caribbean lowlands of the latter three. In Honduras and Nicaragua, there has been considerable mixing between the blacks and the indigenous Garífuna and Miskito peoples, both of which prefer to be known as "Indians." Northwestern Panama contains a large Guaymí Indian enclave, and the Darién of eastern Panama contains sizeable numbers of Cuna Indians along the Caribbean coast and Chocó Indians in the interior. Costa Rica has many blacks and *mulattos*

FIGURE 7.27

These whites of Bogotá, Colombia, are representative of the upper-class European enclaves that have survived in the major urban areas of Latin America.

along its Caribbean coast and a substantial number of *mestizo* Nicaraguan immigrants residing along its northern border. Elsewhere, however, over the past two centuries it has been racially bleached to the point where it now considers the majority of its population to be culturally white, although it is often difficult to physically differentiate between Costa Ricans and their *mestizo* neighbors from Nicaragua and Panama.

South America

Venezuela and Colombia are both *mestizo* nations with large minorities of whites concentrated in the cities of the interior highlands and blacks and *mulattos* along the Caribbean and Pacific coastal lowlands (Figure 7.27). Ecuador, Peru, and Bolivia continue to struggle with great social divisions between the Indian-dominated rural Andean highlands and the *mestizo*- and European-controlled cities and lowlands. Indian influences remain so strong in Peru that Quechua was designated as the second official national language in 1975. Quechua is also the most widely spoken Indian tongue in Ecuador and Bolivia, although there remains in Bolivia a large Aymara-speaking region centered on Lake Titicaca and extending northeastward into the transitional valleys of the Yungas.

FIGURE 7.28

Children of many racial types playing together on a school playground in southern Brazil.

Chile is a thoroughly *mestizo* country with a sizeable Mapuche Indian population in the south. Many of the Mapuche have become *mestizos* over the past several decades, having been forced to migrate from their rural communities into the cities where they have formed a much neglected urban underclass. Argentina and Uruguay are both overwhelmingly white, the former having bleached out its black population and killed virtually all of its Indians in the nineteenth century. Paraguay is an unusual situation: almost all of its population are physical and social *mestizos,* but Guaraní is the most widely spoken and official second language.

Brazil has become one of the most racially mixed nations on earth. Following the collapse of the Indian populations, the Northeast came to be dominated by blacks, the South by whites, and the interior—including Amazonia—by *mestizos* or *mamelucos* (Figure 7.28). Since then a tremendous amount of miscegenation has occurred. Social bleaching is also widespread, with approximately three-fifths of the population now considering itself white. We noted earlier how Guyana and Suriname came to be divided between blacks, called Creoles, and East Indians, with a large Javanese minority in Suriname and a numerically small Amerindian population remaining in interior forest areas. Unless a new spirit of tolerance is fostered, the higher birth rates of the Asian populations in these countries are likely to cause the Creoles to feel even more threatened in the coming decades. French Guiana, a mostly Creole entity with moderate-sized East Asian and European minorities, has largely escaped the bitter racial divisiveness of Guyana and Suriname.

The Caribbean

With the loss of the dense American Indian populations in the early colonial era, the islands of the Greater and Lesser Antilles came to be divided into two general landuse spheres, each of which promoted a certain racial composition. Western Hispaniola, or Saint Domingue, Jamaica, and most of the French- and English-controlled Lesser Antilles were converted into plantation societies. This resulted in these areas becoming almost entirely black and *mulatto* in a physical sense, though, as we have noted, very European in self-perception. On the other hand, eastern Hispaniola, or the Dominican Republic, and Cuba, and Puerto Rico all continued under the control of the Spanish, whose emphasis on extensive animal grazing delayed the large-scale introduction of black slaves until the later development of regional plantation economies. Consequently, the amount of black admixture, while significant, is lower in the Cuban, Dominican, and Puerto Rican populations than in the rest of the Caribbean. Of the three, the Dominican Republic has had the greatest amount of black mixing and Puerto Rico the least. Ironically, antiblack racism is by far the strongest in the Dominican Republic, which, in order to mentally distance itself from its hated black neighbor Haiti, has gone to great lengths to culturally whiten itself by emphasizing the Hispanic aspects of its heritage and bitterly denouncing blacks. Thompson (1997) has observed, for instance, that officially there are no blacks in the Dominican Republic. One is either white, light Indian, dark Indian, *mestizo,* or *moreno* (dark)—but never black.

SUMMARY

The peoples of Latin America and the Caribbean include descendants of American Indians, European whites, African blacks, and Asians. Racial mixing has been so extensive that race is now defined primarily on the basis of culture and social class membership. While prejudice based on physical appearance is not common, discrimination toward members of the lower socioeconomic groups is widespread. This has resulted in a conscious effort by many individuals and societies to adopt European lifestyles, often while simultaneously giving considerable lip service to the glories of their American Indian, African, or Asiatic past.

Having noted the deep social divisions that exist today, it is important to recognize that the peoples of Latin America and the Caribbean do indeed share common values and behaviors that set them apart from the earth's other great culture realms. These values and behaviors are the focus of chapter 8.

KEY TERMS

Chichimecs 143
Aztec 145
Maya 146
Chibcha 148
Inca 148
Araucanian 151
Chono 151
Alacaluf 151
Yahgan 151
Ona 152
Tehuelche 152

Puelche 152
Charrúa 152
Gê 152
Ciboney 152
Guaraní 153
Arawaks 153
Island Caribs 153
tumbeiros 157
manumission 158
Maroon 159

miscegenation 161
mestizo 161
mameluco 161
mulatto 161
zambo 161
"whitening" 163
"bleaching" 163
Social Darwinism 164
indigenismo 166
négritude 166

SUGGESTED READINGS

Alchon, Suzanne Austin. 1991. *Native Society and Disease in Colonial Ecuador.* Cambridge: Cambridge University Press.

Andrews, George Reid. 1995. "The Black Legions of Buenos Aires, Argentina, 1800–1900." In *Slavery and Beyond: The African Impact on Latin America and the Caribbean,* ed. Darién J. Davis, 55–80. Wilmington, Del.: Scholarly Resources.

Aveni, Anthony F. 2001. *Skywatchers.* Austin: University of Texas Press.

Bauer, Brian S. 1992. *The Development of the Inca State.* Austin: University of Texas Press.

Benson, Elizabeth P., ed. 1977. *The Sea in the Pre-Columbian World.* Washington, D.C.: Dumbarton Oaks Research Library and Collections.

Berdan, Frances F. 1982. *The Aztecs of Central Mexico: An Imperial Society.* New York: Holt, Rinehart and Winston.

Borah, Woodrow, and Sherburne F. Cook. 1963. *The Aboriginal Population of Central Mexico on the Eve of the Spanish Conquest.* Ibero-Americana: 45. Berkeley: University of California Press.

Carrasco, David. 1999. *City of Sacrifice: The Aztec Empire and the Role of Violence in Civilization.* Boston: Beacon Press.

———. 1998. *Daily Life of the Aztecs.* Westport, Conn.: Greenwood Press.

Carrasco, Pedro. 1999. *The Tenochca Empire of Ancient Mexico: The Triple Alliance of Tenochtitlán, Tetzcoco, and Tlacopan.* Norman: University of Oklahoma Press.

Carroll, Patrick J. 1991. *Blacks in Colonial Veracruz: Race, Ethnicity and Regional Development.* Austin: University of Texas Press.

Clarke, Colin G. 1986. *East Indians in a West Indian Town: San Fernando, Trinidad, 1930–70.* London: Allen & Unwin.

Cook, Noble David, and W. George Lovell. 1991. *"Secret Judgments of God": Old World Disease in Colonial Spanish America.* Norman: University of Oklahoma Press.

Crosby, Alfred W., Jr. 1972. *The Columbian Exchange: Biological and Cultural Consequences of 1492.* Westport: Greenwood Press.

Curtin, Philip D. 1969. *The Atlantic Slave Trade: A Census.* Madison: University of Wisconsin Press.

Dabydeen, David, and Brinsley Samaroo, eds. 1987. *India in the Caribbean.* London: Hansib Publishing.

Davies, Nigel. 1997. *The Ancient Kingdoms of Peru.* London: Penguin Books.

———. 1980. *The Aztecs.* Norman: University of Oklahoma Press.

Davis, Darién J., ed. 1995. *Slavery and Beyond: The African Impact on Latin America and the Caribbean.* Wilmington, Del.: Scholarly Resources.

Davis, D. D., and R. C. Goodwin. 1990. "Island Carib Origins: Evidence and Nonevidence." *American Antiquity* 55(1):37–48.

De la Cadena, Marisol. 2000. *Indigenous Mestizos: The Politics of Race and Culture in Cuzco, Peru, 1919–1991.* Durham, N.C.: Duke University Press.

de las Casas, Bartolomé. 1951. *Historia de las Indias.* México, DF: Fondo de Cultura Económica.

Denevan, William M., ed. 1992a. *The Native Population of the Americas in 1492.* 2nd ed. Madison: University of Wisconsin Press.

———. 1992b. "The Pristine Myth: The Landscape of the Americas in 1492." *Annals of the Association of American Geographers* 82:369–385.

Dobyns, Henry F. 1966. "Estimating Aboriginal American Population: An Appraisal of Techniques with a New Hemispheric Estimate." *Current Anthropology* 7:395–416.

Elbow, Gary S. 1992. "Migration or Interaction: Reinterpreting Pre-Columbian West Indian Culture Origins." *Journal of Geography* 91:200–204.

Field, Les W. 1994. "Who Are the Indians? Reconceptualizing Indigenous Identity, Resistance, and the Role of Social Science in Latin America." *Latin American Research Review* 29(3):237–248.

Fischer, Edward F., and R. McKenna Brown, eds. 1996. *Maya Cultural Activism in Guatemala.* Austin: University of Texas Press.

Gudmundson, Lowell K. 1986. "De 'negro' a 'blanco' en la Hispanoamérica del siglo XIX: la asimilación afroamericana en Argentina y Costa Rica." *Mesoamérica* 7:309–329.

Harner, Michael J. 1977. "The Ecological Basis for Aztec Sacrifice." *American Ethnologist* 4:117–135.

Harris, Marvin. 1964. *Patterns of Race in the Americas.* New York: Walker and Company.

Hassig, Ross. 2001. *Time, History, and Belief in Aztec and Colonial Mexico.* Austin: University of Texas Press.

Hendrickson, Carol. 1995. *Weaving Identities: Construction of Dress and Self in a Highland Guatemalan Town.* Austin: University of Texas Press.

Herlihy, Peter H. 1997. "Central American Indian Peoples and Lands Today." In *Central America: A Natural and Cultural History,* ed. Anthony G. Coates, 215–240. New Haven: Yale University Press.

———. 1989. "Opening Panama's Darién Gap." *Journal of Cultural Geography* 9(2):41–59.

Illueca, Jorge. 1997. "The Paseo Pantera Agenda for Regional Conservation." In *Central America: A Natural and Cultural History,* ed. Anthony G. Coates, 241–257. New Haven: Yale University Press.

Kaufman, Terrence. 1990. "Language History in South America: What We Know and How to Know More." In *Amazonian Linguistics: Studies in Lowland South American Languages,* ed. Doris L. Payne, 13–67. Austin: University of Texas Press.

Klein, Herbert S. 1986. *African Slavery in Latin America and the Caribbean.* Oxford: Oxford University Press.

Lai, Walton Look. 1993. *Indentured Labor, Caribbean Sugar: Chinese and Indian Migrants to the British West Indies, 1838–1918.* Baltimore: Johns Hopkins University Press.

Lovejoy, Thomas E. 2000. "Amazonian Forest Degradation and Fragmentation: Implications for Biodiversity Conservation." In *Amazonia at the Crossroads: The Challenge of Sustainable Development,* ed. Anthony Hall, 41–57. London: University of London Institute for Latin American Studies.

Lovell, W. George. 1992. " 'Heavy Shadows and Black Night': Disease and Depopulation in Colonial Latin America." *Annals of the Association of American Geographers* 82:426–443.

Lovell, W. George, and Christopher H. Lutz. 1997. "Conquista y Población: Demografía Histórica de los Mayas de Guatemala." In *De los Mayas a la Planificación Familiar: Demografía del Istmo,* eds. Luís Rocero Bixby, Anne Pebley, and Alicia Bermúdez Méndez. San José, Costa Rica: Editorial de la Universidad de Costa Rica.

———. 1995. *Demography and Empire: A Guide to the Population History of Spanish Central America, 1500–1521.* Boulder: Westview Press.

Lowenthal, David. 1972. *West Indian Societies.* London: Oxford University Press.

Malmström, Vincent H. 1997. *Cycles of the Sun, Mysteries of the Moon: The Calendar in Mesoamerican Civilization.* Austin: University of Texas Press.

Mason, J. Alden. 1950. "The Languages of South American Indians." In *Handbook of South American Indians.* Vol. 6, ed. Julian H. Steward, 157–317. Washington, D. C.: Smithsonian Institution Bureau of American Ethnology Bulletin 143.

Newson, Linda. 1987. *Indian Survival in Colonial Nicaragua.* Norman: University of Oklahoma Press.

———. 1995. *Life and Death in Early Colonial Ecuador.* Norman: University of Oklahoma Press.

———. 1986. *The Cost of Conquest: Indian Decline in Honduras under Spanish Rule.* Boulder, Colo.: Westview Press.

Nishida, Mieko. 1993. "Manumission and Ethnicity in Urban Slavery: Salvador, Brazil, 1808–1888." *Hispanic American Historical Review* 73:361–391.

Padden, Robert Charles. 2000. "Cultural Adaptation and Militant Autonomy among the Araucanians of Chile." In *The Indian in Latin American History: Resistance, Resilience, and Acculturation,* rev. ed., ed. John E. Kicza, 71–91. Wilmington, Del.: Scholarly Resources.

Patiño, Victor M. 1963. *Plantas Cultivadas y Animales Domésticos en América Equinoccial.* Cali, Colombia: Imprenta Departmental.

Payne, Doris L., ed. 1990. *Amazonian Linguistics: Studies in Lowland South American Languages.* Austin: University of Texas Press.

Price, Richard. 1996. *Maroon Societies: Rebel Slave Communities in the Americas.* 3rd ed. Baltimore: Johns Hopkins University Press.

Radell, David R. 1992. "The Indian Slave Trade and Population of Nicaragua during the Sixteenth Century." In *The Native Population of the Americas in 1492.* 2nd ed., ed. William M. Denevan, 67–76. Madison: University of Wisconsin Press.

Rogozinski, Jan. 1999. *A Brief History of the Caribbean: From the Arawak and Carib to the Present.* New York: Penguin Putnam.

Rosenblat, Angel. 1954. *La Población Indígena y el Mestizaje en America.* 2 vols. Buenos Aires: Editorial Nova.

Rouse, Irving. 1986. *Migrations in Prehistory.* New Haven, Conn. Yale University Press.

———. 1992. *The Tainos: Rise and Decline of the People Who Greeted Columbus.* New Haven: Yale University Press.

Russell-Wood, A. J. R. 1998. *The Portuguese Empire 1415–1808: A World on the Move.* Baltimore: Johns Hopkins University Press.

Rustomji-Kerns, Roshni, ed. 1999. *Encounters: People of Asian Descent in the Americas.* Lanham, Md.: Roman and Littlefield.

Salzano, Francisco M., and Sidia M. Callegari-Jacques. 1988. *South American Indians: A Case Study of Evolution.* Oxford: Clarendon Press.

Scarborough, Vernon L., and David R. Wilcox, eds. 1991. *The Mesoamerican Ballgame.* Tucson: University of Arizona Press.

Schevill, Margo Blum, Janet Catherine Berlo, and Edward B. Dwyer, eds. 1996. *Textile Traditions of Mesoamerica and the Andes: An Anthology.* Austin: University of Texas Press.

Schmidt-Nowara, Christopher. 1999. *Empire and Antislavery: Spain, Cuba, and Puerto Rico, 1833–1874.* Pittsburgh, Pa.: University of Pittsburgh Press.

Sharer, Robert J. 1994. *The Ancient Maya.* 5th ed. Stanford: Stanford University Press.

Sundberg, Juanita. 1998. "Strategies for Authenticity, Space, and Place in the Maya Biosphere Reserve, Petén, Guatemala." In *Yearbook, Conference of Latin Americanist Geographers.* Vol. 24, eds. David J. Keeling and James Wiley. Austin: University of Texas Press.

Telles, Edward E. 1992. "Residential Segregation by Skin Color in Brazil." *American Sociological Review* 57(2):186–197.

Thompson, Alvin O. 1997. *The Haunting Past: Politics, Economics and Race in Caribbean Life.* Armonk, N.Y.: M.E. Sharpe.

Wagley, Charles. 1971. *An Introduction to Brazil.* Rev. ed. New York: Columbia University Press.

Watts, David. 1987. *The West Indies: Patterns of Development, Culture and Environmental Change Since 1492.* Cambridge: Cambridge University Press.

Whitmore, Thomas M. 1992. *Disease and Death in Early Colonial Mexico: Simulating Amerindian Depopulation.* Boulder, Colo.: Westview Press.

Wright, Winthrop R. 1990. *Café con Leche: Race, Class, and National Image in Venezuela.* Austin: University of Texas Press.

8

Latin America as a Culture Region

Latin Americans belong to a number of racial, ethnic, and socioeconomic subgroups whose lifestyles differ markedly. Nevertheless, an enduring Iberian heritage binds the peoples of the region together and sets Latin America apart as one of the world's most distinctive culture realms. This cultural heritage consists of common values, beliefs, and behavior patterns. Not all Latin Americans, of course, participate equally in each value and behavior. Some Latinos, including many of Indian and African ancestry, exhibit few, if any, of the dominant culture traits, most of which are of Mediterranean origin. The relative influence of these cultural minority groups is steadily declining, however, as many of their members endeavor to advance their socioeconomic standing through racial or cultural whitening. As they do so, they are progressively assimilated into the unifying Hispanic culture that has been variously described as the Latin American way of life (Crist 1968), ethos (Gillin 1955), and tradition (Wagley 1968). Many of these culture traits are timeless, while others are evolving in response to changes from both within and without the region. An appreciation of them helps us understand not only the collective behavior of the people but also its ultimate impact upon the development and underdevelopment of the region as a whole.

HONOR AND SHAME: INDIVIDUAL IDENTITY AND SELF-WORTH

Europeanized Latin Americans have inherited from Spain and Portugal a belief that each person born into the world is given as a birthright from God an inherent, inner goodness that is the essence of human self-worth.

One's individual self-worth is referred to as his or her personal dignity, a quality known as **la dignidad de la persona.** Although all people possess a basic level of human dignity, Latinos believe that an individual should strive to increase his or her self-worth throughout life by being unique and by establishing an identity separate from all others. In the Spanish language, for example, the term *distinguirse* means "to stand out" (*Larousse* 1999) as well as to be "distinguished." It is the person who best plays the role of nonconformist—the individual who has the courage to be visible and different—who is most admired. Gilmore (1990, 36) has noted in this regard that "the excellent man, the admired man, is not necessarily a 'good' man in some abstract moral sense. Rather, he is *good at being a man*" (emphasis by Gilmore). Latin Americans demonstrate that they are good at being men (or women) through gender-based behaviors that bring personal honor.

Machismo

Traditionally, male dignity is expressed through **machismo,** a term derived from the word *macho,* which means simply "the male." To many non-Latinos, the word *macho* connotes a Latin American male whose life is focused wholly on the pursuit and conquest of women. In reality, however, sexual promiscuity is but one part of the *macho.*

The real Latin American *machos* are the men who seemingly have control of every facet of their lives and of the persons and events that shape them. The true *macho* is the epitome of self-confidence. He is a fearless gambler and risk-taker who, under the gaze of his

173

peers, will stride confidently up to the betting table at a cockfight and put down a stack of bills three inches high on his pick, knowing he is going to win, and win every time. The true *macho* is an extrovert, a hard-drinking braggart who always has a witty joke and a quick comeback to anyone who verbally challenges him. Not only can he take an insult but he returns it one better while never losing his composure.

The *macho,* at least in theory, is the supreme master, both at home and at work. His authority and judgment cannot be questioned by his wife, his children, or his underlings. To admit publicly that he has erred is most difficult, for it would bring about a loss of personal honor and self-worth. He is decisive and action-oriented, always seeking fame in the eyes of others. He is totally independent and unencumbered and always successful. He is the consummate winner.

The personal space or sphere of action of the *macho* is the public domain; he lives his life *en la calle,* meaning "on the street." He holds a good job and is involved in community affairs. It would be terribly demeaning to the *macho* to spend large amounts of time at home. To help with domestic chores, such as cleaning the house, washing or mending clothes, preparing meals, or tending children would bring unthinkable shame.

In the political arena, the *macho* is the politician who performs the grand deed and does it with flair. As president or mayor, he is wont to erect heroic monuments to himself. Publicly, he can do no wrong, and compromise with his opponents is difficult to achieve because it implies weakness on his part. Internationally, the *macho* leader will defy with impunity the heads of foreign countries, even if those nations are manifestly more powerful than his own. In this regard, Cuba's Fidel Castro has grown to become, in the eyes of many Latinos, the region's most *macho* politician. He is a man who has defiantly resisted the threats of both American and former Soviet leaders for over forty years, who never has admitted to having erred, and who is still the ruler of his kingdom.

Marianismo

The female counterpart to male *machismo* is **Marianismo,** an expression derived from María, the Spanish name for Mary. The term has reference to Mary, the mother of Jesus Christ, whose perceived life has traditionally been accepted as a model worthy of the emulation of all Latin American women. Just as Mary is said to have set a perfect example in her own life of submissiveness, self-denial, humility, and service, so too the ideal Latin American woman is gentle, kind, loving, patient, and long-suffering. Following Mary's lead, she is a devoted mother who nurtures and serves her husband and children and, through membership in charitable associations, cares for the poor and afflicted. Because women are said to possess these virtues, which males do not have, women are believed to be inherently morally superior to men. Women are the preservers of the family and the strength of society; without them civilization would fail.

Female space is in *la casa,* meaning "the home." It is there that the woman rears and teaches her children and fulfills her divinely appointed role as homemaker. While bearing responsibility for the home, the ideal Latin American woman still recognizes that she must wait upon her husband and sons when they are present. At mealtime, for instance, she serves them first and refrains from speaking when they are engaged in worldly conversation. She also trains her daughters to render similar respect to their father and brothers.

Sex, Reproduction, and Honor

Out of *machismo* and *Marianismo* a double sexual standard has evolved. Men, whose sphere is *la calle,* are expected to pursue every girl or woman who appeals to their lusts, pinching them and in innumerable other ways attempting to physically touch and verbally seduce their prey. It is assumed that if a man and a woman are left alone together, he will attempt to conquer her. For this reason, custom holds that a man and a woman who are not husband and wife should never be alone in a home. Women, on the other hand, are expected to lead virtuous lives. If a girl loses her virginity, some will assume that her only recourse in life is to become a prostitute, reasoning that if she were ever to marry, her husband could never trust her to be true to him.

The ideals of *machismo* and *Marianismo* require that the same man who has a mistress in *la casa chica,* or a small or secondary house, and who has violated or at least attempted to violate countless women, defend by every means possible the honor of his wife, legitimate daughters, sisters, and mother. To fail to do so would ultimately bring him personal shame and a loss of *dignidad.* This ideal of defending male honor has been used successfully on numerous occasions in Brazil to acquit a man for the murder of his unfaithful wife. In 1991, for example, in the southern Brazilian state of Paraná, a court acquitted of murder charges a man who had confessed to having stabbed to death his wife and her lover when he caught them alone together in a hotel room.

One outgrowth or custom historically based on these values, but which recently has begun to change, has been the rigid separation of the sexes after puberty. Secondary school student bodies traditionally were comprised exclusively of either all males or all females. Likewise, courtship was carefully chaperoned, with private expressions between the young woman and young man limited by grated windows.

FIGURE 8.1

This statue of Mary holding the Christ child is found in the main plaza of La Ceiba, Honduras.

It is significant that, in both *machismo* and *Marianismo*, human reproduction is the highest possible achievement. For the male, the fathering of offspring, either legitimate or illegitimate, is the supreme evidence of virility and masculinity. For the female, conception symbolizes, at a mortal level, a duplication of Mary's act of mothering a child of God (Figure 8.1). Thus, to have no children is shameful and to bear a large family is honorable.

In sum, Latin America remains a male-dominated, family-centered society. While there are numerous exceptions to these generalized behavior patterns (Gutmann 1996), the majority of the people continue to be influenced to varying degrees by the images and ideals we have described.

PERSONALISMO

To the Latino, the most visible measure of personal honor and self-worth is the amount of power a person exerts in the lives of others. This extension of one's sphere of influence is called **personalismo,** or person-

alism. It can best be achieved through the establishment of a continually expanding set of personal contacts and ties that bind or obligate others to support or assist one and one's family and friends.

Personalismo permeates every level of Latin American society and is developed through the giving and receiving of personal favors. Where *personalismo* reigns, one's success and advancement in life comes to depend more on whom one knows than on what one knows. Being able to call upon key people for personal favors becomes essential to success. Spanish expressions for having personal connections include *tener palanca* and *tener cuello,* meaning to be able to "pull" or "leverage" or "collar" others. Brazilians use the term *jeito,* which means to deal.

Personalismo is both a contributor to and a product of a broad, pervasive disregard for the law. It is supported by and associated with a lack of trust in government and other public institutions. One of its manifestations is a widespread reliance upon special confidants, variously called *hombres de confianza* (men in whom one can place confidence) or *amigos de carne* (literally, "flesh friends" or bosom buddies). One relies on confidants to ensure the success of one's initiatives, rather than trusting that the merits of a proposal will prove sufficient alone to achieve its goal.

The ritual kinship linkages formed through the Catholic *compadrazgo,* or **godparenthood,** system historically have served as one of the most valued forms of *personalismo.* Under this arrangement, the father of a newborn child will invite a man unrelated to the family by blood to serve as the godfather, or *padrino,* of the infant. If the person agrees, he also becomes the *compadre* (literally "co-father") of the natural father, and the wives of the two men becomes *comadres* ("co-mothers") to each other. From the Church's standpoint, the system provides a set of surrogate parents for the child, who share responsibility with the natural parents for the proper moral upbringing of the youth. One aspect of that responsibility is the sponsorship by the godparents of key sacraments and other rites of passage, such as baptism, confirmation, first communion, and marriage.

To the *compadres,* however, godparenthood represents a mutual support system that, if properly cultivated, will advance the personal material and social interests of each party. Van den Berghe and Primov (1977, 89) have noted that the "choice of godparents is often a coldly calculated business." Parents almost always seek out social equals or superiors to serve as godparents of their children. *Compadrazgo* between equals is a means of consolidating business relationships or friendships or honoring a person one admires. On the other hand, *compadrazgo* between social unequals involves gifts and protection from the superior

Case Study 8.1:   The Hidden Costs of Corruption

Corruption can be defined as the conscious violation of rules or laws for the purpose of realizing personal gain. It can be found among all societies and is often justified by those who engage in it as harmless to others. In reality, however, corruption inflicts enormous long-term, indirect costs upon the community as well as immediate losses upon its intended victims.

While Latin America clearly has no monopoly on corruption, there is a growing realization among Latinos of all walks of life that the perception of high levels of corruption continues to impede the development of the region. Enrique V. Iglesias, president of the Inter-American Development Bank, has described corruption as "one of the greatest evils plaguing the consolidation of democracy in Latin America and the Caribbean," and stated that if "inflation is a tax on the poor," then corruption "is a tax on the entire society." Iglesias went on to note that corruption exacts economic costs "by diverting resources away from development," political costs through "popular disaffection and the weakening of democratic regimes," and social costs by "disintegrating the social fabric, perverting culture and strengthening illegality and clientelism" (Constance 1997, 10). More recently, Oscar Andrés Rodríguez, Roman Catholic Cardinal of Honduras and a leading reform advocate, has spoken of corruption as a "great scourge" that is undermining the foundations of Latin American societies. "Corruption is a cancer that has spread throughout the world," said Rodríguez, "but it has done so in a particular way on our continent, to the point where we have a culture of corruption. The abuse of public office, political kickbacks, omissions, illegal gifts, bribes, tax evasion, fraud—these are our daily bread" (Constance 2002, 1).

Because corruption, by its very nature, is secretive, it is not possible to quantitatively measure its presence within a society. Transparency International has published since 1995, however, an annual Corruption Perception Index (CPI) that is a composite of fourteen data sources derived from seven leading international institutions. By averaging the scores assigned to each country in each of the surveys, the CPI provides a numerical measure of perceived corruption, with 10.0 representing a highly clean country and 0.0 a highly corrupt one. The CPI scores for the Latin American nations listed in the 2001 Global Corruption Report are shown in Case Study Figure 8.1. It is noteworthy that those countries perceived as most corrupt tend, for the most part, to be the economically poorer nations of the region, and those perceived as least corrupt tend, as a group, to be the more prosperous (see chapter 13). Note also the relatively low overall rankings of the Latin American nations compared to other nations of the world.

continued

in exchange for loyalty and gifts in kind, such as free labor, from the inferior. Godparenthood thus has the potential to act as a vehicle of patronage, and it is not uncommon for powerful men to serve as godfather to hundreds of godchildren, or *ahijados.* Chevalier (1965) reported, for example, that the Dominican strongman Rafael Trujillo became, often at his own encouragements, the godfather of thousands of children. The practice of godparenthood has declined somewhat in the urban areas of Latin America. It continues, however, to be used by countless rural landlords and political bosses to perpetuate, through the unique combination of mutual dependence and personal familiarity that it engenders, the age-old structures of social and economic inequality.

Personalismo has contributed to the development of a number of other Latin American culture traits. Together with the rigid social class stratification, which is addressed later in this chapter, it has encouraged the use of much **ceremonial politeness** in conversation. Long, flowery salutations, for example, are still common in written correspondence. Instead of receiving a letter addressed "Dear _____," followed by an "I am writing to inform you that . . ." lead-in to the body, in Latin America it is not unusual for the salutation to be six to eight words long and quite sentimental. An example would be: "My very dearest, most highly esteemed friend _____," followed by an equally personal touch to the start of the body of the letter: "I hope that this finds you and your family enjoying good health and all the other good things of life."

Similarly, a guest entering the home of another person is likely to be told—even if the host does not know the visitor at all: *"Esta es su casa,"* meaning, "this is your home." Other common flowery expressions include *a sus órdenes* ("at your orders"), and *para servirle* (to "serve" or "wait upon you"). Ceremonial politeness can also take the form of the speaker choosing to say what he or she believes the listener wants to hear, rather than telling the truth, in order not to disappoint or hurt the feelings of the other party. This practice commonly affects foreign visitors

Case Study 8.1: The Hidden Costs of Corruption *continued*

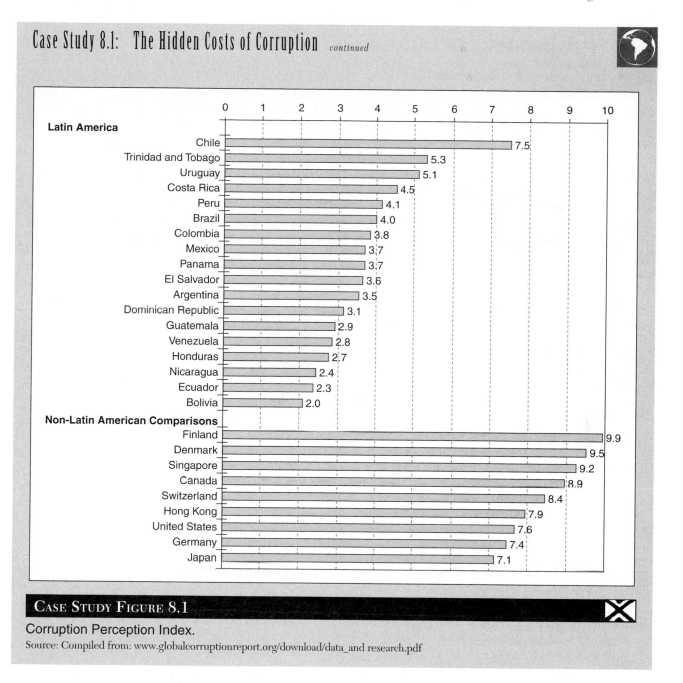

CASE STUDY FIGURE 8.1

Corruption Perception Index.

Source: Compiled from: www.globalcorruptionreport.org/download/data_and research.pdf

to Latin America when they are looking for a certain address or object. Everyone they ask directions of tells them in all seriousness that the address is just a few blocks and turns away. After following five or six of the sets of directions and realizing they are no closer to their destination than when they began their search hours earlier, the visitors may feel considerable anger or frustration at having been knowingly misled. In reality, the persons consulted for assistance likely had no idea where to send them but made up some directions to avoid hurting the visitors' feelings by replying that they do not know the way. *Machismo* also may have contributed to the locals' unwillingness to admit that

they did not have the desired information. Ceremonial or ritual politeness is also related to the widespread use of the titles *dón* or *doña,* as well as to the prolific use of professional titles such as *doctor, profesor,* and *ingeniero* (engineer).

Nepotism, or the practice of giving jobs and appointments to relatives regardless of their qualifications, has obvious ties to *personalismo* and is a grand tradition of great antiquity in Latin America. Chevalier (1965, 37) cited an early case of the president of the Audiencia of Guadalajara, who by 1602 had appointed forty-six relatives and an unspecified number of "dependents" to government positions. These were all alleged to be necessary to assist him in ministering to the needs of 160 households in the community and an unspecified number of persons residing in the surrounding rural zones.

Although the ratio of sycophants to the total population is not quite as high today as in seventeenth-century Guadalajara, nepotism continues to hinder seriously the efficiency and development of Latin American governments, both local and national. For example, a visit to a government office for the purpose of acquiring even the simplest document will almost certainly confirm the existence of what Latinos call *tinterillismo* and *papeleo*, meaning respectively the "ink-bottleism" and "paper shuffling" syndromes. There in the office complex one can expect to be directed to countless clerks or deputy ministers. Each seems determined to force the applicant to acknowledge the importance of his or her office by making the petitioner wait an appropriate length of time before stamping each of the one or two dozen copies of the forms necessary to secure the document.

Not surprisingly, the greater the number of required copies of forms and documents, the less is their individual value and significance. The widely acknowledged Latin American trait of legal-mindedness is thus an unspoken acknowledgment of the lack of confidence Latinos have in written guarantees in a world where most decisions are ultimately made on the basis of *personalismo*.

Yet another outgrowth of *personalismo* is the promotion of corruption, dishonesty, and ineptitude at all levels of Latin American society. Not every Latino official or businessperson is corrupt, of course, but the practice is so pervasive that those who are not are often intimidated and limited in their effectiveness. Bribery, widely known as *la mordida* ("the bite") in Spanish and *presentinho* ("the little gift") in Portuguese, is a standard means of expediting business transactions, from high governmental officials and professionals to the humblest clerks and traffic officers (Little and Posada-Carbó 1996; Constance 1997) (Case Study 8.1). Its effects are both to slow the pace of work and to increase the "indirect costs" of domestically produced goods and services, thereby making them less competitive on the international market. The practice also raises serious issues among present and prospective foreign business leaders who frequently must choose between investing in Latin America or elsewhere.

Personalismo has also contributed to much of the historic political instability of the region through the maintenance of political parties that are sustained principally through the patronage generated by a charismatic leader. *Personalismo* may appear to have

the positive effect of enabling individuals to receive goods or services that would otherwise be difficult or impossible to obtain. But ultimately it has hindered the development of virtually every sector of Latin American society by failing to encourage and reward creativity and achievement and by promoting the short-term rather than the long-term interests of the Latino peoples.

SOCIOECONOMIC CLASSES

The Latin American stress on *personalismo* can also be interpreted as an attempt to circumvent the social rigidity of the region. Most Latin Americans, from pre-Conquest times to the present, have belonged to an exploited lower class whose opportunities for personal advancement have been extremely limited (Figure 8.2). The huge gap in wealth between the elites and the masses has been documented in a number of studies. We will summarize the results of two such studies.

Table 8.1 is derived from studies published annually by the World Bank, which analyze the percentage of household income in various countries earned by percentile groups of households. Of particular significance is the fact that, among those Latin American and Caribbean nations reporting, the wealthiest fifth of the population generally received from one-half to five-eighths of all household income, whereas the poorest fifth of the population received an average only 4.0 percent of the wealth produced.

Table 8.2 gives us a more personal glimpse into Latin America's extreme socioeconomic inequality by focusing on the distribution of land in the two neighboring communities of Juazeiro and Petrolina in the arid backlands of Northeast Brazil. Notice that in Juazeiro, the very poorest class—those who have access to less than 10 hectares (24.7 acres) of land—make up approximately 84 percent of all landholders in the municipality but collectively control only 6 percent of the land. The poverty of these smallholders stands in stark contrast to the wealth of the great *latifundistas*, who form a mere 2.8 percent of the population and yet monopolize over 81 percent of the land. A similar pattern is found in Petrolina. When asked to state whether they belonged to the upper, middle, or lower socioeconomic class, three-fourths of the residents of the two settlements declared themselves part of the lower class, with all but a few of the remaining fourth identifying with the middle class (Table 8.3). These figures are fairly representative of conditions found throughout Latin America. Let us turn now to an overview of the distinct lifestyles of each of the three major socioeconomic classes and to an analysis of how class structure has impacted the development of the region.

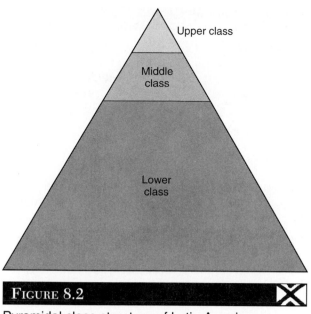

FIGURE 8.2

Pyramidal class structure of Latin America.

The Upper Class

At the top of Latin America's socioeconomic pyramid is the elite upper class. Many rural areas of Latin America continue to be dominated by a handful of great, local patriarchal clans, which alternately feud and intermarry one with another while perpetuating a semifeudal agrarian society in which the serfs or peasants do the bidding of the noble families. These rural elites have proven remarkably resilient, passing on their wealth and power from one generation to the next, regardless of the ebb and flow of national political parties, programs, and ideologies. One reason for their enduring success has been their adaptability to change. Contrary to what might be expected, the rural elites have not been slow to move their residences to the growing provincial and national cities and to invest their wealth in commerce, finance, and industry. Many of Latin America's urban elites are thus extensions of the rural aristocracy.

TABLE 8.1

Percentage Share of Household Income by Percentile Group of Households

COUNTRY	PERCENTAGE SHARE OF HOUSEHOLD INCOME, BY QUINTILE GROUPS OF HOUSEHOLDS				
	Lowest 20%	2nd 20%	3rd 20%	4th 20%	Highest 20%
Latin America					
Bolivia	5.6	9.7	14.5	22.0	48.2
Brazil	2.5	5.5	10.0	18.3	63.8
Chile	3.5	6.6	10.9	18.1	61.0
Colombia	3.0	6.6	11.1	18.4	60.9
Costa Rica	4.0	8.8	13.7	21.7	51.8
Dominican Republic	4.3	8.3	13.1	20.6	53.7
Ecuador	5.4	9.4	14.2	21.3	49.7
El Salvador	3.4	7.5	12.5	20.2	56.5
Guatemala	2.1	5.8	10.5	18.6	63.0
Honduras	3.4	7.1	11.7	19.7	58.0
Jamaica	7.0	11.5	15.8	21.8	43.9
Mexico	3.6	7.2	11.8	19.2	58.2
Paraguay	2.3	5.9	10.7	18.7	62.4
Peru	4.4	9.1	14.1	21.3	51.2
Uruguay	5.4	10.0	14.8	21.5	48.3
Venezuela	3.7	8.4	13.6	21.2	53.1
Anglo America and Europe					
Canada	7.5	12.9	17.2	23.0	39.3
Germany	8.2	13.2	17.5	22.7	38.5
Sweden	9.6	14.5	18.1	23.2	34.5
United States	5.2	10.5	15.6	22.4	46.4

Source: Compiled from *World Development Report 2000/2001*, 2001, 282–283.

TABLE 8.2

Size of Land Holdings in Juazeiro and Petrolina, Northeast Brazil

SIZE OF HOLDINGS (hectares)	HOLDINGS		HECTARES	
	No.	%	No.	%
Juazeiro				
<10	830	83.9	1,733	6.0
11–100	132	13.3	3,660	12.8
101–1,000	25	2.5	7,750	27.1
1,001–10,000	2	0.2	4,700	16.4
10,001–100,000	1	0.1	10,800	37.7
Total	990	100.0	28,643	100.0
Petrolina				
<10	2,348	82.2	8,243	18.3
11–100	480	16.8	12,595	28.0
101–1,000	26	0.9	7,932	17.7
1,001–10,000	3	0.1	16,200	36.0
Total	2,857	100.0	44,970	100.0

Source: Modified from Chilcote 1990, 154.

TABLE 8.3

Social Class Membership in Juazeiro and Petrolina, Northeast Brazil

SOCIAL CLASS	PERCENTAGE IN	
	Juazeiro	Petrolina
Upper	3.2	1.4
Middle	24.6	22.8
Lower	72.2	75.8

Source: Chilcote 1990, 37.

One example of the evolving yet ongoing dominance of the elites over a region is found in the northeastern Brazilian state of Rio Grande do Norte, where the Alves, Maia, and Rosado families have long competed for political control. In the late 1800s, the Rosado and Maia families were united through the marriage of Jeronimo Rosado to Isaura Maia. The couple established their base of power in Mossoró, the state's largest city, and its hinterlands. Today, all of Mossoró's public buildings are painted pink (*rosa*) in honor of the Rosado family, which commanded the mayor's office in every election except one between 1950 and the early 1990s. Dom Jeronimo's numerous sons, named with French numbers because of his great admiration for French culture, include Dix-Sept (Seventeenth) Rosado Maia, who served as state governor, Dix-Huit (Eighteenth)

Rosado Maia, who has held the positions of senator and mayor of Mossoró, the noted industrialist Dix-Neuf (Nineteenth) Rosado Maia, Vingt (Twentieth) Rosado Maia, who was once a federal deputy, and Vingt-et-un (Twenty-first) Rosado Maia, who rose to head the school of agriculture at the state university. Now, Jeronimo's grandchildren are winning posts at the state and municipal levels (DeWitt 1992).

As the rural aristocracy in recent times has expanded its power and influence into Latin America's emerging urban centers, it has often been joined, by marriage or business ties or both, to a new rising generation of *nouveau riche*. The latter are generally of urban and, not infrequently, of foreign origin and have built their wealth upon commerce and industry. The effect of these unions has been to consolidate and perpetuate the power of the elite oligarchy, who continue to dominate, far more than is generally evident, the regions and nations of which they are a part (Case Study 8.2).

The lifestyles of the elites are outwardly gracious, hospitable, and refined. Courtly mannerisms, patterned after those of the European aristocracy from whom most of the elites descend, include the men greeting the women with a kiss on the cheek or the hand and bowing. Manual labor is avoided at all costs, and work itself is often viewed with disdain as an unfortunate evil that detracts from the higher callings of writing, socializing, and conversing with others of equal rank. The homes of the elites appear to the common people as great walled, guarded compounds whose chambers and beautifully landscaped grounds are maintained by ser-

Case Study 8.2:   The Perpetuation of Oligarchical Power in a Central American Nation

Although reforms have begun to alter the social and economic landscapes of Latin America and the Caribbean, many of the traditional ruling families have succeeded in perpetuating and even expanding their wealth and power. The following account of one such family comes from a Central American student attending an American university. The names of the family members have been changed and the identity of their country withheld in order to assure confidentiality.

Juana recognizes her maternal grandfather, Carlos Calderón, as the individual most responsible for building the wealth of the clan. In the late nineteenth century, Carlos, whose parents were Spanish immigrants, began to practice medicine in the capital city of a district situated in the central highlands. Doctor Calderón prospered in his medical practice and

was soon recognized as one of the leading men of the community, which honored him by electing him first as a district judge and, later, as a member of the national legislature. So great did his influence become that even today, years after his death at age 90 in 1965, both the city's main street and hospital carry his name.

Don Carlos' progeny consisted of three legitimate daughters and an estimated one hundred or more illegitimate children whom he fathered by scores of women who resided in the surrounding towns and villages. He attempted to provide for the latter by allowing them, and their families as they married, to practice subsistence agriculture on tiny plots carved out of the great 100,000-hectare (247,000-acre) ranch that he had acquired.

Following Don Carlos' demise, ownership of the ranch passed to

Juana's mother and two aunts. In the succeeding decades, they not only have retained control of the original *latifundio* but have subsequently added a second ranch and a coffee *finca* in the highlands and two large farms in the lowlands. At present, the three sisters control approximately 250,000 hectares (618,000 acres) of prime cattle and farming lands while simultaneously operating a chain of urban pharmacies.

The political mantle of the Calderón family is now being worn by Juana's sister, Patricia, and by Enrique, Juana's cousin. Patricia has served for years as mayor of the country's fifth largest city. Enrique is one of Latin America's most prominent statesmen, having held two national cabinet posts and the chief judgeship of the Supreme Court. He has also served as ambassador to both the Organization of American States and the United Nations.

---

vants (Figure 8.3). Convention dictates that the upper class never carry their own bags or purchases, run errands, or wait in line. They expect special treatment and they almost always get it.

In many ways, the members of the upper class identify more with foreign elites than with the lower-class masses of their native lands. They think nothing of flying to the United States or Europe to shop, obtain medical care, or vacation. Many of them are fluent in English and other foreign languages. Thanks to Cable News Network and other satellite and electronic communication media, they are as knowledgeable about events in foreign lands as about those occurring in their own neighborhoods. Fearful of the political and economic instability of their own countries, they are also inclined to transfer much of their wealth to foreign bank accounts. Their dual status as both the preservers of the old order and creators of the new has contributed to their being extremely conservative in some respects and very liberal in others. Although they strongly resist social change at home, for instance, they are quick to embrace computers and other new technologies. Their children, while trained in the classical disciplines of history, philosophy, law, languages, and the arts, are quick to adopt the latest fashions from around the world.

The Middle Class

Social scientists have long debated whether a middle class exists at all in Latin America. If we use income levels as the sole criterion of socioeconomic class, it will always be possible to identify a group of people who fall in the midrange for a given area. On the other hand, sociologists have long maintained that a social class must consist of people who not only have similar income levels but share a set of goals and aspirations. Using this perspective, it can be argued that until recently most regions of Latin America had no middle class. This is because the members of the economic midgroups embraced upper-class values, to the extent that their financial circumstances permitted. In other words, the middle groups long viewed themselves simply as future elites working their way up, and consequently they failed to develop a distinct group identity.

The ongoing processes that have clearly ended the blurring of the middle and upper classes are urbanization and industrialization. A new salaried, industrial workforce emerging in the cities has created a growing group of wage earners with few ties to, and even less sympathy for, the classical worldview of the traditional elites. The members of the new middle class are primarily skilled

FIGURE 8.3

Upper-class housing of the urban elite, San José, Costa Rica.

manual laborers, many of whom belong to ideologically left-leaning labor unions that tend to view human relations in terms of the management of class conflict. They have been joined in the new middle class by poorly paid lower and midlevel white-collar groups, such as teachers and government bureaucrats, who have been frustrated in their hopes of advancing through education into the upper class. Yet a third component of the middle class is the small-to-midsized merchants, entrepreneurs, and lower-level professionals. They do not share the sense of alienation common to the industrial workers and government employees, yet they recognize that they will never be accepted as equals by the elites.

The size of this new middle class varies widely by region. In some of the larger urban centers, it may include as much as a fourth to a third of the total population. In the smaller provincial cities it is likely to be closer to 10 to 15 percent. In the economically poor, unindustrialized, and less educated rural areas, those who would be grouped as middle class in the cities, such as the teachers and merchants and the local Catholic priest, become, at least in terms of respect, the upper class for that region.

Although the middle class is a very diverse entity, its members are bound together by several shared attributes. These include a strong sense of nationalism and pride in their indigenous cultural heritage. Great value is placed on public education as a perceived vehicle of upward socioeconomic mobility. The members of the middle class are, for the most part, relatively well read and are strongly opinionated about events of national and international importance. Above all, they feel overworked and trapped as they struggle to maintain their lifestyles in the face of relatively low salaries and unrelenting inflation.

The Lower Class

Far beneath the power and opulence of the upper class and the frenzied, vociferous life of the middle class, we find the faceless lower-class masses. These are the tens of millions of rural peasants, agricultural laborers, and artisans, and the even greater numbers of urban have-nots, many of whom are separated from the rural countryside by no more than a single generation. What binds these two groups, rural and urban, together are their lack of marketable urban job skills and poverty. These are the social nobodies who can never expect to get ahead in the peasant villages. Once in the cities, they resort to an endless series of menial jobs in order to feed themselves and their families. These are the shoeshine men waiting for a customer in the plaza, the countless street vendors, the lottery ticket hawkers, the bus and taxi and truck drivers, the beggars, the full- and part-time domestic workers, and the marketplace women (Figures 8.4, 8.5, and 8.6). They are also the vast throngs of the urban unemployed who, by many unofficial estimates, range from a quarter to a half of all those seeking work.

The urban lower class resides in tiny, makeshift homes within the squatter settlements at the edge of the city or in inner-city tenements or government-built housing projects. They spend large amounts of time walking and riding busses and subways. They are not necessarily less happy than the middle and upper classes. Life goes on, with children running and laughing outside their residences while radios and televisions are playing within. The one attribute that most share, however, is a sense of resignation—the feeling that there is little they personally can do to measurably improve their social and material standing. As a taxi driver in Guatemala City once observed: *El que tiene más dinero tiene más valor,* meaning "he who has more money has greater individual worth."

FIGURE 8.4 ✕

Shoeshine men waiting for customers in the main plaza of Otavalo, Ecuador.

FIGURE 8.5 ✕

Many Latin American women supplement the family income through the operation of tiny backyard restaurants and food stands.

CONSEQUENCES OF SOCIAL RIGIDITY

The inability of the majority of Latin America's poor to bring about meaningful changes and improvements in their lives is one factor that has contributed to the strengthening of two interrelated culture traits: a strong sense of fatalism and an equally strong belief in luck.

Fatalism and Luck

Fatalism expresses itself, either overtly or covertly, in a number of ways. The widely held Latin American belief that one's life is mostly preprogrammed and that one can do little to alter one's destiny is evident in the saying: *que será, será,* meaning "what will be, will be."

FIGURE 8.6 ✕

Street vendors at a station along the Puno-Cuzco railroad, highland Peru.

One aspect of the Spanish language that ingeniously accommodates fatalism is the third person impersonal tense. Bad or undesirable occurrences brought about by one's own actions are described in such a manner that the blame or responsibility is shifted from the person relating the incident to the alleged actions of some inanimate object associated with the event. If, for instance, a person accidently drops a glass or plate that shatters upon contact with the floor, he or she might say, *"Se me cayó y se rompió,"* which, translated literally, is "it fell from me and broke itself." Similarly, purses and wallets and other valuables are said to lose themselves and busses go off and leave those who arrive too late at the bus stop.

The Latino belief in luck is nowhere more evident than in the widespread gambling that occurs at all societal levels, from boisterous card and domino games played out nightly in neighborhood *cantinas* and restaurants to betting on the weekly community cockfights and soccer matches. One of the most widespread forms of gambling in Latin America is playing the national lotteries. From a governmental perspective, the lotteries, which were common throughout the region long before they became popular in the United States, are a socially acceptable means of raising money from the lower-class masses, most of whom pay no income taxes.

Another way in which fatalism and luck intertwine in the lives of many Latin Americans is their belief in the ability of certain persons to cast magical spells upon others. For example, the mere mention that someone has placed a curse upon another by burying a black chicken in front of their residence may be enough to cause the victim to develop serious psychosomatic disease symptoms, and through worry bring accidents or other misfortunes upon him or herself.

FIGURE 8.7

The American School of San Salvador, El Salvador. Private schools and universities have served the educational needs of the Latin American elites since colonial times and are currently increasing rapidly in number throughout the region.

Dual Educational Systems

A second outgrowth of Latin America's historic social rigidity has been the perpetuation of a dual educational system (Figure 8.7). The elites have sent their children to expensive private schools, while the children of the poor, if they have attended school at all, have been forced to enroll in poorly funded public schools. This resulted, down into the 1930s and 1940s, in literacy rates as low as 20 to 40 percent in Mexico and most of the Central and South American nations. The only bright spots have been the more Europeanized countries of southern South America, Costa Rica, and the United States-dominated Panama. Only within the last generation or two have most Latin American nations undertaken serious efforts to educate the masses. Now dedicated teachers can be found in even the poorest and most remote areas, often teaching in bare one- and two-room school buildings (Figure 8.8).

The motivation for the literacy campaigns has not been a newfound concern by the elites for the welfare of the poor. Rather, the government has recognized both the economic costs of illiteracy and the capacity of public education to promote the assimilation of ethnic minorities and the inculcation of nationalist ideologies. Regardless of their purposes, the ongoing education programs have now brought most national literacy rates into the middle 80 to upper 90 percent range, and have also encouraged a belated acceptance of the value of technical education (Figures 8.9 and 8.10). One of the reasons why rates of economic growth of many Latin American countries have not kept pace with the dramatic increases in literacy levels is the extremely low

FIGURE 8.8

Teacher and pupils at a village school along the Ucayali River, eastern lowlands of Peru.

FIGURE 8.9

Technical education was long despised by Latin American elites, owing to its association with manual labor. The substantial economic gains derived from industrial technologies, however, are now giving technical education a newfound respectability. This photograph shows the four areas of technical emphasis of Panama's Ministry of Education: industrial, commercial, agricultural, and home economics.

number of years of schooling completed in the Latin American nations (Figure 8.11). Recent studies have found, for instance, that Latin American working adults average only 4.8 years of schooling and that only 15 percent of students from poor families are enrolled by their ninth year of schooling (Bate 1998). This pattern contributes to low functional literacy rates and a mostly unskilled workforce. Nevertheless, the gradual creation of an ever more educated workforce promises a long-term strengthening of egalitarian values and the fuller utilization of the region's human resources.

FIGURE 8.10

Literacy rates (percentage) in Latin American and Caribbean nations.

Source: *United Nations Statistical Yearbook 1998* 2001, 73–74.

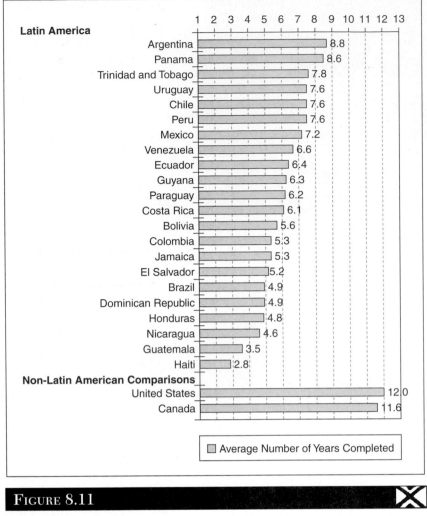

Years of schooling.

Source: Compiled from: http://www.undp.org/hdr2001/indicator/indic_243_1_1.html. Belize and Suriname show no data.

FAMILY LIFE

Families are of great importance in Latin America, where they provide a refuge from the injustice and stress of the outside world. While few, if any, would question the need for strong families in any society, some scholars, including Harrison (1988) and Aptekar (1988), have expressed concern that the loyalty of many Latinos to their immediate families may exceed their loyalty to society. When individuals consistently place the short-term gain of their families ahead of the long-term collective interests of society as a whole, they harm the processes of social and economic modernization. Banfield (1958) has used the term **amoral familism** to describe such societies, where "right" and "wrong" actions come to be defined only on the basis of whether they result in immediate gain or punishment to the individual and his

or her family, rather than on the impact of the actions on the larger community over time. These societies, according to Banfield, are characterized by seventeen conditions. Seven are most relevant to development:

1. No one will work to further the interests of the group or community, except as it is to his or her private advantage to do so.
2. Organizations based on selfless service and sacrifice will be difficult to establish and maintain.
3. The law is wantonly disregarded when there is no fear of punishment.
4. Officeholders take bribes whenever possible; the public assumes that officeholders take bribes, whether they do or not.
5. Claims of persons declaring their desire to render public good are regarded as fraudulent.

TABLE 8.4

Divorce Rates in Latin and Anglo America

COUNTRY	YEAR	RATE[a]
Latin America		
Brazil	1996	0.60
Chile	1998	0.42
Colombia	1994	0.11
Costa Rica	1998	2.04
Cuba	1999	3.54
Dominican Republic	1999	1.17
Ecuador	1998	0.73
El Salvador	1998	0.49
Guatemala	1997	0.13
Jamaica	1998	0.55
Mexico	1998	0.48
Panama	1997	0.65
Suriname	1997	1.23
Trinidad and Tobago	1997	1.00
Uruguay	1998	2.01
Venezuela	1996	0.79
Anglo America		
Canada	1997	2.25
United States	1998	4.19

[a]Rate figures are the number of final divorce decrees granted under civil law per 1,000 midyear population.

Source: Compiled from *United Nations Demographic Yearbook 1999* 2001, 550.

6. There is a shortage of both leaders and followers in public affairs because no one will take the initiative and no one will follow, except as it is to their private advantage to do so.

7. Support for beneficial community projects will come only from those who believe that they will personally gain from them; all others will attempt to defeat the projects.

Amoral familism has obvious ties to *personalismo.* Its presence continues to hinder community development efforts throughout Latin America in at least three ways. The first is that it discourages public planning and service and a willingness to sacrifice in order to invest in the future. The second is that it undermines the trust and cooperation needed from all segments of society if a community is to work for the advancement of the common good. Finally, because the members of amoral familistic societies expect injustice, they condone or tolerate corruption and nepotism. Ironically, in amoral familistic societies, as individuals place their own interests above those of the group in the belief that they are gaining personal advantage, they and their loved ones are actually sustaining great loss through the collective stagnation and/or deterioration of society as a whole.

Street Children

Having noted the negative ramifications of amoral familism, it is important to recognize that families historically have contributed much to the strength of Latin American society. Children are welcomed and loved and even pampered in most Latin American families, and divorce rates are far below those of the United States and Canada (Table 8.4). Emotional and economic support for immediate or extended family members in distress is generally high. Most unfortunate, then, are the increasingly ominous signs that the nuclear family is under attack. The situation is made all the more serious by the scarcity and relative ineffectiveness, in the weakly organized societies of Latin America and the Caribbean, of governmental and civic organizations capable of providing alternative socialization opportunities for the young.

One of the most visible and disturbing evidences of the weakening of the nuclear family is the growing presence and plight of so-called **street children.** The subject is very sensitive and complex, and opinions vary regarding both the scope and nature of the problem and the most effective responses to it. Much of the confusion stems from disagreements over just what constitutes a street child. Does the term apply only to boys and girls who have been totally abandoned by their parent(s), or does it also include youths whose activities center on the streets during the day but who return at night to their homes or a shelter? How do we classify these youths who may leave home for a time and later return? Do we include in our counts children who are working on the streets during the day under the direction of the parent(s) simply to help the family survive economically? On these and many other descriptive questions there is no consensus. Probably the most basic and best definition is that street children are minors who live away from home for part or all of their lives.

The United Nations Children's Fund (UNICEF) has estimated that there are approximately 40 million street children worldwide, with about 25 million, or 63 percent, of those residing in Latin America and the Caribbean (Figure 8.12). While it is unlikely that we will ever have an exact count, two things are clear with respect to the situation in Latin America. The first is that although the greatest numbers of street children are found in Brazil, Colombia, and Mexico, the problem is universal. The second is that although the phenomenon is not new, its breadth and scope are clearly on the increase.

Public perception of the nature of the street children and the lives they live has often been warped by sensationalized media reports and the fears of those who do not know the youths. According to typical newspaper and magazine accounts, the *niños de la calle* or *gamines,* as they are often called, have been totally abandoned by their families and live lives of idleness,

FIGURE 8.12

Many of Latin America's estimated 25 million street children work during the day rather than attend school.

supported by petty thievery, drug running, prostitution, and other forms of deviant social behavior. Interestingly, young street children are generally perceived as cute and worthy of compassion and assistance, whereas older youths are often seen as dangerous and irredeemably lost. The roots of the problem, according to common belief, are found in family abuse and conflict and residence in poor, dilapidated neighborhoods.

Like all stereotypical images, the circumstances described above do have some basis in truth, yet the realities are far more complex. One of the few objective, in-depth, scholarly studies of Latin American street children is Aptekar's (1988) social and psychological analysis of the *gamines* of Cali, Colombia. While the conclusions are far too numerous to all be cited here, and while there are differences in the composition and lifestyles of street children from one community to the next, the report does much to illuminate the phenomenon generally.

Demographically, Aptekar found that 95 percent of Cali's street children are males and that their ages range from 6 to 16 years, with 43 percent being less than 12 years old. Generally speaking, the *gamines* have not been abandoned by their families, keep in regular contact with their parents, and are free to return home at night, although many do so only rarely. Those who cannot return home are usually successful in developing a *personalismo*-like relationship with a benefactor and have regular places where they go to obtain rest, food, and association with playmates. Most of them leave home gradually, in stages, and with the consent of their mothers. By providing for themselves and possibly contributing to the support of their brothers and sisters, the children relieve the economic pressures on the family.

Contrary to public opinion, Cali's *gamines* get along with one another as well as children do elsewhere. Among the younger children, relationships center on a close friendship with a person of the same sex and, later, on gang and boy-girl associations. Petty thievery is common, but drug abuse, prostitution, and homosexuality occur infrequently. The greatest surge in the numbers of street children came with the economic downturn of the 1980s, which led to numerous youths abandoning their education in favor of vending and servicing jobs. Although all of the youths experience emotional and physical trauma to varying degrees, Aptekar found that about three-fourths of the *gamines* cope reasonably well and that most of these eventually become law-abiding, contributing members of society. We must be careful, of course, not to assume that the specific forces at work in the growth of the number of *gamines* of Cali are identical to those forces associated with the proliferation of street children in Brazil, Mexico, and other Latin American nations. What is apparent, however, is that the success of any programs intended to improve the quality of life of the region's street children will hinge on the ability of the program directors to recognize the cultural, social, and economic realities of the families of the youths.

POSSIBLE RESPONSES

Virtually every country in Latin America and the Caribbean has well-intentioned legislation designed to discourage children from leaving home at early ages (Haspels and Jankanish 2000). The problem with these regulations, which include compulsory elementary education and minimum age-to-work laws, is that they are generally unenforceable and thus ineffective. Many observers, including Bequele and Boyden (1988), have concluded that governmental attempts to enforce these laws more strictly would have the unintended effect of pushing the youths into even more secretive and exploitative labor relationships with the outside world. These same observers suggest that if economic realities dictate that the children must work, the wisest approach would be to concentrate efforts on protecting the street children in their employment or on devising forms of alternative education that provide both work and study experiences together with nutritional, health, and recreational programs (Haspels, de los Angeles-Bautista, and Rialp 2000). A successful example of the latter is the Projeto Axé initiative of Salvador, Brazil, which first encourages street children to dream and reflect on what they would like to be if all things were possible. The youths are then offered a combination of literacy and vocational training designed to enable them to earn a little money while preparing to achieve their long-term goals (Bellamy 1997).

Although this and other programs have helped many Brazilian youths and their families, ongoing "death squad" (*grupos de extermínio*) assassinations of street children in Brazil and other expressions of violence toward young people in Latin America provide ample evidence that promoting their well-being remains one of the regions's most urgent social needs (Human Rights Watch 1994a and 1994b).

Status of Women

As we noted previously in our discussion of *Marianismo,* the traditional roles of women in Latin America have been those of faithful mothers, wives, and homemakers. This social ideal, combined with the avoidance of physical labor by the upper class, has contributed to the impression held by many outsiders that most Latin American women do not work to earn a living.

This perception may have some basis in fact among the elite upper class, but in reality the great majority of Latin American women have always played prominent roles in the economic activities of their individual families and communities. To appreciate the pervasive economic contributions of women in Latin America and the Caribbean, one need look no further than the female-dominated public market and street vendors, the women toiling in the tiny fields of the high Andes and Caribbean lowlands, the handicraft sector of the Indian regions, the countless operators of small, corner *tiendas,* or stores, and the ever-growing ranks of domestic servants and workers (Yudelman 1994; Elbow 1995) (Figures 8.13 and 8.14). The income-producing work of these and other Latin American women has often been necessary to support themselves and their children in the absence of a reliable male provider. Such work has also become increasingly common in households where the husband is faithful and hard-working but unable to fully support his family in the midst of rampant inflation and declining levels of living.

Because of the Latin American bias against female activities unrelated to domestic responsibilities, most of the region's working women have been forced to labor in the **informal economic sector.** This consists of jobs performed by persons whose working conditions are not covered by national labor and employment legislation. To be socially acceptable, most women employed in the informal sector have taken jobs traditionally related to feminine responsibilities. These include food and clothing production and preparation, health care, and service to the young and elderly. The women engaged full- or part-time in informal sector jobs are rarely counted in official national employment statistics. As a group, they constitute one of the largest and most neglected and underutilized segments of the regional workforce (Case Study 8.3).

FIGURE 8.13

Women assisting in the harvesting of a bean crop in rural Mexico.

FIGURE 8.14

Female lottery ticket vendors in the plaza of Tegucigalpa, Honduras.

Opportunities for women to engage in activities beyond those of the home and the informal sector have come only gradually and with considerable opposition in most Latin American and Caribbean nations. One indication of the degree of opposition to feminine involvement in the public arena was that until the 1940s and 1950s most Latin American governments refused to grant women the simple right to vote in public elections.

Constitutional guarantees of equal protection and of the equality of the sexes in many personal and family economic issues have been even slower in coming. In some of the Hispanic culture nations, for instance, wives are still prohibited from filing income tax returns

Case Study 8.3:   Women and the Informal Economic Sector

Latin America's informal sector has been defined by Rakowski (1994) as those economic activities that escape standard fiscal and accounting mechanisms and that circumvent government-mandated regulations, benefits, and payment of taxes. It is also often associated with the selective or capricious application of labor and employment legislation by the state. Examples of employment that is commonly associated with the informal economic sector include street vendors and market sellers, domestic servants, tiny home or "cottage"-based industries, and part-time jobs. Other expressions of the informal economic sector often include child laborers, subsistence and semisubsistence agriculturists, and persons engaged in any number of illegal activities ranging from prostitution, drug running, money laundering, and smuggling to the bribery of government officials.

While it is virtually impossible to calculate precisely the number of persons that participate in Latin America's informal economic sector, there is a growing awareness that it is responsible for a very substantial portion of the region's economic output. Recent research by the Inter-American Development Bank (*Annual Report 2000,* 2001), for instance, has shown that "microenterprises" employ more than half the workforce in most Latin American nations. Although some of these businesses fall into the formal economic sector, a large proportion of them do not. Portes, Itzigsoln, and Dore-Cabral (1997), for example, concluded that 92 percent of the Haitian workforce, 54 percent of workers in Guatemala City, and 28 percent of employment in San José, Costa Rica, are associated with the informal sector. Additional studies by Hays-Mitchell (1994) and Cartaya (1994) placed 68 percent of the Peruvian workforce and 32 percent of Venezuela's urban laborers in the informal sector.

Many activities within Latin America's informal economic sector traditionally have been dominated by women. In her survey of marketers in highland Peru, for example, Babb (1998) found that 77 percent were women. Similar findings have been reported by Hays-Mitchell (1994) in her analysis of Peruvian *ambulantes,* or street vendors, and by Ehlers (2000), who studied female economic behavior in Guatemala's western highlands. Portes et al. (1997) have noted the same pattern in the Caribbean Basin, where 68 percent of informal sector vendors and 53 percent of persons engaged in small services and agriculture are women.

While not all persons employed in the informal sector are poor, there is broad agreement that governmental and nongovernmental agencies alike can and should do more to assist informal sector workers. One form of assistance that, until recently, has been sorely lacking has been the provision of loans and other financial services to informal sector entrepreneurs. A great need also exists for an expansion of nonfinancial services, including training, marketing assistance, and technology transfer. As the level of these services is raised, it is anticipated that national governments will derive significant benefit through increased economic productivity and that the status of women and their children will experience a corresponding improvement.

---

separate from those of their husbands. Additionally, children born out of wedlock have no legal claim for support from their fathers nor for a share of their deceased fathers' estates, nor are women entitled to property settlements in cases of divorce. These circumstances have denied Latin American women full equality with men and, in many cases, perpetuated female impoverishment.

Two recent studies that have undertaken the difficult task of quantifying the social, economic, and legal standing of women in the Latin American nations are the Gender-Related Development Index, published annually by the United Nations Development Programme, and an Inter-American Development Bank analysis authored by Bonilla (1990b). By averaging the numerical scores earned in each study, it is possible to gain a general understanding of the comparative status of women among the nations of the region (Figure 8.15).

Owing to the barriers to full female participation in society, Latin American women have only recently begun to move into the traditional masculine domains of politics and the formal economic sector. Levels of feminine involvement in politics have been aided recently in many countries by quota laws, which require that political parties reserve a fixed percentage of their candidate lists for women. As a result, the percentage of female parliament members in the Latin American countries has increased considerably in the last decade and now surpasses the global average (Table 8.5). These changes notwithstanding, it should be noted that only three women, Violeta Barrios de Chamorro of Nicaragua, Mireya Moscoso of Panama, and Mary Eugenia Charles of Dominica have ever served as elected heads of state in Latin America and the Caribbean. Furthermore, in many rural areas, women are effectively excluded from political office by the continuing presence of the *carguero,* or Catholic lay

FIGURE 8.15

Status of Latin American and Caribbean women.

Source: Compiled from: www.undp.org/hdr2001/indicator/pdf/hdr_2001_table_21.pdf; *Bonilla*, 1990b:211.

TABLE 8.5

Women in National Governments

COUNTRY	YEAR WOMEN RECEIVED RIGHT TO VOTE	Female Percentage		
		MINISTERIAL POSITIONS	SEATS IN PARLIAMENT	
			Lower House	Upper House
Argentina	1947	7.3	30.7	33.3
Nicaragua	1955	23.1	20.7	—
Guyana	1953	—	20.0	—
Costa Rica	1949	28.6	19.3	—
Peru	1955	16.2	18.3	—
Suriname	1948	—	17.7	—
Trinidad and Tobago	1946	8.7	16.7	32.3
Dominican Republic	1942	—	16.1	6.7
Mexico	1947	11.1	16.0	15.6
Ecuador	1929	20.0	14.6	—
Jamaica	1944	12.5	13.3	23.8
Chile	1931	25.6	12.5	4.1
Uruguay	1932	—	12.1	9.7
Colombia	1954	47.4	11.8	12.8
Bolivia	1938	—	11.5	3.7
Panama	1941	20.0	9.9	—
Venezuela	1946	0.0	9.7	—
El Salvador	1939	15.4	9.5	—
Guatemala	1946	7.1	8.9	—
Belize	1954	11.1	6.9	37.5
Brazil	1934	0.0	6.8	6.3
Honduras	1955	33.3	5.5	—
Haiti	1950	18.2	3.6	25.9
Paraguay	1961	—	2.5	17.8
Non–Latin American Comparisons				
Canada	1917	24.3	20.6	32.4
United States	1920	31.8	14.0	13.0
Europe	—	—	17.5	14.8
Latin America	—	—	16.7	14.1
World	—	—	14.4	13.7

Sources: Compiled from: www.ipu.org/wmn-e/world.htm and www.undp.org/hdr2001/indicator/indic_237_1_1.html. Data are for late 2001 and early 2002.

religious hierarchy system, which restricts village office holding to males. The secondary political status of women can also be seen in the ongoing failure of many to register to vote.

Female participation in the **formal economic sector** historically has been limited to teaching, nursing, and secretarial-clerical jobs that fit the cultural stereotype. Sanders (1987) has noted that, in Brazil, women constitute 85 percent of all educators, 66 percent of medical services employees, and 58 percent of all social security workers. Over the past three decades a rapidly growing area of female formal-sector employ-

ment has been the so-called off-shore transnational manufacturing and assembly plants (chapter 11). Employers believe that the perceived "feminine" traits of patience, persistence, and docility render women more suited than men to the endless tedium of these jobs. Female employees are also believed to be less supportive than men of labor unions. Women have begun gradually to move into private sector executive and managerial positions. Although their numbers are still small, the increasing participation of females in advanced educational study programs suggests that the number of women in administrative positions will in-

TABLE 8.6

Gender-Based Wage Gaps

COUNTRY	Percentage Difference between Male and Female Compensation for Similar Work	
	PRIVATE SECTOR	PUBLIC SECTOR
Honduras	42.9	−1.0
Ecuador	42.1	15.9
Guatemala	41.3	−1.2
Chile	34.9	27.9
Mexico	34.6	5.3
Dominican Republic	34.6	−8.4
Brazil	32.1	36.3
El Salvador	32.1	0.8
Venezuela	29.2	9.1
Panama	28.7	12.1
Paraguay	28.5	7.6
Uruguay	25.9	30.6
Peru	25.8	8.0
Costa Rica	25.2	7.3
Bolivia	24.8	13.5
Colombia	20.3	10.4
Nicaragua	2.9	2.7

Note: Negative numbers reflect female compensation exceeding male compensation.

Source: Compiled from Panizza 2000.

crease in the coming decades. Such increases are likely to be gradual, however, and those women who do advance are likely to meet with continued male resistance and to be paid less than their male counterparts for equivalent work. Panizza (2000), for example, found that Latin American women employed in the private sector earn, on average, 30 percent less than male workers with similar skills (Table 8.6).

As Latin American women have gradually expanded their involvement and influence in the public arena, they have found that the most effective strategy to advance their causes has been to emphasize the value of the experiences they are perceived to have acquired through motherhood. This approach contrasts markedly with the anti-motherhood positions of many feminist leaders in the United States, where, according to Nader (1986, 390), the women's movement "shifted away from a concern with community to the development of 'self.'" In other words, by stressing the publicly perceived positive values of womanhood—such as compassion, kindness, patience, selflessness, and stability—Latin American women have succeeded in claiming a higher moral authority than their male counterparts.

They are regarded as the guardians of the long-term interests of the community, especially in matters related to education and health and human services.

One result of the pro-motherhood positions of most Latin American female business leaders and government administrators has been a hesitancy to refer to themselves as "feminists." Another has been that women continue to prefer to wear dresses and other articles of clothing that are perceived by the public as feminine. In contrast, many Anglo-American businesswomen have felt compelled, to varying degrees, to imitate male clothing styles, behavior, and customs in order to maximize their opportunities to advance in a male-dominated environment. Yet a third consequence has been the tendency of Latin American female leaders and women to resist the implementation of family planning programs. This resistance has lessened greatly in recent decades, however, and it is now estimated that approximately three-fifths of married women of childbearing age use contraception (*World Development Report 2000/2001* 2001, 287).

It is impossible to determine precisely the number of Latin American women working full-time outside the traditional domestic sphere. However, the Inter-American Development Bank estimates that the total has now reached 65 million or more, or 32.0 percent of the regional workforce. Ironically, increasing women's employment has been associated in many parts of Latin America with a deterioration of male employment (Safa 1995; Burinic and Lycette 1994) (Table 8.7). Regardless of their social rank, women will likely continue to face four common challenges in the male-dominated societies of Latin America.

The first, and most fundamental, of these challenges will likely be the persistent attitude of many Latin American males that women are inherently inferior and less capable than men and, therefore, less deserving of equal opportunity. The second probable challenge will be a perpetuation of the female "double shift," or **"double day" workload.** This occurs when the woman puts in a full day of work at her job outside the home, and then comes home to face the same amount of cooking, cleaning, and washing that she would be doing if she stayed home all day. The increasing amount of work done by Latin American women outside the home has not been matched by a corresponding increase in housework performed by their *macho* husbands. Most Latin American men refuse, either out of a concern for preserving their masculine images or out of insensitivity to the workloads of their spouses, to help with the household chores and childrearing. Bonilla (1990a, 6) found, for instance, that Latin American women employed outside the home still average about 70 hours per week of work at home, whereas men in Latin America are willing to give only 5.25 hours weekly to household tasks.

TABLE 8.7

Labor Force Participation in the Dominican Republic (Percentages)

CATEGORY	Year				
	1950	1960	1970	1980	1990
Female	16.0	9.3	25.1	28.0	38.0
Male	85.1	75.9	72.6	72.0	72.2

Source: Safa 1995, 23.

A third challenge to Latin American working women will likely be continued overt prejudice. Historically, women throughout the region have been treated essentially as minors before the law, which has prohibited or limited in various ways their ability to inherit land, to obtain loans and other banking services, or to qualify for technical assistance in their economic enterprises. These limitations continue to be widespread in Latin America. In some instances they are the result of blatantly discriminatory laws and codes; in others they flow from institutional rigidity or bias. Those female business activities that occur within the informal sector, for example, often either go undocumented or are classified by governments and development agencies as "unproductive," thereby rendering their operators ineligible for assistance.

The fourth major challenge to Latin American women of all regions and subcultures will be to reverse their collective declining economic standing—a process often described as the **feminization of poverty.** In addition to those factors addressed previously, one major cause of the progressive impoverishment of Latin American women is that those who are employed in the informal sector have no supplemental benefits, such as maternity leaves, health insurance, and retirement or social security pensions. Another cause has been widespread rural-to-urban migration, which often has resulted in the gradual abandonment of formerly productive small farms when the women left behind by the migrating males have found themselves unable to keep up with the necessary farm labor (Chaney 1989). Yet a third contributor to female impoverishment has been the increased mechanization of the traditional handicraft sector, particularly textile and cooking ware production. Formerly these were dominated by prosperous, small-scale, female entrepreneurs, many of whom are now being driven out of business by large-scale, male-controlled, mechanized operations (Ehlers 2000). As technology levels continue to increase in the urban business sector, the mounting demand for a better-trained workforce also has had the unintended effect of favoring the employment of men, who until recently were better educated as a group than their female counterparts.

The consequences of adult female impoverishment extend also to children. Mothers are unable to adequately feed and educate the young people whose performance will determine the future productivity of the region. Latin America has thus arrived at a crossroads along the path of human and economic development. In many respects, the opportunities for a broad-based utilization of its human resources, including women, have never been brighter. Yet there also looms the very real possibility that present neglect of the children of the poor will limit the potential contributions of many of the next generation.

OTHER ELEMENTS OF CULTURE

Although space does not permit a detailed treatment, we will touch briefly on the development within Latin America and the Caribbean of five additional components of culture: literature and poetry, art, music, sports, and the mass media. We will then conclude our study of Latin America as a culture region by summarizing the interrelationships of human values, behavior, and regional economic development.

Literature and Poetry

As in all the world's great culture realms, it is possible to find in Latin America literature and written works expressing the full spectrum of human emotions, experiences, and aspirations. Works produced since the European Conquest can be divided into three broad periods. The earliest was the colonial era, which was characterized primarily by historical, geographical, and biographical accounts of the Spanish and Portuguese settlement of the New World. Lima, Mexico City, and Bahia (now Salvador) served as the publication centers of colonial literature written almost exclusively for Europeans and creoles who had little understanding of, and little sympathy for, indigenous American cultures and conditions.

Political independence in the nineteenth century brought a romantic literary period that rejected the European past and idealized native American peoples and physical environments. This Romantic Period was then largely replaced in the late nineteenth and twentieth centuries by a diverse movement called *Modernismo,* or **Modernism,** which has attempted to address the multitude of real-life challenges faced by the Latin American peoples. In its quest for absolute freedom of expression, *Modernismo,* and its contemporary variant, postmodernism, have come to include everything from existentialist musings to social commentary and regionalism (Jrade 1998; Stavans 1997; Williams 1995; Tapscott 1996).

A short list of some of the most prominent regional works of the Romantic and Modernist periods would include Rómulo Gallegos' *Doña Bárbara* (1929), a passionate novel of life on the Venezuelan Llanos, and two powerful portrayals of the Argentine Pampa: José Hernández' epic poem, *Martín Fierro* (1872) and Ricardo Güiraldes' *Don Segundo Sombra* (1926). José Eustasio Rivera's *The Vortex* (1924), Mario Vargas Llosa's *The Green House* (1966), and Alejo Carpentier's *The Lost Steps* (1955) all address human responses to South American rain forest environments. Claude McKay's *Banana Bottom* (1933) is a sensitive account of Caribbean peasant life. Euclides da Cunha's *Rebellion in the Backlands* (1944), J. Lins do Rêgo's *Plantation Boy* (1966), and Jorge Amado's *Gabriela, Clove and Cinnamon* (1962) convey the inner character of Brazil's Sertão, Northeast sugar, and Atlantic cacao regions respectively.

A reading of these and numerous other regional and social commentaries confirms both the close ties between literature and geography and also the strength of the Latin American literary tradition. Worldwide recognition of the latter has included six Latin American and Caribbean recipients of the Nobel Prize for Literature: Gabriela Mistral (Chile, 1945), Miguel Angel Asturias (Guatemala, 1967), Pablo Neruda (Chile, 1971), Gabriel García Márquez (Colombia, 1982), Octavio Paz (Mexico, 1990), and Derek Walcott (Trinidad and Tobago, 1992).

Architecture, Painting, and Sculpture

The development of the visual arts in colonial Latin America was most influenced by the Roman Catholic Church, which viewed architecture, painting, and sculpture as media for the promotion of religious worship. One of the most urgent of the initial needs of the Church was the erection of physical structures to serve not only as gathering places for the faithful but also as symbols of its spiritual and secular authority. The Spaniards thus began the construction, in the sixteenth and seventeenth centuries, of literally thousands of church buildings, which ranged from great vaulted cathedrals in the major urban centers to beautiful, tranquil monasteries, and remote village *capillas,* or chapels. Most of these have endured to the present day; their massive stone exteriors, thick wooden doors, and graceful bell towers form enduring images of the Latin American cultural landscape (Figure 8.16).

FIGURE 8.16

The massive wooden doors of a Catholic church in Zacatecas, Mexico, are characteristic of similar structures erected during the colonial era in many of Latin America's principal urban centers.

When one investigates the histories of these churches, one often finds that they were under construction for 80 to 150 years or more. Most of the labor was supplied by lower-class Indians, blacks, and mixed-race persons who added indigenous elements to the original Gothic and Baroque designs (Figure 8.17). The churches erected in the more prosperous regions were frequently adorned with stained-glass windows, gold-leafed interior ornamentation, and brightly colored exterior paint or glazed tiles. Others, including most of those found in the poor rural areas, are simple and plain in appearance.

As the buildings themselves were completed, the Church often commissioned paintings and sculptures of religious motifs for the most prominent buildings. The most widely acclaimed individual artist of the colonial era was a beloved Brazilian *mulatto* nicknamed Aleijadinho, or "Little Cripple" (Antônio Francisco Lisboa, 1738–1814), who crafted magnificent stone and wood sculpture for the churches of Minas Gerais (Figure 8.18). Ultimately, great numbers of paintings were hung on the walls of dimly lit church interiors throughout Latin America. In many cases, they have remained virtually untouched to the present time. In this respect, many of the principal churches have acquired an almost museumlike character, and no trip to Latin America is complete without visiting these buildings where so much of the history and the culture of the people are to be experienced. In the eyes of the local peoples, the sanctity of the churches has on occasion

FIGURE 8.17

Altar of a small Catholic church, Meseta Central, Costa Rica.

FIGURE 8.18

Church sculpture by Aleijadinho, Brazil's most famous artist of the colonial period.

FIGURE 8.19

The Teatro Colón, of Buenos Aires, Argentina, is one of the greatest architectural achievements of Latin America's neoclassical era.

been augmented by accounts of miraculous events that are alleged to have occurred in connection with a certain painting, carving, or other feature of the structure. When this happens, the favored church will often become a shrine, drawing vast numbers of pilgrims in search of healings and other supernatural experiences.

The domination of the Catholic Church over the visual arts did not begin to lessen until the late nineteenth and early twentieth centuries, when Latin America passed through a neoclassical era that brought, among other accomplishments, the famed Teatro Colón of Buenos Aires and the Palacio de Bellas Artes of Mexico City (Figure 8.19). The neoclassical era was followed by an avant-garde movement, which was greatly influenced by political nationalism and the glorification of native peoples and cultures embodied in *indigenismo* and *négritude.*

The initial inspiration for much of the avant-garde movement was the Mexican Revolution. In the minds of many of the left-leaning Latin American intelli-

gentsia, The Revolution came to symbolize a glorious triumph of exploited indigenous peoples over their former masters (Traba 1994). In 1921, José Vasconcelos, then Secretary of Education in the administration of Alvaro Obregón, persuaded a Mexican artist named Diego Rivera (1866–1957) to return from Europe. Rivera was commissioned to paint a series of murals de-

FIGURE 8.20

Cuban poster art.

picting the struggles and triumphs of The Revolution on the walls of the edifice housing the department. Rivera's work, which presented common, lower-class Mexicans with pronounced Indian features accomplishing heroic deeds on behalf of their fellow countrymen, coincided with the work of José Clemente Orozco (1883–1949) and David Alfaro Siqueiros (1896–1974). Rivera, Orozco, and Siqueiros formed the heart of an emerging Mexican muralist school that came to be recognized internationally in the 1920s and 1930s as the first original Latin American art movement.

The social realism of the Mexican muralist school soon diffused to Ecuador, where beginning in the 1940s Eduardo Kingman and Osvaldo Guayasamín led an indigenist movement that included both muralism and easel painting. Strong nativist and nationalist art schools have also been established in Peru and Bolivia. In Haiti, *négritude* has inspired the black African-influenced primitive or *naif* movement of Caribbean painters. In Marxist Cuba, murals and especially posters have been employed as media of public indoctrination by govern-

ment ideologists (Figure 8.20). The use of architecture to accomplish political or nationalist purposes is also evident in the Oscar Niemeyer–designed government buildings of Brasília and in the buildings of Mexico City's University of Mexico (UNAM). UNAM centers on a central library whose outer walls are covered by murals rendered by Juan O'Gorman. More recently, Latin American art has come to be dominated by abstract and surrealist movements that nevertheless continue to be sensitive to the social and political conditions of the masses (Barnitz 2001).

Music

Of all the arts, Latin America's music, with its capacity to emotionally transcend linguistic and ethnic barriers, has had the greatest cultural impact on the non-Latino world. The Spaniards and Portuguese who first arrived in the New World found that singing, chanting, dancing, clapping, and other forms of musical expression, often accompanied by drums, rattles, flutes, and pan pipes, were present among all the indigenous peoples. This native musical base was subsequently enriched by both European and African elements.

The European contributions to Latin America's musical heritage have taken two principal forms. The first is a strong formal, classical tradition, which centered initially on the composition and performance of sacred church music and has subsequently closely imitated overseas trends and tastes. The classical tradition is supported most strongly today by the Europeanized urban upper classes, but will also be found occasionally among the rural peasantry.

The second of the European musical traditions is Iberian folk music and its modern variants. One popular expression of the folk tradition in Latin America is the flirtatious couple dance, which has evolved into numerous regional forms, including *la marinera* of Peru, *el joropo* of Venezuela, and *la bamba* of Mexico. Each of these involves rapid, staccato footwork between male and female dancers whose body movements symbolize stereotyped gender roles. Other forms of Iberian-derived folk music include the *tango*, whose dual Iberian-black roots have been largely overlooked since its emergence in early twentieth-century Argentina (Chasteen 2000), and *el jarabe tapatío*, the Mariachi-accompanied Mexican Hat Dance. In addition, one cannot walk the streets of most Latin American cities without hearing a seemingly never-ending series of lively *rancherías*, or popular folk songs, blaring from radios and loudspeakers. Equally typical of the impact of Iberian folk music is the sight of a lone guitarist singing and playing in the cool of the evening on a park bench or on the curb of an urban sidewalk.

African contributions to Latin American music have been pronounced in Brazil, which gave the world the *samba* and the *Bossa Nova*. African influences are also prominent in the Caribbean, whose immense musical achievements include *la bamba* of Puerto Rico, *cadence* music of the French Antilles, and the *salsa*, the *mambo*, and the *cha cha chá* of Cuba. Another Caribbean musical form of significance is the calypso, or *kaiso*, of Trinidad, which has come to be associated with bands playing tuned steel pans made from discarded oil drums. The *mento* originated in urban Jamaica in the 1940s as an offshoot of Trinidadian *calypso*, and later spawned *ska* and *reggae*. The intensely interactive character of Caribbean music, its improvisation, and its widespread use as social commentary have all led to its development as a powerful political force within the region (Holton 2000; Manuel 1995; Bilby 1985).

Sports

No overview of Latin American and Caribbean culture would be complete without mention of sports, many of which exhibit pronounced geographical patterns. The most popular of all sports in Latin America is soccer, or *fútbol,* which young boys start to play on city streets and rural pastures from the time they are old enough to walk. The fascination with *fútbol* continues for most Latin American males through adulthood, and late afternoons and evenings are commonly filled with neighborhood pickup matches (Figure 8.21). Throughout Latin America, few events in life surpass in importance the watching of the televised professional matches. Every four years national all-star teams are selected to represent their countries in the international soccer competition known as the World Cup. One of the best indicators of the high level of Latino *fútbol* is the fact that half of the World Cup competitions held since 1950 have been won by Latin American teams. These included the Pelé-led Brazilian teams of 1958, 1962, and 1970, the Diego Maradona-dominated Argentine team of 1986, and the Brazilian teams of 1994 and 2002.

Other sports that are popular throughout Latin America include boxing, wrestling (*lucha libre*), bullfighting, and cockfighting, all of which are closely associated in the Latino mindset with *machismo* (Figure 8.22). Tennis, competitive cycling, basketball, and volleyball are all growing rapidly in popularity among the more affluent urban middle and upper classes. Chess remains a popular pastime throughout the region, with open-air pickup matches being a common sight in many parks and plazas (Figure 8.23).

Of the regional sports, baseball has the widest distribution. The game is believed to have been introduced into Latin America in the early twentieth century by American military personnel stationed in Cuba, Puerto Rico, the Dominican Republic, Haiti, Panama,

FIGURE 8.21

Pickup soccer matches are a common sight throughout Latin America and the Caribbean.

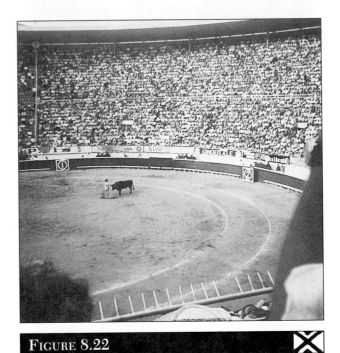

FIGURE 8.22

The traditional Hispanic emphases on cattle grazing and *machismo* are expressed in a ritualistic manner in bullfighting. This scene is from Bogotá, Colombia.

and Nicaragua. It subsequently became popular in Mexico, Colombia, and Venezuela. High-caliber professional winter leagues were established in the mid–twentieth century, and the region has become a leading source of United States major league talent. Cuba provided the greatest number of Latin American major leaguers until Castro established Marxism on the island. While a few Cuban players have successfully fled to the United States in recent years, the Dominican Republic now supplies approximately half of all the Latino major leaguers, with Puerto Rico, Venezuela, and Mexico following in that order.

In track and field, numerous world-class sprinters have developed in the Caribbean and long-distance runners in Mexico. Cricket is very popular in the English-speaking Caribbean, and horse racing is widespread throughout Middle America. Rugby and polo have strong followings in Argentina, and Brazil is currently a hotbed of auto racing, with Emerson Fittipaldi, Nelson Piquet, and the late Ayrton Senna all having won multiple World Formula One Grand Prix championships.

Mass Media and Tempo of Living

One final important aspect of popular Latin American and Caribbean culture is the mass media, which now reach into the most remote rural villages as well as into the poorest urban squatter settlements via newspapers, magazines, radio, and television (Cole 1996; Stevens 1997). The latter has rapidly become a virtual necessity over the past few decades for rich and poor alike, many of whom are as addicted to their soap-operas, called *telenovelas,* sports, and musical programs as are viewers in the United States. On a far more somber note, the abduction, physical beating, and even murder of journalists has become commonplace in some Latin American nations. The rapid expansion of Latin America's mass media, together with the explosive growth of cell phone and Internet usage, is transforming a society once famed for its leisurely tempo of living.

FIGURE 8.23

Open-air chess matches being played in Lima, Peru.

SUMMARY

We will conclude our study of the underlying cultural unity of Latin America and the Caribbean with a brief summary of the linkages between human values and behavior and regional economic development. No culture is inherently better or worse than another. However, the human values and beliefs associated with a given culture invariably influence the behaviors and ways in which the people of a region treat one another. People are ultimately the greatest resource of any nation or region. If the collective effect of a society's values is to promote the fullest possible utilization of human resources, that culture is supportive of long-term economic development. Conversely, if the values and behaviors associated with a given society work to stifle or limit the initiative and creativity of a significant segment of a region's human resource base, that culture is undermining the long-term economic development of the region.

History has shown repeatedly that, regardless of a country's natural resource endowment, those nations that most effectively develop the capacities of all their citizens are those that eventually achieve the highest levels of sustained

development. We have learned in this chapter that the Latin American way of life historically has included certain elements or attributes that have discouraged rather than encouraged the complete utilization of the region's human resources. These attributes include a rigid social class structure, which has systematically repressed the initiative and creativity of the majority of the population. They include a failure to provide adequately for the education and health care of the masses and a failure to promote the full development and utilization of the talents and abilities of women. They also include a historic tolerance of graft, corruption, and nepotism, which has undermined the efficiency of government and discouraged the trust, cooperation, and long-term planning necessary to advance the common good. However, progress in overcoming these obstacles to collective advancement is ongoing.

Culture is not the only component of economic development, but it is one of the most significant and least understood. Underdevelopment is, as Harrison (1988) has noted, in part "a state of mind." To point out the attributes above is not to suggest that the Latin American culture as a whole is inferior to others, for every culture has negative as well as positive qualities, all of which are subject to modification and change. While the Latin American culture has much that is of great worth, the future economic growth of the region will largely reflect its collective willingness or unwillingness to continue to pursue the fullest possible development of its immense human resources.

KEY TERMS

la dignidad de la persona 173
machismo 173
Marianismo 174
personalismo 175
godparenthood 175

ceremonial politeness 176
nepotism 177
amoral familism 186
street children 187
informal economic sector 189

formal economic sector 192
"double day" workload 193
feminization of poverty 194
Modernism 195

SUGGESTED READINGS

Annual Report 2000. 2001. Washington, D.C.: Inter-American Development Bank.

Aptekar, Lewis. 1988. *Street Children of Cali.* Durham, N.C.: Duke University Press.

Babb, Florence E. 1998. *Between Field and Cooking Pot: The Political Economy of Marketwomen in Peru,* rev. ed. Austin: University of Texas Press.

Banfield, Edward C. 1958. *The Moral Basis of a Backward Society.* Glencoe, Ill.: The Free Press.

Barnitz, Jacqueline. 2001. *Twentieth-Century Art of Latin America.* Austin: University of Texas Press.

Bate, Peter, 1998. "Education: The Gordian Knot." *IDB America* (http://www.iadb.org/idbamerica/Archive/stories/1998/eng/e1198e4.htm).

Bellamy, Carol. 1997. *The State of the World's Children.* Oxford: Oxford University Press for the United Nations Children's Fund.

Bequele, Assefa, and Jo Boyden, eds. 1988. *Combating Child Labour.* Geneva: International Labour Office of the United Nations.

Bilby, Kenneth. 1985. "The Caribbean as a Musical Region." In *Caribbean Contours,* eds. Sidney W. Mintz and Sally Price, 181–218. Baltimore: Johns Hopkins University Press.

Bonilla, Elssy. 1990a. "Poor, Female, and Working in Latin America." *The IDB* 17:4–7.

———. 1990b. "Working Women in Latin America." In *Economic and Social Progress in Latin America 1990 Report,* 207–256. Washington, D.C.: Inter-American Development Bank.

Buvinic, Mayra, and Margaret Lycette. 1994. "Women's Contributions to Economic Growth in Latin America and the Caribbean: Facts, Experience, and Options." In *Women in the Americas: Participation and Development,* i–iv. Washington, D.C.: United States Agency for International Development Executive Summaries, Background Papers of the Regional Forum, Guadalajara, Mexico.

Cartaya, Vanessa. 1994. "Informality and Poverty: Casual Relationship or Coincidence?" In *Contrapunto: The Informal Sector Debate in Latin America,* ed. Cathy A. Rakowski, 223–249. Albany: State University of New York Press.

Chaney, Elsa M. 1989. "Scenarios of Hunger in the Caribbean: Migration, Decline of Smallholder Agriculture, and the Feminization of Farming." *International Studies Notes* 14:67–71.

Chasteen, John Charles. 2000. "Black Kings, Blackface Carnival, and Nineteenth-Century Origins of the Tango." In *Latin American Popular Culture: An Introduction,* eds. William H. Beezley and Linda A. Curcio-Nagy, 43–59. Wilmington, Del.: Scholarly Resources.

Chevalier, Francois. 1965. "The Roots of *Personalismo.*" In *Dictatorship in Latin America,* ed. Hugh M. Hamill, 35–51. New York: Knopf.

Chilcote, Ronald H. 1990. *Power and the Ruling Classes in Northeast Brazil.* Cambridge: Cambridge University Press.

Cole, Richard R., ed. 1996. *Communication in Latin America: Journalism, Mass Media, and Society.* Wilmington, Del.: Scholarly Resources.

Constance, Paul. 1997. "Combating a Hidden Scourge: IDB Supports Anti-Corruption Efforts." *IDB America* 24: 10.

———. 2002. "Inside the Beast . . ." *IDB America.* Internet: http://www.iadb.org/idbamerica/English/MAR02E/mar02e1.html.

Crist, Raymond E. 1968. "The Latin American Way of Life." *American Journal of Economics and Sociology* 27:63–76, 171–183, and 297–311.

DeWitt, John. Letter to author, 17 March 1992.

Ehlers, Tracy B. 2000. *Silent Looms: Women and Production in a Guatemalan Town,* rev. ed. Boulder, Colo.: Westview Press.

Elbow, Gary S. 1995. "Marketing in Latin America: A Photo Essay." *Journal of Cultural Geography* 15(2): 55–77.

Gillin, John. 1955. "Ethos Components in Modern Latin American Culture." *American Anthropologist* 57:488–500.

Gilmore, David D. 1990. *Manhood in the Making: Cultural Concepts of Masculinity.* New Haven, Conn.: Yale University Press.

Gutmann, Matthew C. 1996. *The Meanings of Macho: Being a Man in Mexico City.* Berkeley: University of California Press.

Harrison, Lawrence E. 1988. *Underdevelopment Is a State of Mind: The Latin American Case.* Lanham, Md.: Madison Books and The Center for International Affairs, Harvard University.

Haspels, Nelien, Feny de Los Angeles-Bautista, and Victor Rialp. 2000. "Alternatives to Child Labour." In *Action against Child Labour,* eds. Nelien Haspels and Michele Jankanish, 145–184. Geneva: International Labour Office of the United Nations.

Haspels, Nelien, and Michele Jankanish, eds. 2000. *Action against Child Labour.* Geneva: International Labour Office of the United Nations.

Hays-Mitchell, Maureen. 1994. "Streetvending in Peruvian Cities: The Spatio-Temporal Behavior of Ambulantes." *Professional Geographer* 46:425–438.

Herzfeld, Michael. 1980. "Honor and Shame: Problems in the Comparative Analysis of Moral Systems." *Man,* n.s.15:339–351.

Holton, Graham E. L. 2000. "Oil, Race, and Calypso in Trinidad and Tobago, 1900–1990." In *Latin American Popular Culture: An Introduction,* eds. William H. Beezley and Linda A. Curcio-Nagy, 201–212. Wilmington, Del.: Scholarly Resources.

Human Rights Watch. 1994a. *Final Justice: Police and Death Squad Homicides of Adolescents in Brazil.* New Haven: Yale University Press for Human Rights Watch.

Human Rights Watch. 1994b. *Generation Under Fire: Children and Violence in Colombia.* New Haven: Yale University Press for Human Rights Watch.

Jrade, Cathy L. 1998. *Modernismo, Modernity, and the Development of Spanish American Literature.* Austin: University of Texas Press.

Larousse Diccionario Español-Inglés/English-Spanish Dictionary. 1999. México, D.F.: Larousse-Bordas.

Little, Walter, and Eduardo Posada-Carbó, eds. 1996. *Political Corruption in Europe and Latin America.* New York: St. Martin's Press.

Manuel, Peter. 1995. *Caribbean Currents: Caribbean Music from Rumba to Reggae.* Philadelphia: Temple University Press.

Nader, Laura. 1986. "The Subordination of Women in Comparative Perspective." *Urban Anthropology* 15:377–395.

Panizza, Ugo. 2000. "The Public Sector Premium and the Gender Gap in Latin America: Evidence from the 1980s and 1990s." *Inter-American Development Bank Working Paper 431* (http://www.iadb.org/OCE/working_papers_list.cfm?CODE-WP-431).

Portes, Alejandro, Jose Itzigsoln, and Carlos Dore-Cabral. 1997. "Urbanization in the Caribbean Basin: Social Change during the Years of the Crisis." In *The Urban Caribbean: Transition to the New Global Economy,* eds. Alejandro Portes, Carlos Dore-Cabral, and Patricia Landolt, 16–54. Baltimore: Johns Hopkins University Press.

Rakowski, Cathy A. 1994. "Introduction: What Debate?" In *Contrapunto: The Informal Sector Debate in Latin America,* ed. Cathy A. Rakowski, 3–10. Albany: State University of New York Press.

Ravicz, Robert. 1967. "*Compadrinazgo.*" In *Handbook of Middle American Indians,* vol. 6, ed. Manning Nash, 238–252. Austin: University of Texas Press.

Safa, Helen I. 1995. *The Myth of the Male Breadwinner: Women and Industrialization in the Caribbean.* Boulder, Colo.: Westview Press.

Sanders, Thomas G. 1987. "Brazilian Women in Politics." In *Universities Field Staff International Reports: Latin America,* 1–16. Indianapolis: Universities Field Staff, No. 14.

Stavans, Ilan, ed. 1997. *The Oxford Book of Latin American Essays.* New York: Oxford University Press.

Stevens, Donald F., ed. 1997. *Based on a True Story: Latin American History at the Movies.* Wilmington, Del.: Scholarly Resources.

Tapscott, Stephen, ed. 1996. *Twentieth-Century Latin American Poetry: A Bilingual Anthology.* Austin: University of Texas Press.

Traba, Marta. 1994. *Art of Latin America 1900–1980.* Baltimore: Johns Hopkins University Press.

United Nations Demographic Yearbook 1999. 2001. New York: United Nations.

United Nations Statistical Yearbook 1998. 2001. New York: United Nations.

Van den Berghe, Pierre L., and George P. Primov. 1977. *Inequality in the Peruvian Andes: Class and Ethnicity in Cuzco.* Columbia: University of Missouri Press.

Wagley, Charles. 1968. *The Latin American Tradition: Essays on the Unity and Diversity of Latin American Culture.* New York: Columbia University Press.

Williams, Raymond Leslie. 1995. *The Postmodern Novel in Latin America: Politics, Culture, and the Crisis of Truth.* New York: St. Martin's Press.

World Development Report 2000/2001. 2001. New York: Oxford University Press for the World Bank.

Yudelman, Sally W. 1994. "Women Farmers in Central America: Myths, Roles, Reality." *Grassroots Development* 17–18:2–14.

ELECTRONIC SOURCES

Transparency International Global Corruption Report: www.globalcorruptionreport.org/download/data_and_research.pdf

United Nations Development Programme Human Development Report 2001:
www.undp.org/hdr2001/indicator/indic_237_1_1.html
www.undp.org/hdr2001/indicator/hdr_2001_table_21.pdf
www.undp.org/hdr2001/indicator/indic_243_1_1.html

Inter-Parliamentary Union: www.ipu.org/wmn-e/world.htm

9

Religion

Religion is one of the most talked about and least understood aspects of Latin American and Caribbean culture. For example, most Anglo Americans would place "Roman Catholic" at or near the top of their list of Latino culture traits. In so doing, they would be recognizing both the prominent role that religion has played and continues to occupy in the lives of most Latin Americans and also the enormous historical influence of the institutional Catholic Church in the social, political, and economic development of the region.

Very few of those same Anglo Americans, however, if asked to describe in more detail the nature and spatial distributions of the different forms of Catholicism that are practiced in Latin America, would have even the slightest notion of what we were speaking about. Similarly, if we were to ask them about the presence of non-Catholic faiths in the region, most would be unable to provide us with the information we were seeking and many would likely express surprise at the suggestion that not all Latinos are Catholic.

One reason for the widespread ignorance of the religious geography of Latin America and the Caribbean has been the difficulty of collecting valid data on a sensitive issue. This, however, should not deter us from studying the religious behavior of the region, for it is only through the study of religious feelings and practice that we can fully understand the inner essence or character of peoples and the lands they occupy. We can no more come to know Latin America and its culture without studying Latin American Catholicism and Protestantism than we can appreciate the lands and peoples of the Middle East without knowing of Islam or the culture of India without learning about Hinduism. As we now begin our study of the role of religion in Latin America and the Caribbean,

we do so in a nonjudgmental way, recognizing that this aspect of regional culture, like many others, is changing at a historically unprecedented pace.

RELIGIOUS PRACTICE IN INDIGENOUS AMERICA

We learned in chapter 7 that the American Indians varied widely in group social structure and political organization, as well as in the physical habitats they occupied on the eve of the Spanish and Portuguese conquests. While the indigenous belief systems also differed in the specific names and attributes of their gods, they were nevertheless similar in that each was polytheistic. The sun, the moon, and the stars, as well as countless physical features of the earth, such as mountains and caves, rivers and springs, rocks and trees, all became sacred, venerated either as gods themselves or as the homes of the spirits who controlled the destinies of men and women alike. Because nature was frequently unpredictable and indiscriminate in the harm it inflicted, the gods were viewed as capricious beings who could be influenced or appeased only through repeated pilgrimages or offerings.

Yet another common attribute of the native belief systems was a willingness of the conquering tribes to admit into their pantheons as secondary gods the deities of the subject tribes, so long as the conquered peoples professed to worship the head gods of their masters. Religious allegiance under these circumstances was not the product of a search for absolute truth, as historically has been the case in the Christian, Islamic, and Jewish traditions. Rather it was largely a function of social and political expediency, with the tributary peoples showing outward homage toward the gods of their rulers while simultaneously continuing to supplicate the age-old gods of their ancestors.

MEDIEVAL IBERIAN CATHOLICISM ON THE EVE OF THE CONQUEST

At the same time that the relative standings of the New World Indian deities were evolving in response to changing cultural and political circumstances, the objects of Catholic worship in medieval Spain and Portugal were also changing. Christian (1981) and Sallnow (1987) have both noted that four forms of Catholic devotion had developed in Iberia by the sixteenth century. The first to appear on the peninsula was the veneration, and even worship, of the alleged bodily remains or relics of persons considered by the common people to be saints or holy beings. Such body parts as bones or locks of hair were accepted as possessing miraculous powers capable of healing those who were physically or spiritually afflicted. So strong was the faith of the masses in the power of these relics that it became common practice to place them in the altars of the churches so that those who prayed therein could have direct physical contact with them. There arose numerous cults centering on the relics. One of the most influential was that of Santiago de Compostela, which began in northern Spain in the eleventh century with the claim of the discovery of the remains of Saint James the Greater. The militant, mounted figure of Santiago became the emotional focus of the Christian Reconquest and, later, a leading religious symbol of Spanish dominion in the New World (Figure 9.1).

The second form of medieval Hispanic Catholic devotion was that of venerating the images of saints, a practice that spread into Spain from the Byzantine regions of the Middle East in the centuries immediately preceding the discovery of America. Images enjoyed an advantage over relics in that images were not site-specific; that is, multiple copies of images could be produced and transported at will, whereas relics remained at a single location that often became a pilgrimage shrine. In addition to individuals seeking specific blessings of a given saint, it was common for entire villages or towns to be named in honor of a patron saint who was believed to have intervened to bless the community at a time of great need.

The third and most rapidly expanding form of Catholic devotion in those portions of medieval Spain reclaimed from the Moors during the Reconquest was the worship of the Virgin Mary. Because popular belief held that Mary had ascended bodily into heaven (a concept enunciated as official Catholic dogma in 1950), it was theologically unacceptable for relics of her to be discovered. On the other hand, belief in her bodily As-

FIGURE 9.1

This representation of a militant, mounted St. James the Greater, or Santiago, killing with his sword a dark-skinned Moor, is found under an image of the crucified Christ at the head altar of a Costa Rican Catholic church.

sumption lent itself well to accounts of multiple appearances, apparitions, and images, most of which were alleged to have occurred to poor, marginalized members of society while visiting rural, uninhabited sites near hills, springs, and other natural landscape features (Poole 1995). The most prominent of the Iberian Marian shrines at the time of the New World Conquest was that of the Virgin of Guadalupe in central Spain.

The fourth and final form of medieval Catholic devotion was the veneration of images of Christ himself. Frequently, the images depicted Christ experiencing great agony and pain as he hung dying while nailed to the cross. As time passed, reports surfaced of persons having seen an image miraculously bleeding or sweating in a manner similar to the sufferings of Christ recorded in the Bible. These accounts led to the formation of flagellant brotherhoods, whose members would annually submit themselves to whippings, carrying crosses, and even being nailed upon crosses to demonstrate the depth of their devotion.

As the popular expressions of Catholicism evolved in medieval Spain, the institutional Church in Rome was granting unprecedented powers to the emerging Spanish state, which had at last been unified through the marriage of Ferdinand and Isabella in 1469. While extremely devout and pious in their personal lives, the new Spanish monarchs were also political pragmatists who saw the Church and its leading officials as potential threats to the authority of the state, especially in the southern regions recently reclaimed from the Moors through the Christian *Reconquista.* Accordingly, they undertook a series of maneuvers aimed at limiting the power of the Church. Two of the most significant, both initiated in 1478, were a reform and purging of the existing clergy and the establishment of the Holy Office of the Inquisition. The latter, which has been greatly misunderstood by Anglo Americans, was designed not only to cleanse the Church of Jewish and Islamic influences but also as a mechanism through which the Crown could regulate the actions of the clergy (Lea 1922; Greenleaf 1969).

These initiatives were followed, in 1486, by Ferdinand and Isabella's supreme pre-Conquest achievement—the obtaining of the **royal patronage.** This power, which in 1508 was extended by Pope Julius II to all of Spain's New World possessions, conceded to the Crown the authority to appoint all clergy, to control their assignments and movements, to redraw in the interests of the state both diocesan and parish boundaries, to collect and administer tithes, to preside over all Church councils and synods, and to oversee all Papal communication to the American colonies (Sallnow 1987; Padden 2000). The Church that was charged with ministering to the spiritual needs of the native Americans at the time of the Conquest was thus a thoroughly politicized institution whose officers served at the pleasure of the Crown. Ironically, the privileged and protected status of the Church would prove in the end to be its greatest liability in colonial Spanish America. Controlled by the government, it came to be viewed by many as an arm of imperial conquest and administration. Likewise, the clergy came to be seen as civil as well as ecclesiastical officers whose actions were frequently based on political rather than ethical considerations. It was little wonder that both the sword and the cross became symbolic, in the minds of the American Indians and their descendants, of the Spanish Conquest of the New World.

THE SPIRITUAL CONQUEST OF THE NEW WORLD

As the military conquest of Latin America and the Caribbean proceeded, the Crown took seriously its responsibility to Christianize the native peoples. The importance of this aspect of the Conquest was reemphasized by Pope Paul III, who resolved a heated controversy by confirming in 1537 that the "Indians" were indeed humans and therefore deserving of spiritual salvation. The obligation of indoctrinating the natives was shared between the institutional Church and the *encomenderos,* or Spanish lords, who theoretically were accountable for seeing that their Indian wards were baptized in the holy faith and taught Spanish (the Christian language). Once a frontier area was pacified militarily, it was generally assigned to one of the monastic orders, for proselytizing purposes. The most active of these orders were the Franciscans, Dominicans, Augustinians, and Jesuits. The monastic or regular clergy were purposely chosen by the Crown to perform the initial labors of the Church because they had vowed, both individually and collectively, not to accumulate property or other forms of wealth. Notwithstanding their self-described status as mendicants, or beggars, the orders quickly acquired considerable wealth in the New World, where they justified their possessions on the grounds that the Church was not providing them with the support necessary to accomplish their assigned tasks. The most disciplined and, ultimately, the most successful of the monastic orders in a worldly sense was the Jesuits (Society of Jesus). Their holdings extended from a vast, almost autonomous, communally organized kingdom among the Guaraní Indians of the upper Paraguay and Paraná river basins to a series of fourteen missions in Baja California (Saeger 2000; Pérez de Ribas 1999; Merino and Newson 1995) (Figure 9.2). The order was equally successful in its educational endeavors, through which it founded many of Latin America's leading centers of higher education. The influence and independence of the Society eventually grew to the point where in certain parts of Latin America its presence threatened the authority of the Crown. After failing in numerous attempts to reduce the power of the order, the Portuguese and Spanish governments finally took the extreme steps in 1759 and 1767 of expelling the Jesuits from their American colonies.

The spectacular rise and fall of the Jesuits was but one dimension of the role of the monastic orders in taking the Christian faith to the natives residing within the frontier regions of the empire. Once the members of a given order were dispatched to an area, their first task was to resettle those Indians living in dispersed hamlets and farmsteads into nucleated communities called *reducciones* or *congregaciones.* The purpose of gathering the natives was to facilitate both their religious instruction and the use of their labor on farms and in shops, schools, and hospitals. Those Indians who were already residing in nucleated settlements within the more established areas of the empire were generally assigned to the care of the parish-supported secular, or diocesan, clergy.

FIGURE 9.2

Remains of a Jesuit mission in northeastern Argentina.

Both the regular and secular clergy endeavored to establish Catholicism among the natives through a two-pronged strategy. The first consisted of destroying as many as possible of the native shrines and placing crosses and images of the Virgin Mary and the saints in as many of the ancient holy places as they could discover. The impossibility of totally achieving this goal is evident from a seventeenth-century report of a Peruvian clergyman cited in Sallnow (1987, 51). The dedicated Spaniard reported that he and a companion had discovered and destroyed, in one town alone, 603 "principal" *huacas,* or sacred places, 3,418 household shrines, 189 farmstead shrines, and 617 ancestral mummies. Needless to say, despite what they perceived to be their heroic efforts, the two clergy probably missed more Indian worship sites than they identified. Furthermore, the Spaniards soon realized that despite their much-published campaigns of "extirpation," the Indians continued to practice their ancient rites and ceremonies long after the Europeans had declared an area cleansed of so-called pagan beliefs (Mills 2000; Bauer 1998).

The second strategy of the Spaniards in their attempts to convert the Indians to Catholicism was to try to teach the natives the principles of the Christian gospel and to facilitate their participation in the sacraments. Because many, if not most, of the secular priests could not communicate initially in the language of the local Indians, and because the Spanish *encomenderos* felt unqualified to preach the doctrines of the holy faith, responsibility for the teaching of the natives was often delegated to a layman indoctrinator, or *doctrinero.* Typically the local Indians would all be brought together in the plaza in front of the church, and the *doctrinero* would offer a few prayers and read some passages of scripture. He would then pronounce the natives prepared for baptism, ignoring the fact that few, if any, of the Indians had understood even a single concept he had taught; nor had they been given time to ponder the implications of their impending "conversions." The *doctrinero* would then turn the natives back to the care of the presiding cleric, who would perform a mass baptism of his captive congregation. The Indians were then informed that their souls were saved and, following a few concluding admonitions, were discharged to return to their homes. The *encomendero,* having thus outwardly fulfilled his spiritual obligations toward his serfs, then proceeded to live off of their tribute and labor; and the clergy, of whom there were never a sufficient number to service the rural areas, returned as often as not to the comforts and refinements of their residences in the distant urban centers. For their part, the newly baptized Indians continued, virtually unaltered, the beliefs and practices that they and their ancestors had followed for time immemorial.

In laboring to facilitate the conversion of the natives, Church officials throughout Latin America were guided by an 800-year-old policy of missionary accommodation and tolerance. The policy indicated that whatever was good or indifferent in the indigenous culture of a convert people should be accepted by the Church, so long as it was not expressly forbidden in the Bible. Unencumbered, then, by the need to bring about substantive change in the customs and behaviors of the subject peoples prior to their baptisms, the early Church leaders reported great successes in bringing the Indians, and later the blacks, outwardly unto Christ. One Mexican-based monk named Motolinía, for example, indicated in his records that over 5 million Indians had been baptized in the colony between 1524 and 1536 (Arciniegas 1967, 145). Bishop Zumárraga of Mexico City, in his ecclesiastical report of 1531, attributed more than 1 million Indian baptisms to the Franciscans alone (Poblete 1970, 34–52). In other regions of Indian America, the numbers of converts achieved among the native peoples were equally impressive. The black slaves imported from Africa primarily into Brazil and the greater Caribbean were also forced to accept a Catholic baptism and then, like the Indians, were allowed to practice their ancestral rites and ceremonies.

RELIGIOUS SYNCRETISM

So intent were most of the early Church leaders on fulfilling the outward letter of the law and on stressing their statistical accomplishments that few paused to consider that the Indians neither understood nor were committed to the doctrines and practices of the institutional Church. The type of Catholicism that emerged among the native peoples was thus a religious fusion, or mix, which consisted of little more than a thin veneer of medieval Iberian customs and ceremonies superimposed over a base of ancient pre-Conquest Indian practices.

This mixing, or **religious syncretism,** began with the Europeans endeavoring to promote in the minds of the natives an association between the Indian and Catholic gods. The Inca, Maya, and Aztec culture heroes of Viracocha, Kukulkán, and Quetzalcoatl each became identified with the Christian godhead and the lower-level Indian deities with a Catholic saint responsible for a similar service or function (Figure 9.3).

The blending of the Catholic and Indian belief systems continued with the integration of native religious festivals with the annual round of Catholic *fiestas,* or celebrations. In the Peruvian Andes, for example, the Christian observance of Corpus Christi was substituted for the great pre-Conquest Indian Sun God festival of *Inti-raymi,* and Christ was frequently referred to by the Catholic clergy as "Young Lord of the Sun" (Métraux 1969, 191). In southern Mexico, Christ was reported to

FIGURE 9.3

Religious syncretism is evident in this Guatemalan church, which contains adjoining images of Iberian Catholic saints and ancestral Mayan dieties.

have appeared in a cave near the village of Chalma. Because the same cave had previously served as a sanctuary of the Aztec god Huitzilopochtli, the Christian pilgrimage that developed in honor of Christ's appearance was timed to culminate on the identical *fiesta* date used earlier to honor the Indian deity. Meanwhile, Christ's titles in the region came to include Lord of Chalma, Lord of the Cave, and Lord of the Mountain, the same as those given previously to Huitzilopochtli.

European efforts to promote the Indian assimilation of Iberian Catholicism extended to doctrinal issues. The Catholic clergy were quick to point out to the Andean Indians the similarities between the Incan and Christian creation accounts, both of which included a destruction of the first race of humans in a flood because of their disobedience and the subsequent separation of humankind into geographically distinct language groups (Sallnow 1987, 33). Prior to the conquest, the Aztecs, for their part, had practiced rites of baptism and confession, the latter when facing death (Table 9.1). They also believed in a god (Huitzilopochtli) who had been born of a virgin mother (Tonantzín).

The Spaniards, ever aware of the power of symbolism, even went so far as to construct their churches on the tops of the ancient pyramids and temples and other Indian holy places. In so doing, they unwittingly created the conditions under which the native peoples could continue to worship the same gods in the same holy places as had their ancestors, all while appearing to venerate the Catholic deities. The placement of the Catholic shrines over the remains of Indian worship structures also reminded the Indians, in a most visible and dramatic manner, of the technological and social superiority of the European gods and peoples over their indigenous coun-

TABLE 9.1

Similarities between Iberian Catholicism and Pre-Conquest Aztec Religious Belief and Practice

IBERIAN CATHOLICISM	AZTEC EQUIVALENT
Baptism	Bathing ceremony for infants, four days after birth
Confirmation	Dedication of children to temple schools to become priests
Penance	Prayers of penance to the goddess Tlazolteotl
Eucharist (bread changed into the flesh of Christ through transubstantiation before communion)	Figurines of the gods Huitzilopochtli and Omacatl which were fashioned from amaranth dough and then eaten
Extreme Unction	Ceremonial parallels
Religious Orders	Holy Orders of priests dedicated to specific gods
Holy Matrimony	Marriage celebrations on portentous days and festivals of the gods

Source: Derived from Hassig 2001, 160–161.

terparts. Two of the finest surviving examples of this practice are the Church of Santo Domingo, built literally on the foundations of the Inca Temple of the Sun in the sacred city of Cuzco, Peru, and the beautiful Catholic chapel constructed on the top of the great pyramid of Tepanapa in the central Mexican ceremonial center of Cholula, Puebla (Figure 9.4).

In addition, tens of thousands of small Catholic shrines were erected to guard the sacred springs and mountains and caves that housed the Indian nature gods. Many of these sites long had been the destinations of native pilgrimages, which continue to the present day in the form of popular Catholicism. In all their efforts to integrate the native and European belief systems, the Spaniards were unknowingly replicating the practices of the ruling Indian tribes of the past, who attempted to assimilate culturally their subject peoples by accepting their gods as lesser deities.

Marianism and the Cult of the Virgin Mother

Of all the events that facilitated the outward conversion of the Indians to Roman Catholicism, none was more powerful and binding emotionally than the numerous apparitions of the Virgin Mary. In Peru and Guatemala, Mary partially assumed the role of the Indian earth mother goddesses Pachamama and Ixchel among the Incan and Mayan peoples.

Perhaps the best known of all the Marian apparition accounts dates from December of 1531 when a dark-complected Virgin Mary was alleged to have appeared as the **Virgin of Guadalupe** to a lowly, newly converted Indian named Juan Diego on a hill called Tepeyac a short distance north of Mexico City. According to Mexican tradition, the Virgin requested that Juan go to Bishop Zumárraga and inform him that she wished a sanctuary to be built in her honor at the site of her apparition. Juan

FIGURE 9.4

Spanish construction during the colonial period of a Catholic church on the top of the great pyramid of Tepanapa in the central Mexican ceremonial center of Cholula, Puebla, was a symbolic act designed to reinforce in the minds of the Indians the superiority of the European gods and peoples.

FIGURE 9.5 ✖

Image of the Mexican Virgin of Guadalupe.

FIGURE 9.6 ✖

These stone calvaries, or sculptured representations of the crucifixion of Christ, are situated on a hill overlooking Lake Titicaca near Copacabana, Bolivia.

dutifully complied with Guadalupe's instructions but became deeply discouraged when the cleric told him that he would not act on the Virgin's request unless Juan could offer physical proof that the vision had indeed occurred. Juan is reported then to have returned in great despair to the site of the apparition where he was favored with a second visit from the Virgin. She told him not to worry but to go to a nearby spot where he would find flowers miraculously blooming where only cactus normally grew. Juan was told that he was to pluck the flowers and to carry them under his cloak until he could show them to the bishop. Upon arriving the second time at the bishop's residence, Juan opened his cloak to reveal not only the flowers but also a loving, smiling image of the Virgin herself imbedded permanently within the fabric of the garment (Figure 9.5). The skeptical bishop was now convinced and ordered a church erected on the site. Most significantly from the perspective of religious syncretism, the vision of the Catholic Virgin Mother had occurred at the very place where the Indians had, for centuries, journeyed to worship Tonantzín, the Aztec virgin mother goddess.

News that a dark-skinned Virgin of Guadalupe had appeared to the humble, dark-skinned Juan Diego contributed immensely to the conversion of the Indians to Catholicism. Additionally, because Guadalupe was the most prominent of the Marian shrines of Spain, her appearance also served to legitimize Spanish colonial rule (Taylor 1987). As time passed, Juan Diego came to be revered by the Indians and, in 2002, was elevated to sainthood by Pope John Paul II during a visit to Mexico. The Virgin of Guadalupe became, in turn, the protectress of Mexico's poor Indian and *mestizo* masses and, ultimately, the very symbol of Mexican Catholicism.

The Virgin Mary is also reported to have appeared in numerous other parts of Latin America, both in the colonial and modern periods. These include the Virgin of Copacabana of highland Bolivia (Salles-Reese 1997) and the Virgins of the Rosary and of Santa Rosa in Lima and the surrounding central Peruvian lowlands (Figure 9.6). The Virgin of Chapi is preeminent near Arequipa in southern Peru, while the Virgin of Guadalupe is venerated in northern Peru. In Ecuador, the Virgin of Quinche is worshiped, while the Virgin of Luján has become the national patroness of Argentine Catholics. Venezuelan loyalties are divided among the Virgin of the Valley, whose shrine on the Isla de Margarita is the spiritual focus of Catholicism in the East; the Virgin of Chiquinquirá, who is favored in the western zones centering on the Maracaibo Basin; and the Virgin of Coromoto, whose cult is present throughout the country. The dark-skinned virgins of Aparecida and of the Rosary have gained widespread devotion in Brazil, and the Virgin of San Juan de los Lagos has become the symbol of northern Mexican Catholicism. The Virgin of the Remedies has been considered since

FIGURE 9.7

This grotto, centered on a spring believed to contain healing waters, adjoins the Catholic basilica of Cartago, Costa Rica.

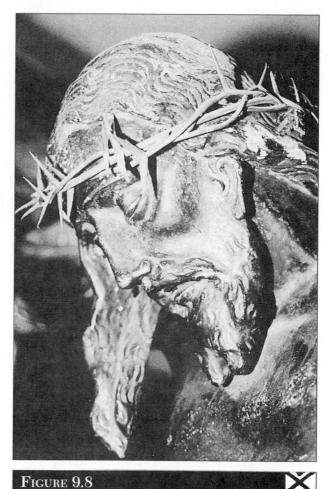

FIGURE 9.8

The Black Christ of Esquipulas sculpture has become the central object of a shrine that annually attracts millions of religious pilgrims to the highlands of eastern Guatemala.

the time of the Spanish Conquest to have a special interest in the well-being of the European elites of Mexico City (Curcio-Nagy 2000). These are but a few of the Marian cults of Latin America, each of which includes a sacred shrine to which pilgrims come annually, often in the millions, to seek divine intervention on their behalf (Figure 9.7).

The conversion of the Indians was also aided by numerous reports of non-Marian miracles (Case Study 9.1). One of hundreds of regional cults that have developed in connection with such miracles is that of the Black Christ of Esquipulas. Esquipulas is today a small agricultural municipality in the eastern highlands of Guatemala near the Honduran and Salvadoran borders. The region is endowed with sulfur springs and edible clays whose curative features contributed to its use as a pre-

Columbian Mayan pilgrimage site centering on the veneration of *ek-chuah*, a Yucatec deity of merchants and cacao who was generally painted black (Kendall 1991). In 1740, Guatemalan Archbishop Fray Pedro Pardo de Figueroa experienced a miraculous cure after venerating a darkened or black crucifix housed in a Catholic hermitage that had been established in the valley in the late sixteenth century. The archbishop's remains were later interred in the new basilica, which he had ordered to be constructed near the hermitage. The crucifix, which by the late 1800s had come to be known variously as the Black Christ, the Miraculous Christ, and the Lord of Esquipulas, replaced *ek-chuah* as the principal dark-colored object of devotion for the 1 million to 2 million pilgrims who visit the shrine annually (Figure 9.8). While in Esquipulas, many of the visitors purchase small

Case Study 9.1:   Cuzco's Lord of the Earthquakes

In the mid–sixteenth century, a small stick-and-straw mannequin, wrapped in layers of cloth after the manner of the imperial Andean mummies, was fashioned into a crucifix that came to be placed in the cathedral of Cuzco. We know little of the uses made of the crucifix until the year 1650, when a devastating earthquake inflicted major damage upon the community. With the aftershocks continuing, the clergy removed the crucifix and paraded it through the streets of the city with the grief-stricken townspeople following behind dressed in ashes and sackcloth as a token of their repentance. When the tremors finally stopped, the crucifix—now called *El Señor de los Temblores*, or Lord of the Earthquakes—was credited with having protected the city from further harm.

As the reputation of the statue for preventing or tempering earthquakes grew in the succeeding years, the complexion of the Christ image itself became ever darker from the accumulation of layer upon layer of smoke from the candles lit at its base. Although the darkening of the crucifix carried no spiritual significance to the European aristocracy, the Indian masses came to view it as a sign that the image was their personal guardian. When the bishop ordered the statue cleaned following the earthquake of 1834, the townspeople refused to permit it, arguing that altering its color would cause it to lose its miraculous power.

With the passage of time, the physical appearance of the image was upgraded as emeralds replaced iron nails, a crown of gold the wreath of thorns, and a richly embroidered skirt the original rag loincloth. In modern times, it has become traditional to parade the ornate *Señor de los Temblores* on a great silver litter through the streets of Cuzco on the Monday before Easter. As the bier passes, the townspeople throw red flower petals in its path in memory of the blood of Christ. The image is then returned to the cathedral to await the joy of another Holy Week or the anguish and suffering of another earthquake.

Sources: Gade 1970; Sallnow 1987; Radin 1969.

cakes of the same edible white clays consumed by Mayan pilgrims thousands of years ago but which are now marketed as *pan del Señor,* meaning the "Lord's bread." They also drink holy water from the same springs and, in many cases, visit the nearby mountains and caves at which their ancestors worshiped. By the late 1990s the cult had diffused widely throughout mainland Middle America, with nearly 200 secondary shrines established from central Mexico southward to Panama (Horst 1999 and 1990).

Challenges to Colonial Catholicism

Within a few decades of the Conquest, then, almost all of Latin America had become at least outwardly, or nominally, Catholic. Catholic churches, from massive metropolitan cathedrals to humble frontier missions, dotted the landscape from California to Tierra del Fuego. Crosses marked the sites of highway fatalities, and the socially prominent competed for the most prestigious burial plots within the courtyard cemeteries of the churches. Images of the Virgin and the saints were to be found in almost every home, business establishment, and government office. From the perspective of the Protestant-dominated regions of Anglo America and northern Europe, Latin America appeared to be universally Catholic.

Yet the very universality of Catholicism in Latin America masked and, indeed, contributed to significant challenges to the institutional Church. The first of these was that religious allegiance came to be viewed by many of the masses more as a culture trait than as a uniform set of spiritual beliefs and practices. To be Latin American came to be synonymous with being Catholic. Catholicism provided a common social bonding and sense of regional belonging and identity to peoples of all stations. Beyond that, however, it failed to require acceptance of and adherence to a common set of beliefs and practices. In so doing, it ultimately would be confronted with its members' widespread ignorance, and even rejection, of its teachings.

A second threat to the Church that stemmed from its universality was the development of a false sense of security among many of the institutional hierarchy. While complimenting themselves on their outward success in Christianizing the natives, the hierarchy often failed to instill a high level of moral discipline in their own members. Many of the secular clergy, in particular, were allowed to live their lives in a manner that was inconsistent with the strict standards of the Church. This situation frequently promoted widespread feelings of cynicism and sarcasm among the general populace toward the priests and, indirectly, toward the Church itself—a condition known as **anticlericalism.**

The third risk to the Church derived from its historic sponsorship by the state. Because civil officers controlled their appointments, the clergy felt beholden to some degree to the government authorities. The decisions of the priests and bishops were thus on occasion politicized and based more on short-term expediency than on long-term moral principle. This led the masses to view the institutional Church as an arm of a government, a self-serving regulatory body only distantly concerned with the welfare of the common people and often exploitative in its actions toward them. Failing to obtain from the clergy the emotional comfort and sustenance they desired, many Latin Americans sought spiritual fulfillment through personal and group activities unrelated to the rituals and ceremonies of the formal Church.

During the critically important early decades of the colonial era, the failure of the institutional Church to gain the allegiance of many of the people was compounded by its refusal to ordain Indians, blacks, or persons of mixed racial ancestry to the priesthood. Initially, the Church justified this policy on the grounds that the non-Europeans failed to meet the criteria laid down in the old doctrine of *limpieza de la sangre*, or purity of the bloodline, which required all candidates for the priesthood to prove that they had no heretics or nonbelievers in their ancestry for four generations prior to themselves. Although the law was occasionally ignored or overlooked—*mulattos* admitted to the priesthood in early nineteenth-century Brazil were simply classified as "white" and three Indians even rose to the office of bishop in colonial Mexico—the masses quickly came to view it as a vehicle for the perpetuation of the power of the aristocracy, from whose ranks the upper clergy were obtained.

Recognizing belatedly the harm created by the ordination practice, and desiring also to counter the growing influence of the proindependence creole clergy, the Crown finally rescinded the policy in the late 1700s and began actively to seek Indian and *mestizo* candidates for the priesthood. By then, however, it was a case of too little too late. The damage had been done; and Indians, blacks, and persons of mixed racial heritage would never look in large numbers to the Church as a vocation. The result has been a severe, chronic **shortage of clergy** throughout Latin America, which has led, among other things, to great numbers of rural communities going without resident priests for extended periods of time and to others being serviced largely by imported foreign clergy whose native culture is non-Hispanic.

THE ECONOMIC AND POLITICAL POWER OF THE CHURCH IN THE COLONIAL AND REPUBLICAN PERIODS

From the beginning of the Conquest onward, the privileged status of the Church resulted in its acquisition of great wealth and worldly influence. In the early colonial period, the Crown often gave free of charge, or sold at only a small fraction of their true value, *latifundios* and other income-generating properties to the religious and secular clergy. To this base, the Church steadily added additional possessions acquired through a variety of means.

One of the most common fundraising strategies was to cultivate among the devoted the notion that it was a sign of weak faith not to include the Church in one's will. In addition to receiving inheritances from conscience-stricken elites, the Church also received a portion of the tithes and "first fruits" offerings collected on its behalf by the government. The Catholic prohibition against clerical marriage resulted in the Church receiving, at the death of the clergy, all their individual possessions that otherwise might have been passed on to their families. Much income was also derived from sale of the products of Church-owned mines, quarries, factories, and shops. Church-owned private schools and universities also contributed considerable income, since the expenses for faculty salaries were minimal. Lastly, and very importantly, the Church became a leading money-lending institution. Nunneries were particularly active in this endeavor, often using the proceeds to finance hospitals, clinics, and orphanages. So pervasive did ecclesiastical lending become that it is estimated that at the end of the colonial era two-thirds of all the capital in circulation in New Spain was controlled by the Church (Parkes 1970, 111).

The cumulative, long-term result of these and other practices was that on the eve of the wars of independence, the Church had become the single wealthiest entity in Latin America, controlling, according to Mecham (1966), approximately half the arable lands and from 40 to 60 percent of all urban real estate. Combined with its dominance of education and its monopoly of the vital birth, marriage, and death registries, the Church's influence had clearly passed beyond the spiritual realm. With its continued power dependent on preserving the status quo, the institutional Church represented, in the minds of many at the end of the colonial era, a leading obstacle to social and economic change. It was not surprising, therefore, that after the defeat of Spain in the early nineteenth century, Latin American society quickly divided into two great conflicting camps, the remnants of which have continued

FIGURE 9.9

A notice painted on the outside wall of a church in Puebla, Mexico, indicating that the building was the property of the federal government.

to compete to the present time. Arrayed on the **conservative** side were those elements that favored the continued, unrestrained economic, social, and political influence of the Church. Opposing them were the **liberals,** who while often devoutly Catholic in their personal religious beliefs, nevertheless were convinced that socioeconomic restructuring and political democratization and pluralism could occur only through a separation of church and state.

While the religious history of each Latin American nation has varied during the republican era, the general pattern has been for liberal regimes to enact legislation limiting Church ownership of property and clerical involvement in politics and education (Figure 9.9). In extreme situations, liberal governments have also claimed authority, through modern interpretations of the old *patronato real*, to control the appointments and numbers of clergy functioning within a given country. Ironically, the desire to regulate the Church more completely has led many liberal governments to reestablish Roman Catholicism as the official state religion, in order to have a clear claim to control it, while simultaneously encouraging the preaching of non-Catholic faiths.

For their part, conservative governments tend to enforce less strictly those laws limiting Church involvement in earthly matters and often cite the need for a strong Catholic presence to combat perceived moral and political decadence. In most nations, an unwritten understanding has evolved between the Church and the state in which each attempts to avoid undue involvement in the other's sphere of influence.

THE CURRENT STATUS OF LATIN AMERICAN CATHOLICISM

In order to appreciate the nature of Latin American Catholicism, it is helpful to view it as a modified extension of southern European, or **Mediterranean, Catholicism.** As such, it differs significantly in a number of ways from Anglo American Catholicism, which emanated from Ireland, Germany, and other regions of northern Europe.

One of the most visually obvious differences between northern and southern European Catholicism is a lower level of attendance at Mass in Mediterranean and Latin American lands. Precise figures vary from place to place, often in response to the quality of pastoral care, but most observers in and out of the Church agree that for Latin America as a whole, only 10 to 15 percent of all Catholics attend Mass in a given week. The figure is often somewhat higher in conservative regions and lower in areas that historically have favored liberal causes. It also tends to be higher in urban zones, where priest-to-member ratios are more favorable to the Church, and lower in rural communities. Thousands of the latter have no resident priests of their own, and when a visiting cleric does come to officiate at a Mass it is not uncommon to see the church empty except for a few older women dressed in black and a grandchild or two brought along for the experience.

A second characteristic of Latin American Catholicism is a low level of sacramental participation. Because baptism, confirmation, and marriage are associated with the institutional Church and its representatives, neither of which historically has been viewed with trust by the masses, large numbers of Latinos have chosen not to seek these and other sacraments. The inclination not to participate has been especially pronounced among the poor, who have frequently viewed the fees levied by the clergy for such services as an unwarranted tax. The cumulative result is that baptism and confirmation are widely neglected, and in many parts of Latin America formal marriage "has become an almost unknown institution . . . reserved to the minorities of the socially prominent and a few white-collar workers" (Latorre Cabal 1978, 152). The abundance of common law unions has produced illegitimacy rates well above 40 percent in many nations (Table 9.2).

A third attribute of Hispanic Catholicism is a high level of anticlericalism, the origins of which were noted earlier. One interesting consequence of Latin American anticlericalism is the capacity of individual Catholics to criticize the clergy most severely while simultaneously holding the institutional Church itself blameless for the actions of its priesthood. For in-

TABLE 9.2   ⊠

Latin American Illegitimacy Rates

NATION OR TERRITORY	PERCENTAGE OF BIRTHS OUT OF WEDLOCK
St. Lucia	85.8
French Guiana	80.0
El Salvador	72.8
Martinique	65.9
Guadeloupe	63.0
Belize	59.3
Venezuela	53.0
Costa Rica	49.0
Peru	42.2
Mexico	27.5

Source: Compiled from *Britannica Book of the Year 2002*, 2002, 545–760.

[handwritten: De alguna forma no hay respeto pero si miedo]

stance, in jokes as well as in more somber conversation, it is not uncommon for many Latinos to openly accuse the local priest of the most serious violations of the Church's moral code and yet never themselves consider leaving the Church over the issue. This ability to differentiate between man and institution contrasts markedly with the Anglo American Protestant tradition, in which groups of members leave a religious body to form their own church when they become disillusioned by the behavior or teachings of their former pastor or minister. Yet another offshoot of anticlericalism is the tendency of many Latin American Catholics to ignore the teachings of the Pope and his representatives without feeling the least bit troubled or threatened in their Church membership.

A fourth and final characteristic of Hispanic Catholicism is a focus on what Willems (1967, 35) has called the "cult of the saints." This contrasts with Anglo American Catholicism, which historically has been more Christ-centered. Closely related to the cult of the saints in Latin America is the cult of the Virgin Mary, who in the minds of many Latinos stands as an intercessor between the saints and Christ, just as the saints are believed to act as intermediaries between her and those who are alive on the earth.

One by-product of the cults of the saints and the Virgin Mother is that images of Christ figure far less prominently in Latin American churches and homes than those of the saints and Mary. This pattern tends to be most pronounced in the old colonial-period structures, whereas the newer churches tend to be less ornate and image oriented.

Subtypes of Latin American Catholicism

Latin American and Anglo American Catholicism differ to such an extent that some have wondered if they should be classified as two separate faiths. Theologians have frequently vacillated between the two positions. Coleman (1958, 1), for example, has argued that "Catholics are literally one, no matter how great their variety or diversity." In the same study, however, he identified three principal subtypes of Latin American Catholicism: formal, nominal, and folk.

Formal Catholicism in Latin America consists of support for and participation in the teachings and programs of the institutional Roman Church. It is, in many ways, an extension of northern European and Anglo American Catholicism. Formal Catholics attend Mass regularly and make a conscious effort to learn and implement in their lives the doctrines of the Church as given through the priests and bishops. Anticlericalism is less frequent among formal Catholics than among the population at large, and families pride themselves on having a son or nephew who is a priest. Formal Catholics believe that spiritual salvation can be obtained only through participation in the sacraments (Figure 9.10).

Formal Catholicism is thus, in many respects, a culture trait of the affluent upper classes who reside in the regional and national urban centers. To the women and adolescent girls of these elite groups, it stresses obedience, salvation, piety, and the commitment of considerable amounts of time to devotional societies, charities, and social clubs. The men may join the Knights of Columbus or another religious brotherhood, if they so choose, and are taught that they have a responsibility to provide generous financial support for the programs of the Church.

The proportion of Church members who are formal Catholics varies by region (Figure 9.11). It is strongest in the old conservative mining and ranching communities of the interior highlands of western Latin America, beginning on the north with central and western Mexico and continuing southward through the volcanic basins of Central America into the high Andes of Venezuela, Colombia, Ecuador, Peru, and Bolivia. Secondary regions of formal Catholic strength include the Andean foothill settlements of northwestern Argentina, the Paraguay–Rio de la Plata Basin of Paraguay and Argentina, and the interior backlands of Northeast Brazil. The latter is also strongly influenced by Indian and African spiritist cults, which will be addressed later in this chapter. While formal Catholicism continues to exert great influence in Latin America, owing to the allegiance of the elites, it is, in reality, a minority religion that claims no more than 10 to 20 percent of the Church membership in conservative areas and even less in liberal regions.

FIGURE 9.10

Formal Latin American Catholicism, as practiced by the urbanized, European elites, stresses participation in the rituals and sacraments of the institutional church. This wedding was performed in Mexico City's national cathedral.

That's me

Nominal Catholics are defined by Coleman (1958, 2) as "believing but not practicing" members of the Church. They constitute the majority of the urban middle and lower classes, whose interests historically were opposed by the institutional Church. Nominal Catholicism, mixed with traces of relic folk Catholicism, is also predominant among the rural peasantry, many of whom have never known a resident priest.

Nominal Catholics rarely, if ever, attend a Mass or other Church ceremony. The majority of the men, in particular, can count on the fingers of one hand the number of times they have entered a church and have little inclination for personal spiritual renewal. When asked whether they still consider themselves to be Catholic, most will reply, *"sí, pero a mi modo,"* meaning "yes, but in my own way." Another common response is *sólo de nombre,* meaning "in name only."

Most nominal Catholics participate in few, if any, of the Church sacraments and consider anticlericalism a mark of a mature, sophisticated mind. Nominal Catholicism is easily the most common of the Catholic subtypes. Even in the conservative western highland regions where formal Catholicism maintains a significant presence, 60 to 70 percent of the Catholic population will usually fall into the nominal category. In the remainder of Latin America, including the Caribbean and Gulf coastal lowlands of Middle America, Chile, southern Argentina, Uruguay, southern Brazil, and almost all of the Amazon and Orinoco river basins, nominal Catholicism either reigns unchallenged or has mixed with folk Catholicism (Figure 9.11).

Latin America's third Catholic subtype, **folk Catholicism,** includes the beliefs and rituals of the isolated Indian peoples and also those religious practices that are largely of African origin (Figures 9.12 and 9.13). Both the Indian and African forms of folk Catholicism consist primarily of pre-Conquest ceremonies and experiences that historically have been viewed by the formal Roman Church as paganistic at best and Satanic at worst. The oppressed Indians and blacks soon discovered,

FIGURE 9.11

Latin American Catholic subtypes.

FIGURE 9.12

Folk Catholicism in Guatemala: a Maya Indian Easter offering of foods and flowers on the stone floor of the great cathedral of Antigua. Foods used include multicolored maize and beans, oranges, manioc, carrots, and cucurbits.

FIGURE 9.13

Folk Catholicism in Mexico: a Totonac headdress positioned at the head altar of a Catholic church in Cuetzalan, Puebla.

however, that if they wrapped their ancestral rites in an outer covering of European Catholicism, the institutional Church would tolerate the perpetuation of their traditional practices, no matter how unrelated they were to formal Catholicism.

In the case of the American Indians, the formal outer covering has frequently consisted of the use of European symbols and terminologies in the perform-

ance of the traditional ancestral rites and ceremonies. For instance, in describing the folk Catholicism of Zinacantán, a Tzotzil-speaking Maya community in the highlands of Chiapas, Mexico, Vogt (1970) noted the presence of hundreds of wooden cross shrines scattered across the mountainous landscape. To the casual observer, the shrines—called *kalvarios* after Calvary hill where Jesus Christ was crucified—made Zinacantán appear outwardly to be one of the most intensely Catholic settlements on earth. Reinforcing the physical appearance, the national census showed Zinacantán to be overwhelmingly Catholic. Yet, after living among the people for an extended time, Vogt learned that the *kalvarios* were considered by the Indians to be doorways or entrances to the abodes of their ancestral nature gods rather than Christian representations of Christ's crucifixion. Not surprisingly, the rituals performed at the *kalvarios*, including the placement of pine boughs and flowers on the crosses, the burning of incense, and the offering by shamen of black chickens, white candles, and liquor, go back to the dawn of Mayan history. Although listed by the formal Church as Catholics, Vogt (12) observed that the Zinacantecos were, in reality, "Maya tribesmen with a Spanish Catholic veneer."

While folk Catholicism survives today among the least Europeanized Indian groups, it is clearly a small remnant religion whose overall status is diminishing as the tribal Indians are progressively integrated into the cultural mainstream. Rural literacy campaigns, improved transportation and communication networks, modern medicine, and racial whitening all threaten the long-term survival of Indian folk Catholicism. Watanabe's (1992) recent study of Santiago Chimaltenango, a highland Maya community of western Guatemala, documents well the dissolution of folk Catholicism at the village level. Here, where folk Catholicism flourished as recently as a generation ago, the Chimaltecos no longer believe in the spirits of mountains and other natural objects. Medical doctors are now preferred over shamen as healers, and the once-thriving *carguero*, or folk religious hierarchy system, is being steadily secularized as formerly appointive positions are opened by the national government to popular voting.

In contrast to the progressive decline of folk Catholicism among the American Indians, black African cults are flourishing in many areas of the Caribbean Basin and Brazil. The origins of African folk Catholicism are diverse, and the various branches have formed and evolved in the New World in response to changing social and cultural circumstances. To appreciate how they came into being, it is necessary to remember that the blacks who were brought to Latin America and the Caribbean as slaves came from distinct geographical and ethnic backgrounds. One of the most influential source regions of New World African religions was the area in and around the modern nations of Togo, Benin, and Nigeria, which supported the highly advanced Yoruban and Dahoman civilizations. Many other slaves, however, came from the Congo Basin of central Africa and from the Bantu-speaking areas of Angola and Mozambique.

These and other African peoples transported against their wills to the New World brought with them their own religious beliefs and practices. Although the religions differed one from another, most included strong elements of nature and ancestor worship. In some cases, these features were combined by attributing to a certain mythical ancestor power over a part of nature, such as the rains, an animal, or even a substance such as iron. Furthermore, in many of the most prominent of the African religions, worshipers believed that the ancestral spirit gods had the power to communicate with their descendants by entering or "mounting" the physical bodies of their devotees. When this occurred, the possessed humans assumed completely the personality, desires, and physical attributes of the god who had mounted them. The possessed persons also frequently received knowledge of future events and instructions on how to prepare for or respond to them.

The African belief systems were generally peaceful and contributed much to the promotion of morality, justice, individual self-worth, and group identity among the oppressed blacks of the New World. However, the formal Catholic Church rejected as categorically evil the practice of spirit possession. The African custom of sacrificing animals in order to obtain blood to link the physical and spiritual realms was also repugnant to the Europeans. As threatened as they were by these African practices, nevertheless the Church fathers were required to see that the slaves, like the Indians before them, were taught the rudiments of Christianity and baptized for the salvation of their souls. The blacks were thus forced to become official members of the very Church that was attempting to destroy key elements of their culture.

As a survival mechanism, the slaves pretended to accept Christianity. Following their forced baptisms, they erected small Catholic altars against the walls of their living quarters and assigned to each of their gods the name(s) of the Catholic saint(s) or godhead member whose attributes and functions corresponded most closely with those of the African deity. Having thus satisfied their European masters, they returned to their ancient rites, dancing night after night before the candle-lit altars upon which were placed the images of ancestral gods masquerading as Catholic saints. As the slaves learned that the whites lacked the power to deprive them of their beliefs and worship, the African re-

ligions came to be viewed as expressions of black resistance and, in a very real sense, the essence of black identity in plantation America. In particular, they have influenced the development of art, music, medicine, and folklore throughout the region (Voeks 1993 and 1997). The association of African folk Catholicism with black cultural survival in the Caribbean and Brazil is a principal reason for the present strength of these belief systems.

FORMS OF AFRICAN FOLK CATHOLICISM

As the various religious traditions have mixed and adapted over time to local conditions in Latin America, a number of distinct forms of African folk Catholicism have evolved. Each form in its present state consists of a mixture of African, American, and European influences. The most publicized of these folk Catholic forms among Anglo Americans and Europeans is *voodoo*, which is also known among academics as *vodoun*, *voudoun*, *vodun*, and *vodu*. The term "voodoo" is derived from the Fon language of Dahomey (now Benin) and means "god" or "spirit." It is practiced primarily in Haiti and the French Antilles, where the African gods are called *loas*. *Voodoo* was used as a unifying force by Toussaint L'Ouverture in promoting the great Haitian slave uprising of the early nineteenth century and has since become the basis of culture and nationalism among Haiti's repressed masses. It is rooted in the rural peasantry and has been the object of numerous purge campaigns by the westernized urban Haitian elites, who, as we noted in chapter 7, have rejected their African heritage (Greene 1993; Desmangles 1992). Haitian voodoo presently consists of two cults, whose altars are generally placed in separate rooms of the *houmfort*, or temple (Heusch 1989). The more dominant cult is called *rada* and focuses on communion with ancient Fon deities, most of whom are peaceful and beneficial. The smaller *petro* cult is of mixed Congo and Creole heritage and is often associated with the use of violent gods to accomplish evil purposes and, on occasion, with the practice of zombification (Davis 1986).

The most widespread form of African folk Catholicism among the Spanish-speaking Caribbean peoples is *Santería*, a term which translates as "the worship of the saints." Although *Santería* focuses outwardly on veneration of the images of Catholic saints, to the *santero*, or practitioner of *Santería*, the images are really representations of ancient Yoruban gods called *orishas* (Fernández Olmos and Paravisini-Gebert 1997; Brandon 1993; González-Wipper 1973). The spirits of the *orishas*, like the Haitian *loas*, have the power to possess the believer, who may engage in violent dancing and other strenuous physical behavior for hours before collapsing when the *orisha* departs. *Santería* is also associated with much sorcery and animal sacrifice. By acting

Spanish speaking Caribbeans

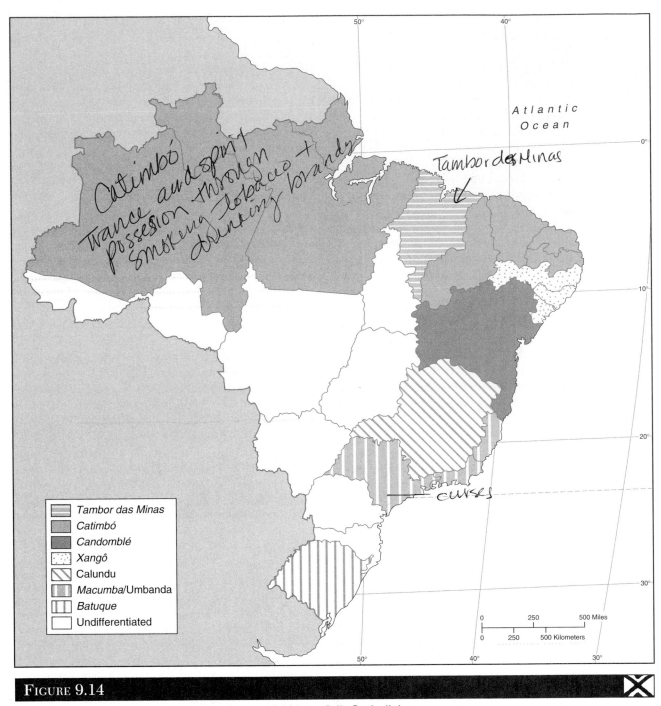

Catimbó
trance and spirit
possesion through
smoking tobacco +
drinking brandy

Tambor das Minas

curses

Tambor das Minas
Catimbó
Candomblé
Xangô
Calundu
Macumba/Umbanda
Batuque
Undifferentiated

FIGURE 9.14

Generalized distribution of Brazilian forms of African folk Catholicism.

Portuguese speaking

out a desired beneficial or malevolent outcome, *santeros* believe that they can cause the event to happen.

In Portuguese-speaking Brazil, African folk Catholicism has taken a number of forms. With the exception of the state of Maranhão, which is dominated by the worship of Dahoman deities belonging to a belief system called *Tambor das Minas* (Ferretti 2001), most of northern Brazil, from Amazonas to Paraíba, is character-

ized by a mixed American Indian–African folk religion known as *Catimbó*, or *santidade* (Figure 9.14). In contrast to the more purely African cults, *Catimbó* rejects dancing and drumming in favor of spirit possession achieved through trances induced through the smoking of tobacco and the drinking of brandy.

The leading form of African folk Catholicism for Brazil as a whole is *Candomblé*, which originated in the

Case Study 9.2:   Jamaican Rastafarianism

The beginnings of Rastafarianism, one of the Caribbean's most unique religious movements, can be traced to the twentieth-century teachings of two Jamaicans: Marcus Garvey and Leonard Howell. Born in 1887, Garvey grew up with a strong sense of pride in his African heritage and an enduring belief in the goodness of all black peoples. He became a wealthy and highly respected and charismatic leader of Jamaican blacks in his adult years, and often prophesied that a black king would be crowned in Africa, through whom God would deliver repressed blacks the world over. Garvey's belief in Africa as a divinely appointed black homeland was expressed in his slogan "Africa for the Africans—At Home and Abroad." It also led him to found a movement that he called the Universal Negro Improvement Association, and to organize an ill-fated steamship company, named the Black Star Line, which he intended to use to transport blacks back to their promised land of Africa.

The ascension of a young Christian prince named Ras Tafari as Emperor Haile Selassie of Ethiopia in 1930 was seen by Garvey as the fulfillment of his prophecies concerning the rise of a black savior. Although Garvey spent much of his adult life in the United States, his teachings were adopted and modified by the more radical Leonard Howell. Howell, a Jamaican seaman, had previously resided in the United States where he had been influenced by a black sect that called itself the "Israelites." In 1940 he established a commune after the manner of his Maroon ancestors. Howell declared that all whites were evil and inferior and should be hated and mistreated by blacks as revenge for the past wrongs inflicted by the whites upon God's chosen black people. He further taught that Haile Selassie was the Living God, descended from King Solomon and the Queen of Sheba, whose rise to power represented the promised Second Coming of Jesus Christ. As such, it was the religious duty of all Jamaican blacks to return to Ethiopia, meaning Africa.

Howell's commune was raided by Jamaican police on two occasions in the 1940s and 1950s, and his followers, now calling themselves Rastafarians after Ras Tafari, migrated into Kingston where the movement became urban-based. Rastafarianism, with its use of Protestant hymns, a form of dancing called "nyabinghi," dreadlocks, and marijuana ("ganja") to achieve spiritual awareness, peaked as an anti-establishment black power movement in the 1970s. The deaths of both Haile Selassie and Bob Marley, whose reggae music had become closely identified with the group, together with the rise of fundamentalist pentecostalism on the island, have left Rastafarianism in a weakened condition. Although condemned by most middle- and upper-class Jamaicans, the movement continues to provide a sense of self-worth to some members of the urban underclass.

Sources: Chevannes 1998; Besson 1998; Simpson 1970; Gullick 1983.

state of Bahia. *Xangô*, a faith closely related to *Candomblé*, is widely practiced in Bahia's bordering states of Sergipe, Alagôas, and Pernambuco. Other *Candomblé*-based folk religions in Brazil include calundu or *canjere* in Minas Gerais and *batuque* in Rio Grande do Sul. In the twentieth century, the Yoruban-based *Candomblé* has mixed with Bantu influences among the urban poor of Rio de Janeiro, Espírito Santo, and São Paulo to form a new movement called *macumba.* Recently, the cult has become highly commercialized, with the priests using black magic and burning incense and sacrificing animals more for their artistic and entertainment value than for worship purposes. One of the principal uses of *macumba*, by blacks and whites alike, is to have curses placed on rivals in sports, politics, and love. Yet another variant of African folk Catholicism, which is strongest in the urban centers of São Paulo and Rio de Janeiro, is a spiritist movement called Umbanda, which has historical ties both to *macumba* and to East Indian Hinduism (Motta 2001).

In addition to the leading forms of African folk Catholicism discussed above, numerous secondary syncretistic movements have emerged and will likely continue to evolve. These include the María Lionza cult of Venezuela, which is of African, American Indian, and Roman Catholic origins, and Jamaican Rastafarianism and Procomania, both of which have significant Protestant infusions (Case Study 9.2). The Trinidadian Orisha (Shango) movement consists of a blending of five religions: African, Roman Catholic, Hindu, Protestant, and Kabbalah (an esoteric belief system of European origins). The intensely fluid and changeable nature of the Orisha and other syncretistic movements was illustrated well by Houk (1995, 61), who described attending a fourteen-hour worship service that included the beating of drums and singing of songs to an Orisha god, offering prayers to Hindu deities, singing hymns to Baptist powers, and reciting incantations to Kabbalah entities. Because there exist so many possible degrees of involvement in African beliefs and practices, it is often impossible to distinguish between an inactive nominal Catholic and an African folk Catholic. While there are far fewer folk Catholics than nominal Catholics for Latin America as a whole, many observers estimate that upwards of 30 percent of all Brazilians, and an even higher percentage of Caribbean peoples, practice some form of African-American religion.

NON-CATHOLIC FAITHS IN LATIN AMERICA AND THE CARIBBEAN

To this point, we have learned that Latin America and the Caribbean became universally Catholic in the early colonial era owing to the interrelationships of church and state in Spanish and Portuguese New World settlement policies. The lack of doctrinal understanding and institutional loyalty among many Catholic converts, together with the Church's internal tolerance of non-European beliefs and practices, led to the development of the Catholic subtypes and forms that presently characterize the region. To complete our study of Latin America's religious geography, we will now turn to an overview of the non-Catholic faiths that have established themselves within the region.

Hinduism and Islam

Hinduism and Islam, two of the Old World's most dominant belief systems, are found today as small minority faiths in portions of the Caribbean Basin. Hinduism was established in the New World by thousands of indentured South Asian laborers who were taken by the British primarily to Guyana, Suriname, Trinidad and Tobago, and Jamaica during the nineteenth and early twentieth centuries. Islam diffused to the Caribbean in two small waves. The first consisted of indentured laborers transported by the Dutch from Java and the Indian subcontinent to Suriname. The second wave was composed of Turks, Arabs, Syrians, Lebanese, and other Middle Easterners who fled the crumbling Ottoman Empire in the late 1800s.

When the Hindus and Muslims first arrived in the New World, they endeavored to perpetuate the religious beliefs and customs of their ancestors. However, their inability to maintain regular contact with the main body of Old World believers, together with their constant exposure to new religious philosophies, resulted eventually in widespread changes in their observances and, in some cases, schisms and loss of members. One of the best known of the Muslim divisions occurred over the simple issue of personal prayer, which the faithful are taught should always be offered while facing Mecca, the Islamic holy city. When the Javanese Muslims first were brought to the Guianas, they continued the oriental custom of praying while facing west, which is the direction of shortest distance between Java and the Arabian peninsula. Later, a more geographically literate newcomer pointed out that the shortest distance between northern South America and Mecca was obtained by facing east. The division that soon arose over the issue became so heated that the members of the opposing groups refused to intermarry or even to speak to one another (Lowenthal 1972). As this and other controversies weakened their internal cohesiveness, the Guiana Muslims also faced mounting economic and political opposition from the black Creole majority populations of the region. The combination of doctrinal dilution and growing persecution eventually led to the partial assimilation of many of the oriental Muslims into the Hindu ranks. Meanwhile, many of their Middle Eastern brethren were abandoning altogether their Islamic identity through intermarriage and absorption into the Caribbean upper class.

While Islam has thus largely disintegrated in the greater Caribbean Basin, Hinduism has also struggled and changed. Many of the rural Hindus have become Christians. Those who have not now practice a Christianized form of Hinduism that rejects caste while mimicking Christianity in holding weekly weekend worship services, employing priests who minister to the faithful, adopting standardized worship services and rituals, and opposing the practice of spirit possession (Vertovec 1992; Lowenthal 1972).

Hinduism and Islam have thus evolved over time from distinct Old World religious enclaves into a Caribbean ethnic grouping that is generically referred to as "East Indian." Owing to its minority status in the black- and mulatto-dominated Guiana nations, East Indianness is now associated primarily with a non-Creole lifestyle whose most visible aspects include forms of personal greeting, dress, diet, and entertainment. Trinidadian East Indians of all religious persuasions, for example, exhibit their Indianness when greeting each other by pressing the palms of their hands together as though in prayer (Clarke 1986). East Indian women wear saris; others do not. East Indians also use much rice, curry, and spinach in their diet. Indian movies and novels are popular not only because of their intrinsic merits but because they serve to set their users apart culturally.

As a combined East Indian ethnic grouping, Hindus, Muslims, and other Asian-Americans now constitute between 30 and 50 percent of the populations of Trinidad and Tobago, Guyana, and Suriname. There are smaller enclaves in Jamaica and in other portions of the English- and Dutch-speaking Caribbean realms. Hinduism and Islam are virtually nonexistent within the American Indian, Hispanic, and black populations of Latin America.

Judaism

One of the least understood aspects of Latin American religious geography has been the continuous presence of Judaism within what, for centuries, was widely perceived as an exclusively Catholic region. To appreciate the nature of the Jewish experience in Hispanic America, it is necessary first to review briefly some key aspects of Old World Jewish history.

The Jewish presence in Latin America can be traced to the destruction of Jerusalem by Roman legions in A.D. 70 and the subsequent Diaspora, or scattering, of the survivors throughout the known world. Those Jews who obtained refuge in eastern Europe, and who came to speak Yiddish, became known as **Ashkenazim.** Those who migrated into northern Africa and western Europe, including Iberia, became known as **Sephardim.** Despite the early Christian prejudice toward them, Sephardic Jews had come to occupy a prominent place in Spanish commerce, education, and industry at the time of the Moorish conquest in the early eighth century. As Muslims, the Moors accepted the Jews and encouraged them to remain in Iberia for three reasons. The first was that the Jews and Arabs shared a common ancestry from the patriarch Abraham and respected each other's beliefs. The second was that the Jews contributed much to the economic and cultural vitality of the region. The third was that the Jewish achievements served to limit Christian influences. With the beginning of the Christian reconquest of Iberia, the Jews experienced renewed persecutions, which eventually culminated in their being forced to choose between death, abandonment of their homes and businesses, or conversion to Roman Catholicism. Although the majority of the Iberian Jews opted for expulsion, large numbers chose to accept a Catholic baptism. Following their baptisms, however, many of these *Marranos,* or Christianized Jews, continued to practice Judaism in secret. They made up a significant portion of the so-called New Christians that the old, established Spanish Christian families attempted to exclude from power through the doctrine requiring a four-generation purity of bloodline.

An undetermined but substantial number of the Spaniards and Portuguese who came to Latin America in the early colonial era were New Christians of Jewish or part-Jewish ancestry who had minimal prospects for personal advancement in the Old World. Most of these were males. Once in the New World, some took Indian wives and allowed their offspring to be raised as Catholics. Others, however, attempted to pass their Jewish heritage on to their children. Those that were caught were tried by the Inquisition. If unrepentant, they were sentenced in a ceremony known as an *auto de fe* to be burned publicly at the stake. The descendants of those Jews who escaped detection attempted to perpetuate clandestinely the faith of their fathers as best they could (Liebman 1970). One rural Jewish congregation has recently surfaced near Pachuca, Mexico, and it is possible that other Sephardic remnants will yet be discovered (Case Study 9.3).

Large numbers of Jews were also present among the early Portuguese and Dutch settlers of Northeast Brazil. An estimated half of the European population of Recife was Jewish in 1645 (Eliade 1987). When Holland lost Brazil in 1654, many of these Jews accompanied the Dutch to Suriname and the Caribbean Islands, where they played major roles in sugar cane farming, banking, and commerce down into the twenty-first century. Jews were so influential in nineteenth-century Jamaica that it was common for the House of Assembly to adjourn its legislative session for Yom Kippur, the Jewish Day of Atonement. Jews also dominated the commerce of Curacao for much of its history. As prominent as the Sephardic Jews were, however, most eventually married into local, upper-class Christian families and ceased to practice their faith. At the turn of the twentieth century, fewer than 10,000 practicing Jews remained in all of Latin America, many of those in small agricultural colonies (Cohen 1971).

The most recent wave of Jewish immigration into Latin America, which lasted from 1889 into the mid-1940s, was dominated by Ashkenazim fleeing the Nazi Holocaust, the spread of Communism, and other threatening developments in eastern Europe (Avni 1991; Mirelman 1990; Levine 1993). The vast majority of these settled in the urban centers of Argentina and Brazil with secondary clusters in Mexico, Uruguay, Venezuela, and Chile (Table 9.3). Since World War II, most Latin American nations have maintained low Jewish population levels through strict immigration restrictions. It appears that the number of Latin American Jews is now increasing only modestly, with secularism, internal disunity, intermarriage, and assimilation offsetting moderate birth rates.

Protestantism

Protestantism, the largest and most rapidly expanding of Latin America's non-Catholic religious movements, was outlawed throughout the colonial era by the Spanish and Portuguese governments, who viewed it as an arm of the rival English and Dutch empires. Its introduction into most of the Latin American countries occurred in the nineteenth century as liberal leaders sought out English and American educators and businessmen with the expectation that their presence would promote the economic development of the newly independent republics. Expressive of the attitudes of many liberal reformers was Benito Juárez'

Case Study 9.3   Venta Prieta, a Jewish Community in Mexico

Contrary to popular perception, the Spaniards who colonized Latin America were not a religiously unified body. While many came from families that had been committed Catholics for generations, others were descended from Moorish or Jewish ancestors who had outwardly converted to Christianity in order to escape exile or death during the religious purges associated with the *Reconquista*. Many of the converted Jews, who were derisively called *conversos*, meaning "New Christians," or *Marranos*, meaning "swine," by members of the Romanized elite, clung privately to the tenets of their Hebrew faith.

With the discovery and settlement of the New World, many Jewish *conversos* chose to come to America, where they often sought residence in physically isolated regions that afforded them the opportunity to practice their religion with a minimum of disruption from the Spanish authorities. However, the continuing purges directed by the Office of the Inquisition, together with their poverty and small numbers, resulted in most Jewish congregations having to worship without the guidance of a trained rabbi. While this led, over time, to a certain amount of religious syncretism, the fundamental beliefs and practices of the faith often remained intact to a remarkable degree.

One rural Jewish congregation that has survived to the present day is found in Venta Prieta, a tiny settlement situated on the outskirts of Pachuca City in the eastern mountains of central Mexico. According to an excellent study by Gardy (2000), the Venta Prieta congregation can be traced back to a Jewish family surnamed Tellezgirón that arrived in Veracruz in the sixteenth century and escaped to Zamora, Michoacán, before fleeing to the highlands of the state of Hidalgo under the shortened name of Téllez. By the 1850s, the Jewish portion of the Téllez family had been reduced to a single couple who founded Venta Prieta. Because some of their descendants intermarried with Catholic neighbors, one branch of the family became Catholic while the other preserved its Jewish heritage. Those who chose to be Jewish held their religious services in the room of a home until they were able to construct a small temple, which was completed in 1923. Fearing increased persecution, the members tore the temple down during World War II, after which a larger edifice was dedicated in 1967. All the while, the Venta Prieta Jews felt compelled to hide their beliefs, and it was not until the 1960s that they openly acknowledged their faith.

Today, the small Jewish community of Venta Prieta has an average attendance of around thirty worshippers at its Sabbath-day services. As do other Jews, the members abstain from eating pork. The temple, which can accommodate about 100 persons, contains Stars of David and a Sacred Ark, which houses the Torah. An eternal flame burns in a bowl filled with olive oil. Notwithstanding their devotion to their faith, expressions of religious syncretism are also evident. The members, for example, make the sign of the cross when touching the Torah's cover with their hands. Recently, ties have been established with the large Jewish synagogue in Mexico City where the children of the Venta Prieta Jews now go to be prepared for their bar mitzvahs (coming of age ceremonies) and to be married.

TABLE 9.3

Latin American and Caribbean Jewish Populations

NATION	JEWISH POPULATION 1982	1998	PERCENTAGE OF TOTAL LATIN AMERICAN/CARIBBEAN JEWISH POPULATION 1998
Argentina	242,000	230,000	45.17
Brazil	110,000	130,000	25.53
Mexico	35,000	40,700	7.99
Uruguay	40,000	30,000	5.89
Venezuela	17,000	30,000	5.89
Chile	25,000	21,000	4.12
Panama	n.a.	7,000	1.38
Colombia	7,000	5,650	1.11
Peru	5,000	3,000	0.59
All Others	12,200	11,840	2.33
TOTALS	493,200	509,190	100.00

Sources: Himmelfarb and Singer 1982, 276; Beker 1998, various pages.

comment that he was encouraging Protestant proselytizing efforts in Mexico because the missionaries would teach the Indians to read rather than "to waste time lighting candles to the saints," and the Argentine Juan Bautista Alberdi's description of Protestantism as a "symbol of cultural progress" (Damboriena 1962–63, Vol. I, 18). The efforts of occasional liberal reformers notwithstanding, very few northern Europeans and Americans chose to reside in Latin America during this turbulent period, and those who did seldom attempted to share their beliefs with their Latino associates. At the turn of the century, there were fewer than 100,000 Protestants in all of Latin America, most of whom resided in isolated urban enclaves or in remote, foreign agricultural colonies (Sawatzky 1971; Tullis 1987). The Anglican, Lutheran, Presbyterian, Baptist, and other mainline European denominations represented seemed to have little appeal to the native peoples.

The beginnings of a truly indigenous Latin American Protestantism occurred in the late 1800s among Methodist congregations in central Chile. There, United States missionaries began introducing fundamentalist practices, including healings, visions, prophesying, and speaking in tongues. News of these so-called charismatic gifts, and the spiritual trances called *tomadas de espíritu* that often accompany them, greatly alarmed Methodist authorities in the United States, who responded by replacing the missionaries and condemning the pentecostal practices. That action, in turn, provoked the charismatic Chilean Methodists to break with the mother Church and to form, in 1909, an independent denomination known as the Pentecostal Methodist Church (Bastian 1990; Vergara 1962; Hoover 1948). Meanwhile, other, mostly North American–based charismatic Protestant and quasi-Protestant faiths, including the Assemblies of God, Church of God, Seventh-Day Adventist, The Church of Jesus Christ of Latter-day Saints (Mormon), Church of the Nazarene, and the Jehovah's Witnesses, were being established throughout Latin America. Many of these have continued to splinter into countless independent evangelical bodies, and it has become increasingly evident that a native Latin American Protestantism now exists whose explosive growth merits a more detailed analysis.

CHARACTERISTICS OF AND CHALLENGES TO LATIN AMERICAN PROTESTANTISM

Latin American Protestantism today is characterized by seven attributes. The first is that its appeal, like that of early Christianity within the declining Roman Empire, has been greatest among the rural peasantry and urban lower classes, both of which have remained economically dispossessed and socially alienated despite the progressive collapse of the old order. Having noted the broader social context within which most Protestant conversions occur, we must also stress that the individual decision to change religions remains a momentous one. In Latin America, it is seldom made in haste and is preceded almost invariably by extensive social and doctrinal assimilation. The old notion was that the peasantry is attracted to Protestantism because it offers an escape from the financial burdens of the folk Catholic *carguero* or *fiesta* system. But that explanation can no longer be considered valid in light of the decline of folk Catholicism and the heavy financial obligations, including the payment of a full tithe, associated with most of the Protestant denominations.

The second attribute of current Latin American Protestantism is that it is dominated by fundamentalist, charismatic beliefs and practices. The most charismatic and fastest growing of all the denominations tend to be the Assemblies of God, Church of God, and other pentecostal faiths whose followers are widely referred to as *evangélicos.* Such groups generally comprise two-thirds to three-fourths of the Protestant population of most Latin American countries. In the middle of the ideological spectrum are the Seventh-Day Adventists, Jehovah's Witnesses, and Mormons, and the least charismatic are the older, mainline denominations of European origin. Even these latter churches, however, tend to be much more charismatic in Latin America and the Caribbean than their counterparts in the United States and Canada. Martin (1990), for example, has noted that as pentecostalism expanded in Jamaica, Anglican congregations on the island became more charismatic, adopting among other practices the evangelical customs of baptism by immersion and the laying on of hands. Simultaneously, some of the island's Presbyterians began speaking in tongues.

Of the many so-called gifts of the spirit, Latin American *evangélicos* most value that of healing (Figure 9.15). All the other gifts are greatly esteemed, however, as is the emotional euphoria achieved as the Holy Spirit fills the soul of the believer. Bastide (1978) has made the argument that the pentecostal *tomada de espíritu* is very similar to the African folk Catholic process of spirit mounting. In both instances, the believer assumes a different personality and is endowed with powers not otherwise available.

A third feature of Latin American Protestantism is its emphasis on unpaid church service. This stress on volunteerism is a reflection of the movement's highly fluid and egalitarian nature in which church positions are shared and rotated regularly among the faithful. This has resulted in the development of strong leadership and management skills within Protestant ranks in such areas as organizational administration, public speaking, planning, reporting, and budgeting. As a consequence, many Protestants have strong, dormant managerial inclinations and abilities and are quick to move into the entrepreneurial sector when given the opportunity.

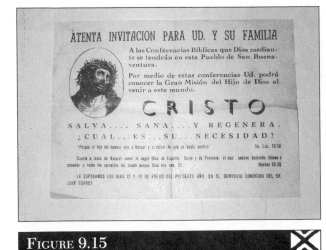

FIGURE 9.15

Pentecostal flyer advertising the gift of healing to be manifested at a forthcoming worship service in a rural Mexican village.

FIGURE 9.16

The indigenous and charismatic character of most Latin American Protestantism is manifest in this evangelical revival conducted by native ministers on the main plaza in front of the Catholic cathedral of Tegucigalpa, Honduras. Those participants who support the teachings of the preachers have raised their arms and bowed their heads in prayer.

A fourth characteristic of Latin American Protestantism is its small church organizational structure. Although exceptions exist, the great majority of congregations number no more than 200 to 300 members and many are as small as 30 to 50. The Protestant church is thus small enough to function effectively as the extended family of the members, many of whom were rejected by their birth family when they joined the new faith. Those converts who remain active soon find that their lives are wholly reoriented toward shared experiences with their fellow members, whom they now address as *hermanos y hermanas,* "brothers and sisters." The congregation also becomes one's social and economic security net, serving, in effect, as a mutual aid society capable of providing employment, physical health, and self-esteem. The Catholic *compadrazgo* system, with its expectations of reciprocal favors, is generally abandoned. Women, in particular, are attracted to the economic and emotional security of Protestantism, whose values are in many ways almost the opposite of the traditional *macho*-dominated Hispanic worldview (Eber 2000; Sherman 1997; Brusco 1995; Smilde 1994). The sense of family and belonging associated with Protestantism contrasts with most Latin American Catholic parishes, the majority of whose members have more limited associations one with another.

A fifth quality of Latin American Protestantism is that it is largely indigenous and independent of foreign control. Not only is this the case within the small, independent, local evangelical churches, but it is also surprisingly valid among the Assemblies of God, Mormons, Jehovah's Witnesses, Methodists, Baptists, and other denominations that were initially introduced by Americans or Europeans. Without exception, these faiths are now sustained almost entirely by local ministers and local missionaries (Figure 9.16). Moreno (1997), for instance, reports that 99 percent of Latin American Pentecostal congregations are led by native pastors. Martin (1990, 52) has observed further that "North American personnel may be present but they are for the most part not in control." In an interesting historical twist, Latin American Protestants, who have often been accused in the past by threatened Catholic leaders of being agents of United States economic and military interests, are now responding that it is Catholicism that is most dominated by foreign personnel and ideology.

The sixth attribute of Latin American Protestantism is that it is extremely fragmented. Intense rivalry prevails among the different churches, whose ministers and believers frequently do not hesitate to verbally attack and proselytize one another. The seventh, and last, feature is its rapid growth. While exact membership figures are impossible to obtain, owing to poor record keeping, frequent double-counting, desertions, and definitional differences, Protestantism's growth rate has clearly outstripped overall population growth over the past several decades, and the movement now claims more than one of every eight Latinos (Table 9.4). One example of this explosive growth is Rio de Janeiro city, where 673 new evangelical churches and 214 spiritist centers were established from 1990 to 1992, as compared to one new Catholic parish (Table 9.5) (Azevedo 1992, 83). Given the low levels of Catholic attendance at Mass and the

TABLE 9.4

Protestant Growth in Latin America

YEAR	TOTAL PROTESTANT MEMBERSHIP	PERCENTAGE OF TOTAL LATIN AMERICAN POPULATION
1916	123,000	.15
1937	1,219,000	1.02
1961	10,128,000	5.12
1990	40,000,000	9.30
2000	70,758,000	13.63

Sources: Compiled from Johnstone and Mandryk 2001; Minard 1990; Rycroft and Clemmer 1963.

TABLE 9.5

New Ecclesiastical Units Established in the City of Rio de Janeiro, Brazil: 1990–1992

	1990	1991	1992	TOTAL
Evangelical churches	187	262	224	673
Spiritualist centers	50	81	83	214
Catholic parishes	0	1	0	1

Source: Azevedo 1992, 83.

high levels of Protestant member involvement, it is possible that a number of Latin American nations, including Chile and the countries of Central America, now have more practicing Protestants than practicing Catholics (Figures 9.17 and 9.18). Brazil is reported to have more full-time Protestant pastors than Catholic priests (Smith 1995, 1; Martin 1990, 50). So rapidly has Protestantism diffused that almost every rural village and urban *barrio* now has one or more *capillas*, or chapels. These have become almost as common a landscape feature as the colonial Catholic churches (Figures 9.19 and 9.20).

Some scholars have suggested that Latin America may now be experiencing a belated Protestant Reformation. Like the European Protestant Reformation of the sixteenth through eighteenth centuries, the Latin American Reformation has appeared at a time of great social and economic change brought on, in part, by rapid industrialization and urbanization (Martin 1990; Norman 1981). As with the Protestant Reformation of Europe, the ongoing Latin American Protestant Reformation is associated with a strict morality and with charismatic Christianity. Whether or not these attributes will continue to characterize the movement is impossible to predict, as is the ultimate extent of Protestant growth. Current growth patterns do suggest that certain nations and regions may eventually have a Protestant majority.

Perhaps more important than statistical projections are the issues of what Protestantism represents in Latin America and what changes it is likely to encourage. In attempting to answer these questions, we must begin with the recognition that Latin American Protestantism is supremely individualistic. Just as spiritual salvation is achieved only through the individual transforming himself or herself by abstaining from previous *vicios*, or vices, such as drinking alcohol, carousing, gambling, and cursing, so too the Latin American Protestant sees individual self-improvement and hard work as the vehicles to material salvation in this life. Consequently, he or she places great stress on education, thrift, honesty, and integrity. The result, as numerous studies have documented, beginning with Willems (1967) and continuing with Hinshaw (1975), Muratorio (1981), and Sherman (1997), is a clear multigenerational pattern of upward socioeconomic mobility among Latin American Protestants. Guatemalan *evangélicos* refer to this process as *del suelo al cielo,* meaning to move from dirt floor to heaven or from poverty to prosperity (Annis 1987).

While some pentecostal pastors may employ authoritarian models of leadership in their relations with their followers (Bastian 1992; Williams and Peterson 1996), Protestantism as a whole is clearly contributing to the transformation not only of Latin America's religious landscape but of some aspects of its fundamental Hispanic culture as well. Its values are generally opposed to those of *personalismo* and *machismo.* Whether Protestantism is the result of, or the cause of, the far-reaching changes now sweeping the region, it is a reform movement that has become widely associated with the causes of political pluralism and social egalitarianism. Protestants have served as elected heads of state in Brazil and Guatemala and are becoming increasingly involved in the political arena in many Latin American nations (Steigenga 2001; Cleary and Stewart-Gambino 1997; Bonicelli 1997). Protestantism's growth has also contributed to a number of changes in the modern Latin American Catholic Church.

FIGURE 9.17

Protestant percentage of population.

Source: Data derived from Johnstone and Mandryk 2001.

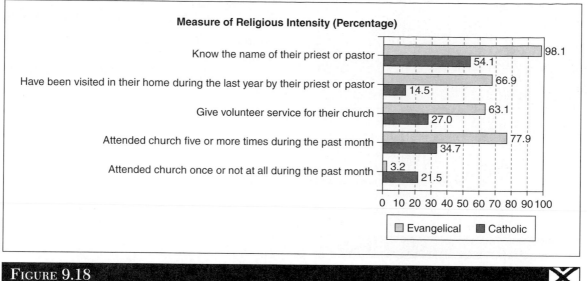

Measure of Religious Intensity (Percentage)

	Evangelical	Catholic
Know the name of their priest or pastor	98.1	54.1
Have been visited in their home during the last year by their priest or pastor	66.9	14.5
Give volunteer service for their church	63.1	27.0
Attended church five or more times during the past month	77.9	34.7
Attended church once or not at all during the past month	3.2	21.5

0 10 20 30 40 50 60 70 80 90 100

☐ Evangelical   ■ Catholic

FIGURE 9.18

Catholic-Evangelical measures of comparative religious intensity: Costa Rica and Guatemala.
Source: Compiled from Steigenga 2001, 42–44.

FIGURE 9.19

Assemblies of God chapel in rural Pernambuco, Brazil.

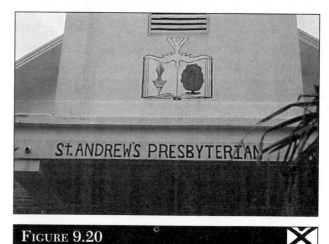

FIGURE 9.20

Protestantism in the English-speaking Caribbean realm is strongly influenced by mainline denominations of European origin. Shown here is St. Andrew's Presbyterian Church of Belize City, Belize.

The Catholic Counter-Reformation

Just as the Protestant Reformation of the sixteenth century was followed by attempts to reform the European Catholic Church, the Protestant Reformation of Latin America has been associated with a number of changes in the Latin American Catholic Church. These changes, which are ongoing, can be viewed collectively as a modern-day Counter-Reformation. While the individual initiatives are too numerous to mention, they fall into three general groupings.

The first has been an attempt to increase the level of involvement of ordinary lay members in Church programs and rituals that were formerly reserved for the clergy and social elites. The earliest of these initiatives

was a program called Catholic Action, which was introduced with some success in the 1930s. The further opening of the Church to the common people was one of the driving motivations behind the reforms of Vatican Council II (1962–1965). Those included approval to say Mass in the language of the local people, encouraging Church members to read the Bible for themselves, and granting authority for lay members to perform certain rites that formerly had been reserved for the priesthood. More recently, Pope John Paul II introduced a "New Evangelization" initiative that attempts to revitalize the Church through greater responsiveness to member

FIGURE 9.21

Modern Catholic pilgrims in Quito, Ecuador.

TABLE 9.6

Ratio of Catholic Priests to Members

COUNTRY	NUMBER OF MEMBERS PER PRIEST	
	1968	2002
Argentina	4,066	5,619
Bolivia	5,953	6,671
Brazil	7,766	8,718
Chile	3,114	4,968
Colombia	3,679	4,979
Costa Rica	4,983	4,075
Cuba	9,310	16,748
Dominican Republic	12,252	9,424
Ecuador	3,400	6,445
El Salvador	8,067	8,400
Guatemala	11,100	9,737
Haiti	7,570	10,503
Honduras	15,899	14,933
Mexico	5,773	9,368
Nicaragua	7,773	10,556
Panama	7,482	6,246
Paraguay	4,150	6,700
Peru	7,257	8,371
Uruguay	4,109	5,054
Venezuela	5,371	9,807

Sources: Bunson 2002; *Annuario Pontificio* 1984; Landsberger 1970.

needs. Interestingly, one outcome of the program has been the growth of a Catholic charismatic movement which bears numerous similarities to evangelical Protestantism. Catholic lay involvement has also taken the form of increased pilgrimage activity (Bauer and Stanish 2001; Horst 1999; Salles-Reese 1997; Crumrine and Morinis 1991; Nolan 1991) (Figure 9.21).

The second thrust of the Latin American Catholic Counter-Reformation has been an attempt to make the institutional Church more responsive to the historic social inequities that many Catholic leaders feel have fueled Protestant growth. One expression of this response, in many countries in the mid–twentieth century, was the emergence of reform-minded Christian Democratic political parties which, although not directly tied to the Church, drew heavily on Catholic values. Another has been a reform movement known as **liberation theology,** which draws on both Christian and Marxist thought (Garrard-Burnett 2000; Dussel 1981; Dahlin 1981; Dodson 1979). Liberation theologians perceive poverty as a structural rather than as an individual problem, and define sin as oppression of the poor by the elites and righteousness as helping the poor overcome their exploited condition. Many of these theologians have consequently devoted themselves to the reformation of Latin America's traditional hierarchical class structure. The instrument used by liberation theologians to organize the masses at the grassroots level is the Basic Ecclesial Community, or CEB (Torres and Eagleson 1981; Gibellini 1979). By the early 1980s, Brazil alone contained over 80,000 CEBs, (Barreiro 1982). More recently, Smith (1995, 11) reported 150,000 to 200,000 CEBs in all Latin America, with an average of fifteen to forty members each. For a time there was concern among the Roman hierarchy that the movement might evolve into a splinter church. This, however, has not occurred and the

past decades have witnessed a progressive reorientation of the objectives of many CEBs from the pursuit of radical political reform to the promotion of individual religious devotion and community development. These changes have yielded widely varying results in membership trends, ranging from stagnation in São Paulo, Brazil (Hewitt 1996) to rapid growth in El Salvador (Williams and Peterson 1996).

The third thrust of the Catholic Counter-Reformation has been a multifaceted effort to strengthen the formal institutional Church. A series of Latin American Episcopal Conferences have been held in Rio de Janeiro, Brazil (1955), Medellín, Colombia (1968), Puebla, Mexico (1979), and Santo Domingo, Dominican Republic (1992). Papal visits to the region have increased in frequency, and the Church has become more outspoken in its opposition to Protestantism. While these and other strategies have succeeded in bringing about some increases in lay member involvement, long-term prospects of the formal Latin American Church continue to be clouded by growth of secularism in the urban areas and the ever-worsening shortage of Catholic clergy. In most Latin American and Caribbean nations, the ratio of priests to members now averages between one to 5,000 and one to 11,000 or more (Table 9.6).

CHAPTER 9 Religion   **229**

SUMMARY

Latin America and the Caribbean are experiencing, for the first time since the Spanish and Portuguese Conquests, a broadly based religious transformation. Latin American Catholicism was imposed upon the masses from the top down during the early colonial era. While exerting vast institutional influence, Catholicism resulted in relatively little change in the personal religious behavior of the people, most of whom have continued as nominal or folk Catholics.

Twentieth-century Protestantism, on the other hand, has emanated primarily from the masses. Because it is individualistic and fragmented in character, it has exerted limited institutional influence to date. It is possible, however, that in the long term, its grassroots nature will impact the personal religiosity of the masses more than has formal institutional Catholicism. Regardless of what the future may bring, it is evident that Latin America has now become a religiously pluralistic society.

KEY TERMS

royal patronage 204
religious syncretism 206
Virgin of Guadalupe 207
anticlericalism 211
shortage of clergy 211
conservative 212

liberals 212
Mediterranean Catholicism 212
formal Catholicism 213
nominal Catholics 214
folk Catholicism 214
voodoo 217

Santería 217
Ashkenazim 221
Sephardim 221
evangélicos 223
liberation theology 228

SUGGESTED READINGS

Annis, Sheldon. 1987. *God and Production in a Guatemalan Town.* Austin: University of Texas Press.

Annuario Pontificio. 1984. Citta del Vaticano: Libreria Editrice Vaticana.

Arciniegas, Germán. 1967. *Latin America: A Cultural History.* Trans. Joan MacLean. New York: Knopf.

Avni, Haim. 1991. *Argentina and the Jews: A History of Jewish Immigration.* Tuscaloosa: University of Alabama Press.

Azevedo, Eliane. 1992. "Fé Explosiva: Censo Protestante Revela: Fiéis abrem quase un templo evangelico por dia no Rio de Janeiro." *Veja* (16 de Dezembro): 82–83.

Barreiro, Alvaro. 1982. *Basic Ecclesial Communities: The Evangelization of the Poor.* Trans. Barbara Campbell. Maryknoll, N.Y.: Orbis Books.

Bastian, Jean-Pierre. 1990. *Historia del Protestantismo en América Latina.* Mexico City: Casa Unida de Publicaciones.

———. 1992. "Protestantism in Latin America." In *The Church in Latin America 1492–1992;* ed. Enrique Dussel, 313–350. Maryknoll, N.Y.: Orbis Books.

Bastide, Roger. 1978. *The African Religions of Brazil.* Trans. Helen Sebba. Baltimore: Johns Hopkins University Press.

Bauer, Brian S. 1998. *The Sacred Landscape of the Inca: The Cuzco Ceque System.* Austin: University of Texas Press.

Bauer, Brian S., and Charles Stanish. 2001. *Ritual and Pilgrimage in the Ancient Andes: The Islands of the Sun and the Moon.* Austin: University of Texas Press.

Beker, Avi, ed. 1998. *Jewish Communities of the World: 1998–1999 edition.* New York: Lerner Publications for the Institute of the World Jewish Congress.

Besson, Jean. 1998. "Religion as Resistance in Jamaican Peasant Life: The Baptist Church, Revival Worldview, and Rastafari Movement." In *Rastafari and Other African-Caribbean Worldviews,* ed. Barry Chevannes, 43–76. New Brunswick, N.J.: Rutgers University Press.

Bonicelli, Paul J. 1997. "Testing the Waters or Opening the Floodgates? Evangelicals, Politics, and the 'New' Mexico." *Journal of Church and State* 39(1997): 107–130.

Brandon, George. 1993. *Santería from Africa to the New World: The Dead Sell Memories.* Bloomington: University of Indiana Press.

Britannica Book of the Year 2002. 2002. Chicago: Encyclopaedia Britannica.

Brusco, Elizabeth E. 1995. *The Reformation of Machismo: Evangelical Conversion and Gender in Colombia.* Austin: University of Texas Press.

Bunson, Matthew, ed. 2002. *Catholic Almanac 2002.* Huntington, Ind.: Our Sunday Visitor.

Chevannes, Barry. 1998. "New Approaches to Rastafari." In *Rastafari and Other African-Caribbean Worldviews,* ed. Barry Chevannes, 20–42. New Brunswick, N.J.: Rutgers University Press.

Christian, William A., Jr. 1981. *Local Religion in Sixteenth Century Spain.* Princeton: Princeton University Press.

Clarke, Colin G. 1986. *East Indians in a West Indian Town: San Fernando, Trinidad, 1930–70.* London: Allen & Unwin.

Cleary, Edward L., and Hannah W. Stewart-Gambino, eds. 1997. *Power, Politics, and Pentecostals in Latin America.* Boulder, Colo.: Westview Press.

Cohen, Martin A., ed. 1971. *The Jewish Experience in Latin America.* 2 vols. New York: Ktav Publishing House.

Coleman, William J. 1958. *Latin American Catholicism: A Self Evaluation.* Maryknoll, N.Y.: Maryknoll Publications.

Crumrine, N. Ross, and Alan Morinis, eds. 1991. *Pilgrimage in Latin America.* New York: Greenwood Press.

Curcio-Nagy, Linda A. 2000. "Native Icon to City Protectress to Royal Protectress: Ritual, Political Symbolism, and the Virgin of the Remedies." In *The Church in Colonial Latin America,* ed. John F. Schwaller, 183–208. Wilmington, Del.: Scholarly Resources.

Dahlin, T. et al., eds. 1981. *The Catholic Left in Latin America.* Boston: G.K. Hall.

Damboriena, Prudencio. 1962–63. *El Protestantismo en América Latina.* 2 tomos. Friburgo, Suiza: Oficina Internacional de Investigaciones Sociales de FERES.

Davis, Wade. 1986. *The Serpent and the Rainbow.* New York: Simon and Schuster.

Desmangles, Leslie. 1992. *The Faces of the Gods: Vodou and Roman Catholicism in Haiti.* Chapel Hill: University of North Carolina Press.

Dodson, Michael. 1979. "Liberation Theology and Christian Radicalism in Contemporary Latin America." *Journal of Latin American Studies* 11:203–222.

Dussel, Enrique. 1981. *A History of the Church in Latin America: Colonialism to Liberation.* Trans. Alan Neely. Grand Rapids, Mich.: Wm. B. Eerdmans.

Eber, Christine. 2000. *Women and Alcohol in a Highland Maya Town: Water of Hope, Water of Sorrow,* updated edition. Austin: University of Texas Press.

Eliade, Mircea, ed. 1987. *The Encyclopedia of Religion.* Vol. 8. New York: Macmillan.

Fernandez Olmos, Margarite, and Lizbeth Paravisini-Gebert, eds. 1997. *Sacred Possessions: Vodou, Santería, Obeah, and the Caribbean.* New Brunswick, N.J.: Rutgers University Press.

Ferretti, Sergio F. 2001. "Religious Syncretism in an Afro-Brazilian Cult House." In *Rewriting Religions: Syncretism and Transformation in Africa and the Americas,* eds. Sidney M. Greenfield and André Droogers, 87–97. Lanham, Md.: Rowman and Littlefield.

Gade, Daniel W. 1970. "Coping with Cosmic Terror: The Earthquake Cult in Cuzco, Peru." *The American Benedictine Review* 21:218–223.

Gardy, Allison. 2000. "Emerging from the Shadows: A Visit to an Old Jewish Community in Mexico." In *On Earth as It Is in Heaven: Religion in Modern Latin America,* ed. Virginia Garrard-Burnett, 63–69. Wilmington, Del.: Scholarly Resources.

Garrard-Burnett, Virginia, ed. 2000. *On Earth as It Is in Heaven: Religion in Modern Latin America.* Wilmington, Del.: Scholarly Resources.

Gibellini, Rosino, ed. 1979. *Frontiers of Theology in Latin America.* Trans. John Drury. Maryknoll, N.Y.: Orbis Books.

González-Wipper, Migene. 1973. *Santería: African Magic in Latin America.* New York: Julian Press.

Greene, Anne. 1993. *The Catholic Church in Haiti: Political and Social Change.* East Lansing: Michigan State University Press.

Greenleaf, Richard E. 1969. *The Mexican Inquisition of the Sixteenth Century.* Albuquerque: University of New Mexico Press.

Gullick, Charles J. M. R. 1983. "Afro-American Identity: The Jamaican Nexus." *Journal of Geography* 82:205–211.

Hassig, Ross. 2001. *Time, History, and Belief in Aztec and Colonial Mexico.* Austin: University of Texas Press.

Heusch, Luc de. 1989. "Kongo in Haiti: A New Approach to Religious Syncretism." *Man,* n.s. 24:290–303.

Hewitt, W.E. 1996. "The Changing of the Guard: Transformations in the Politico-Religious Attitudes and Behaviors of CEB Members in São Paulo, 1984–1993." *Journal of Church and State* 38:115–136.

Himmelfarb, Milton, and David Singer, eds. 1982. *American Jewish Yearbook 1983.* New York: The American Jewish Committee and Jewish Publication Society of America.

Hinshaw, Robert E. 1975. *Panajachel: A Guatemalan Town in Thirty-Year Perspective.* Pittsburgh: University of Pittsburgh Press.

Hoover, W. C. 1948. *Historia del Arivamiento Pentecostal en Chile.* Valparaíso: Imprenta Excelsior.

Horst, Oscar H. 1999. "Building Blocks of a Legendary Belief: The Black Christ of Esquipulas, 1595–1995." *The Pennsylvania Geographer* 41:6–19.

———. 1990. "The Diffusion of the Cult of the Guatemalan Black Christ of Esquipulas." Paper presented at the meetings of the Association of American Geographers, Toronto.

Houk, James T. 1995. *Spirits, Blood, and Drums: The Orisha Religion in Trinidad.* Philadelphia: Temple University Press, 1995.

Johnstone, Patrick, and Jason Mandryk. 2001. *Operation World,* 21st-century edition. Carlise, Cumbria, U.K.: Paternoster Publishing.

Kendall, Carl. 1991. "The Politics of Pilgrimage: The Black Christ of Esquipulas." In *Pilgrimage in Latin America,* eds. N. Ross Crumrine and Alan Morinis, 139–156. New York: Greenwood Press.

Landsberger, Henry A., ed. 1970. *The Church and Social Change in Latin America.* Notre Dame, Ind.: Notre Dame University Press.

Latorre Cabal, Hugo. 1978. *The Revolution of the Latin American Church.* Trans. Francis K. Hendricks and Beatrice Berler. Norman: University of Oklahoma Press.

Lea, Henry Charles. 1922. *The Inquisition in the Spanish Dependencies.* New York: Macmillan.

Levine, Robert M. 1993. *Tropical Diaspora: The Jewish Experience in Cuba.* Gainsville: University of Florida Press.

Liebman, Seymour B. 1970. *The Jews in New Spain.* Coral Gables, Fla.: University of Miami Press.

Lowenthal, David. 1972. *West Indian Societies.* London: Oxford University Press.

Martin, David. 1990. *Tongues of Fire: The Explosion of Protestantism in Latin America.* Oxford: Basil Blackwell.

Mecham, J. Lloyd. 1966. *Church and State in Latin America.* Chapel Hill: University of North Carolina Press.

Merino, Olga, and Linda A. Newson. 1995. "Jesuit Missions in Spanish America: The Aftermath of the Expulsion." In *Yearbook, Conference of Latin Americanist Geographers, 1995,* ed. David J. Robinson, 133–148. Austin: University of Texas Press for the Conference of Latin Americanist Geographers.

Métraux, Alfred. 1969. *The History of the Incas.* Trans. George Ordish. New York: Schocken Books.

Mills, Kenneth. 2000. "The Limits of Religious Coercion in Midcolonial Peru." In *The Church in Colonial Latin America,* ed. John F. Schwaller, 147–180. Wilmington, Del.: Scholarly Resources.

Minard, Lawrence. 1990. "The New Missionaries." *Forbes,* 14 May, 41–42.

Mirelman, Victor A. 1990. *Jewish Buenos Aires, 1890–1930: In Search of an Identity.* Detroit: Wayne State University Press.

Moreno, Pedro C. 1997. "Pentecostals Redefine Religion in Latin America." *The Wall Street Journal,* 29 August, A-11.

Motta, Roberto, 2001. "Ethnicity, Purity, the Market, and Syncretism in Afro-Brazilian Cults." In *Rewriting Religions: Syncretism and Transformation in Africa and the Americas,* eds. Sidney M. Greenfield and André Droogers, 71–85. Lanham, Md.: Rowman and Littlefield.

Muratorio, Blanca. 1981. "Protestantism, Ethnicity, and Class in Chimborazo." In *Cultural Transformations and Ethnicity in Modern Ecuador,* ed. Norman E. Whitten, Jr., 506–534. Urbana: University of Illinois Press.

Nolan, Mary Lee. 1991. "The European Roots of Latin American Pilgrimage." In *Pilgrimage in Latin America,* eds. N. Ross Crumrine and Alan Morinis, 19–49. New York: Greenwood Press.

Norman, Edward. 1981. *Christianity in the Southern Hemisphere: The Churches in Latin America and South Africa.* Oxford: Clarendon Press.

Padden, Robert C. 2000. "The Ordenanza del Patronazgo of 1574: An Interpretive Essay." In *The Church in Colonial Latin America,* ed. John F. Schwaller, 27–47. Wilmington, Del: Scholarly Resources.

Parkes, Henry Bamford. 1970. *A History of Mexico.* Boston: Houghton Mifflin.

Pérez de Ribas, Andrés. 1999. *History of the Triumphs of Our Holy Faith,* trans. Daniel T. Reff, Maureen Ahern, and Richard K. Danford. Tucson: University of Arizona Press.

Poblete, Renato. 1970. "The Church in Latin America: A Historical Survey." In *The Church and Social Change in Latin America,* ed. Henry A. Landsberger, 39–52. Notre Dame, Ind.: Notre Dame University Press.

Poole, Stafford. 1995. *Our Lady of Guadalupe: The Origins and Sources of a Mexican National Symbol, 1531–1797.* Tucson: University of Arizona Press.

Radin, Paul. 1969. *Indians of South America.* New York: Greenwood Press.

Rycroft, W. Stanley, and Myrtle M. Clemmer. 1963. *A Factual Study of Latin America.* New York: United Presbyterian Church in the U.S.A.

Saeger, James Schofield. 2000. *The Chaco Mission Frontier: The Guaycuruan Experience.* Tucson: University of Arizona Press.

Salles-Reese, Veronica. 1997. *From Viracocha to the Virgin of Copacabana: Representation of the Sacred at Lake Titicaca.* Austin: University of Texas Press.

Sallnow, Michael J. 1987. *Pilgrims of the Andes: Regional Cults in Cuzco.* Washington, D.C.: Smithsonian Institution Press.

Sawatzky, Harry L. 1971. *They Sought a Country: Mennonite Colonization in Mexico.* Berkeley: University of California Press.

Sherman, Amy L. 1997. *The Soul of Development: Biblical Christianity and Economic Transformation in Guatemala.* New York: Oxford University Press.

Simpson, George Eaton. 1970. *Religious Cults of the Caribbean: Trinidad, Jamaica, and Haiti.* Rio Piedras: University of Puerto Rico Institute of Caribbean Studies.

Smilde, David A. 1994. "Gender Relations and Social Change in Latin American Evangelicalism." In *Coming of Age: Protestantism in Contemporary Latin America,* ed. Daniel R. Miller, 39–64. Lanham, Md: University Press of America.

Smith, Christian. 1995. "The Spirit and Democracy: Base Communities, Protestantism, and Democratization in Latin America." In *Religion and Democracy in Latin America,* ed. William H. Swatos, Jr., 1–25. New Brunswick, N.J.: Transaction Publishers.

Steigenga, Timothy J. 2001. *The Politics of the Spirit: The Political Implications of Pentecostalized Religion in Costa Rica and Guatemala.* Lanham, Md.: Lexington Books.

Taylor, William B. 1987. "The Virgin of Guadalupe in New Spain: An Inquiry into the Social History of Marian Devotion." *American Ethnologist* 14:9–33.

Torres, Sergio, and John Eagleson, eds. 1981. *The Challenge of Basic Christian Communities.* Trans. John Drury. Maryknoll, N.Y: Orbis Books.

Tullis, LaMond. 1987. *Mormons in Mexico.* Logan: Utah State University Press.

Vergara, Ignacio. 1962. *El Protestantismo en Chile.* Santiago: Editorial del Pacífico.

Vertovec, Steven. 1992. *Hindu Trinidad: Religion, Ethnicity and Socio-economic Change.* London: Macmillan Caribbean.

Voeks, Robert. 1993. "African Medicine and Magic in the Americas." *The Geographical Review* 83:66–78.

———. 1997. *Sacred Leaves of Candomblé: African Magic, Medicine, and Religion in Brazil.* Austin: University of Texas Press.

Vogt, Evon Z. 1970. *The Zinacantecos of Mexico: A Modern Maya Way of Life.* New York: Holt, Rinehart, and Winston.

Watanabe, John M. 1992. *Maya Saints and Souls in a Changing World.* Austin: University of Texas Press.

Willems, Emilio. 1967. *Followers of the New Faith.* Nashville, Tenn.: Vanderbilt University Press.

Williams, Philip J., and Anna L. Peterson. 1996. "Evangelicals and Catholics in El Salvador: Evolving Religious Responses to Social Change." *Journal of Church and State* 38:873–897.

Part III

ECONOMIC PATTERNS

Having studied the physical and cultural geography of Latin America and the Caribbean, we now proceed to analyze patterns of economic activity and the consequences of that activity on individual well-being. We begin, in chapter 10, with an overview of the principal forms of agriculture. That is followed, in chapter 11, with an analysis of the past and present contributions of mining, manufacturing, and tourism. Chapter 12 addresses the effects of the interrelated processes of urbanization, population growth, and migration on the economic development of the region. We conclude, in chapter 13, with evaluations of the influences of international economic cycles and organizations and of the role of disease in the development and underdevelopment of the Latin American and Caribbean peoples.

10

Agriculture and Agrarian Development

The agricultural geography of Latin America and the Caribbean reflects, perhaps more faithfully than any other dimension of the human landscape, the diverse cultural and economic forces that have interacted within the region since the time of the Conquest. The general public has often been led to believe that the encounter of Old and New World peoples and technologies was essentially one of extreme contrast between advanced and primitive peoples. If we reflect, however, on the great indigenous civilizations and dense native populations that characterized much of the region prior to the arrival of the Spaniards and Portuguese, we will realize that some of the American Indians must surely have possessed advanced agricultural production techniques. We noted in chapter 7 that one of the greatest and most lasting achievements of the pre-Colombian peoples was their domestication of a great variety of food, medicinal, and fiber crops, many of which continue to rank, both nutritionally and economically, among the world's leading agricultural commodities (Table 7.2).

A second major agronomic achievement of the native Americans was the development of sophisticated hydrologic and land modification systems that enabled them to produce abundant harvests in harsh physical environments, environments that are often utilized far less intensively today than they were prior to the Conquest. One widely recognized example of this skill was the capacity of the highland peoples of Middle and South America to construct and maintain thousands upon thousands of terraced fields on slopes so steep that they seemingly defy human occupation (Knapp 1991; Mathewson 1984; Treacy 1987; Williams 1990) (Figure 10.1). As if that accomplishment were not enough, many of the terraced fields were also irrigated.

FIGURE 10.1

The terraced fields of southern Peru are one evidence of the advanced agriculture that prevailed in much of Latin America prior to the Conquest.

The native Americans also excelled at the utilization of wetlands and other bottomlands whose soils were saturated for all or large parts of the year. The cumulative evidence accruing from the research of Denevan (2001; 1966), Parsons (1969), Siemens (1998), Turner and Harrison (1983), Knapp (1982), Gómez-Pompa (1982), Sluyter (1994), and others suggests that many of Latin America's river floodplains were utilized intensively in pre-Columbian times through various forms of drained and/or raised field agriculture. Remnant forms of raised or ridged field farming practiced today in the high intermontane basins and valleys of western Latin America include the so-called floating gardens, or *chinampas*, of the Valley of Mexico (West and Armillas 1950) and the *wachu* (Zimmerer 1991b) and *waru waru* agriculture of the central Andes (Case Study 10.1). While it would be

Case Study 10.1:   Waru Waru Agriculture

Waru waru is a specialized form of raised field agriculture developed some 3,000 years ago on over 122,000 hectares of Lake Titicaca wetlands on the Peruvian and Bolivian Altiplano. Simple in its basic design, it involved taking dirt dug from shallow, parallel canals and using it to form elevated beds or platforms that could then be sown with staple food crops.

Four significant advantages were inherent in the system. The first was microclimatic tempering. During the day, the water in the canals absorbed considerable solar heat, which was then radiated out during the cold night hours, thereby minimizing frost damage to the plants. A second advantage of waru waru agriculture was its capacity to regulate water supplies to the crops. When excess rain fell, flooding was avoided as the water drained off the platforms and into the surrounding canals. Conversely, water from the canals worked its way up from the zone of saturation to the roots of the plants during dry periods, thereby protecting them from drought. The third principal advantage of waru waru was the maintenance of high levels of soil fertility through the annual addition to the platforms of fertile muck sediments dredged from the canal bottoms. Yet another benefit of waru waru agriculture was its capacity to generate supplemental fish harvests at no additional cost.

Recent tests of this labor-intensive, low-technology form of farming have resulted in potato yields two to seven times greater than those currently achieved in nearby conventionally farmed, chemically fertilized fields. Waru waru is but one of a growing number of traditional Latin American agricultural systems whose simplicity and productivity offer sound alternatives to many economically expensive and environmentally threatening forms of modern industrial agriculture.

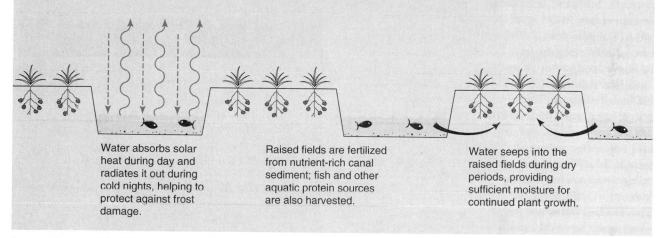

Water absorbs solar heat during day and radiates it out during cold nights, helping to protect against frost damage.

Raised fields are fertilized from nutrient-rich canal sediment; fish and other aquatic protein sources are also harvested.

Water seeps into the raised fields during dry periods, providing sufficient moisture for continued plant growth.

Sources: Denevan 2001; Kolata 1996; Popenoe and King 1989.

naive to assume that every form of pre-Columbian agriculture was more productive than modern farming, in some cases it was. Whitmore (1991, 474) has noted, for instance, that the *chinampas* of central Mexico produce almost three times the maize yields of irrigated lands nearby and over eight times those of neighboring dry-farmed plots.

A third invaluable accomplishment of the indigenous agriculturists of the New World tropics was the development of intensive multicropping, or polycultural cropping, systems, which were, and continue to be, capable of producing much higher yields than those generated subsequently through Iberian single crop, or monocultural, farming techniques. We will discuss the advantages of these traditional polycultural cropping systems later in this chapter. It is sufficient to note here that the agricultural practices of the pre-Conquest American Indians were sophisticated enough not only to sustain dense populations but to feed them well, often with considerable surplus production. The relatively well nourished conditions of these overwhelmingly rural Indian peoples were dramatically altered, however, by the arrival of the Spanish and Portuguese conquerors.

IMPACT OF THE EUROPEAN CONQUEST

When the Iberians arrived in the New World, they brought with them not only their social, political, and religious structures but their material culture as well. A significant portion of that material culture consisted of the crops, tools, and animals necessary to produce the foods

TABLE 10.1

Selected Old World Plants and Animals Introduced by the Spanish and Portuguese into the Americas

COMMON NAME	BOTANICAL NAME
Plants	
wheat	*Triticum* spp.
barley	*Hordeum* spp.
rice	*Oryza sativa*
olives	*Ceratonia siliqua*
grapes	*Vitis vinifera*
figs	*Ficus carica*
citrus	*Citrus* spp.
chickpea	*Cicer arietum*
sugar cane	*Saccharum officinarum*
bananas	*Musa* spp.
coconut	*Cocos nucifera*
mango	*Mangifera indica*
coffee	*Coffea* spp.
nutmeg	*Curcuma longa*
castor bean	*Ricinus communis*
cotton	*Gossypium* spp.
Animals	
horses	*Equus przewalskii caballus*
donkeys	*Equus asinus asinus*
cattle	*Bos primigenius taurus*
sheep	*Ovis ammon aries*
goats	*Capra aegagrus hircus*
swine	*Sus scrofa domestica*
chickens	*Gallus gallus*

FIGURE 10.2

The Mediterranean scratch plow continues to be used by smallholder farmers in much of Latin America and the Caribbean, especially on steeply-sloped, thin-soiled fields.

to which they had become accustomed in Spain and Portugal (Gade 1992). Accordingly, they introduced wheat, grapes, and olives for the production of bread, wine, and cooking oil, as well as numerous commercial crops that were soon to become the basis of export plantation agriculture (Table 10.1). To these were added chickens and Old World meat and draft animals, including horses, cattle, donkeys, sheep, goats, and pigs. The Mediterranean scratch plow was also introduced (Figure 10.2).

As the triumphant conquerors established themselves in their new kingdoms, the Indians were forced to flee their farms and to seek safety in more remote and less desirable lands. One of the consequences of this forced relocation, and the catastrophic population collapse that accompanied it, was a widespread reduction in the intensity of native agriculture. Countless terraces were abandoned, and vast stretches of drained and raised fields fell into disrepair. Latin America as a whole became one of the emptiest of all the earth's culture spheres, characterized by low population densities and seemingly boundless tracts of little-used land.

Ironically, the depopulation of much of rural Latin America during the colonial era was accompanied by ever-increasing land hunger, owing to the unwillingness of the dominant European upper class to allocate New World lands in a relatively equitable manner. We can only imagine the positive economic, social, and political good that would have come to Latin America had the Iberian elite been willing to distribute medium-to-large parcels of prime farmland to members of the lower and middle classes. Such was not the case, however, and most rural Latinos became either serfs on the great European-controlled *latifundios* or subsistence *minifundistas*, condemned in either case to living their lives devoid of any realistic hope of future advancement for themselves or their children.

Thus, when rapid population growth finally did come in the mid–twentieth century to the neglected, poverty-stricken rural countryside, it forced the Indians to divide and redivide their tiny farms to the point where millions upon millions felt compelled to seek a better life for themselves in the distant urban centers. While Latin America's rural to urban migration has contributed to explosive population growth in the cities, continuing high birth rates have kept rural populations at record high levels (Figure 10.3). It would thus be a serious misjudgment to presume that full economic development can ever come to Latin America and the Caribbean without a meaningful improvement of rural living conditions. To help understand the conditions and issues that must be addressed if rural development is ever to be achieved

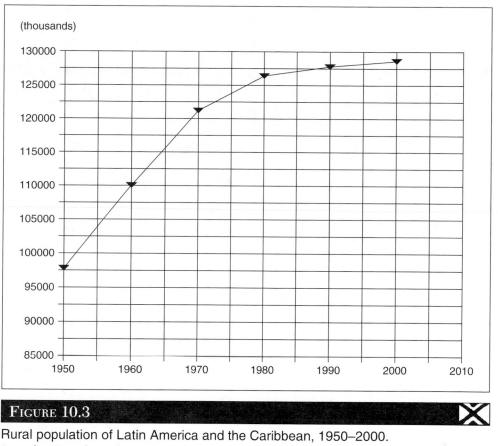

(thousands)

FIGURE 10.3

Rural population of Latin America and the Caribbean, 1950–2000.
Source:http://apps.fao.org/lim500/wrap.pl?Population.

within the region, we will now turn to an analysis of specific agricultural landuse systems.

ENCLAVE ECONOMIES

As the Spaniards and Portuguese carved out for themselves huge *latifundios* in the New World, they initiated a pattern of uneven, or dual, development between large and small landholdings that has plagued Latin America to the present time. Because the *latifundios* historically have existed as virtual islands of European technologies and products, surrounded at every hand by an almost endless sea of traditional agriculture, the largeholdings have often been described as socioeconomic **enclaves.** Largeholder agriculture in Latin America occurs in three principal forms: *haciendas, estancias,* and plantations.

Haciendas

The form of Latin American largeholder agriculture that most embodies traditional hispanic culture values and aspirations is the **hacienda.** The stereotypical Latin American *hacienda* is a huge cattle ranch owned by a great *caballero,* or *hidalgo,* who lives in a distant urban center and who periodically visits his estate to survey his vast domain and to bask in his status as *patrón* to his rural serfs. Because the principal purpose of owning an *hacienda* is the acquisition and maintenance of prestige by the blue-blooded aristocracy, the estates tend to be neglected by their owners and their productivity notoriously low. Although the *hacendado,* or estate owner, is of course not adverse to realizing a profit on his estate, historically he has been reluctant to invest in technological improvements.

Life on the *hacienda* is inward looking and tradition bound. Most are situated on the isolated plateaus and basins of Latin America's western highlands. They export virtually nothing overseas, and the shipment of meat products to the far away urban centers is a low priority. Aside from small plots of land that the mostly Indian peasantry is allowed to work for the cultivation of subsistence crops, the estate produces almost nothing of economic value. Animal stocking levels are low, and one can often travel mile after mile across the estate without observing more than a few tiny clusters of cattle, or sheep, or donkeys grazing in the distance.

Socially, the *hacienda* is an anachronism—a remnant, as it were, of the medieval past when the serfs were bound to the estate for life through debt peonage and

FIGURE 10.4

The great manor house of an abandoned *hacienda* in Yucatán, Mexico.

FIGURE 10.5

Cattle grazing on an *estancia* of the Argentine Pampa.

networks of social dependency such as *compadrazgo.* Physically, the *hacienda's* complex usually consists of a great manor house occupied most of the year by the overseer, a store, a small Catholic chapel, or *capilla,* and housing for the serfs (Figure 10.4). While the *haciendas* are obviously prime candidates for land expropriation and other forms of agrarian reform, a surprising number have survived relatively intact to the present owing to the political and social prominence of their owners.

Nevertheless, change is gradually altering the traditional landscape in most *hacienda* regions. In highland Ecuador, for example, renting, and a form of sharecropping called *aparcería,* are currently replacing the old feudalistic *huasipungo* system, which required the peons to work as many as six days per week on the lands of the *patrón* in exchange for use of a small plot of land (Lawson 1988). It is not unusual in many parts of Latin America to find some of the old *haciendas* evolving into modern agricultural enterprises, such as dairies, cattle ranches, and even farms. In such cases, the estate often retains the term *"hacienda"* in its name for tradition's sake, but its economic and social character has been greatly altered. In another potentially confusing twist, one occasionally finds agricultural entities that combine features of both *haciendas* and plantations, such as the coffee estates of highland Central America. For this reason, it is often fruitless to attempt to depict a given zone on a map as being dominated by either one or the other.

Estancias

The second, and least widespread, form of largeholder agriculture in Latin America is the **estancia.** *Estancias* are the great cattle ranches of the Argentine and Uruguayan Pampa (Figure 10.5). Although similar to the *hacienda* in its focus on animal grazing, the *estancia* qualifies as a unique estate type for a number of reasons. One of the most physically obvious is that the *estancias* are found on a humid, lowland plain rather than in the interior highlands that harbor most *haciendas.* Another difference between *haciendas* and *estancias* is that the *gauchos,* or laborers, of the *estancias* are primarily of European rather than American Indian descent. The most significant difference, however, is that the *estancia* is an export, profit-oriented entity. This has led to the adoption over the past 150 years of a number of new products and technologies, including improved pastures, genetic upgrading of animal stocks, and refrigerated packaging and canning of the great quantities of beef and mutton produced throughout the region.

Plantations

The most economically significant and politically controversial form of largeholder agriculture in Latin America and the Caribbean is the **plantation.** Unlike the *haciendas* and *estancias,* both of which are found only within the Spanish and Portuguese culture spheres, plantations are common to all of the region, including the English, Dutch, and French-speaking Caribbean realms.

The New World plantation was born of the ambition of the early European colonists to profit from the exportation of tropical agricultural produce to Europe and Anglo America. The selection of which tropical crops to ship to overseas markets had nothing to do with the size of the estate or the nature of the labor force that produced them. It was, instead, a purely economic decision based, first, on which tropical crops were in demand abroad for use as food or industrial components, and second, on the perishability of a given product and its ability to withstand lengthy periods of shipment. In other words, the only true test of what, then or now, constituted a plantation crop was whether it was a tropical product grown for export overseas.

The history of plantation agriculture in Latin America and the Caribbean can be divided into three principal phases. The earliest was the establishment of family-owned estates along the coastal lowlands of the Caribbean islands, Mexico, Venezuela, Ecuador, and Brazil. Because most of these frontier zones experienced a rapid decline of their American Indian populations, dependence upon the use of imported African slaves became increasingly common. The number of slaves utilized varied widely, according to the labor demands of a given crop, world market conditions for the product, and the management skills of the plantation owner. Often, the number of slaves was low, and the social and physical relationships of whites and blacks were intimate.

The era of family-dominated plantations ended in the late nineteenth century with the rise in the United States and Europe of trading and shipping companies specializing in the distribution of tropical produce. As these firms became involved in the Latin American tropics, they came to believe that they could increase their profits by expanding into the production end of the business. Their economic and technological resources were so much greater than those of the old family-operated plantations that in most regions they soon drove the latter out of business. In the process, management-labor strife replaced the personalistic security of the past, and the influence of the transnational corporations in national politics frequently reached high levels.

The era of corporate plantation dominance lasted only into the mid–twentieth century, when nationalistic sentiment reached a level in most nations sufficient to begin to challenge the previously unrestrained corporate influences. The reaction of the transnational firms to the changing political realities has varied by company and by country according to local circumstances. In many instances, it has led to a third, ongoing phase of plantation agriculture, which has generally included a renewed emphasis by the transnational corporations on shipping and marketing, with much of the production coming from native, independent, commercial farmers.

Plantation agriculture in Latin America and the Caribbean is thus quite diverse. There often exists in a given region a unique blend of family, corporate, and postcorporate production units and technologies (Table 10.2). No two plantation regions are identical, so descriptions of their common attributes are of little value. Plantation crops are now grown from the cool

volcanic basins of Central America to the selvas of lowland Bolivia. They are produced by labor of all ethnic origins on estates ranging from large to small. Their sole common characteristic is that they are tropical crops grown for export overseas. Because they include many of the dominant products of the region and have exercised, and continue to exert, immense influence on the socioeconomic development of many countries, we will now analyze in more detail the most prominent plantation crops and their impact on selected regions.

SUGAR CANE IN THE CARIBBEAN

Sugar cane (*Saccharum officinarum*), a member of the grass, or *Gramineae*, family of plants, is native to Southeast Asia and was introduced into Iberia by the Moors. It was one of the first crops taken by the Spaniards to the Caribbean islands and by the Portuguese to Northeast Brazil. Both of these regions offered almost ideal growing conditions, with fertile volcanic, limestone, or alluvial soils, warm- to hot-temperatures throughout the year, and alternating wet and dry seasons that encourage rapid plant growth and high sucrose content, respectively. Sugar cane is propagated vegetatively by taking cuttings, called ratoons, of existing stalks and rooting them in the ground. Like other grasses, when the stalks of the plants are harvested, the roots will send up new growth that will be ready for cutting about a year later (Figure 10.6). The number of harvests that can be obtained from a single planting of ratoons varies considerably, according to the fertility of the soil and the care given the plants. Generally, yields of sugar are highest with the first cutting and decline gradually but steadily with each successive harvest, so that the farmer eventually decides that there is more to be gained in increased harvests by replanting than to be lost by the expense of sowing the field anew.

Although sugar cane was widely cultivated in the Caribbean during the sixteenth century, the plant was considered less profitable as an export crop than was tobacco. The rise of sugar cane as the preeminent Caribbean plantation crop was tied to the introduction of improved production techniques by the Dutch fleeing Brazil in the mid-1600s and to the simultaneous emergence of Virginia and other British Anglo American colonies as superior tobacco-producing regions (Galloway 1989).

The initial burst of sugar cane cultivation in the Caribbean was sustained by small, family-operated farms, which extracted the juice from the cut stalks by hand-feeding the canes into an animal-powered grindstone mill called a *trapiche*. The process was extremely slow and inefficient and the sugar that was eventually produced from the juice was dark brown in color and

TABLE 10.2

Characteristics of Latin American and Caribbean Plantation Crops

CROP	PRINCIPAL PRODUCTION ZONE(S)	PRINCIPAL PRODUCER GROUP
sugar cane	all areas	largeholders
coffee	Southern Brazil	largeholders
	Central America	mixed
	Colombia	smallholders
bananas	Ecuador	smallholders
	Central America	mixed
	Lesser Antilles	smallholders
cotton	Eastern Brazil	largeholders
	Northern Argentina/Southeastern Paraguay	largeholders
	Northwestern Mexico	largeholders
pineapple	Northeast Brazil	largeholders
	Caribbean Basin	largeholders
oil palm	Colombia, Ecuador	largeholders
	Central America	largeholders
	Brazil	largeholders
cacao	Northeast Brazil	mixed
	Caribbean Basin	mixed
	Ecuador	mixed
sisal	Northeast Brazil	largeholders
	Middle America	largeholders
jute	Amazonia	smallholders
coca leaf/cocaine	Eastern Andean foothills of Bolivia, Peru, and Colombia	mixed

FIGURE 10.6

Sugar cane is the dominant plantation crop of the Latin American and Caribbean lowlands.

filled with little pieces of leaves and other impurities. The number of Caribbean *trapiches* peaked at about 3,900 in the late 1700s and has been in decline ever since. A few still survive in the more remote areas, producing a rich-tasting loaf sugar called *panela*.

As mechanization came to the Caribbean, the *trapiches* were replaced by water- or wind-powered mills called *ingenios,* which in turn were supplanted in the nineteenth and twentieth centuries by even larger, often corporate-owned, steam-powered complexes called *centrales* (Figure 10.7). Ironically, the adoption of improved milling technologies has been accompanied over the past 200 years by a number of economic, environmental, and cultural developments that have greatly weakened the Caribbean sugar industry and the economies of the nations that depend upon it. Economically, sugar cane must now compete not only against sugar beets but also against corn syrup and low-calorie artificial sweeteners that can be produced at a cost far

FIGURE 10.7

A large sugar *centrale* in the state of Veracruz, Mexico.

lower than cane sugar. Environmentally, many of the Caribbean soils, some of which have been producing sugar cane continuously for as long as 380 years, are now exhausted, and increasingly production is sustained by chemical fertilizers that are both expensive and harmful to the soil texture when used for extended periods of time. Soil erosion is also worsening. Culturally, Caribbean sugar cane production has always struggled with labor supply problems, it being difficult to recruit sufficient workers during the dry season harvest period, called the *zafra*, who must then be laid off during the long rainy season. Mechanized harvesting only increases unemployment levels. These and other problems have contributed to the Caribbean and Latin America's other sugar-producing regions becoming areas of endemic social unrest, deep poverty, political ferment, and outmigration (Beckford 1972; DeWitt 1989). These conditions will likely prevail until sugar cane production, as a way of life, is finally abandoned in favor of a more diverse agricultural base.

COFFEE IN COLOMBIA, CENTRAL AMERICA, AND BRAZIL

Coffee, like sugar cane, is native to the Old World, where three commercially important varieties are found. *Coffea arabica*, the most widely grown and consumed of the three, was domesticated in the highlands of the Ethiopian massif and the nearby southwestern Arabian peninsula. It was widely adopted by the Arabs and was introduced into Iberia by the Moors. Its arrival in the New World, however, can be traced to the Dutch settlements in the Guianas, from whence it diffused to the Atlantic coast of Brazil in the early 1700s. Simultaneously, it was introduced by the French into the Caribbean, where it soon proved to be an ideal smallholder upland complement to the largeholder lowland sugar estates.

The collapse of the Haitian coffee economy in the aftermath of the great slave uprising of the early 1800s led to increased plantings in Central America and the Eastern Cordillera of Colombia in the early nineteenth century, and to the establishment in 1861 of the first *cafetales,* or coffee farms, in the Antioquia region centering on Medellín. There, on the steep volcanic mountain slopes overlooking the lower and middle Cauca River Valley, the small family farmers encountered almost ideal coffee growing conditions, including cool but frost-free temperatures, abundant, evenly distributed precipitation, considerable mist and cloud cover, and deep, well-drained acidic soils.

Initially, coffee was the end product of frontier expansion driven by rapid population growth. Upon finding the best lands already taken in their home districts, young Antioqueño married couples would migrate up the Cauca Valley to where virgin forested lands could be claimed. They would then clear the vegetative undergrowth while leaving enough larger, older trees to create a semiopen woodland that provided shade for the young coffee bushes. Subsistence food crops were then raised between the rows of coffee trees until the shrubs began to bear at three to five years of age. Meanwhile, a similar type of shade-grown smallholder coffee cultivation was evolving on family farmsteads, called *fincas,* in Central America.

Coffee farming in Brazil was initially centered on the mountain slopes overlooking the Atlantic coast from Rio de Janeiro southward to Santos. From there, it spread slowly in the early nineteenth century into the interior Paraíba River Valley, where it was cultivated on large, inefficient slaveholding estates called *fazendas.* Although slavery was not abolished in Brazil until 1888, its high social and economic costs prompted experiments using European tenant farmers, called *colonos,* beginning in the 1840s. Under this arrangement, the *colonos* contracted with the *fazendero* to clear an area completely of its original forest cover and then to plant the land in coffee bushes. The *colono* was allowed to grow subsistence crops among the young trees until they began to bear, at which time the tenant farmer was required to relinquish the farm and to move on. The arrangement proved immensely popular, for it allowed poor European peasants, many of Italian and German ancestry, to establish a new life, and the *fazendero* was able to plant his vast holdings to a profitable export crop at minimal cost and personal labor. All told, over 3 million *colonos* came to Brazil between 1886 and 1936, helping to create a great westward-moving coffee frontier across the *terra roxa* soils of the Paraná Plateau portions of São Paulo and Paraná states.

There thus evolved in Latin America two principal coffee-producing regions: a mixed smallholder-largeholder, shade-grown zone extending from southern Mexico southward through Central America into Colombia; and the mostly largeholder, sun-grown

FIGURE 10.8

A smallholder coffee farm of the Colombian highlands. Note the use of taller trees to shade the coffee bushes.

fazenda zone of southern Brazil (Roseberry, Gudmundson, and Kutschback 1995)(Figure 10.8). The shade-grown mountain coffees of Colombia and Central America are widely considered to be superior in flavor and aroma to the Brazilian lowland types. This superiority, if it indeed exists, is not attributable to the shade or superior upland climate or soils but rather is a result of the Central American and Colombian practice of hand-picking each coffee berry at the time of optimal ripeness, as opposed to the mechanized, mass-harvesting of berries of varying degrees of ripeness by the Brazilians. There is currently a trend in Central America and Colombia toward replacing the shade-tolerant varieties with new high-yielding, sun-loving strains. Recent research suggests that while the sun-oriented varieties do outproduce those grown in shade, they also require more chemical fertilization, are more costly to produce, result in greater soil erosion, and yield inferior quality beans (Rice 1998). The traditional shade-grown varieties of Colombia and Central America also allow for the harvesting of numerous secondary tree crops, including citrus, mango, cinchona, banana, and avocado, as well as a variety of wood products.

BANANAS IN CENTRAL AMERICA AND ECUADOR

Eating bananas (*Musa sapientum*) belong to a family of large, fast-growing, broad-leafed plants that are often mistakenly referred to as "trees." Unlike true trees, which are woody and branched and take years to ma-

ture, the banana plant has a branchless, soft fleshy trunk and is capable of producing hundreds of fruits approximately a year from the time of planting. Once the fruit has been harvested, the plant that bore it immediately begins to die, only to be replaced by new shoots emerging from the roots of the old. Bananas are thus capable of propagating themselves indefinitely following the initial planting.

As with so many other crops, bananas were introduced by the Moors into the Spanish-held Canary Islands and transported from there to the New World in the early colonial era. The tastiness of the fruit, its rapid growth and high yields, and its capacity to continue producing year after year combine to make it an outstanding smallholder subsistence crop. One of the most common sights in rural Latin America is a tiny, isolated dwelling set amidst clusters of lush, dark-green banana plants, with barefoot children playing in the yard and a few chickens and pigs scavenging freely nearby.

The utilization of the banana as a plantation crop in Latin America and the Caribbean was closely tied to the development of railroads and refrigerated ships in the late nineteenth century. Although Jamaica and other Caribbean islands contributed to the initial banana trade with the United States, production soon came to focus on smallholder exports from the fertile Caribbean river valleys of Central America and the southern Mexican state of Tabasco. The turning point in the establishment of foreign-owned, largeholder banana operations occurred in 1870 when the Costa Rican government contracted with an American railroad builder named Henry Meiggs to construct a line from the coffee-producing areas of the Meseta Central to the Caribbean port of Limón. Meiggs passed the contract on to his nephew, Henry Meiggs Keith, who in turn arranged for his younger brother, Minor, to do the work.

While overseeing the completion of the railroad, Minor Keith persuaded the Costa Rican government to allow him to establish some large-scale banana plantations on sparsely settled Caribbean lowlands. The success of this and similar endeavors along the eastern river floodplains of Panama, Nicaragua, Honduras, Guatemala, and British Honduras led Keith to form the United Fruit Company in 1899. A rival consortium, called the Standard Fruit and Steamship Company, was created in 1924. Together, the United and Standard fruit companies transformed millions of acres of previously underutilized Caribbean lowlands into efficient, highly productive largeholder plantations. In the process, they acquired numerous tax, land, and transportation concessions and became renowned for the influence—both positive and negative—that they wielded at local and national levels of government (Dosal 1993; Schlesinger and Kinzer 1982; Williams 1986; Bourgois 1989).

FIGURE 10.9

Banana plantation lands on the northern coastal plain of Honduras.

Even as the economic and political power of the transnational fruit companies was expanding, however, the plantations themselves were being attacked by a root fungus, called Panama disease, for which an effective treatment has yet to be developed. The plague became so severe in the 1930s and early 1940s that the only recourse was to abandon altogether the Caribbean operations and to establish new ones along the Pacific coast of Central America. The fruit companies also began to take a greater interest at this time in marketing Colombian and, especially, Ecuadorian smallholder production. Meanwhile, a second dreaded disease, called Sigatoka leaf blight, appeared in the late 1930s. Copper sulphate spray eventually proved effective in controlling Sigatoka, but when the Panama disease fungus caught up to the Pacific coast plantings, the growers responded by dropping altogether the cultivation of the tasty but disease-prone Gros Michel variety of banana in favor of less flavorful but more disease-resistant Cavendish strains. The hardiness of the Cavendish bananas, together with repeated applications of fungicidal and insecticidal sprays, has made it possible for the growers gradually to reopen over the past several decades a number of their Caribbean operations (Figure 10.9).

Yet another change that has begun to reshape the Central American banana scene has been a modest scaling back by the transnational fruit corporations of some of their largeholder production units in favor of a renewed emphasis on the marketing of produce purchased from local smallholder farmers. The economic viability of this approach has long been evident in Ecuador, which has maintained its position as the world's leading exporter of bananas since the early 1950s on the basis of smallholder production from the Guayas and other Pacific coast river floodplains. Another ongoing plantation trend in Central America and

FIGURE 10.10

Oil palm cultivation has expanded rapidly in the Latin American lowlands. The palm, which was introduced from West Africa, is raised for the high quality vegetable oil that is obtained from the nuts.

elsewhere is for the banana companies to attempt to diversify their production through plantings of pineapple and other, lesser-known tropical fruits such as carambola and guava.

THE PLANTATION CONTROVERSY

In addition to the areas described above, there are numerous other plantation zones in Latin America and the Caribbean. Some of the more prominent of these include cacao along the Atlantic coast of Bahia in northeastern Brazil and in the Guayas River floodplain of Ecuador; cotton in eastern Brazil along a belt extending from Paraná northeastward to Ceará, and also in northern Argentina and neighboring southeastern Paraguay as well as in the irrigated river valleys of northwestern Mexico and coastal Peru; oil palm in the lowlands of Colombia, Ecuador, Central America, and the interior savannas of Brazil (Figure 10.10); sisal in Yucatán and other arid regions of Middle America; and jute from

FIGURE 10.11

The loss over the past several decades of countless coconut palms in the greater Caribbean Basin to a disease called lethal yellowing is a highly visible and tragic example of the risks of catastrophic disease in monocultural agriculture.

portions of the Amazon Basin. Market demand for and plantings of sisal and jute, both grown for their natural fiber, have declined steadily in recent decades owing to competition from synthetic materials and to increasing reliance on bulk handling of agricultural commodities that were formerly transported in bags.

As we noted previously, Latin America's plantation crops are produced on landholdings of all sizes in widely varying physical environments and by peoples of distinct ethnic backgrounds. The general public, frequently ignorant of these distinctions, often criticizes all plantations generally for the perceived evils and excesses of foreign-owned, largeholder agriculture. The list of common criticisms can be broken down into two general categories: sociocultural and environmental.

Under the sociocultural heading, one often finds accusations that the big companies underpay and treat their workers harshly, that the firms are insensitive to local cultures, and that the wealth of the multinational corporations has corrupted the morals of both the plantation officials and the government authorities. Environmentally, critics have argued that plantation agriculture is not sustainable over the long haul owing to its reliance on **monoculture,** or the growing of a single crop, which requires massive chemical spraying in order to stave off attacks of catastrophic disease (Figure 10.11).

The response of the leaders of the multinational corporate plantations to these attacks has been to argue that their employees are paid much higher salaries than nonplantation workers in the area and are provided generous housing, health care, and educational and retirement benefits. While admitting that bribes and other unscrupulous activities have occurred in the past, they

declare that their dealings with government officials are now honorable and that, indeed, the economic and social development of the host countries is largely dependent upon the foreign exchange earnings, salaries, and taxes flowing from the transnational corporations. Regarding the alleged poisoning of the environment, they insist that its seriousness has been greatly exaggerated.

In truth, both the opponents and proponents of largeholder plantation agriculture are partially correct and partially mistaken, with each side showing a willingness, at times, to manipulate facts in an attempt to advance its own agenda. Interestingly, both sides have largely ignored what may be the most significant of all issues, which is the questionable wisdom of using a large portion of a small country's best farmland for the production of export crops while unemployment and malnutrition persist among the masses who are kept alive only through the consumption of imported foodstuffs. As local populations continue to grow amidst mounting nationalist sentiments, pressures may build to convert the largeholder plantations into more agriculturally diversified smallholder production units. These forces may be responsible in part for the recent modest growth and, in some cases, actual declines in plantation crop production in Latin America and the Caribbean (Table 10.3).

SMALLHOLDER AGRICULTURE

While Latin America's *latifundistas* have focused on the acquisition of prestige and the production of food for foreign markets, most rural Latin Americans continue to struggle to feed themselves on small plots of marginal fertility. In order to appreciate how they manage to do so, we will first summarize the basic characteristics of all Latin American and Caribbean smallholder agriculturists and then analyze in more detail the two most widespread production systems: slash and burn agriculture and peasant farming.

Common Attributes of Smallholder Agriculturists

The most fundamental and perhaps most easily overlooked characteristic of all Latin American smallholder farmers is that they are poor. Rather than being caused by inefficient farming methods, as has often been assumed by outside agricultural development personnel, this poverty is attributable primarily to the fact that the smallholder farmers simply do not have access to enough land to produce a large surplus for sale elsewhere. The poverty born of this limited resource base is frequently intensified by the uncertain land tenure status of the smallholder, who may be sharecropping or renting or who may have been prevented from obtaining a clear title to land that is rightfully his. If any of

TABLE 10.3

Plantation Crop Production in Latin America, 1980–2000

CROP	1000s METRIC TONS			PERCENTAGE CHANGE 1980–2000
	1980	1990	2000	
sugar	26,425	27,812	37,412	+41.6
bananas	16,216	18,945	24,211	+49.3
coffee	2,987	3,700	4,262	+42.7
pineapple	1,862	2,071	3,840	+106.2
cotton	1,635	1,666	1,079	−34.0
cocoa beans	553	672	466	−15.7
palm oil	190	631	1,306	+587.4
sisal	352	260	249	−29.3
tea	50	60	72	+44.0
rubber	49	58	153	+212.3
jute	102	46	31	−69.6
Totals	50,421	55,921	73,081	+44.9

Sources: Compiled from http://apps1.fao.org/servlet/XteServlet...ction.Crops; and *FAO Production Yearbook 1990* 1991.

these widespread conditions apply to a given farmer, his or her productivity may be limited by an unwillingness to invest in long-term improvements to the farm out of a fear that he or she will not be able to work the land long enough to reap the potential benefits.

The poverty of the smallholder is often complicated further by the fact that the two or three acres of land that he or she works usually consist of a number of small, widely scattered, discontiguous fields that require a great deal of travel time to service. While such a situation would be considered terribly inefficient in the more industrialized nations, the ingenious Latin American smallholder generally turns it to his or her advantage by matching crops or crop varieties to the exact microenvironment of a given field. It is not uncommon in the Andes, for example, for a single farmer to work fifteen to twenty minuscule plots, each situated at a different altitude or exposure to the sun and possessing distinct soil and microclimatic conditions (Figure 10.12).

A second common characteristic of Latin American smallholder farmers is that they have acquired, through trial and error, an intimate knowledge of how to modify and/or create microenvironments in order to temper the harshness of nature. Techniques such as organic composting and the formation of tiny ridges and valleys or mounds and depressions are often employed to increase the soil moisture or the temperature of a certain spot of land according to the needs of the crop at that moment (Wilken 1987; Horst 1989; Denevan 1984; Vargas Carranza 1990; Clawson 1987).

FIGURE 10.12

The cultivation by Latin American smallholder agriculturists of multiple tiny farm plots, each varying slightly from the others in altitude, exposure to the sun, degree of surface slope, and other microenvironmental conditions, is a common practice in the western highland regions. This scene is from Ecuador.

The third shared attribute of the region's smallholder farmers is that they are more concerned with harvest security than with realizing a monetary profit from their crops. Because they have little land and no means of obtaining food other than by growing it themselves, smallholders' first concern is to make sure that

FIGURE 10.13

Tuber-dominated polycultural farming in northeastern Jamaica.

FIGURE 10.14

The banana- and allspice-dominated upper story of a polycultural plot located near the field shown in Figure 10.13.

the crops or crop varieties they grow will not fail to produce at least something. Because their surplus is usually very small anyway, they are hesitant to experiment with a new agricultural product or technology until it has been proven reliable by others. In recent years, numerous well-intentioned agricultural development schemes have failed in Latin America because government leaders, bankers, and extension personnel have been either unwilling or unable to acknowledge and factor into their plans the fundamental conservatism of smallholder farmers (Hewitt de Alcantara 1973–1974; Clawson and Hoy 1979). This said, we should emphasize that smallholders are perfectly willing to adopt new innovations, but only when they have proven to offer increased security when utilized under the environmental conditions faced by the farmers in their own fields.

A fourth, and very important, characteristic of Latin American and Caribbean smallholders is that they cultivate simultaneously a wide array of crops and crop varieties within their fields. The practice of growing a number of different crops together in shared space is called **polyculture,** or **interspecific agriculture,** and is especially common in the *tierra caliente,* where high temperatures and abundant rainfall provide a year round growing season (Figures 10.13 and 10.14). In these areas, it is not unusual to find a plot no more than thirty by fifty feet in size that contains twenty to thirty different crops, starting with coconut, banana, and papaya within the upper story, followed at the next level by sugar cane, maize, and cassava, and dropping down to tomatoes, beans, and peppers, before encountering yams, sweet potatoes, squashes, and other vine crops and assorted herbs running along the ground itself (Figure 10.15). The principal advantage of polyculture to the smallholder farmer is that it maximizes har-

vest security by providing the family with a continuous supply of various foods throughout the year.

Polyculture is also attractive environmentally because the diversity of crops reduces the buildup of harmful insects and diseases to such a degree that toxic chemicals rarely need to be used. Yet another advantage of polyculture is that it requires very little weeding because the good plants tend to choke out the undesirable ones. The natural recycling of fallen leaves and other organic debris also improves the texture and nutrient levels of the soil and eliminates the need for costly and harmful chemical fertilizers. Finally, research by Mead and Willey (1980), Hildebrand (1976), Innis (1997), and others indicates that polyculture yields a greater total harvest in the course of a year than does monoculture, owing to the more complete utilization of solar energy, water, and soil nutrients by plants whose leaves and roots occupy complementary spatial niches.

In sum, polyculture is a highly productive, low cost, environmentally sustainable form of agriculture. It has been criticized by proponents of commercial export agriculture for being ill-suited to large-scale mechanization and to the production of cash crops. The former criticism is physically correct but culturally insensitive and irrelevant. As for cash crop production, there is nothing to prevent polycultural smallholder farmers from selling their surplus if they so desire; and, indeed, many commercial farms in the Latin American tropics could increase their output significantly if they were to adapt polycultural principles to their own specific cropping patterns.

Latin American and Caribbean smallholders not only utilize interspecific diversity or polyculture, they also usually cultivate multiple varieties of their most important crops, a practice known as **intraspecific diversity.** The farmers of Mexico, Central America, and

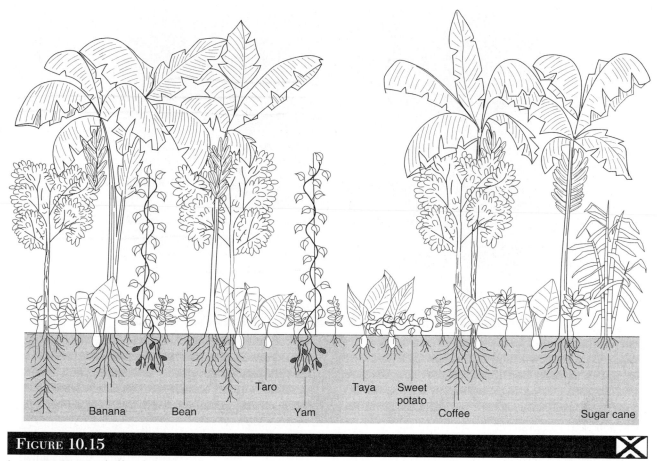

Banana   Bean   Taro   Yam   Taya   Sweet potato   Coffee   Sugar cane

FIGURE 10.15

Interspecific tropical polyculture.

Source: Adapted with permission from Donald W. Innis, "The Future of Traditional Agriculture," *Focus,* 30 January/February 1980. Copyright © 1980. The American Geographical Society.

highland South America, for example, typically plant four or five varieties of maize in their fields, each having a different color, taste, texture or feel to the tongue, and length of growing season (González 2001; Louette, Charrier, and Berthaud 1997; Anderson 1952; Clawson 1985). Similarly, smallholder Andean farmers often plant as many as twenty or thirty varieties of potatoes in a single tiny field, each varying ever so slightly from the others in its physical attributes (Zimmerer 1998, 1991a; Brush, Carney, and Huamán 1981; Johns and Keen 1986). The purpose of the traditional farmer who cultivates multiple varieties of food crops in his or her tiny fields is simply to increase dietary variety and harvest security. One of the valuable side benefits of the practice, however, is that the farmers' plots function as thousands upon thousands of genetic field laboratories where vigorous new crop varieties evolve through natural mutation and crossing and where priceless and irreplaceable plant attributes can be preserved. Many of these qualities, such as resistance to a certain strain of disease or a tolerance to extremes of temperature or soil moisture or alkalinity or acidity, are of significant value to modern industrialized commercial farming. It

can therefore be argued that, in the present age of the earth's dwindling biodiversity, it is in the interest of all peoples, be they urban or rural dwellers living in rich countries or poor, to find a way to strengthen and preserve the long-term vigor of traditional Latin American and Caribbean smallholder agriculture.

Slash and Burn Agriculture

Because traditional Latin American and Caribbean smallholder agriculture has adapted through time to site-specific physical and cultural conditions, it can presently be found in an almost endless variety of forms. While these forms differ one from another in the specific combinations of crops and crop varieties grown, in their spacing and timing, and in countless other details, most of them fall within two general types of production: slash and burn agriculture and peasant farming.

Slash and burn agriculture is found throughout the earth's humid tropics wherever forest cover remains. Its practice in Latin America predates recorded history and is known by a wide variety of names, including *milpa* agriculture in Mexico and Central America,

FIGURE 10.16

A slash and burn *milpa* near Huatusco, Mexico, with the fallen trees lying upon the ground prior to burning.

conuco agriculture in the Caribbean realm and in Venezuela, and *roza* and *chacra* agriculture in much of mainland South America.

The basic steps involved in practicing slash and burn agriculture are quite simple. Having selected a small area at the beginning of the local dry season, the farmer first proceeds to cut down, usually with nothing more than a large *machete* or an ax, as much of the vegetation as possible. Those trees that are too large to be readily felled are allowed to remain standing. As the cut trees and branches fall at random upon the ground, they begin to dry and decay (Figure 10.16). This process continues to the end of the dry season, which may last from as little as a few weeks to as long as five or six months. Then, just before the anticipated arrival of the new rains, the farmer burns the clearing as thoroughly as possible. Occasionally, when the cut vegetation is relatively small and there has been ample time for it to dry, the burn is so complete that nothing remains on the ground except the ash. More often, however, the trunks and stumps of the larger trees are not completely consumed, so that the plot is left with numerous charred logs and a very disorganized appearance (Figure 10.17).

The principal advantage of burning is that many of the nutrients, especially phosphorus and potassium, that were present in the living trees prior to their being cut and burned, are returned in relatively large quantities to what is often deeply leached and infertile soil. Burning also eliminates plants that would otherwise compete with the farmer's crops for scarce soil nutri-

ents; drives out rats, snakes, insects, and other harmful creatures; and cleanses the soil of fungi and other disease organisms.

Having thus prepared the clearing, the farmer will then proceed to plant his field with a wide array of crops in traditional polycultural fashion. Among the most widely sown crops are plantains, a type of banana that is eaten boiled or fried rather than raw, and cassava, an almost indestructible tropical root crop whose tubers will remain edible for two to three years if left in the ground. Cassava is also known as yuca or manioc (see Figure 3.4). As these and the other plants establish themselves, the plot begins to resemble a miniature rain forest—a condition that has been described by some scholars as a "food forest."

To the farmer, the beauty of the system is that, at least initially, he receives relatively abundant harvests from a minimum of personal labor. The plants grow rapidly and produce well in the enriched soil, and there are few weeds, pests, and diseases. The system begins to break down, however, as the soil is depleted of its nutrients through leaching and the continuing demands of the crops themselves, and as weeds, pests, and disease begin to take their toll. Usually, by the end of the second or the beginning of the third year, the farmer will search out another spot of land to clear and begin the entire process anew. The abandoned plot may continue to yield a few tubers or fruits for a year or two after it has been left but will quickly be reclaimed by the forest from which it was temporarily removed.

FIGURE 10.17

The disorganized appearance of many slash and burn plots is evident in this field, situated along a highway in Campeche, Mexico. Note how the dominant crops of maize and squash have been planted among large limestone surface boulders and young palm trees that survived the burning of the preexisting forest vegetation.

The only major disadvantage to slash and burn agriculture is the limited number of people it can support. If the population of a given area is low enough that a cleared parcel of land can be left unused until the new forest growth has matured within it, there are no long-term negative consequences to the physical environment. If, on the other hand, population pressure increases to the point where the parcel is reused before the new vegetation has had time to mature, soil depletion and erosion and vegetative deforestation will ensue. It is thus not the slash and burn cropping system itself that is inadequate; rather, cultural forces, such as the lack of access for the poor to more favorable lands, are behind the long-term environmental degradation of much of rural Latin America and the Caribbean. Pressure on the land has become so great in many regions that it is now commonplace to find steep mountain slopes literally checkered with a patchwork pattern of tiny slash and burn clearings, while the fertile valley bottoms at the base of the slopes are sown to tropical export crops.

Peasant Agriculture

In contrast to slash and burn agriculture, which supports only sparse populations living at a subsistence level in remote and isolated regions, peasant agriculture has for millennia nourished Latin America's greatest civilizations. There are both physical and cultural differences between slash and burn and peasant agricultural systems. Physically, slash and burn farming is seminomadic, with the farmer and his family being forced to relocate to a new plot every few years. In contrast, peasant agriculture exhibits an extraordinary sense of permanence, with the farmer owning and working lands, often with the assistance of draft animals such as oxen or mules, that he has come over time to reverence and that he will eventually pass on as an inheritance to his children.

Culturally, slash and burn farmers live alone, and their harvests are consumed almost exclusively by the nuclear family. Peasants, on the other hand, are

FIGURE 10.18

Peasant farmers harvesting maize in the highlands of central Mexico.

FIGURE 10.19

Broad-based agrarian reform has come only to a handful of Latin American and Caribbean nations and may or may not prove to be lasting. Shown here is a small, privately owned farm in Cuba.

semisubsistence farmers who live in a village that is situated close enough to a larger town or city that they can sell their agricultural surplus in the nearby urban marketplace. Peasants thus live out their lives in two culture worlds, one being the rural village, which often has changed remarkably little in form and function since the time of the Conquest, and the other the ever-growing urban centers where the peasants, as farmers, are accustomed to being treated as nobodies by the proud European and *mestizo* elites.

The specific crops grown by Latin American and Caribbean peasants vary by physical environment and cultural preference. In the high Andes from Chile and Argentina northward through Ecuador and Colombia, peasant agriculture focuses on the cultivation of indigenous, cold-weather tubers such as potatoes, oca, and ulluco and on native grains, including quinoa, kaniwa, and kiwicha (see Table 7.2). European grains such as wheat and barley are also grown in the more favored locales, and animal grazing is widely practiced. In contrast, the classic peasant crop trilogy of highland Mexico and Central America continues to be maize, beans, and squash (Figure 10.18). Animal grazing, in these regions, is relatively unimportant. Maize, rice, beans, plantains, and lowland tubers, such as cassava, yams, and sweet potatoes, form the basis of peasant agriculture throughout the Latin American lowlands. Wherever they reside and whatever plants they cultivate, the Latin American and Caribbean peasants are sophisticated, highly productive farmers who possess detailed knowledge of their local environments and crops. Without them, many of the region's urban elites would starve and the earth's biodiversity would be greatly diminished.

AGRARIAN REFORM

Agrarian reform, or the redistribution of land from the wealthy few to the masses of poor who have little or no land, might today be considered a politically radical and economically destructive concept in the relatively stable and prosperous countries of Anglo America and Europe. Yet, in the poverty-stricken and socially stratified nations of Latin America and the Caribbean, it is often seen, even by politically conservative interests, as essential to the achievement of substantive change leading to the social and economic development of the rural sector. The clear need for land reform has for centuries inspired the participation of the Latin American and Caribbean masses in countless revolutions and political campaigns that promised a more just and equitable distribution of land. Fear among United States and Latin American political leaders that land hunger might fuel the fires of Communist insurgency throughout the region led them to include support for agrarian reform in the Alliance for Progress program established in 1961.

With this degree of consensus and level of support, one might suppose that by now meaningful progress toward agrarian reform would have been realized in most Latin American and Caribbean nations. The continuing opposition of national elites, however, has limited broad-based land reform in the past century to Mexico, Bolivia, and Cuba—and recent changes to Mexico's constitution allowing *ejidos* (redistributed parcels of land) to be sold or rented may ultimately lead once again to a concentration of land in the hands of a few (Randall 1996; Cornelius and Myhre 1998) (Figure 10.19). More limited and less enduring pro-

TABLE 10.4

Agrarian Reform in Latin America

COUNTRY	REFORM PERIOD COVERED	PERCENT OF RURAL HOUSEHOLDS BENEFITTING	TYPE OF ORGANIZATION[a]
Argentina	1940–1968	0.8	Indv[b]
Bolivia	1953–1975	78.9	Indv
Brazil	1964–1969	0.4	Indv
Chile	1962–1973	20.0	*Asentamientos*
counter-reform	1973–1975	4.0	Indv
Colombia	1961–1977	8.0	Indv; Coop
Costa Rica	1961–1979	13.5	Indv; Coop
Cuba	1959–1981	100.0	State; Indv; Coop
Dominican Republic	1962–1986	19.2	Indiv; Coop
Ecuador	1964–1976	9.1	Indv; Coop
El Salvador	1980–1983	12.0	Indv; Coop
Guatemala	1952–1954	33.0	Indv; Coop
counter-reform	1954–1969	3.0	Indv
Haiti	1809–1883	80.0	Indv
Honduras	1962–1980	10.4	Indv; Coop
Mexico	1917–1980	52.4	*Ejidos*
Nicaragua	1979–1983	30.0	State; Indv; Coop
Panama	1963–1969	2.7	Indv; Coop
Paraguay	1963–1969	6.8	Indv[b]
Peru	1967–1979	21.3	*Empresas Asociativas*
Uruguay	1948–1969	0.5	Indv[b]
Venezuela	1959–1975	25.4	Indv; Coop

[a]Indv = Individual; Coop = Cooperative
[b]Colonization projects only

Source: Modified from Meyer 1989, 4.

grams have been attempted during periods of politically leftist dominance in Peru, Nicaragua, Guatemala, and Chile, but none of these initiatives has proven to be politically sustainable. Land reform in the remaining nations has been extremely modest and, in some cases, purely cosmetic (Thiesenhusen 1995; Powelson and Stock 1990) (Table 10.4). Latin America as a whole has yet to experience the socioeconomic leveling and development achieved through the agrarian reform programs of many Asian nations.

Current statistics on regional landholding patterns indicate clearly that the inequalities initially imposed by the Spaniards and Portuguese through the *latifundia-minifundia* complex have continued largely unchanged. King (1977, 78–82) has estimated that 90 percent of the land of Latin America is presently controlled by 10 percent of the population, that fully half of all agricultural lands are held by less than 2 percent of the landowners, and that 40 percent of all farm families are completely landless. Of those farmers who do have access to land,

less than half have legal title to the plots they are working (Jaramillo and Kelly 1999). Scholars widely believe that Latin America and the Caribbean have the most unequal landholding pattern of any of the world's culture realms.

While conditions such as these raise serious long-term ethical issues, one of their most troubling short-term impacts is their role in reducing regional agricultural output. Unequal land holdings depress Latin American food production in two ways. First, large-holders in Latin America do not use anywhere near as great a proportion of their land for food production as do smallholders. Second, smallholders work their lands more intensively and achieve a higher output per unit of land worked than do largeholders (Cornia 1985; Harwood 1979). In Brazil, for example, over 76 percent of the land belonging to farms smaller than five hectares in size is worked in a given year; whereas less than 7 percent of the land belonging to farms larger than fifty hectares is cultivated (Table 10.5). Another

TABLE 10.5

Relationship of Size of Landholding to Proportion of Land Farmed in Latin America and the Caribbean

| PERCENTAGE OF AREA CULTIVATED[a] | | | |
FARM SIZE (HECTARES)	COLOMBIA	BRAZIL	PERU
Less than 5	58.6	76.4	55.5
5–50	32.0	35.6	28.6
Over 50	8.1	6.5	2.6

[a]Temporary and permanent crops, excluding fallow land

Source: Adapted from *Economic and Social Progress in Latin America 1986* 1986.

FIGURE 10.20

Mexico's Papaloapan River Basin Commission continues to act as a quasi-governmental entity in the administration of one of Latin America's oldest and largest planned lowland colonization projects.

way of expressing these same figures would be to say that the amount of *latifundista* land used to produce food would increase almost twelvefold if Brazilian largeholders would use the same proportion of their land for farming as do that nation's smallholders. In famine-stalked Northeast Brazil, there are three hectares of *latifundia* lands not being used for any agricultural purpose whatsoever for every hectare cultivated by a *minifundista* (*Economic and Social Progress* 1986). A recent study by Navarro (1993) reported that farms under 100 hectares in size accounted for only 20 percent of all privately held land under cultivation in Brazil, yet were responsible for half of all agricultural output by value. In contrast, farms larger than 1,000 hectares in size controlled 45 percent of all productive land yet generated only 16 percent of national harvests. When we realize that similar conditions prevail in most of Latin America and the Caribbean, it becomes clear that the absence of broadly implemented land reform continues to be one of the principal impediments to the economic and social development of the region.

New Land Colonization

As the Latin American and Caribbean elites have wrestled with the dilemma of how to make, or at least how to appear to be trying to make, more land available to the poor without taking any away from themselves, they have frequently turned to the idea of allowing the peasantry to colonize distant, sparsely settled lands. These areas are situated for the most part within the humid and semihumid lowlands, whose development has been impeded historically by the presence of malaria, yellow fever, and other life-threatening disease organisms. Their gradual, ongoing occupation over the past cen-

tury has been driven primarily by the deepening poverty and growing sense of hopelessness among the highland peasantry. Other motivations for colonization may include the desire of a national government to occupy more effectively and thus strengthen its claim to disputed frontier territories, to develop more completely the natural resources of these regions, or to defend an area from foreign encroachment.

Controversy has long existed among Latin American government planners and development personnel over the relative merits of planned versus voluntary colonization. Planned colonization entails the screening of potential colonists by ministry officials and, at least in theory, the provision by the government of services and financial aid. It also results in the government placing restrictions on the colonist, including an assignment of where he or she must settle and what crops must be grown. By contrast, voluntary colonists are free to go where they wish and to raise what crops they may. There is little or no government interference in their lives, and likewise few services are provided.

One of the first of Latin America's planned, large-scale colonization schemes centered on the Papaloapan River Basin of the Veracruz coastal plains in southern Mexico. Established in the late 1940s after the pattern of the Tennessee Valley Authority of the United States, it included the construction of penetration roads and highways, dams for flood control and hydroelectric power generation, schools, and health clinics (Poleman 1964) (Figure 10.20). It was followed in the 1960s and 1970s by a host of other government-directed projects, extending from Las Majaguas on the Venezuelan Llanos to Tingo María-Tocache in eastern

TABLE 10.6

Staple Food Crop Production in Latin America, 1980–2000 (1,000s Metric Tons)

CROP	1980	1990	2000	PERCENTAGE CHANGE 1980–2000
Grains				
maize	45,058	50,090	76,088	+68.9
wheat	15,091	20,801	23,712	+57.1
rice	16,444	15,521	23,060	+40.2
Tubers				
cassava	29,699	32,713	31,290	+5.4
potatoes	10,489	11,695	16,189	+54.3
sweet potatoes	2,015	2,341	1,671	−17.1
Beans, dry	3,681	4,306	5,077	+37.9
Totals	122,477	137,467	177,087	+44.6

Sources: Compiled from http://apps1.fao.org/servlet/XteServlet...ction.Crops; and *FAO Production Yearbook 1990* 1991.

Peru (Eidt 1975; Wesche 1971; Stewart 1965). Planned Brazilian occupation of its Amazon Basin territories peaked in the 1970s and 1980s with disastrous consequences for the tropical forests that formerly covered much of the area (Ozório de Almeida, Luíza, and Campari 1995; Hecht and Cockburn 1989; Ozório de Almeida 1992). Although the government-directed schemes have fulfilled their initial objectives to some degree, they have been plagued by inadequate planning and resource inventories, poor service delivery, corruption, inappropriate settler selection procedures, low agricultural productivity, and large cost overruns. Voluntary colonization, on the other hand, has been generally successful (Parsons 1968; Crist and Nissly 1973; Nelson 1973).

Foreign colonization has also been promoted, or in some cases tolerated, by Latin American governments over the years, often in the hope that the newcomers would be successful in developing a certain area where little economic activity previously had occurred. The composition of such groups has included United States Confederates in Brazil, Venezuela, and Middle America; Japanese in numerous South American nations; Quakers in Costa Rica; Mormons in Mexico; and Mennonites in Mexico, Paraguay, Bolivia, and Belize. While most of these groups have achieved some degree of economic success, their methods have seldom been adopted by the local farmers, who tend to attribute the foreigners' prosperity to preferential government treatment and to good luck. Because foreign colonization is increasingly offensive to nationalist political sentiment, it is unlikely to play a prominent role in future development schemes.

AGRICULTURAL MODERNIZATION

When most people speak of "modern" agriculture, they are actually referring to the use of industrial technologies and products to produce food, fiber, or wood. Modern industrial agriculture generally includes the use of tractors and other mechanized farm implements, large-scale irrigation systems where needed, and chemical fertilizers, insecticides, and herbicides. It also relies on the use of hybrid seeds that have been bred to produce higher yields by utilizing greater quantities of water and plant nutrients. The term coined to describe this package of industrial inputs is **Green Revolution** agriculture.

The Green Revolution has profoundly altered Latin American and Caribbean agriculture over the past fifty years, both for better and for worse. Because smallholders rarely have a need for tractors or mechanized irrigation systems and because they cannot financially afford to purchase expensive chemical fertilizers and sprays, they have largely rejected the Green Revolution. This should not be interpreted as meaning that they are opposed to progress or are unwilling to change, but rather that many of the Green Revolution products are, in their present form, culturally and economically incompatible with smallholder agriculture.

Latin America's largeholders, on the other hand, generally have been quick to embrace Green Revolution technologies because of their capacity to increase the farmers' net profits by reducing labor demands and increasing yields. As a result of these changes, production of most staple food crops has increased modestly in recent years for Latin America as a whole (Table 10.6).

When the rapid population growth of the region is factored into the equation, however, a mixed picture emerges. Food production increases in most of the Caribbean nations (including Suriname and French Guiana) have not kept pace with the rate of population growth whereas they have exceeded population growth in most of the mainland countries (Table 10.7). One of the socioeconomic consequences of the Green Revolution, then, has been a growing income gap between the wealthy largeholders and the poor smallholders.

The Green Revolution has also spawned a host of environmental concerns. One of the most threatening to human health has been atmospheric, groundwater, and soil contamination by the residues of toxic insecticides and herbicides (Grossman 1998; Murray 1994; Thrupp 1991). While the degree of contamination and the extent of its impacts vary by region, they are often found at much higher levels than in the Anglo American and Western European nations where environmental regulations are more strict and are enforced more effectively. Many of Latin America's most productive agricultural zones, such as the former cotton-producing lowlands of western Guatemala, El Salvador, and Nicaragua and the fruit and vegetable-dominated Central Valley of Chile, have become entrapped in a vicious cycle of having to make greater and greater use of increasingly toxic chemicals in a seemingly futile attempt to stay ahead of mutating insect populations. Beneficial wildlife populations may then become threatened, as has been the case in the Cachapoal Valley of central Chile, which has been described as a "Valley Without Birds" (Breslin 1988).

The present and future damage to human health brought about by the continued use of pesticides, some of which are so toxic that their consumption has been banned in the United States, is compounded in Latin America by widespread farmer ignorance of the risks of usage and of the need to minimize human exposure to the poisons. Reports abound, for example, of unsuspecting farmers storing drinking water or food in spent pesticide containers, of stirring a mix of concentrate and water with their bare arms and hands, and of breathing, for hours on end without protective face gear, air filled with chemical mists (Figure 10.21).

A second major environmental concern stemming from the Green Revolution is its harmful long-term impact on the soil. This harm is caused primarily by the use of chemical fertilizers that gradually kill off beneficial soil organisms, lower soil organic matter levels, and make the soil more acidic. It is also caused by the use of tractors and other heavy farm equipment that results in undesirable soil compaction.

TABLE 10.7

Per Capita Food Production in Latin America and the Caribbean

COUNTRY	1979–1981	1989–1991 Annual Average = 100.0	1999–2001
		1989–1991	
Caribbean Islands			
Antigua and Barbuda	94.8	100.0	95.1
Barbados	115.9	100.0	95.9
Cuba	98.5	100.0	59.0
Dominica	57.5	100.0	85.5
Dominican Republic	105.6	100.0	92.8
Grenada	123.8	100.0	90.9
Guadeloupe	129.4	100.0	103.7
Haiti	128.1	100.0	84.6
Jamaica	95.6	100.0	109.3
Martinique	88.4	100.0	115.3
Puerto Rico	110.3	100.0	75.6
St. Kitts and Nevis	143.5	100.0	103.3
St. Lucia	72.5	100.0	64.0
Trinidad and Tobago	114.6	100.0	107.8
Mexico and Central America			
Belize	89.4	100.0	139.6
Costa Rica	97.2	100.0	112.3
El Salvador	99.3	100.0	95.4
Guatemala	87.1	100.0	103.5
Honduras	120.3	100.0	85.1
Mexico	103.3	100.0	112.5
Nicaragua	155.3	100.0	108.6
Panama	105.2	100.0	89.5
South America			
Argentina	106.2	100.0	125.6
Bolivia	87.8	100.0	114.3
Brazil	84.4	100.0	126.6
Chile	84.0	100.0	118.7
Colombia	92.8	100.0	99.7
Ecuador	99.8	100.0	126.7
French Guiana	52.3	100.0	85.6
Guyana	133.4	100.0	189.6
Paraguay	82.2	100.0	105.5
Peru	96.1	100.0	144.4
Suriname	104.9	100.0	73.9
Uruguay	92.8	100.0	127.1
Venezuela	103.7	100.0	99.4
Latin America and the Caribbean	96.7	100.0	117.1

Source: Compiled from: http://apps1.fao.org/servlet/XteServlet.jrun?Areas

FIGURE 10.21 ☒

This young highland Mexican farmer is applying by hand insecticidal dust to his recently harvested maize crop. Many Latin American farmers are unaware of the serious health risks posed by prolonged exposure to agricultural chemicals.

A final environmental issue associated with the Green Revolution centers on the loss of genetic or biodiversity brought on by the expansion of hybrid-dependent industrial agriculture. Although the hybrid varieties are, in some but not all cases, more productive than the indigenous food crop strains, their adoption by many Latin American farmers has led to the reduced or discontinued use of literally thousands of native food crop cultivars, which possess genetic qualities of great value to plant breeders (Zimmerer 1996; Thrupp 1998). Efforts to arrest this ongoing process of "genetic erosion" have centered on two strategies. Proponents of the first, called off-site or *ex situ* **conservation,** have generally assumed that traditional Latin American smallholder agriculture either cannot or should not be saved and that the best hope for preserving the threatened loss of indigenous food crop varieties is to collect samples of as many of them as possible. The germplasm samples are then placed in long-term cold storage facilities maintained by a network of International Agricultural Research Centers (Bebbington and Carney 1990) (Table 10.8).

Critics of this "seed bank" approach acknowledge that *ex situ* conservation of the endangered crop varieties is needed but contend that it is not enough because it does not allow for the ongoing, natural evolution of new food crop varieties. These critics also point out the inability of the Research Centers to preserve vegetatively propagated tuber crops for extended periods of time. They suggest, instead, that priority be given to a strategy called on-site or *in situ* **conservation,** which consists of encouraging farmers to raise as many varieties of food crops as possible. Proponents of *in situ* conservation believe that the region's biodiversity can best be preserved over time by strengthening traditional smallholder agriculture, a concept that leads to the understanding that traditional agriculture has a valuable contribution to make in modern Latin America.

EXPANSION OF COMMERCIAL AGRICULTURE

Commercial agriculture is increasing much more rapidly in Latin America and the Caribbean than is basic food crop production (compare Table 10.6 to Table 10.9). As Brazil, for example, has become one of the world's leading producers of soybeans for overseas markets, it has found itself having to import great quantities of common black beans, a dietary staple throughout the country. Likewise, Mexico now imports annually millions of tons of maize to cover local production shortages brought on by corresponding increases in the cultivation of sorghum used as a component in exported animal feeds (Barkin and De-Walt 1988). Notwithstanding these contradictions, commercial agriculture continues to expand in Latin America and the Caribbean in the following five areas: plant crop cultivation, animal husbandry, tree farming, fishing, and narcotic drug production.

Plant Crop Production

Over the past thirty years, Latin America has developed a number of economically significant commercial crop production zones. One of the most successful, when measured in terms of growth of overseas market share, has been the Central Valley of Chile, where many former *haciendas*, called *fundos*, have been transformed into one of the world's leading centers of fruit, nut, table grape, and wine production (Crowley 2000).

Commercial farming in Argentina centers on the Pampa, which consists of a humid, eastern inner zone used mostly for maize, soybeans, and mixed agriculture and an arid, western outer zone dominated by wheat and extensive cattle ranching. Grape cultivation for wine production is centered on the Andean foothills from Mendoza northward, and large quantities of peaches and apples are harvested in the river valleys of northern Patagonia. Farming in southern Brazil is also highly commercialized and includes soybeans and mixed agriculture in much of São Paulo and Paraná and a rapidly growing Atlantic citrus belt in the states of Rio de Janeiro and Espírito Santo. Much of Brazil's citrus, like that of Florida,

TABLE 10.8

Ex Situ Germplasm Conservation of Human Food Crops in Latin America, 1989–1990

CROP	NUMBER OF ACCESSIONS	INTERNATIONAL AGRICULTURAL RESEARCH CENTER WHERE STORED[a]
wheat	66,931	CIMMYT
bean, common	35,950	CIAT
maize	10,700	CIMMYT
cassava	4,600	CIAT
potato	4,165	CIP
sweet potato	3,212	CIP
squash	1,093	CATIE

[a]CIMMYT is the International Maize and Wheat Improvement Center of Chapingo, Mexico; CIAT is the International Center for Tropical Agriculture of Cali, Colombia; CIP is the International Potato Center of Lima, Peru; and CATIE is the Tropical Agriculture Center for Research and Teaching of Turrialba, Costa Rica.

Source: Modified from Clawson 1992.

TABLE 10.9

Commercial Crop Production in Latin America, 1980–2000 (1,000s metric tons)

CROP	1980	1990	2000	PERCENTAGE CHANGE 1980–2000
Fruits				
citrus	19,270	27,597	33,716	+75.0
grapes	5,523	5,727	5,776	+4.6
apples	1,690	2,789	3,318	+96.3
peaches, nectarines	727	894	927	+27.5
pears	307	444	933	+203.9
plums	177	246	337	+90.4
Vegetables, melons	17,901	21,937	32,209	+79.9
Soybeans	19,814	33,273	57,454	+190.0
Meat	15,835	19,080	31,169	+96.8
Milk	34,888	40,863	57,981	+66.2
Eggs	2,585	3,760	5,126	+98.3
Totals	118,717	156,958	228,946	+92.9

Sources: Complied from: http://apps1.fao.org/servlet/XteServlet...ction.Crops; and *FAO Production Yearbook 1990* 1991.

is processed into orange juice concentrate. Brazil's São Francisco River Valley now produces large quantities of irrigated melons, table grapes, and tropical fruits (Caviedes and Muller 1994), while large areas of east-central Brazil have been planted to aluminum-resistant varieties of soybeans and rice. Cashew and Brazil nuts are harvested commercially from the Brazilian Amazon.

Peru's irrigated coastal river valleys produce cotton, bananas, and other fruits and vegetables. Farmers in highland Ecuador, Colombia, and Central America are turning in increasing numbers to the production of fresh flowers, which are cut, boxed, and shipped under refrigeration to lucrative overseas markets. The temperate uplands of Central America are also becoming

FIGURE 10.22

One of the most rapidly expanding forms of commercial agriculture in Central America is the cultivation of high quality, cool-weather vegetables. Shown here is a planting of garlic in the highlands of Costa Rica.

FIGURE 10.23

A modern dairy farm near Cayambe in the Ecuadorian Andes.

important exporters of broccoli, asparagus, snow peas, and other cool-weather vegetables (Figure 10.22).

Mexico's leading commercial plant crop zones include the irrigated river valleys of the arid northwest, which produce great quantities of winter vegetables, melons, cotton, and oil seed crops for export to the United States, the northeast sorghum and citrus areas; and Yucatán, where irrigated citrus is fast replacing henequen on many former *hacienda* lands. Citrus is also an important crop in Belize. Secondary commercial farming zones are scattered through Latin America and the Caribbean, many of them specializing in fresh produce for nearby urban populations.

Animal Husbandry

Commercial animal husbandry in Latin America today occurs in two forms. The first, which tends to be oriented almost exclusively toward domestic urban markets, consists of modern poultry and pig farms and dairies. These can now be found near every major city, supplying the demand of the middle and upper economic classes for relatively expensive, high-protein foods such as meat, eggs, milk, and cheese (Figure 10.23). Many of the dairies are operated by the descendants of nineteenth- and twentieth-century immigrants from the alpine zones of Germany and northern Italy. The high quality, safe-to-drink milk that they produce often stands in stark contrast to contaminated municipal water supplies.

The second form of modern Latin American and Caribbean animal husbandry centers on improved methods of sheep and cattle ranching. The first region to

move in this direction was the Argentine and Uruguayan Pampa, whose coarse native prairie grasses, known as *pasto duro,* or tough pastures, were utilized throughout the colonial era for the extensive grazing of horses and *criollo* cattle. The principal products of the latter were hides, tallow, and a dried and salted beef called *tasajo.* The introduction in the 1820s and 1830s of improved breeds of European sheep and shorthorn cattle combined with the advent in the 1870s of refrigerated shipping to create a large market for chilled and canned meat from the Pampa. That in turn triggered the importation of millions of mostly Italian indentured laborers, who were used to convert the native *pasto duro* grasses into alfalfa and other *pasto tierno,* or tender pastures, which now predominate on most of the *estancias.*

The introduction of improved breeds of cattle in the remainder of Latin America and the Caribbean has been greatly facilitated by the diffusion of imported African pasture grasses, an ongoing process that Parsons (1970) described as the "Africanization of the New World tropical grasslands." Those that have proven most durable in the heat and humidity of the *tierra caliente* include a tall clump-former called Guinea grass (*Panicum maximum*), a runner-type pasture called Pará grass (*Brachiaria mutica [Panicum purpurascens]*), and Jaraguá grass (*Hyparrhenia rufa*), which is extremely tolerant of the drought and fires common to the tropical savanna. Much of Latin America's temperate *tierra templada* is now overrun by Molasses grass (*Melinis minutiflora*). Yet another African import, Pangola grass (*Digitaria decumbens*), is found from the *tierra caliente* into the lower reaches of the *tierra fria,* with much of the latter now dominated by Kikuyu grass (*Pennisetum clandestinum*).

FIGURE 10.24   ✖

Zebu or Brahman cattle now form the basis for much of the lowland Latin America's commercial animal husbandry.

Just as the Latin American and Caribbean pastures have been upgraded through the diffusion of foreign or exotic grasses, the quality and vigor of the cattle herds have also been raised through the recent introduction of additional Old World breeds. The cattle originally brought by the Spaniards and Portuguese to the New World were Mediterranean strains belonging to the *Bos taurus* family. Once established in the New World, they evolved into a number of lines or breeds whose most valued asset was their sheer toughness and ability to survive, often in a wild state. Productivity, whether measured in meat or milk production, was not a major concern in most of Latin America until the latter half of the nineteenth century, when new breeds were introduced. These included European shorthorns, which are now raised for both meat and milk production in the cooler, temperate zones, and the hump-necked Zebu, or Brahman (*Bos indicus*), a product of the Indian subcontinent, whose tolerance of heat, humidity, insects, and diseases has resulted in its becoming, over the last 100 years, the dominant breed of the lowland tropics (Figure 10.24).

The rapid expansion of cattle ranching and the pastures that support it has been described by some as the **grassification** of the American tropics. While there is nothing inherently evil in cattle ranching, the practice has come under increasing attack in recent years by those who see its growth as the leading cause of **tropical deforestation.** Nair, Follis, and Murphy (1991), for instance, reported that approximately 65 percent of the deforestation that occurred in Latin America and the Caribbean during the 1980s was attributable to land clearing by cattle ranchers, whose efforts were frequently heavily subsidized by national governments bent on increasing exports of beef (Figures 10.25 and 10.26). They went on to note that most of the remaining deforestation was caused by smallholder farmers, with mining, commercial logging, and fuel wood operations being only minor contributors. Deforestation, with its attendant loss of wood products, wildlife, and plant biodiversity, is presently proceeding at an average rate of 1.7 percent annually for all of Latin America and the Caribbean. Interestingly, while the Latin American country that is losing the most land in absolute terms to deforestation continues to be Brazil, the highest annual rates of deforestation are found in Haiti, El Salvador, Nicaragua, and the other smaller nations of Central America and the Caribbean (Table 10.10). Barring widespread natural reforestation, only a few local instances of which have been documented (Rudel, Perez-Lugo, and Zichal 2000; Rudel, Bates, and Machinguiashi 2002), the natural forests of these countries will almost completely disappear within the next twenty to forty years (Figure 10.27).

Tree Farming

Concern over the threatened loss of Latin America's tropical rain forests and other woodlands has led some to experiment with tree farming, a strategy advocated by its proponents as a means of relieving pressure on the remaining natural forests (Lorentzen 2001; Fisher 1999; Evans 1992). Tree farming can be divided into two broad categories. The first, **plantation forestry,** consists of planting large areas to a single tree species in the hope of reaping commercially sustainable harvests of wood, food, or industrial products. Attempts at monocultural plantation forestry have a long, and often spectacular, history of failure in the humid lowlands of Latin America, particularly in Amazonia, where numerous large-scale rubber- and fiber-producing schemes have collapsed under the onslaught of disease and mismanagement. Despite such experiences, plantation forestry expanded from approximately 800,000 hectares (1,984,000 acres) in 1965 to almost 10,000,000 hectares (24,619,000 acres) in 1995, with most of it devoted to the cultivation of pine, teak, eucalyptus, gmelina, and *Acacia mangium* on degraded grazing lands (Table 10.11; Figure 10.28). Plantation forestry is not limited to tropical rain forest environments. Central and southern Chile, for example, now have over 1,500,000 hectares (3,700,000 acres) of Monterrey pine (*Pinus radiata*) plantations which have been planted at the expense of native *Nothofagus* species (Veblen, Hill, and Read 1996).

FIGURE 10.25

Burning of rain forest vegetation near Rio Branco in the state of Acre in the far western regions of Amazonian Brazil.

FIGURE 10.26

Former rain forest lands converted to pasture for animal grazing in Acre, Brazil.

Nonplantation forestry strategies include a wide range of small-scale development approaches. One of these, **agroforestry,** advocates the selective planting and harvesting of forest products without altering the basic polycultural composition and structure of the forest (Smith 1999; Faizool and Ramjohn 1995; Reed 1995; Nair 1990; Hiraoka 1989). Another, called the **harvest strip system,** permits small-scale clear-cutting of narrow strips that are naturally reseeded by

TABLE 10.10

Annual Deforestation Rates for Latin American and Caribbean Countries, 2000

COUNTRY	PERCENTAGE OF TOTAL FORESTED AREA LOST ANNUALLY
Haiti	5.7
El Salvador	4.6
Nicaragua	3.0
Guatemala	1.7
Panama	1.6
Jamaica	1.5
Ecuador	1.2
Mexico	1.1
Honduras	1.0
Costa Rica	0.8
Argentina	0.8
Trinidad and Tobago	0.8
Paraguay	0.5
Peru	0.4
Colombia	0.4
Brazil	0.4
Venezuela	0.4
Bolivia	0.3
Chile	0.1
Cuba	−1.3
Uruguay	−5.0

Note: negative numbers represent rates of reforestation.

Source: http://devdata.worldbank.org/data-query/

FIGURE 10.27

Tropical hardwood trees cut for timber in the Caribbean lowlands of Costa Rica.

FIGURE 10.28

Seedlings of *Acacia mangium* growing in a nursery in the uplands of central Panama prior to being transplanted into a plantation forestry project.

TABLE 10.11

Plantation Forestry in Latin America: 1965–1995

COUNTRY	1000 Hectares		
	1965	1980	1995
Brazil	500	3,855	4,805
Chile	10	23	1,747
Argentina	—	182	830
Venezuela	2	112	589
Cuba	143	243	470
Peru	20	127	349
Uruguay	—	—	348
Colombia	16	95	300
Costa Rica	2	4	129
Ecuador	4	9	120
Guatemala	1	3	68
Others	100	281	208
Totals	798	4,934	9,963

Sources: Adapted from: www.earthtrends.wri.org; and Evans 1992, 34–35.

the rain forest, thereby preventing a long-term decrease in forest cover and biodiversity (Case Study 10.2).

Fishing

Commercial fishing historically has remained a poorly developed food source in most of Latin America and the Caribbean. The causes of this underdevelopment are not physical, for the coastal waters are populated with abundant stocks of both cold and warm water marine species. Rather, the causes are cultural and include a traditional dietary bias in favor of beef products, low levels of scientific management knowledge, and the establishment of misguided regulatory policies by government officials.

A classic example of the mismanagement of fishing stocks is that of Peru's cold offshore current. Owing to the upwelling of nutrient-rich sediments, the current is endowed with a vast supply of tiny marine organisms called phytoplankton, which once sustained a seemingly inexhaustible supply of anchovy. While not important as a food fish for humans, the small, bony anchovies have served as food to the hundreds of millions of pelicans, gannets, cormorants, and other seabirds, whose nitrogen-rich droppings, called *guano*, have been viewed for almost two centuries as one of the earth's finest natural fertilizers. Yet another use for the anchovies, that of ground fishmeal added as a high-protein component of Western European and Anglo American livestock feeds, expanded rapidly in the 1950s, bringing with it a series of northern and central coastal boomtowns focusing on the catching and processing of the fish. Between 1965 and 1971, Peruvian anchovy harvests made up approximately one-fifth of the entire world fish catch and over 80 percent of Latin America's marine harvests. As the profits poured in, the Peruvian government encouraged the expansion of the industry, while ignoring, at least publicly, repeated warnings from its own marine biologists that gross overfishing was occurring. The inevitable collapse,

Case Study 10.2   Harvest Strip Forestry

Lumbering, or the harvesting of trees for their wood or fiber, is accomplished through either selective cutting or clear-cutting. Selective cutting involves the felling of single mature trees that are often located far from one another. Because it is costly to remove individual trees from a forest for processing, most commercial forestry employs the clear-cutting technique in which each and every tree in a designated area is felled at the same time. Although clear-cutting is economically more profitable than selective cutting, it also has more harmful environmental consequences because it results in reduced biodiversity and habitat loss in the affected area.

Tropical foresters have long sought a lumbering strategy that would combine the economic advantages of clear-cutting with the envi-

ronmental protections of selective cutting. One such approach that was tested in the Palcazú Valley of eastern, Amazonian Peru is called the harvest strip system. It entails dividing the tropical rain forest up into long, narrow strips of land whose width is no greater than the height of the tallest trees of the forest. By clear-cutting a given strip only once every thirty to forty years, the mature forest left standing on the neighboring uncut strips reseeds the harvested strip, maintaining the biodiversity of the entire region.

In the Palcazú Valley experiment, the larger logs were transported by animals to a centrally located, on-site, cooperative mill where they were cut into lumber. The remaining, smaller-diameter materials were converted into utility poles, posts, or charcoal.

Power for the processing plant was obtained in part from the burning of unprocessed biomass.

Although the Palcazú forestry project was finally abandoned in 1989, owing to increased guerrilla activity in the region and negative economic returns brought about by low timber prices and high marketing costs, it nevertheless demonstrated that high timber yields can be sustained in tropical rain forests without inflicting long-term environmental damage. While it is recognized that no single management system is suitable to all tropical forests, the harvest strip approach presents a viable alternative to plantation forestry, cattle ranching, or largeholder agriculture, each of which has resulted in the long-term degradation of the rain forest environment.

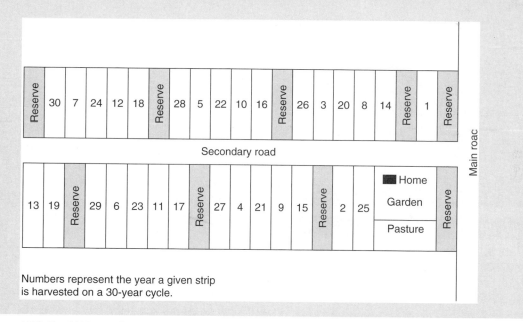

Numbers represent the year a given strip
is harvested on a 30-year cycle.

Sources: Southgate 1998; Tosi 1991; Hartshorn, 1989.

which began in 1972, was triggered by an invasion of warm tropical waters linked to the El Niño Southern Oscillation (ENSO). Successive Peruvian governments first nationalized and then abandoned the battered fishing industry, which finally began to recover par-

tially in the 1980s on the basis of sardine, mackerel, pilchard, and hake catches. Chile also has emerged as a leading fishing nation, having managed wisely its stocks of swordfish, salmon, king crab, and shrimp (Table 10.12; Figure 10.29).

TABLE 10.12

Latin American and Caribbean Fish Catches, 1953–2000

COUNTRY	Total Catch (1000s Metric Tons)			
	1953	1967	1977	2000
Argentina	77	241	370	1,130
Brazil	161	419	749	855
Chile	107	1,053	1,319	3,558
Colombia	16	93	64	213
Cuba	10	66	185	100
Ecuador	9	47	434	457
Guyana	3	14	22	57
Mexico	67	350	611	1,222
Panama	n.a.	72	208	219
Peru	165	10,134	2,534	4,346
Uruguay	3	11	48	141
Venezuela	63	107	146	517

Sources: Compiled from *Statistical Yearbook 1998* 2001, 355–364; *Statistical Yearbook 1993* 1995, 347–355; and *Statistical Yearbook 1968* 1970, 460–472.

FIGURE 10.29

Fishing boats at port in Antofagasta, a coastal city of northern Chile that fronts the cold, nutrient-rich Peru Current.

Other commercially important fisheries include the Guayas lowlands of Ecuador, where one of Latin America's few successful mariculture enterprises focuses on shrimp; the northern Pacific coast of Mexico, which yields large catches of tuna, sardine, and shrimp; the Atlantic coasts of both Brazil and Mexico, which produce lobster, red snapper, grouper, bonito, shrimp, and other warm water species; and the Pacific coasts of Central America and Colombia, which are worked for

FIGURE 10.30

Terrace cultivation of coca in the transitional valleys of the Yungas of northern Bolivia. Most of the leaves on these plants have been harvested.

shrimp. Sport fishing is locally important in much of the Caribbean and in both the Argentine and Chilean lake districts. While all of these are of some value to the local inhabitants, fishing as an occupation and fish products as dietary components are unlikely to assume, in the near term, the importance in Latin America that they have historically occupied in the developing nations of East and Southeast Asia.

Narcotic Drug Production

The final category of Latin American and Caribbean commercial agriculture is narcotic drug production. Although cannabis, or marijuana, has long been cultivated throughout Middle America and opium poppy farming began to appear in the 1980s, the most economically significant of Latin America's drug crops is cocaine. The coca plant (*Erythroxylon coca*) is a bushy, perennial shrub indigenous to the eastern Andean foothill country of Bolivia, Peru, and Brazil (Figure 10.30). Its use by the Indians reaches back to prehistory. It was viewed as a sacred plant and employed as a bartering medium by the Inca, who endeavored to limit its use to the ruling caste. Following the Conquest, coca use was deregulated by the Spaniards and the habit of chewing the leaves with a lime solution expanded rapidly among the enslaved Indians, who relied on the practice as a means of dispelling the effects of thirst, hunger, and fatigue at high altitudes. It is important to recognize that coca's historic use among the Andean peoples has been, and continues to be, as a stimulant—not as a hallucinogenic. For this reason, the dried, unprocessed leaves have been available as freely in the marketplaces as potatoes or quinoa.

TABLE 10.13

Andean Coca and Cocaine Production: 1999

COUNTRY	HECTARES OF ILLICIT COCA CULTIVATION	COCA LEAF HARVEST (METRIC TONS)	COCAINE PRODUCTION (METRIC TONS)	PEOPLE INVOLVED	TURN OVER (US $ MILLION)	GDP SHARE (%)
Bolivia	9,800[a]	22,800	70	51,300	86	2.2
Colombia	122,500	195,000	520	200,000	2,200	2.3
Peru	51,000	95,600	240	200,000	325	0.5
Totals	183,300	313,400	830	451,300	2,611	—

[a]Excludes 12,000 hectares cultivated legally for traditional purposes.

Source: Compiled from the Bolivia, Colombia, and Peru country profiles at: www.undcp.org

Until recently, the principal coca production zones were Peru's Huallaga Valley and the Chaparé and Yungas regions of Bolivia. These patterns have changed significantly in recent years, however, and Colombia now far exceeds Peru and Bolivia, both in the amount of land given over to illicit coca cultivation and in coca leaf production (Table 10.13). These changes are attributable to the success in Bolivia and, to a lesser degree, in Peru of alternative development programs that have paid farmers as much as US $2,500 per hectare (2.47 acres) per year to substitute food crops for coca while simultaneously investing in roads, schools, medical clinics, and other projects designed to improve the quality of life of the poverty-stricken rural farmers. The reduction in levels of guerrilla activity in Peru has also freed much of the peasantry from forced alliances with organized drug traffickers and enabled the government to invest more heavily in regional development projects. In contrast, the high levels of violence and corruption and the low levels of governmental presence in much of Colombia have provided a setting in which drug producers have encountered little state interference and in which alternative development programs have been largely absent. As a result, approximately half of Colombia's coca leaf is now brazenly produced on large plantations up to 80 hectares (198 acres) in size.

While the geographical center of coca cultivation has shifted from Peru and Bolivia to Colombia in recent years, there is no evidence that the total amount of cocaine flowing out of the Andean region has diminished. Until such time that foreign demand for the product is reduced and Colombian interdiction efforts prove effective, Andean coca production is unlikely to decline.

SUMMARY

As we have studied in this chapter the agricultural geography of Latin America and the Caribbean, we have come to understand that it consists of far more than tables of important crops or maps showing dominant cropping systems. Rather, it is a microcosm of all the complex cultural, social, and physical forces that have interacted within the region over time. This landuse complexity is often so great that we have avoided, in our analysis, the common practice of assigning to a given area on a map a single farming system category. Central America, for example, is characterized by a mix of banana and sugar cane plantations, peasant farming, and commercial ranching operations in the lowlands, and coffee plantations, *haciendas*, commercial dairying, vegetable and flower production, peasant villages, and slash and burn agriculturists in the highlands.

Just as it would be unrealistic to attempt to generalize about the cropping patterns of Central America, so too is it impossible to list here all the political, economic, and social variables that have contributed to the agricultural development or underdevelopment of each individual area. What can and must be noted, however, is that the traditional Hispanic perception of urban areas and peoples being superior to their rural counterparts continues unchallenged in most of Latin America. The result of those values has been, and continues to be, the enactment of policies that inhibit the development of the countryside in favor of the urban, industrial sector. These policies have resulted in an increasing gap in levels of living between the urban and rural zones. This gap is now threatening, in the form of uncontrolled migration and social and political unrest, the very urban development used by the elites to justify the exploitation of the agrarian sector.

It is vitally important to the present and future welfare of all the Latin American peoples, both rural and urban, that agriculture be strengthened through enlightened government policies that recognize the contributions of smallholder as well as largeholder agriculturists. Only then will the land be used to its fullest potential and the socioeconomic inequalities that have so plagued the region begin to be rectified.

KEY TERMS

enclaves 237
hacienda 237
estancia 238
plantation 238
monoculture 244
polyculture 246

interspecific agriculture 246
intraspecific diversity 246
Green Revolution 253
ex situ conservation 255
in situ conservation 255

grassification 258
tropical deforestation 258
plantation forestry 258
agroforestry 259
harvest strip system 259

SUGGESTED READINGS

Anderson, Edgar. 1952. *Plants, Man, and Life.* Boston: Little, Brown.

Barkin, David, and Billie R. DeWalt. 1988. "Sorghum and the Mexican Food Crisis." *Latin American Research Review* 23:30–59.

Bebbington, Anthony, and Judith Carney. 1990. "Geography in the International Agricultural Research Centers: Theoretical and Practical Concerns." *Annals of the Association of American Geographers* 80:34–48.

Beckford, George L. 1972. *Persistent Poverty: Underdevelopment in Plantation Economies of the Third World.* New York: Oxford University Press.

Bourgois, P. I. 1989. *Ethnicity at Work: Divided Labor on a Central American Banana Plantation.* Baltimore: Johns Hopkins University Press.

Breslin, Patrick. 1988. "The Valley Without Birds." *Grassroots Development* 12:24–29.

Brush, S. B., H. J. Carney, and Z. Huamán. 1981. "Dynamics of Andean Potato Agriculture." *Economic Botany* 35:70–88.

Caviedes, César N., and Keith D. Muller. 1994. "Fruticulture and Uneven Development in Northeast Brazil." *Geographical Review* 84:380–393.

Clawson, David L. 1992. "Conservation of Food Crop Genetic Resources in Latin America." In *Benchmark 1990*, ed. Tom L. Martinson, 11–17. Auburn, Ala.: Conference of Latin Americanist Geographers.

———. 1985. "Harvest Security and Intraspecific Diversity in Traditional Tropical Agriculture." *Economic Botany* 39:56–67.

———. 1987. "Teaching Traditional Tropical Agriculture." *Journal of Geography* 86:204–210.

Clawson, David, and Don R. Hoy. 1979. "Nealtican, Mexico: A Peasant Community That Rejected the 'Green Revolution.' " *American Journal of Economics and Sociology* 38:371–387.

Cornelius, Wayne A., and David Myhre, eds. 1998. *The Transformation of Rural Mexico: Reforming the Ejido Sector.* San Diego: Center for U.S.–Mexican Studies, University of California, San Diego.

Cornia, Giovanni A. 1985. "Farm Size, Land Yields and the Agricultural Production Function: An Analysis for Fifteen Developing Countries." *World Devlopment* 13:513–534.

Crist, Raymond E., and Charles M. Nissly. 1973. *East from the Andes.* Gainesville: University of Florida Press.

Crowley, William K. 2000. "Chile's Wine Industry: Historical Character and Changing Geography." *Yearbook: Conference of Latin Americanist Geographers* 26:87–101.

Denevan, William M. 2001. *Cultivated Landscapes of Native Amazonia and the Andes.* Oxford: Oxford University Press.

———. 1984. "Ecological Heterogeneity and Horizontal Zonation in the Amazon Floodplain." In *Frontier Expansion in Amazonia,* eds. M. Schmink and C. H. Wood, 311–336. Gainesville: University of Florida Press.

———. 1966. *The Aboriginal Cultural Geography of the Llanos de Mojos of Bolivia.* Berkeley: Ibero-Americana No. 48, University of California Press.

DeWitt, John. 1989. "Sugar Cane Cultivation and Rural Misery: Northeast Brazil." *Journal of Cultural Geography* 9 (2):31–39.

Dosal, Paul J. 1993. *Doing Business with the Dictators: A Political History of United Fruit in Guatemala, 1899–1944.* Wilmington, Del.: Scholarly Resources.

Economic and Social Progress in Latin America 1986. 1986, Washington, D.C.: Inter-American Development Bank.

Eidt, Robert C. 1975. "Agrarian Reform and the Growth of New Rural Settlements in Venezuela." *Erdkunde* 29:127–139.

Evans, Julian. 1992. *Plantation Forestry in the Tropics.* 2nd ed. Oxford: Clarendon Press.

Faizool, Sheriff, and Robert K. Ramjohn. 1995. *Agroforestry in the Caribbean.* Santiago, Chile: Food and Agriculture Organization of the United Nations Regional Office for Latin America and the Caribbean.

Fisher, Richard F. 1999. "Forest Plantations in the Tropics." In *Managed Ecosystems: The Mesoamerican Experience,* eds. L. Upton Hatch and Marilyn E. Swisher, 202–211. New York: Oxford University Press.

Food and Agriculture Organization Production Yearbook 1990. 1991. Rome: United Nations.

Gade, Daniel W. 1992. "Landscape, System, and Identity in the Post-Conquest Andes." *Annals of the Association of American Geographers* 82:460–477.

Galloway, John H. 1989. *The Sugar Cane Industry: An Historical Geography from Its Origins to 1914.* Cambridge: Cambridge University Press.

Gómez-Pompa, Arturo et al. 1982. "Experiences in Traditional Hydraulic Agriculture." In *Maya Subsistence: Studies in Memory of Dennis E. Puleston,* ed. K.V. Flannery, 327–342. San Diego: Academic Press.

González. Roberto J. 2001. *Zapotec Science: Farming and Food in the Northern Sierra of Oaxaca.* Austin: University of Texas Press.

Grossman, Lawrence S. 1998. *The Political Ecology of Bananas: Contract Farming, Peasants, and Agrarian Change in the Eastern Caribbean.* Chapel Hill: University of North Carolina Press.

Hartshorn, Gary S. 1989. "Sustained Yield Management of Natural Forests: The Palcazú Production Forest. "In *Fragile Lands of Latin America: Strategies for Sustainable Development,* ed. John O. Browder, 130–138. Boulder, Colo.: Westview Press.

Harwood, R. R. 1979. *Small Farm Development.* Boulder, Colo.: Westview Press.

Hecht, Susanna B., and Alexander Cockburn. 1989. *The Fate of the Forest: Developers, Destroyers, and Defenders of the Amazon.* New York: Verso.

Hewitt de Alcantara, Cynthia. 1973–1974. "The 'Green Revolution' as History: The Mexican Experience." *Development and Change* 5:25–44.

Hildebrand, Peter E. 1976. "Multiple Cropping Systems are Dollars and 'Sense' Agronomy." In *Multiple Cropping,* 348–371. Madison, Wis.: American Society for Agronomy Special Publication Number 27.

Hiraoka, Mario. 1989. "*Ribereños'* Changing Economic Patterns in the Peruvian Amazon." *Journal of Cultural Geography* 9 (2): 103–119.

Horst, Oscar H. 1989. "The Persistence of Milpa Agriculture in Highland Guatemala." *Journal of Cultural Geography* 9(2):13–29.

Innis, Donald Q. 1997. *Intercropping and the Scientific Basis of Traditional Agriculture.* London: Intermediate Technology Publications.

———. 1980. "The Future of Traditional Agriculture." *Focus* 30:1–8.

Jaramillo, Carlos Felipe, and Thomas Kelly. 1999. "Deforestation and Property Rights." In *Forest Resource Policy in Latin America,* ed. Kari Keipi, 111–133. Washington, D.C.: Inter-American Development Bank.

Johns, T., and S. L. Keen. 1986. "Ongoing Evolution of the Potato on the Altiplano of Western Bolivia." *Economic Botany* 40:409–424.

King, Russell. 1977. *Land Reform: A World Survey.* Boulder, Colo.: Westview Press.

Knapp, Gregory. 1991. *Andean Ecology: Adaptive Dynamics in Ecuador.* Boulder, Colo.: Westview Press.

———. 1982. "Prehistoric Flood Management on the Peruvian Coast: Reinterpreting the 'Sunken Fields' of Chilca." *American Antiquity* 47:144–154.

Kolata, Alan L., ed. 1996. *Tiwanaku and Its Hinterland: Archaeology and Paleoecology of an Andean Civilization,* Vol. I: *Agroecology.* Washington, D.C.: Smithsonian Institution Press.

Lawson, Victoria A. 1988. "Government Policy Biases and Ecuadorian Agricultural Change." *Annals of the Association of American Geographers* 78:443–452.

Lorentzen, Erling. 2001. "Aracruz Celulose." In *Footprints in the Jungle: Natural Resource Industries, Infrastructure, and Biodiversity Conservation,* eds. Ian Bowles and Glenn T. Prickett, 134–144. Oxford: Oxford University Press.

Louette, Dominique, Andre Charrier, and Julien Berthaud. 1997. "In Situ Conservation of Maize in Mexico: Genetic Diversity and Maize Seed Management in a Traditional Commuinity." *Economic Botany* 51:20–38.

Mathewson, Kent. 1984. *Irrigation Horticulture in Highland Guatemala: The Tablón System of Panajachel.* Boulder, Colo.: Westview Press.

Mead, R., and R. W. Willey. 1980. "The Concept of a 'Land Equivalent Ratio' and Advantages from Intercropping." *Methodology of Experimental Agriculture* 16:217–228.

Meyer, Carrie A. 1989. *Land Reform in Latin America: The Dominican Case.* New York: Praeger.

Murray, Douglas L. 1994. *Cultivating Crisis: The Human Costs of Pesticides in Latin America.* Austin: University of Texas Press.

Nair, P. K. R. 1990. *The Promise of Agroforestry in the Tropics.* Washington, D.C.: The World Bank.

Nair, P. K. R., Mark B. Follis, and Timothy P. Murphy. 1991. "Agroforestry and Sustainable Development in the Humid Tropical Lowlands of Latin America and the Caribbean." In *Proceedings of the Humid Tropical Lowlands Conference: Development Strategies and Natural Resource Management,* vol. I, ed. Dennis V. Johnson, 55–72. Bethesda, Md.: Development Strategies for Fragile Lands.

Navarro, Zander. 1993. "Rural Poverty and the Promise of Small Farmers in Brazil." *Grassroots Development* 17(1):20–24.

Nelson, Michael. 1973. *The Development of Tropical Lands: Policy Issues in Latin America.* Baltimore: Johns Hopkins University Press for Resources for the Future.

Ozório de Almeida, Anna Luíza. 1992. *The Colonization of the Amazon.* Austin: University of Texas Press.

Ozório de Almeida, Anna Luíza, and João S. Campari. 1995. *Sustainable Development in the Brazilian Amazon.* Oxford: Oxford University Press.

Parsons, James J. 1968. *Antioqueño Colonization in Western Colombia.* Rev. ed. Berkeley: University of California Press.

———. 1969. "Ridged Fields in the Rio Guayas Valley, Ecuador." *American Antiquity* 34:76–80.

———. 1970. "The 'Africanization' of the New World Tropical Grasslands." *Tubinger Geographische Studien* 34:141–153.

Patiño, Victor M. 1963. *Plantas Cultivadas y Animales Domésticos en América Equinoccial.* Cali, Colombia: Imprenta Departamental.

Poleman, Thomas T. 1964. *The Papaloapan Project: Agricultural Development in the Mexican Tropics.* Stanford: Stanford University Press.

Popenoe, Hugh, and Steven R. King et al. 1989. *Lost Crops of the Incas.* Washington, D.C.: Board on Science and Technology for International Development, National Research Council, National Academy Press.

Powelson, John P., and Richard Stock. 1990. *The Peasant Betrayed: Agricultural and Land Reform in the Third World.* Washington, D.C.: Cato Institute.

Randall, Laura, ed. 1996. *Reforming Mexico's Agrarian Reform.* Armonk, N.Y.: M.E. Sharpe.

Reed, Richard K. 1995. *Prophets of Agroforestry: Guaraní Communities and Commerical Gathering.* Austin: University of Texas Press.

Rice, Robert A. 1998. "A Rich Brew from the Shade." *Américas* 50(2):52–59.

Roseberry, William, Lowell Gudmundson, and Mario Samper Kutschback, eds. 1995. *Coffee, Society, and Power in Latin America.* Baltimore: Johns Hopkins University Press.

Rudel, Thomas K., Diane Bates, and Rafael Machirguiashi. 2002. "A Tropical Forest Transition: Agricultural Change, Out-Migration, and Secondary Forests in the Ecuadorian Amazon." *Annals of the Association of American Geographers* 92: 87–102.

Rudel, Thomas K., Marla Perez-Lugo, and Heather Zichal. 2000. "When Fields Revert to Forest: Development and Spontaneous Reforestation in Post–War Puerto Rico." *Professional Geographer* 52: 386–397.

Schlesinger, S., and S. Kinzer. 1982. *Bitter Fruit: The Untold Story of the American Coup in Guatemala.* New York: Doubleday.

Siemens, Alfred H. 1998. *A Favored Place: San Juan River Wetlands, Central Veracruz, A.D. 500 to the Present.* Austin: University of Texas Press.

Sluyter, Andrew. 1994. "Intensive Wetland Agriculture in Mesoamerica: Space, Time, and Form." *Annals of the Association of American Geographers* 84:557–584.

Smith, Nigel J. H. 1999. *The Amazon River Forest: A Natural History of Plants, Animals, and People.* New York: Oxford University Press.

Southgate, Douglas. 1998. *Tropical Forest Conservation: An Economic Assessment of the Alternatives in Latin America.* New York: Oxford University Press.

Statistical Yearbook 1998, 1995, 1993 and *1968.* 2001, 1997, 1995, 1970. New York: United Nations.

Stewart, Norman R. 1965. "Migration and Settlement in the Peruvian Montaña: The Apurimac Valley." *Geographical Review* 55:143–157.

Thiesenhusen, William C. 1995. *Broken Promises: Agrarian Reforms and the Latin American Campesino.* Boulder, Colo.: Westview Press.

Thrupp, Lori Ann. 1998. *Cultivating Diversity: Agrobiodiversity and Food Security.* Washington, D.C.: World Resources Institute.

———. 1991. "Long-term Losses from Accumulation of Pesticide Residues: A Case of Persistent Copper Toxicity in Soils of Costa Rica." *Geoforum* 21:1–15.

Tosi, Joseph A., Jr. 1991. "Integrated Sustained Yield Management of Primary Tropical Wet Forest: A Pilot Project in the Peruvian Amazon." In *Proceedings of the Humid Tropical Lowlands Conference: Development Strategies and Natural Resource Management,* vol. III, ed. Joshua C. Dickinson III, 47–64. Bethesda, Md.: Development Strategies for Fragile Lands.

Treacy, John M. 1987. "Building and Rebuilding Agricultural Terraces in the Colca Valley of Peru." *Yearbook, Conference of Latin Americanist Geographers* 13:51–57.

Turner, B. L., II, and P. D. Harrison. 1983. *Pulltrouser Swamp: Ancient Maya Habitat, Agriculture, and Settlement in Northern Belize.* Austin: University of Texas Press.

Vargas Carranza, Jorge Luís. 1990. "Prácticas Agrícolas Indígenas Sostenibles en Areas del Bosque Tropical Húmedo en Costa Rica." *Geoistmo* 4:1–94.

Veblen, Thomas T., Robert S. Hill, and Jennifer Read, eds. 1996. *The Ecology and Biogeography of Nothofagus Forests.* New Haven: Yale University Press.

Wesche, Rolf. 1971. "Recent Migration to the Peruvian Montana." *Cahiers de Geographie de Quebec* 15:251–266.

West, Robert C., and P. Armillas. 1950. "The Chinampas of Mexico." *Cuadernos Americanos* 50:165–182.

Whitmore, Thomas M. 1991. "A Simulation of the Sixteenth-Century Population Collapse in the Basin of Mexico." *Annals of the Association of American Geographers* 81:464–487.

Wilken, Gene C. 1987. *Good Farmers: Traditional Agricultural Resource Management in Mexico and Central America.* Berkeley: University of California Press.

Williams, Lynden S. 1990. "Agricultural Terrace Evolution in Latin America." *Yearbook, Conference of Latin Americanist Geographers* 16:82–93.

Williams, R. G. 1986. *Export Agriculture and the Crisis in Central America.* Chapel Hill: University of North Carolina Press.

Zimmerer, Karl S. 1996. *Changing Fortunes: Biodiversity and Peasant Livelihood in the Peruvian Andes.* Berkeley: University of California Press.

———. 1998. "Disturbances and Diverse Crops in the Farm Landscapes of Highland South America." In *Nature's Geography: New Lessons for Conservation in Developing Countries,* eds. Karl S. Zimmerer and Kenneth R. Young, 262–286. Madison: University of Wisconsin Press.

———. 1991a. "Managing Diversity in Potato and Maize Fields in the Peruvian Andes." *Journal of Ethnobiology* 11:23–49.

———. 1991b. "Wetland Production and Smallholder Persistence: Agricultural Change in a Highland Peruvian Region." *Annals of the Association of American Geographers* 81:443–463.

ELECTRONIC SOURCES

Food and Agriculture Organization of the United Nations: http://apps.fao.org/lim500/wrap.pl?Population.

Food and Agriculture Organization of the United Nations: http://apps1.fao.org/servlet/XteServlet...ction.Crops.

Food and Agriculture Organization of the United Nations: http://apps1.fao.org/servlet/XteServlet.jrun?Areas

United Nations Office for Drug Control and Crime Prevention: www.undcp.org

World Bank World Development Indicators database: http://devdata.worldbank.org/data-query/

World Resources Institute: www.earthtrends.wri.org

11

Mining, Manufacturing, and Tourism

Although the majority of the economically active population of Latin America and the Caribbean historically supported itself through agriculture, one of the principal sources of regional wealth—from colonial times to the present—has been mining. Mining was dominated during the colonial era by the pursuit of gold, silver, and precious gems. Today it is centered on the production of numerous industrial ores whose use is essential to Latin America's emerging manufacturing sector. In this chapter, we will analyze the changing economic, social, and cultural impacts of mining, manufacturing, and tourism, which collectively have come to constitute one of the most rapidly growing components of the regional economic landscape.

PRE-CONQUEST MINING AND MANUFACTURING

Mining and manufacturing were practiced widely among the South and Middle American Indians prior to the Conquest, often with remarkably impressive results. Among the Inca, Maya, and Aztec peoples, the quarrying of massive stones for construction purposes sustained architectural accomplishments that were equal or superior to those of the Old World at comparable times in history. The use of jade, obsidian, and other precious stones as ornaments was widespread in both highland and lowland America prior to the Conquest. Although iron mining was absent, metallurgy also reached advanced levels throughout the region, with gold and copper jewelry serving as common items of tribute. Pre-Conquest manufacturing focused primarily on the production of textiles, whose patterns

could be exceedingly intricate and beautiful. Pottery, baskets, implements of war, large boats, and countless other articles were also produced, each adapted to the needs and resources of the local peoples. Mining and manufacturing were thus present throughout much of the New World on the eve of the Spanish and Portuguese conquests. What most differentiated the Old and New World civilizations that were soon to be placed into direct contact with one another was not the presence of mining and manufacturing within the Old World and its absence within the New World. It was, instead, the greater overall cultural and economic value placed by the Europeans on the generation of mineral wealth and the more advanced industrial technologies of the Spaniards and Portuguese. Many of these technologies had diffused to Iberia from northern and central Europe, while others had originated within the Islamic lands of Africa and the Middle East.

MINING AND MANUFACTURING IN COLONIAL LATIN AMERICA

The great importance placed on mining by the Portuguese, and especially by the Spanish, conquerors of the New World can best be summarized by reminding ourselves of the threefold objectives or purposes of the individual *conquistadores:* the acquisition of wealth, service to God, and the achievement of personal honor and glory. One task to which many of the early settlers devoted themselves with great enthusiasm was the search for precious metals, particularly gold. This was accomplished primarily through two distinct strategies. The first, and easier, means of accumulating gold was

267

to steal it from the Indians, many of whom, as we noted previously, had access to large quantities buried in ancestral tombs or in use as jewelry and other bodily ornaments.

The second, and more laborious, means of acquiring gold was to search for it, as the Indians had for thousands of years, among the many small streams and rivers that drained the highly mineralized interior highlands of the Greater Antilles and the American mainland. The Spaniards, ever conscious of their self-proclaimed superior social standing, generally sought to force the Indians to work the placer deposits for them, utilizing shallow wooden bowls and other native American implements. In addition to bestowing wealth and high social standing, the possession of gold was also of great practical value during the early decades of the colonial era. It provided the means through which the European settlers paid for numerous expensive Old World foods and manufactured items, the cultivation or production of which was prohibited under the terms of the prevailing mercantile systems. These items included iron and steel products of all kinds, fine textiles, wines, and olive oil.

Silver Mining

As the alluvial gold deposits of the early settlement sites dwindled in the beginnings of the sixteenth century, the Spaniards' insatiable need for additional revenues led them to intensify their search for alternative precious metals, the foremost of which was silver. In contrast to recent times, when the value of silver has typically averaged only about one-fiftieth to one-one hundredth the value of an equivalent amount of gold, the value of silver in early sixteenth-century Europe was approximately one-tenth to one-eleventh that of an equivalent quantity of gold. This reflected, in part, the presence of relatively stable and abundant gold supplies at a time when the rise of the Ottoman Empire in the eastern Mediterranean Basin was resulting in the partial cutoff of traditional central Asian silver sources (Céspedes 1976).

The search for silver in Latin America was thus an activity of the highest priority in the first decades of the colonial era. By the 1530s, the Spanish had discovered numerous mines of small to modest size worked previously by the native peoples in the western highlands, from central Mexico southward into the Peruvian Andes. This was followed, in the mid-1540s, by the chance discoveries of what would ultimately prove to be two of the richest mineral deposits in the history of the world. Their subsequent development was to significantly shape the social fabric and settlement and landuse patterns of Latin America's most influential regions (Figure 11.1).

By far the more spectacular of the two finds was the discovery in 1545 of rich surface deposits of silver on a barren, windswept, red-colored mountain whose sugarloaf-shaped peak rose to over 4,900 meters (16,000 feet) in the Western Cordillera of Upper Peru or Bolivia. The name given to the mountain, and to the mining town that was laid out two years later at its base, was Potosí (Figure 11.2). During the last half of the sixteenth century, fully half the world's output of silver originated from the legendary Potosí mines, whose successful functioning became one of the highest priorities of the Spanish Crown.

One year after the discovery of Potosí, a second great silver deposit was uncovered thousands of miles to the north at Zacatecas in the Western Sierra Madre of central Mexico. This was soon followed by the development of additional major deposits at Sombrerete, Guanajuato, Durango, Guadalajara, Parral, and numerous other sites in western Mexico and at Pachuca and San Luís Potosí in the Eastern Sierra Madre (Figure 11.3). Because most of the Mexican mines were situated at elevations ranging from 1,500 to 2,400 meters (5,000 to 8,000 feet), their physical environments were not nearly as harsh as that of Potosí.

Initially, the combined silver output of the mines of New Spain was far exceeded by that of Potosí, whose very name became synonymous in the early colonial era with wealth and opulent lifestyles. The Mexican production, however, ultimately proved to be more stable and substantial. The reasons for Mexico's greater long-term yields were both physical and political. Physically, Mexico was blessed with a large number of major mines, whereas Peruvian production, despite the existence of scattered secondary operations at Cerro de Pasco, Oruro, and elsewhere, was concentrated at a single site. Given the fact that output fluctuated considerably over time at each mine, the greater number of Mexican operations resulted in a more consistent long-term production. Potosí's long-term decline was also related to the early exhaustion of its rich surface ores and to subsequent dependence on its less accessible subsurface veins, whose silver content was considerably lower than that of the Mexican ores. Then, too, the incredible 4,200 to 4,900 meter (14,000 to 16,000 foot) altitudes at which the Potosí mines were situated took a far greater toll on men, animals, and machinery than was the case in Mexico. Potosí's altitude also meant that all of its food and supplies had to be imported at great cost from distant regions, whereas farming and animal husbandry were never far from the Mexican mines.

FIGURE 11.1

Principal mining and manufacturing centers of colonial Latin America.

Mexico's higher grade ores and lower production costs ultimately resulted in greater long-term Spanish investment. One of the most significant ways in which Spain favored Mexico over Peru was its decision, in 1548, to require Mexican mine operators to pay only a tenth, or *diezmo,* of their output to the Crown rather than the customary 20 percent, or *quinto real,* which continued to be exacted from the Peruvians until 1736. Peruvian sales taxes and import and export duties were also higher than their Mexican counterparts, and Peru's government was exceptionally corrupt and inefficient, even by colonial standards (Lynch 1992). Overall, it is estimated that Mexican production costs averaged approximately 20 percent of the silver's value over the course of the colonial era compared to a 35 percent overhead on Peruvian production. All of these challenges notwithstanding, Garner (1988, 898) concluded that, from 1560 to 1810, Spain's American colonies produced between 93.3 and 108.9 billion grams (3.0 and 3.5 billion ounces), or 100,000 tons, of silver, most of which was transported to Spain rather than invested in the colonies.

IMPACT OF AMALGAMATION

In addition to the physical and political challenges to the production of silver discussed above, the colonial mines also had to deal with numerous on-site obstacles that required the utilization of what were then the highest levels of European mechanical and chemical technology. These obstacles included the necessity of tunneling deeper and deeper into the earth in pursuit of the narrow, ore-bearing veins, or *vetas,* the difficulty of transporting the ore back to the surface, and the need to control water seepage into and flooding of the shafts themselves.

FIGURE 11.2

The fabled mines of Potosí provided half the world's silver output during the early colonial era. Most of the mines were situated at extremely high altitudes, making labor very strenuous.

FIGURE 11.3

Sombrerete is one of a number of silver mining communities in Mexico's Western Sierra Madre whose output sustained the colonial economy of New Spain.

Even more troubling than the mechanical obstacles were the low metallic content of the vein ores and the inefficient smelting techniques of the period. Because of the latter, in the early years of mining much of the silver that was brought to the surface went unrecovered. The most significant technological breakthrough associated with colonial silver mining occurred in 1544, when Bartolomé de Medina applied to the Viceroy of New Spain for a concession, or *solicitud de privilegio,* on a mercury amalgamation process that he had adapted from Old World practices (Assadourian 1992). The purpose of the **amalgamation process** was to extract silver from ores whose content was so low that they could not otherwise be profitably smelted. Its principal steps included crushing the silver ore into a fine dust or powder called *harina*, mixing the *harina* with water, salt, and mercury to the consistency of a thick slime, treading the mix for two to three months in a large stone paved open yard or *patio* while adding additional water and mercury as needed, washing the sludge in vats and settling troughs where much of the heavier silver-mercury amalgam sank as particles and was recovered, pressing the amalgam into lead molds, heating the molds to release the mercury as distilled gas or vapor, which condensed on the ceiling and walls of the chamber and was recovered for reuse, and finally removing the almost pure silver bars from the cooled molds (Bakewell 1971).

The introduction of amalgamation, or the *patio* process as it was often called, thus created a tremendous demand for mercury, or "quicksilver." Significant mercury deposits never were located in Mexico, which was supplied throughout the colonial era from mines at Almadén and Idria in Spain and, through the Manila galleon that disembarked at Acapulco, with some mercury mined in the Orient. Potosí, on the other hand, was supplied primarily from a rich deposit discovered at Huancavelica in central Peru in 1563. Land transport of the mercury over the 1,300 kilometers (800 miles) of mountain trails separating Huancavelica from Potsoí was so difficult that it was usually shipped first to the Peruvian port of Chincha, from where it was routed southward by boat to Arica and then loaded onto the backs of mules and llamas for the tortuous climb by pack train back up the Andes (see Figure 11.1). Amalgamation remained the dominant, but not exclusive, treatment of lower-grade Latin American silver ores until the late nineteenth century, when it began to be replaced by advanced flotation processes.

LABOR SUPPLY

The greatest obstacle facing the European mine owners was the provision of an adequate labor supply. The problem was particularly acute in Peru where high production costs led the mine owners to rely on Indian draft labor. Indian laborers were supplied initially to the mines by the *curacas,* or local Indian chiefs, who were compelled by the mine owners to provide an assigned number of men from each *ayllu,* or village (chapter 7). In addition to those forced to work in the mines, a fair number of free Indians, called *yanaconas,* also migrated voluntarily to Potosí during the midsixteenth century, where many became skilled craftsmen. Work in the mines was so demanding physically, and the Spaniards treated the Indians so harshly, however, that by the 1560s most *curacas* were refusing to send additional men to Potosí, knowing that large numbers of those previously drafted had never returned to their families.

Facing the prospect, then, of losing much of his labor supply at the very time that larger numbers of Indians were needed to work the deeper vein deposits, Peruvian Viceroy Toledo resorted in 1571–1572 to an even more oppressive form of draft labor, which was given the innocent-sounding name of **mita minera,** meaning "turn in the mines." Under this arrangement, all Indian villages within a designated radius of the mines were required to provide annually one-seventh of their tribute-paying males to work the mines for a period of one year. During his assigned year, each Indian was expected to work every third week. In theory, the drafted Indians, called *mitayos,* were to be paid for their services and lovingly cared for by the mine owners. In reality, worker compensation was minimal and great numbers of the *mitayos* died from overwork while countless others succumbed to the ravages of European diseases.

Dreading their forthcoming "turns" in the mines, many of the remaining *ayllu* Indians fled over time from their ancestral villages. Their most common destinations were non-*mita* communities or, ironically, the highland *haciendas,* whose labor-starved owners were more than willing to protect their newly gained serfs from conscription in the mines in exchange for a lifetime of labor on the great estates. The mounting rates of Indian mortality and village abandonment led, from the seventeenth century onward, to such severe labor shortages in the mines that the owners began accepting cash payments from the *curacas* as substitutes for laborers. The owners then used the payment money to hire free Indian laborers, called *Indios mingados,* or *mingas.* Eventually, the owners of many of the less productive mines found that they could realize greater profits by accepting tribute from the *curacas* than by using *mitayos* in their mines. This led to the closing of numerous smaller mines and to the evolution of the meaning of the term *mita* to refer primarily to an annual forced cash payment from a *curaca* to a mine owner. The *curaca,* of course, obtained his tribute by extorting the Indians under his control. The *mita* had thus been transformed from its original character of a labor tax intended to benefit the Crown to a mostly cash tribute that enriched the creole mine owners (Lynch 1992).

The vicious cycle of ever-worsening labor shortages eventually led to the *mitayos* being forced to work every other week rather than every third week as in the beginning. From 1740 onward, *mitayos* were required to labor continuously with no weeks off, and debt servitude at the mines reached the point where the children of indebted *mitayos* were obligated from birth to labor in the mines when old enough to do so (Dore 1988). Although the *mita minera* was officially rescinded in 1812, it continued in practice in Peru and Bolivia into the midtwentieth century, leaving in its wake a lingering legacy of widespread death, suffering, and social and economic dislocation throughout the central Andean highlands.

As was the case in Bolivia and Peru, Mexican silver mining was dependent in the beginning on drafted Indian laborers whose work obligations, called *servicio personal*, arose from the *repartimiento* system. However, the greater profitability of the Mexican mines, together with their lower elevations and less stressful physical environments, soon resulted in the mine owners of New Spain finding that large numbers of free Indians were eager to be hired as salaried laborers. So rapidly did the transformation from draft to salaried labor take place in the Mexican mines that by the 1590s, 68.5 percent of all mine laborers were *naborías*, or free Indian workers, with only 17.7 percent being *repartimiento* Indians and the remaining 13.8 percent being slaves (Garner 1988, 928). Free Indian labor has thus sustained the Mexican mining industry from the late sixteenth century to the present.

Gold and Diamond Mining in Colonial Brazil

While the early Portuguese settlers of colonial Brazil were as eager as their Spanish counterparts to enrich themselves from mineral wealth, little of any consequence was discovered either in the Northeast, where life came to center on the great sugar estates, or in São Paulo and other southern outposts. Brazil, as a whole, remained a mining backwater zone until the late 1600s, when armed Paulista Indian slave-hunting bands, called *bandeiras,* began to notice small amounts of gold in the beds of streams draining the mineralized uplands of São Paulo and southern Minas Gerais. These discoveries were so small, however, that the region attracted relatively little permanent settlement until 1689 onward, when major alluvial deposits, first of gold and later of diamonds, began to be reported in the southern Serra do Espinhaco, some 320 kilometers (200 miles) inland from Rio de Janeiro (see Figure 11.1). These were followed, in the early 1700s, by additional gold and diamond discoveries of major proportions in southern Goiás and in eastern Mato Grosso.

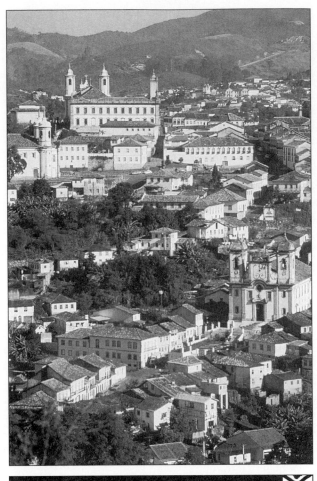

FIGURE 11.4

Ouro Prêto was the economic and cultural center of southern Brazil's gold and diamond mining districts during the boom period of the late seventeenth and eighteenth centuries.

By the early eighteenth century, a massive influx of settlers was pouring into the region. Its capital city, first named Vila Rica de Albuquerque and later changed to Ouro Prêto ("Black Gold"), soon exceeded Salvador in wealth as the focus of the Brazilian economy shifted from sugar farming to mining. Financing for the new mining enterprises came primarily from banking houses in Rio de Janeiro, which became Brazil's dominant city. Then, as is so often the case in mining ventures, the boom ended as suddenly as it had begun. By the early nineteenth century, the Minas mining districts were in visible decline and the economic center of the nation stood poised to shift to the coffee and industrial estates of São Paulo state. Ouro Prêto, whose regional population had peaked in 1760 at around 250,000, including some 100,000 slaves, is now a small tourist center whose numerous historical and architectural monuments, including Aleijadinho's Church of St. Francis of Assisi, serve as silent visual reminders of the wealth and influence of a bygone era (Figure 11.4).

Other Colonial Mining Centers

The enormous fortunes gained from the mining of silver and gold in the mountains of Mexico and Bolivia and in the uplands of southeastern Brazil inspired countless additional prospecting efforts throughout the colonial era. Many of these were undertaken in regions of great physical and social isolation. While most failed to generate significant wealth, a few were moderately successful. Chief among these secondary mining regions were Colombia's Antioquia and Chocó regions, whose streams yielded gold from the 1530s onward (West 1952), and Honduras, where silver deposits were developed in the interior uplands around Tegucigalpa and where placer gold operations occurred along the banks of the Guayape River (see Figure 11.1). Other small silver and gold mines, too numerous to mention, were opened in the highlands of northwestern and southern Mexico, the Greater Antilles, and in scattered locations throughout the Andes (West 1998 and 1993). A few of these have continued in use down to the present day, but most were worked only briefly before being abandoned by men whose visions of wealth were never realized.

One of the most widespread, and yet least studied, forms of colonial mining was highland salt mining—an activity that in many parts of Latin America was and continues to be found traditionally within the female domain (Pomeroy 1988). In addition to being available locally, in contrast to sea salt, which had to be brought up from coastal zones, the salt produced from highland salt springs was generally high in iodine. This had the effect, for reasons possibly not understood by the Indians, of reducing the occurrence of goiter and cretinism. Highland salt was thus an extremely valuable commodity during the pre-Conquest and colonial periods, when it was regularly traded for gold, cotton, and staple food products.

Impacts of Mining

Mining's impact on the land and its occupants thus varied widely from place to place in colonial Latin America. Where small mining operations functioned either permanently or temporarily, their influence on the dominant agrarian landscape was minimal. The development of the larger mining centers, on the other hand, led to major geographic changes. Most obvious of these was the establishment of large urban centers, which housed both the European and creole elites as well as the *mestizo* and Indian working classes, and the creation of vast regional trade and service networks. The networks included *caminos reales,* or royal highways, agricultural supply zones, secondary towns and cities, and the distant ports through which silver and gold were exported in exchange for foreign products (Case Study 11.1).

Beyond the development of new cities and transportation networks, mining was also associated with most of the technological advancements that impacted Latin America during the colonial era. On the negative side of the ledger, it must be noted that mining was frequently associated with deforestation, as local timber resources were exhausted to provide charcoal and lumber, and with the poisoning of soils and streams through the indiscriminate use and disposal of mercury and other toxic substances. It also contributed in no small way to the massive Indian population declines that swept the region during the sixteenth and seventeenth centuries and to the establishment of unequal social and racial relationships that have endured to the present day.

While it would be inaccurate, then, to suggest that the impacts of mining on colonial Latin America were either all good or all bad, they were certainly extensive. Perhaps most regrettable is that after having supplied the world with an enormous quantity of silver and gold throughout the colonial era, most of Latin America was poverty stricken as the wars of independence approached. We can only guess at how different the economic and social history of the region would have been had even a modest amount of the wealth generated from mining been used to educate and otherwise better the circumstances of the masses rather than having been squandered by Spain on endless wars and by the creole elite on showy living.

MANUFACTURING IN COLONIAL LATIN AMERICA

As we noted earlier, manufacturing, or the production of items made either by hand or by machinery from raw materials, was present in much of Latin America and the Caribbean on the eve of the Iberian Conquest. The majority of indigenous manufacturing was of the cottage, or home-centered, type and focused on the production of cotton or woolen clothing and articles used in the cultivation, preparation, storage, and transportation of food (Figure 11.5).

The arrival of the Spaniards and Portuguese in the Americas brought the introduction of new forms of manufacturing (Figure 11.6). Chief among these were large woolen textile factories called **obrajes,** which were first established in central Mexico in the 1530s and in the Andean highlands in the late sixteenth and early seventeenth centuries. The *obrajes* were complex production centers where Castilian techniques of carding, warping, weaving, stripping, burling, and fulling were all integrated at a single site. As in the mines, forced Indian labor was used, with the majority of the work force being women, older men, and boys below the age of tribute. Black slave labor was also occasionally employed, often in the manufacture of hats and

Case Study 11.1: Mining and Regional Development in Colonial Latin America

The immense wealth generated by Latin America's larger mining camps, variously called *reales de minas* or *asientos de minas* (royal mines or mining centers), resulted in many of them growing into cities of great stature during the colonial period. The population of Potosí during the late sixteenth and early seventeenth centuries stabilized at around 120,000 persons, making it for a time the largest city in the Americas. The eighteenth-century gold rush of southern Minas Gerais attracted over 800,000 immigrants, with the Ouro Prêto region alone accounting for some quarter of a million people in the mid-1700s. Zacatecas, Guanajuato, and other leading Mexican mining communities, while they did not become as large as the South American mining centers, ranked among the largest cities of New Spain during the colonial era.

Because the occupants of the mining districts were limited either by the cold mountain climates or by upper-class social norms from producing their own food and other necessities, the great mining cities soon developed vast regional supply networks that reached far beyond their immediate hinterlands. Food consumed by the Zacate-can urban elite, for example, included wheat from the fertile basins of the Bajío, maize and beans from Aguas-calientes, and wine from Parras. Sugar was imported mostly from Michoacán, shrimp from the Pacific coast, and cacao or chocolate from as far away as Guatemala and Venezuela. The trading network of Potosí was even more extensive than that of Zacatecas and included grains, fruits, cotton, and melons from the transitional valleys of Tucumán in northcentral Argentina, wheat from Chile, sugar from Cuzco, and sheep and cattle from distant Paraguay and Buenos Aires. Fish caught in Pacific waters off the coast of central Peru were taken immediately up into the high Andes, where they were frozen overnight or packed in snow for shipment to Potosí. Food for Ouro Prêto came largely from São Paulo, Paraná, and other points far to the south of the city itself, with the São Francisco River Valley of interior Bahia serving as a secondary supply region. Imported goods of all kinds, from European manufactures to Oriental silks, porcelains, and spices, were brought by ship to the major ports for transshipment to the interior mining centers.

Since trucks and trains did not exist, the supplies were transported either through human portage or, most often, on the backs of tens of thousands of mules and llamas. Where roads were adequate, small wooden, animal-drawn carts called *carretas* and larger, heavier wagons called *carros* were also used to carry goods and produce. With income levels far greater than in most of Latin America, the mining cities represented the most affluent markets in the New World, with prices typically running five to six times greater than for the same products sold elsewhere. The wealth of the mining districts also sustained a thriving contraband trade, which, in the case of Potosí, reached as far as the Atlantic coast of Brazil.

The mining cities thus functioned not only as major urban areas in their own right but also as the centers of regional development in much of Latin America. As time passed and the output of the mines declined, the populations of Potosí, Zacatecas, and Ouro Prêto also fell. Although none is as prominent today as in former times, each continues to function as a center of conservative, upper-class culture and as a showpiece of regional and national history.

Sources: Cobb 1949; Bakewell 1971; *Brazil Trade and Industry* 1986.

---

indigo. While the location of the *obrajes* was not restricted by law, many of those established in New Spain were situated in and around the city of Puebla. Those of the Andes were centered on the high intermontane basins of central and northern Ecuador, where intensive textile production predated the Spanish Conquest. Both of these regions have continued to the present day as centers of woolen textile manufacturing (Bebbington 2000).

The *obrajes* were limited to the production of coarse woolen clothing, which eventually replaced native cotton manufactures as the dress of the lower-class Indians in much of the western highlands. The production of high quality clothing that might compete with imported Spanish articles was expressly prohibited by the Crown in secret instructions sent to the American viceroys in 1568 (Assadourian 1992).

Closely associated with the textile *obrajes* were factories that specialized in the processing of dyes. The two most significant of these were cochineal and indigo, or *añil*. Cochineal was a bright red dye obtained from scale insects found on the leaves of highland nopal or prickly pear cactus. Indigo was a deep blue dye extracted from a small herbaceous plant native to the *tierra caliente*. Because both were highly esteemed but unsuitable for cultivation in Iberia, the Spaniards encouraged their production in the New World, where they often ranked as the leading nonmineral exports. Leather goods, wooden furnishings, iron agricultural implements and tools, and various foods and alcoholic beverages were also manufactured at scattered locations throughout colonial Latin America.

New World manufacturers were impressive in the areas of mining, textiles, and foodstuffs. Yet in assessing

FIGURE 11.5

The woolen textile products of highland Ecuador are an example of indigenous manufacturing that has endured down to the present day with the addition of new production techniques.

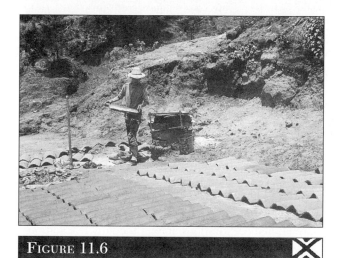

FIGURE 11.6

This Guatemalan tile production site is an example of the new and sometimes simple forms of manufacturing introduced into the New World by the Spaniards and Portuguese following the Conquest.

the overall strength of industry in the Latin American colonial economy, it is essential to recognize that manufacturing was widely repressed by the Spanish and Portuguese mercantile systems. As the wars of independence drew near, Latin America and the Caribbean were overwhelmingly agrarian in character and, with the exception of certain textiles and foods, dependent on foreign supplies for virtually every manufactured commodity. Those few industries that did exist stood, in effect, as isolated islands or enclaves of development in a vast sea of rural poverty and isolation.

MINING AND MANUFACTURING DURING THE EARLY REPUBLICAN PERIOD

Although political independence in the early nineteenth century brought an end to the despised mercantile system for most of Latin America, little structural change occurred in the regional economy as power passed from the *peninsulares* to the elite creole class. Indeed, mining output declined and manufacturing remained largely undeveloped in most of Latin America until the mid-1800s. One reason for this was that Latin America and the Caribbean had been deliberately denied access to and even, in many cases, knowledge of the emerging industrial technologies of northwestern Europe by their overseas masters during the preceding four hundred years. Other forces that worked against the development of a strong industrial base in the newly freed Latin American nations were the absence in most areas of a domestic market large enough to support local manufacturing, the chronic political instability of the region, and a traditional Hispanic value system that idealized the indolent lifestyle of many of the landed aristocracy.

The limited manufacturing that did exist in the early 1800s, then, was primarily a continuation of the scattered enclave mining enterprises and modest textile and agricultural processing operations of the colonial era. Those manufactured articles required by the elites in their business and personal affairs generally were imported at considerable cost from abroad, while the poverty-stricken masses, for their part, simply did without.

BEGINNINGS OF MODERN MANUFACTURING

As unlikely as it appeared in the early republican period that manufacturing conditions would ever change for the better, during the mid to late nineteenth century modern manufacturing emerged in many of Latin America's most prominent regions. The growth of industry in these areas was made possible, in part, by the development in Europe and Anglo America of a number of new technologies, including labor-saving machinery, electricity, refrigeration, and the use of steam power in long-distance rail and water transportation. The collective effect of these advances was to greatly stimulate the growth in Latin America and the Caribbean of industries associated with the large-scale production and processing of agricultural products and nonferrous minerals for which there was strong overseas demand.

Not coincidentally, the last half of the nineteenth century was also a period in much of Latin America of

Case Study 11.2:   "Big Bill" Greene and American Industrialization in Northern Mexico

Industrialization in late nineteenth- and early twentieth-century Latin America was frequently associated with foreign financial interests, whose leaders were able to amass vast short-term fortunes through questionable business practices. One of the most flamboyant of the foreign operators in turn-of-the-century Mexico was an American speculator named William "Big Bill" Greene, who built a spectacular, if short-lived, mining and ranching empire in northern Sonora and Chihuahua during the waning years of the Díaz regime. Greene's Mexican ventures began in 1896 when, as a poor Arizona rancher, he obtained from the widow of a prominent Sonoran family a 99 year lease on four mines in the vicinity of Cananea. Within three years, Big Bill had incorporated his holdings as the Cananea Consolidated Copper Company, S.A., the stock of which he then sold to many of America's leading financial tycoons to finance the operations of the Greene Gold-Silver Company and the Greene Consolidated Gold Company. He also founded the Sierra Madre Land and Timber Company, which supplied cut timber to the mines, and the Rio Grande, Sierra Madre, and Pacific Railroad, which linked his domain to the Gulf of California. With the "dividends" from these companies, Greene purchased over 240,000 hectares (600,000 acres) of prime Mexican ranchlands on which he grazed close to 40,000 head of pure-bred Hereford cattle.

In order to house his mine workers, Greene founded a company town that he called La Cananea and that by 1906 included several office buildings, two sawmills, and a hospital, in addition to a restaurant and hotel. Big Bill resided in a huge two-story mansion in which he entertained Mexican and American dignitaries at lavish parties. The mine workers themselves lived in small rental properties and were compelled to purchase food and other necessities on credit at the company store.

Anglos employed by Greene's companies were paid 50 percent higher wages than Mexicans doing the same work and were given superior housing. Two of Big Bill's closest associates were the governor of Sonora, who ordered a state road constructed across Greene's private landholdings at no cost to the American, and the governor of Chihuahua, who arranged for a 1,400,000 hectare (3.5 million acre) concession of timberland to the Sierra Madre company.

Greene's public image as an American exploiter and his ties to the corrupt Díaz regime eventually led in 1906 to a number of opposition-inspired labor actions, which served as a prelude to Mexico's tumultuous Revolution. Greene's troubles culminated in a nervous breakdown in 1908, and he died in 1911.

Source: Adapted from Raat 1981.

relative political calm, which served to attract millions of Italian, German, and other European migrants from their impoverished and war-torn homelands. It was largely through the entrepreneurial initiatives and capital investments of these new immigrants, who were joined in some cases by aspiring members of the Hispanic landed elite, that the foundations of modern Latin American manufacturing were laid.

In addition to the application of new technologies and to the arrival of the southern and eastern European immigrants, Latin American industrialization was also encouraged by freewheeling *laissez-faire* economic philosophies that attracted considerable American, British, French, and German investment in the agricultural and mining sectors. One example of this type of foreign investment, already noted in chapter 10, was the United and Standard Fruit companies operations in Central America. Other commodities invested in heavily by foreign interests included Argentine and Chilean beef and wheat, Peruvian sugar and minerals, Brazilian, Colombian, and Central American coffee, Chilean nitrates, Cuban and Puerto Rican sugar, and Mexican copper. Exports of all of these increased dramatically around the turn of the century on the basis of unprecedented levels of foreign investment. The arrival during this period of aggressive, well-financed multinational corporations among the largely tradition-bound peoples of Latin America and the Caribbean proved to be both a blessing and a curse. On the positive side, advanced mining and agricultural production techniques were introduced, and rail systems, highways, and port facilities were upgraded. Other benefits included the creation of thousands of new, relatively high-paying jobs and the emergence, for the first time in post-Conquest Latin America, of a small but growing urban middle class.

Unfortunately, these gains were offset, to some degree, by the isolated enclave nature of the new economic enterprises, which had the effect of perpetuating, and in many cases even exacerbating, the historic gaps between the wealthy and the poor. This in turn led to growing feelings of resentment by local populations toward the foreign elites, who in far too many instances were guilty of meddling in the internal political affairs of the host nations (Case Study 11.2). Yet another negative by-product of the new multinational activity was the growth of labor-management conflicts, which have proven in the twentieth century to be particularly divisive and enduring in the mining and plantation sectors (Harner 1998).

These problems notwithstanding, by the early twentieth century, large-scale industrialization, while far from universal, was clearly under way in the more technologically advanced regions of Latin America and the Caribbean. Two of the leading manufacturing zones at this time were the areas centered on Buenos Aires, Argentina, and Monterrey, Mexico. By 1905, industrial output had risen to constitute 18 and 14 percent of the gross domestic product of Argentina and Mexico, respectively (Gwynne 1986, 22). Other emerging industrial regions included São Paulo and the neighboring areas of southern Brazil, Medellín and the Antioquia region of Colombia, southeastern Uruguay, Lima and the irrigated coastal valleys of Peru, Santiago and the Central Valley of Chile, and the petroleum-rich Maracaibo Lowlands of Venezuela (Figure 11.7). With the notable exception of Monterrey, where a thriving steel industry was well established by the 1920s, manufacturing in each of these regions continued to center on the production of minerals, textiles, or agricultural products for export overseas. The proceeds from the sales of these commodities were then used by the elites to purchase machinery and other manufactures not available locally.

With industrial jobs and profits thus increasing steadily in the 1920s, free trade appeared to be serving well the interests of the Latin American middle and upper socioeconomic classes. Those interests were heavily dependent, however, on a precariously balanced world financial system that was soon to crumble.

THE GREAT DEPRESSION AND IMPORT SUBSTITUTION INDUSTRIALIZATION

The collapse of the world's financial markets in 1929 and the Great Depression that followed in the 1930s inflicted tremendous harm on Latin America's youthful industrial sector. Overseas demand for South and Middle American minerals and agricultural products, which had formerly sustained Latin America's rapidly expanding industrial work force, plummeted. Widespread unemployment and mounting social unrest spread among the incipient urban middle class. The elites, too, became increasingly disillusioned with the progressive erosion of their earnings and their inability to purchase the foreign manufactured products to which they had become accustomed.

As the sense of despair and hopelessness deepened with each passing year, Latin Americans began to question many long-held beliefs and assumptions. Among these were the *laissez-faire* economic policies of the previous century, which had held that governments should intervene as little as possible in the activities of individuals and corporations. Those policies were now seen, by increasing numbers of alienated Latinos, as being responsible for the loss of their jobs and for the enduring postcolonial poverty of their homelands.

This anger and frustration eventually came to form the basis of two related academic theories that attempted to explain the persistent poverty and underdevelopment of Latin America and other less industrialized regions in terms of world trade relationships. These theories, entitled the **core-periphery model** and the **dependency model,** perceived the countries of the world as belonging to one of two great multinational groupings that were distinguished on the basis of levels of industry and technology. The more industrialized and technologically advanced nations of Europe and Anglo America were said to form the world's economic core, and the less industrialized nations of Latin America, Africa, and Asia the economic periphery. According to these theories, the principal cause of underdevelopment among the nations of the periphery was their general lack of industry, which resulted in their always having to trade for high-cost manufactured products from the nations of the core. Because the nations of the periphery had little except low-value primary commodities to offer in exchange for their industrial imports, and because the prices of the periphery's raw materials were controlled by the power-brokers of the core, the core-periphery theorists argued that the peoples of the periphery were involuntarily dependent on the peoples of the core. In other words, the peoples of the periphery were being intentionally kept in a state of never-ending economic underdevelopment by the peoples of the core, whose own prosperity was derived from the poverty of the Third World nations.

If the underdevelopment of Latin America and the Caribbean were indeed caused primarily by so-called neocolonialist trade relationships, it followed that the only path to economic development and true political liberation lay in the promotion by national governments of internal industrial development. The strategy that emerged in the 1940s and 1950s as the means through which the Latin American and Caribbean nations were going to achieve internal industrialization was called **import substitution industrialization,** or ISI. The main idea behind ISI was that any country, no matter how small or impoverished it might be, could eventually become industrialized and prosperous if government leaders would rechannel the bulk of the nation's spending into the development of a domestic industrial sector. The products of local industries would then replace imported manufactured goods, saving the country vast sums of money that could be used to raise the levels of living of its citizens.

FIGURE 11.7

Principal industrial regions of early twentieth-century Latin America.

Recognizing that most Latin American nations had a limited industrial base, ISI theorists suggested that industrial self-sufficiency would have to be accomplished in four stages. The first, which had already been achieved by most countries, was the manufacture of clothing, pharmaceuticals, and processed foodstuffs. Once a nation became self-reliant in the production of these commodities, it would then move, according to the ISI model, into the production of durable consumer items, including stoves, radio and television sets, and even automobiles. Stage Two nations that did not have the capacity to manufacture every part of a given product, such as a stove or an automobile, were allowed by the theorists to import the needed components. Thus, the importation of industrial parts was permitted but that of assembled manufactured goods was not. With the industrial foundation gained in the second stage, the Latin American nations would hypothetically be prepared to move into Stage Three, the development of so-called intermediate industries, which specialized in the manufacture of the parts and materials needed in the first two phases. The fourth, and final stage, emphasized the production of steel and chemicals needed in the manufacture of heavy machinery, at which point it was assumed that the country had become fully developed.

The only way for any of this to occur, according to ISI proponents, would be for national governments to protect local industry during its formative period from foreign competitors, who, because of their head start, were capable of underselling local manufactures. The most common ways in which the Latin American and Caribbean governments chose to protect their new industries were the establishment of import quotas, which limited the number of competing foreign goods allowed into a country, the imposition of very high tariffs or duties on those products that were permitted into the country, and the maintenance of overvalued currencies, which had the effect of discouraging the sale abroad of industrial raw materials. Yet another common government response was the outright nationalization of public utilities, airlines, railroads, mines, steel mills, and other basic industries.

There is no question that the import substitution industrialization policies of the midtwentieth century did result in a short-term increase of Latin American and Caribbean manufacturing. It is also clear that these gains were achieved at very high long-term costs to the peoples and nations of the region. The sheltering of Latin American firms from foreign competition, for example, led to a proliferation of inefficient, state-subsidized industries whose products were both more expensive and of lower quality and reliability than those available overseas. This in turn contributed both to chronically high inflation rates and to an upper-class tradition of spending huge sums on overseas shopping trips, with everything from electronic appliances to perfume being purchased and then shipped back into the country with the aid of generously tipped customs agents.

Import substitution industrialization also contributed to the growth of the already bloated bureaucracies of government regulatory agencies and to a reduction of foreign investment and technology. Of great damage to the long-term development of the Latin American and Caribbean nations was ISI's systematic exploitation of the rural peasantry. This occurred as national governments, in an attempt to provide cheap food for the industrial labor force in the cities, imposed artificially low prices on crops sold by the farmers at the same time that agricultural production and marketing costs were soaring (Grindle 1986). Uneven development also occurred as large-scale industrial agriculture expanded at the expense of smallholders, which resulted both in reduced local food supplies and in a flood of rural migrants into towns and cities ill-prepared to accommodate them. Perhaps the most damning indictment of import substitution industrialization is that it failed in its avowed purpose of reducing levels of imported manufactured items. Indeed, by the substitution of parts and assembling equipment for imports of finished products, dependence on industrial imports actually increased rather than decreased in most Latin American and Caribbean nations. This increase brought, in its cumulative wake, the unmanageable foreign debt burdens of the 1980s and 1990s (Thorp 1992) (chapter 13).

In sum, the overall effect of import substitution industrialization was, at best, mixed. Those nations with relatively large populations and markets, such as Brazil, Mexico, and Argentina, emerged from the experiment with the capacity to produce, albeit at high unit costs, many of their own industrial products. Regrettably, for the remaining Latin American and Caribbean nations, whose domestic markets were too small to support most forms of local manufacturing, ISI did far more harm than good by siphoning off, in the form of inefficient industrial subsidies, monies that otherwise could have been invested in projects of long-term worth to the populace.

GROWTH POLES AND THE SPATIAL CONCENTRATION OF INDUSTRY

As the shortcomings of import substitution industrialization became more and more apparent during the late 1950s and 1960s, many Latin American and Caribbean officials began to turn to an alternative industrial development strategy known as the **growth pole model.** As conceived originally by French regional planners and economists, the term growth pole referred to an urban area that contained a large number of interrelated industries whose presence created a demand for other industries and businesses within the immediate vicinity of the lead industries (Richardson and Richardson 1975). As the related businesses and industries grew and prospered along with the lead firms, it was hypothesized that the entire urban area would advance to an economic takeoff stage, after which growth would be self-sustaining. The ongoing growth and prosperity of the urban pole would then diffuse to the surrounding rural countryside through a network of input and output linkages, which in turn would bring economic development to the entire region or nation.

The key to development, according to growth pole advocates, was thus the spatial concentration or agglomeration of industry in critical geographical areas. Because the determination of which areas were to be favored by government investment was influenced to a great extent by the location of preexisting industries, growth pole economics had the effect of benefiting, in most instances, the larger urban areas rather than the towns and villages. While acknowledging openly that the practice of concentrating scarce resources in a few favored areas was unfair, growth pole proponents insisted that it was necessary in the short term in order to achieve the universal long-term development of Latin America and the Caribbean.

Fortified by this rationalization, planners from Mexico to Argentina were soon occupied in drawing up grandiose regional development schemes. Because each nation was endowed with a distinct set of physical resources and historic needs and aspirations, specific strategies varied widely from country to country. Rather than listing the details of all the national plans, most of which have subsequently been discarded, we will now focus on selected aspects of the Chilean, Brazilian, and Venezuelan experiences in order to illustrate both the most common types of approaches used and the challenges inherent in each.

Chile and the Arica Automobile Manufacturing Experiment

One of the most ambitious of Latin America's growth pole strategies was devised during the 1960s in Chile, where government planners chose to combine multilevel spatial and functional approaches. Spatially, three tiers of growth poles were identified. The lowest tier was the regional primate cities found within each of twelve geographic regions. The middle was the multiregional poles of Antofagasta on the north, Valparaíso in the center, and Concepción on the south (Figure 11.8). The highest order was the national growth pole of Santiago. In addition, a few cities, designated as functional growth poles, were chosen to be the centers of certain types of industry. These included Punta Arenas in the far south, which was assigned the production of petrochemical products; Concepción, which was to manufacture the nation's steel; and the far northern port of Arica which, despite its extreme isolation and aridity, was selected to become Chile's automobile and electronics manufacturing center (Conroy 1973).

FIGURE 11.8

The port of Antofagasta was designated the northernmost of Chile's three multiregional industrial growth pole sites during the 1960s.

While Arica's selection was grounded in geopolitical considerations, it made little sense economically or geographically to build automobiles in an area far removed from the components, labor, and markets necessary to assemble and distribute the final product. After years of subsidizing the Arica experiment, the remnants of the Chilean automobile industry were finally returned to Santiago. In like manner, Chile's other growth pole experiments eventually gave way, first to the Marxist planners of the Allende era and finally to the economic liberalism and free trade policies of the Pinochet and subsequent regimes. With the benefit of hindsight, it can now be stated that Chile's growth pole experiment did more to reinforce preexisting patterns of industrialization than to alter them.

Brazil: Large-Scale Regional Development and Underdevelopment

Two of the greatest historic challenges to the economic development of Brazil have been the sheer size of its territory and the difficulty of achieving uniform levels of human well-being among regions that vary greatly in their physical and cultural attributes. In the early colonial era, a wide gap in human living levels soon evolved between the more affluent plantation region of the humid coastal plains of the Northeast and the arid, poverty-stricken interior backlands of the sertão. As time passed, the coastal plantation zone itself failed to adapt to technological and social change, and the entire Northeast emerged as a chronically underdeveloped region, one of whose principal long-term exports to the more developed South was uneducated human laborers. Meanwhile, the vast and isolated region encompassing the heart of the Amazon Basin, which Brazilians came to call the North, attracted little permanent European settlement and development.

By the time, then, that the growth pole industrial development model became fashionable in the late 1950s, Brazilians were keenly aware of the extreme differences in living levels between the more prosperous and rapidly industrializing South and the tradition-bound, largely agrarian Northeast and North. The desire to bridge that gap through growth pole industrialization led to the formation of the two regional development agencies that were most responsible for directed change in the latter twentieth century (Figure 11.9). The first, the Superintendency for the Development of the Northeast, or SUDENE, was created in 1959 as a successor to a long line of Northeast regional development commissions that had included the Board of Public Works against Drought (1909) and the São Francisco Valley Commission (1948) (Kleinpenning 1971a). The second, the Superintendency for the Development of Amazonia, or SUDAM, evolved in the 1960s as the successor to a body called the Superintendency for the Economic Valorization of Amazonia, which had been formed in 1953 (Despres 1991).

While SUDENE and SUDAM have functioned independently of each other for over three decades, they have a common objective of promoting the industrialization of their respective regions. Industrialization has been encouraged by SUDENE in the Northeast through legislation that exempted certain factory equipment from import duties and through laws that allowed Brazilian firms to use 50 percent of their federal income tax liability to invest in approved manufacturing projects within the region. In like manner, the much publicized highway projects within Amazonia initially were undertaken for the interrelated purposes of providing employment for jobless laborers from the Northeast, supplying minerals and other raw materials to the factories of the Northeast, and serving as a market for the finished industrial products of the Northeast (Kleinpenning 1971b).

Assessments of the effectiveness of Brazil's growth pole industrialization efforts have varied widely. Supporters point to the massive highway projects and mining operations of Amazonia as evidence of success (Figure 11.10). They likewise point with pride to the expansion of industry in the Northeast, particularly the rejuvenation of old textile and leather factories, and the growth of branch assembly plants in and near the large cities of Recife, Salvador, and Fortaleza. The expansion of large-scale industrial agriculture is also seen as evidence of economic progress and modernization. Critics, on the other hand, decry the environmental degradation of Amazonia and note that the North and Northeast, notwithstanding the recent changes, continue to lag far behind the South in every measure of development (Savedoff 1995; Barham and Coomes 1996).

Venezuela: Growth Pole Development in a Frontier Setting

One of the most unusual and, in many ways, successful of all the growth pole industrialization projects undertaken in Latin America has been the building of Ciudad Guayana, Venezuela, into one of South America's leading steel, aluminum, and heavy manufacturing centers. The process began during World War II when two American firms, United States Steel and Bethlehem Steel, initiated large-scale iron ore mining operations at Cerro Bolívar and El Pao. The mines were situated on either side of the Caroní River, which drains the eastern Guiana Highlands before plunging to its junction with the Orinoco River along the southern edge of the Llanos.

FIGURE 11.9

Regional development agencies of northern Brazil.

Although the area was one of the most remote in all Venezuela, studies undertaken during the 1940s and 1950s revealed a great array of natural resources within the region (Figure 11.11). These included, in addition to the world-class iron ore deposits at Cerro Bolívar and El Pao, economical waterborne access to the Atlantic by way of the Orinoco and an abundance of energy sources. The latter included extensive petroleum and natural gas deposits under development in the eastern Llanos to the north and, even more importantly, an immense hydroelectric potential along the lower Caroní. The inexpensive electricity could justify the establishment of both a steel industry and an aluminum industry, even if the bauxite, or aluminum ore, had to be imported from the Caribbean or Brazil.

FIGURE 11.10

The development of Brazil's Trans-Amazon highway network is intended, in part, to support the industrialization of the Northeast.

Armed with this knowledge, Venezuelan officials formed, in 1960, a state-owned holding company named the **Corporación Venezolana de Guayana,** or CVG, and set out to convert the tiny river port of Santo Tomé de Guayana into one of Latin America's leading industrial centers. The process through which this was accomplished included three thrusts. The first, the development of a Venezuelan steel industry, was achieved through the formation in 1962 of the state-owned steel corporation, Siderúrgica del Orinoco (SIDOR), the absorption of SIDOR by the CVG in 1964, and the erection in the 1970s and 1980s of a huge steel complex that was fed by local iron ores and imported coal and limestone. Production centered initially on seamless steel pipe for use in the Llanos oil fields but has now expanded to include a wide array of rolling stock.

The second thrust, the building of an aluminum industry, began with the CVG and Reynolds Aluminum agreeing, in 1962, to form a joint company called ALCASA. After subsequent disagreements be-

tween the principals, the CVG formed a state-owned subsidiary in 1970 called VENALUM and built a smelter that by 1979 had become the largest in the world. This was followed in 1983 by the establishment of a new CVG subsidiary called INTERALUMINA and the development, by yet another CVG subsidiary called BAUXIVEN, of a huge 500 million-ton high-grade bauxite deposit along the upper Orinoco at a site called Los Pijiguaos. Venezuela has thus developed, over the past three decades, Latin America's first fully integrated aluminum industry, whose principal components are the mining of bauxite, the conversion of bauxite into alumina, and the production of aluminum from alumina.

The third, and in many ways most essential, thrust was the building of the massive Guri Dam and hydroelectric complex along the lower Caroní River a few miles upstream from Ciudad Guayana. Begun in 1965, the dam was built in three stages. By the 1990s it was generating 10,000 megawatts of power for Ciudad Guayana and much of Venezuela. With steel, aluminum, and power all in place, Fiat, John Deere, Mack, and numerous other multinational manufacturers built plants of their own within the region. Although the development of satellite industries was slower than initially anticipated, by 2002 Ciudad Guayana had developed into a modern city of over 700,000 persons, fully linked by ground and air transportation to the rest of the country. A second major hydroelectric power plant, called Caruachi, is now under construction along the lower Caroní River and is scheduled to be fully operational by 2006. The success of Ciudad Guayana can thus be attributed to a fortuitous combination of factors that include proximity to some of the earth's richest iron ore and bauxite deposits, abundant inexpensive electrical power, water transportation to the outside world, an absence of competing governmental agencies, and a national government capable of and willing to commit billions of dollars to the project.

STATE OWNERSHIP AND PRIVATIZATION

While not rejecting the growth pole industrialization model altogether, most of the leaders of the Latin American and Caribbean nations had decided, by the mid-1970s, that it was not working as well as had been hoped. To many, it smacked of the old core-periphery and dependency development models, with the growth poles taking on the role of foreign capitalist enclaves. Politically, it was becoming increasingly difficult for politicians to justify the favoring of one area and people over another, and the resultant widening income

FIGURE 11.11

Ciudad Guayana regional development.

gaps between the rich and the poor. Further, many of the growth pole sites had been selected on the basis of political rather than geographic considerations and had been poorly managed after their completion. Most troubling to many was the evidence that the long-term effect of the growth pole strategy had been to spur the growth of the dominant cities, whose populations were now exploding under a never-ending flow of rural migrants, most of whom failed to secure the industrial jobs they sought.

Faced with these challenges and influenced by a rising tide of antiforeign sentiments, many Latin American and Caribbean governments resorted to **nationalizing** the holdings of the larger transnational mining and industrial firms remaining within their borders. In-

cluded in the wave of expropriations were the Mexican petroleum industry in 1938, Bolivian tin in 1952, the foreign bauxite industry of Guyana in 1971, Chilean copper in 1969–1970, Peruvian silver, tin, and iron ore operations in 1974–1975, and Venezuelan petroleum in 1976. Yet another expression of the nationalist fervor of the day were the actions taken by the Brazilian government during the early 1980s to force Daniel K. Ludwig, a publicity-shy American shipping tycoon, to sell to Brazilian interests a huge 1,600,000 hectare (4 million acre) farming, mining, and pulp and paper project situated along the Jari River in the lower Amazon Basin.

Mines and agricultural enterprises were not the only targets of the nationalist era. Public utilities, transportation interests, banks, and heavy metallurgical plants

FIGURE 11.12

The nationalization of banks and other financial institutions has been a common development strategy of many Latin American and Caribbean countries since the midtwentieth century. Shown here is the National Bank of Panama building, Panama City, Panama.

TABLE 11.1

Value of Latin American and Caribbean Privatizations, 1990–1999

COUNTRY	TOTAL (MILLIONS OF U.S. DOLLARS)
Brazil	69,607.7
Argentina	44,588.0
Mexico	28,593.0
Peru	8,134.4
Venezuela	6,072.0
Colombia	5,979.5
Chile	2,138.4
Panama	1,427.3
Guatemala	1,351.2
El Salvador	1,070.1
Cuba	706.0
Dominican Republic	643.4

Source: www.worldbank.org/data/wdi2001/pdts/tab5_1.pdf

were all vulnerable to public takeover (Figure 11.12). The extent to which state ownership of leading industries was pursued varied widely from nation to nation and, often, from one political administration to another within a given country. Although some state-owned industries were well managed, many were not, and over the years they came to require ever-larger government subsidies in order to remain operative.

The subsidies had to come from somewhere, and by the 1980s it had become all too clear that the bulk of the money had been obtained from foreign borrowings, which most of the Latin American and Caribbean governments could not repay. This eventually led to debt restructuring negotiations, one aspect of which was an insistence by the International Monetary Fund and other international lenders that Latin American nations sell off many state-owned enterprises to private firms. This process, called **privatization,** was embraced most vigorously during the 1980s by Mexico and Chile, both of which sold off hundreds of government-owned companies to the private sector. Argentina, Brazil, and most of the other Latin American and Caribbean nations adopted the practice, with varying degrees of enthusiasm and success, during the 1990s (Table 11.1) (see Case Study 13.2). While privatization led, in some instances, to increased investment, greater access to technical expertise, more efficient management, and renewed economic competitiveness, in other cases it was implemented without adequate preparation and few, if any, fiscal safeguards. In these latter instances,

private monopolies were substituted for their public predecessors and the proceeds from the sales of the state-owned enterprises were misused by inadequately trained or corrupt government officials. The results of privatization thus have varied widely, with the citizens of some nations realizing improved services at reduced costs and those of other countries experiencing simultaneously rising prices for basic services and the loss of social programs previously provided by the government.

Privatization is the latest in a long series of Latin American and Caribbean industrialization development models, which have included nationalization, growth poles, import substitution, core-periphery and dependency and, before them, *laissez-faire* economics and colonialism. Viewed historically, each model has evolved as a response of the elite to the most pressing issues of their day, only to be replaced when a new set of priorities emerges. None of these industrial development models has proven capable by itself, however, of generating prolonged and widespread economic development. This is because each has failed to address the cultural roots of underdevelopment considered in chapter 8, and also because each, in its own way, has neglected the agrarian sector. What we presently find, then, in every Latin American and Caribbean nation, is a mix of past industrialization models and projects, all of which are set in the midst of pervasive rural poverty and widening income gaps between the wealthy and the poor.

EXPORT PROMOTION MANUFACTURING: AN ALTERNATIVE DEVELOPMENT MODEL

The failure of the various Latin American and Caribbean industrialization models to substantially raise the living levels of the masses during the past six decades has prompted a number of Latinos to investigate alternative industrial development strategies practiced by other Third World nations. In doing so, many came to be impressed with the rapid economic growth achieved since the 1960s by South Korea, Hong Kong, Taiwan, Singapore, and the other newly industrializing countries of East and Southeast Asia. Although each of the Asian nations has operated within its own physical and cultural parameters, they have all followed an industrialization development model known as **export promotion manufacturing.** In contrast to the inward-looking Latin American and Caribbean models, whose emphasis has been primarily on the redistribution of existing internal resources, export promotion manufacturing focuses on expanding a nation's resource base by increasing foreign sales of locally fabricated products, some of which may be produced by domestically owned firms and others by branch plants of transnational corporations.

In attempting to persuade foreign industrial firms to locate within their homelands, the leaders of the Latin American and Caribbean nations have sought to make their countries financially and culturally attractive to the corporations they were courting. The incentive package offered the multinationals typically includes some combination of subsidized land, buildings, and utilities, relaxed or modified government regulation of imported raw materials and exports of finished products, low or deferred taxes, permission to freely move money and personnel in and out of the host country, and, most importantly, a docile, well-trained, and low-cost work force. While it is clear that such export promotion arrangements carry a great risk of foreign meddling and exploitation, their potential economic benefits have inspired a number of Latin American and Caribbean nations to adopt the strategy in one form or another. Following are some representative examples.

Puerto Rico's Operation Bootstrap

One of the earliest and most widely publicized Latin American and Caribbean export promotion manufacturing projects was begun in Puerto Rico in the early 1950s. Prior to that time, Puerto Rico's economy and social structure had changed little from the Spanish colonial period, despite the fact that American control during the twentieth century had resulted in a widespread upgrading of the commonwealth's schools,

roads, and utilities. Facing what they considered to be an urgent need to stimulate the island's economy, President Luis Muñoz Marín and other leaders of the ruling Popular Democratic Party (PPD) decided to aggressively recruit United States investors. Accordingly, they developed a package of tax and infrastructural incentives that proved to be attractive to labor-intensive apparel and textile firms headquartered on the mainland. Although Puerto Rico's native monied class was harmed by the new competition, **Operation Bootstrap,** as the program was named, resulted in the addition of more than 140,000 new, relatively high-paying industrial jobs, which contributed to significant increases in per capita income over the next two decades. While never approaching average income levels of those on the United States mainland, Puerto Rico came to be widely viewed as an economic miracle and Muñoz Marín as a wise, visionary leader. The PPD, for its part, solidified its power by representing itself as the political party that was truly concerned with the welfare of the common people.

The key to the early success of Operation Bootstrap was the formation of what Cabán (1989) has called a "social compact" between the PPD-dominated government and the Puerto Rican labor unions, many of which chose to affiliate with the North American–based AFL-CIO. In brief, the compact consisted of union support of the party in exchange for the PPD endorsing the unions' attempts to recruit members, the inclusion of union leaders in the party's ruling councils, and the passage of a minimum wage law and a broad range of social legislation. The latter featured state-subsidized housing, medical care, free education, and worker training programs.

Initially, the arrangement appeared to benefit everyone. Living levels of the industrial workers increased dramatically over those of the steadily shrinking rural peasantry, the unions were able to claim credit for the workers' gains, and PPD rule continued virtually unchallenged. By the mid-1960s, however, Puerto Rico was losing its comparative advantage as Asian competitors entered the American marketplace with goods produced from less expensive labor. This in turn created intense pressure on the PPD-union coalition to suppress worker compensation in an attempt to make Puerto Rican goods more competitive. When that failed to prevent the Asian capture of the American textile market, widespread layoffs occurred, outmigration to the United States mainland increased, and the popularity of both the AFL-CIO and PPD leadership plummeted. In an attempt to preserve its power, in the 1970s the coalition abandoned its commitment to the apparel industry and began recruiting more mechanized, capital-intensive firms specializing in the manufacture of petrochemical and pharmaceutical products, specialty foods, and electronic appliances.

In the meantime, Puerto Rico's agricultural sector floundered, and dependence on United States government welfare assistance increased to the point that food stamps had become a leading source of income to two-thirds of all Puerto Rican families by 1974 (Cabán 1989). The Commonwealth itself also became deeply indebted as it spent huge sums in an attempt to continue funding the programs to which the industrial work force had become accustomed. When the social compact eventually collapsed, the PPD was swept from power, and Puerto Rico's economy stagnated. By the late 1990s, Puerto Rico could best be described as an industrialized yet, in many ways, underdeveloped Caribbean island whose economy was highly dependent on benefits derived from its special political relationship with the United States.

The Mexican Maquiladora Industry

A second large-scale example of Latin American export promotion manufacturing is Mexico's **maquiladora,** or Border Industrialization Program. The term *maquila* is derived from the Spanish verb *maquilar,* which means "to mill or to process." It is used in Mexico to refer to an industrial plant that hires native workers to assemble manufactured items, usually made of imported materials. The finished products are then returned for sale to the foreign market.

The roots of the *maquiladora* program reach back to 1942 when the United States and Mexico agreed to allow Mexican agricultural laborers to work for American farmers, many of whom were experiencing severe labor shortages as a consequence of World War II. Although the *bracero* program, as it came to be called, was initially designed to terminate when the war was ended, the arrangement proved so attractive to both nations that it was extended until 1964. In that year, the United States unilaterally cancelled the agreement and expelled the Mexican farmworkers, leaving Mexico searching desperately for a way to create jobs for tens of thousands of former *braceros* living along the border. The solution came in 1965 with the establishment of the Border Industrialization Program, which offered government-subsidized land, utilities, and tax holidays to any foreign company that would establish an assembly plant in the Mexico–United States border zone. As was the case in Puerto Rico, both the imported parts and the exported finished products would be allowed to cross the border freely, and the Mexican government would guarantee political stability and peaceful labor relations.

The number of *maquiladoras* grew rather slowly at first but gained momentum as regulations governing the program were liberalized during the 1970s and 1980s. By 2001, over 3,700 plants employing 1.3 million people were registered with the Mexican government

TABLE 11.2

Growth of the Mexican *Maquiladora* Industry

YEAR	NUMBER OF MANUFACTURING PLANTS (*MAQUILAS*)	NUMBER OF EMPLOYEES
1968	79	17,000
1974	455	75,974
1981	605	130,973
1988	1,400	369,489
1990	2,042	486,000
1996	2,398	750,000
2001	3,750	1,264,390

Sources: South 1990; Wilson and Mather 1990; Sklair 1993; MacLachlan and Aguilar 1998; www.maquiladirectory.com/statistics/

(Table 11.2). Two of the most significant of the regulatory changes were the adoption of policies permitting full foreign ownership of *maquiladoras* and the establishment of plants virtually anywhere in the country. The latter has led to many of the lower technology plants, especially those producing textile products and packaged food, being located in interior sites as far south as Oaxaca and Yucatán, where there exists an abundant supply of low-cost, unskilled laborers (MacLachlan and Aguilar 1998). The northern border zone has come to specialize in the assembly of electronic appliances and heavy equipment and offers higher wages to its skilled and semiskilled workforce. While these wages are extremely low by American standards, with most Mexican workers receiving no more for a day's labor than their United States counterparts earn in half an hour, they are nevertheless equal to or higher than locally prevailing rates.

Initially, the great majority of the plants were owned by American corporations, with a modest Japanese presence in Tijuana (Wilson and Mather 1990). By 2001, however, 29 percent of all *maquiladoras* were controlled by Mexican investors, with 62 percent being controlled by companies headquartered in the United States. Mexican majority ownership is most prevalent in the lower technology plants situated in the interior, while American corporate ownership continues to dominate in the border region.

Many of the world's largest transnational corporations are now operating within the program, including International Business Machines, American Telephone and Telegraph, Westinghouse, General Electric, Phillips, Zenith, Motorola, and Honeywell in the electronic and computer sectors. The *maquiladora* program has now replaced tourism as Mexico's leading source of foreign exchange earnings, and its export earnings are nearly four times greater than those of petroleum.

These stunning economic statistics hide, however, a host of social and environmental ills. One of the great ironies of the program, for instance, was that women below the age of 25 initially accounted for 70 to 80 percent of the workforce (Kopinak 1996, 12; Sklair 1993, 167). The apparent causes of this pattern included the old Hispanic perception that women are more patient than men when performing tedious, repetitive tasks and the belief that female employees are more reliable and less likely to contribute to labor unrest. The program thus resulted in a situation where hundreds of thousands of young, single women found themselves supporting households headed by their unemployed fathers and grandfathers (Cravey 1998). The physical and emotional stress of the work was so great that the average tenure of the laborers was less than three years. One of the most significant changes of the past two decades, however, has been the progressive masculinization of the labor force in the northern border region.

The Mexican government has also been criticized for being either unwilling or unable to enforce strict environmental legislation (Corral 2000; Peña 1997; Lord 1992; South 1990). The millions of migrants who have been attracted to the northern border region over the past three decades have overloaded already inadequate sewage treatment plants to the point that over 80 percent of municipal raw sewage passes untreated into local rivers, streams, and open cesspools. Outbreaks of typhoid fever, hepatitis, cholera, polio, salmonella, and intestinal parasites are common and have the potential to threaten American populations through contaminated imported food products. Great quantities of toxic chemical discharges, including industrial solvents, acids, heavy metals, and PCBs, also go unmonitored, raising the risks of birth defects and other serious health complications. Eye stress is so great on the workers assigned to assemble small components that many experience partial blindness before the age of 30.

Even the economic performance of the *maquiladora* program has been criticized for taking jobs from American workers and for failing to generate spinoff employment through linkages to other sectors of the Mexican economy. By 2001, for example, 97 percent of all raw materials, equipment, and supplies used by the manufacturing plants came from United States suppliers (www.solunet-infomex.com). The *maquilas* are functioning, then, largely as industrial enclaves. While there is no question that the program has become vital to Mexico's economic well-being, it is also equally clear that its potential for lifting overall levels of living within the region has yet to be fully realized.

FIGURE 11.13

This *maquiladora* plant is one of dozens situated in or near San Pedro Sula on the north coast of Honduras.

Other Export Promotion Initiatives

Numerous other export promotion initiatives have been undertaken throughout Latin America and the Caribbean. One of the most unlikely of these, in a geographical sense, was the creation in 1967 of the Manaus Free Trade Zone (ZFM). By the late 1990s, it had attracted over 500 domestic and foreign firms to the heart of the Amazon Basin through a package of trade and tax concessions. Over 90 percent of the companies, which include many of the world's leading electronic appliance, motorcycle, computer, and optical equipment manufacturers, rely on the assembly of imported components (Despres 1991). Mining firms and timber exporters are also represented. One of the principal side benefits of the ZFM has been an upsurge of economically upscale tourists, who now mix shopping trips for discounted electronic goods with jungle riverboat cruises.

Several other export promotion programs have been developed by Caribbean and Central American nations (Figure 11.13). These include apparel-dominated ventures in Barbados (Long 1983), the Dominican Republic (Safa 1995; Portes, Itzigsohn, and Dore-Cabral 1994; Thoumi 1991), Haiti (DeWind and Kinley 1988), and Honduras, information processing in Jamaica (Mullings 1999), and the so-called offshore financial services industry of the Cayman Islands, the Bahamas, Belize, and Bermuda (Case Study 11.3). The Intel microprocessor manufacturing plant that began operations in San José, Costa Rica, in 1998 is an advanced technology variant of export promotion manufacturing.

Case Study 11.3: Offshore Caribbean Financial Services Industries

One of the most successful export promotion industries of the English-speaking Caribbean nations is the provision of specialized financial services not generally available to companies and individuals in their home countries. These so-called offshore, or foreign, services cater to clients from around the world, whose patronage is predicated on their transactions being processed rapidly and with absolute anonymity and confidentiality by the Caribbean service centers. The services are available to any person or company willing to pay for them, regardless of the client's legal standing at home or abroad.

The services offered can be grouped into three broad categories, the first of which is banking. Offshore Caribbean banks are renowned for their willingness to process, almost instantaneously, huge deposits, withdrawals, or transfers made by drug lords, organized crime figures, and other "high net worth individuals" seeking to launder their money and/or evade the payment of taxes. The offshore banks also allow seemingly reputable foreign banks to establish Caribbean subsidiaries to which the parent foreign banks sell bad loans at discounted rates in order to circumvent mainland regulations requiring the maintenance of adequate reserves to cover loan losses.

The second financial services category consists of the "captive" insurance market. Under this arrangement, foreign firms are allowed to set up wholly owned subsidiary insurance companies that then assume risks that the parent company could not afford to obtain coverage for from legitimate insurance interests operating on the mainland. The most common types of insurance offered by the captive subsidiaries include medical malpractice, product liability, and workers' compensation.

The third common service available in the offshore financial centers is the registration of ships and dummy companies. Foreign ship owners can frequently save large amounts of money by registering their vessels under the flag of a Caribbean or Central American nation whose fees are significantly lower and whose safety and environmental regulations are much more relaxed than those of the owner's home country. In a similar vein, the establishment of tax-exempt Caribbean companies is a common means of hiding large amounts of flight capital from the mainland.

Of all the Caribbean nations active in the offshore financial services market, the Cayman Islands have been the most successful of late. In October of 2001, for example, there were 570 commercial banks and trust companies licensed on the islands, whose total population was only 35,530. This averaged out to a ratio of one bank or trust company for every 62 persons living on the islands. At the same time, there were more than 40,000 companies registered on the islands, or a ratio of one company for every 0.89 persons! Bermuda and Panama, two entities on the periphery of the Caribbean, have been the historic leaders in the captive insurance and ship registration sectors, respectively. In 1999, for instance, the capacity of Panama's registered merchant shipping fleet was by far the largest in the world, totaling over 105 million tons. This compared to a capacity of 12 million tons for the United States and 2.5 million tons for Canada. Numerous other nations, including the Bahamas, Belize, Barbados, the Netherlands Antilles, and the Turks and Caicos Islands, have also offered offshore financial services to varying degrees (Case Study Figure 11.1).

The offshore financial services industry is seen by many Caribbean peoples as a golden path to wealth and economic development. It is often described as the ideal, upscale, high-technology, white-collar complement to the lower-paying jobs generated by tourism. Critics of the industry argue, however, that it is notoriously unstable and thus a very poor foundation on which to base long-term development plans.

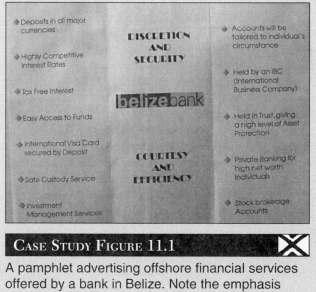

CASE STUDY FIGURE 11.1

A pamphlet advertising offshore financial services offered by a bank in Belize. Note the emphasis placed on discretion and security for high net worth individuals.

Sources: Blackman 1991; Roberts 1990; www.state.gov/r/pa/ei/bgn5286.htm; *Statistical Yearbook 1998* 2001.

While export promotion manufacturing has unquestionably contributed to short-term increases in industrial employment in those areas where it has been established, it has also been criticized for its general failure to form functional linkages to the local manufacturing sector and for contributing to the decline of native agriculture. Other concerns center on its tendency to increase, rather than decrease, dependence on foreign investment and on its footloose, transitory nature. Its capacity to promote the long-term economic development of Latin America and the Caribbean is also limited, like the other industrialization models that have come and gone, by its failure to address the cultural roots of underdevelopment.

CURRENT INDUSTRIALIZATION PATTERNS

As we now assess the overall geographical impacts of the various industrialization models practiced in Latin America and the Caribbean, two clear patterns emerge. The first is that Latin American industry continues to be concentrated, as it has been from the time of the European Conquest, in the major urban centers and neighboring provincial capitals (Figure 11.14). The São Paulo–Belo Horizonte–Rio de Janeiro industrial triangle of Brazil, the Toluca–Mexico City–Puebla corridor of Mexico, the Buenos Aires–Rosario–Santa Fé lower Río de la Plata region of Argentina, the Santiago–Rancagua–Valparaíso nucleus of Chile, the Lima–Callao area of Peru, and the Caracas–Maracay–Valencia axis of Venezuela are each responsible for between 50 and 75 percent of the manufacturing output of Latin America's leading industrial nations. Even in Latin America's smaller, less industrialized countries, the principal cities usually have an assortment of soft drink and beer bottling plants, automobile and truck dealerships and repair shops, cement and lumber companies, and home-centered sewing and food-preparation businesses (Figures 11.15 and 11.16). Recent research ("Industry's Tiny Titans" 1998) suggests that these small and medium-sized enterprises (SMEs) account for approximately 90 percent of manufacturing companies, half of manufacturing employment, and one-third of industrial output in most Latin American countries. Latin America's rural villages, by contrast, have almost no industrial activity whatsoever. The historically persistent **urbanization of industry** in Latin America and the Caribbean is an eloquent testimony to the capacity of a culture trait to endure in the midst of changing economic and political ideologies.

The second geographical pattern of Latin American industry is its limited regional distribution. Although national political leaders and development personnel seem always to be extolling the industrial growth of their home countries, the stark truth of the matter is that over four-fifths of all Latin American and Caribbean manufacturing occurs in just three countries—Brazil, Mexico, and Argentina (Figure 11.17). Another 12.5 percent is generated in Chile, Venezuela, Colombia, and Peru. The remaining nations produce collectively a mere 5 percent of the region's industrial output. This trend is equally evident in the production of steel, pig iron, aluminum, cement, paper and paperboard, and automobiles—the key components of modern manufacturing (Table 11.3).

CURRENT MINING DEVELOPMENTS

The Latin American and Caribbean mining industry has experienced three long-term historical phases. The first was associated with the Colonial Period and was characterized by the pursuit of silver, gold, and gemstones such as diamonds and emeralds. We have noted previously that mining was the dominant force in colonial industry and exerted an enormous influence on race and labor relations and on regional settlement and trading patterns. Yet another significant attribute of colonial mining was the forced removal of large portions of the ores and gems to Spain and Portugal as part of an exploitative mercantile system that left the New World colonies largely underdeveloped at the close of the eighteenth century.

Mining's second historical phase, which might be called the Republican Period, lasted from the achievement of independence in the early nineteenth century into the mid to late 1800s. It was, in general, a time of economic and technological stagnation associated with widespread political turmoil and weak foreign demand for Latin American minerals.

The Modern Period of Latin American and Caribbean mining began in the late 1800s and continues to the present. It was associated initially with the arrival of European and American investors, who did much to upgrade the technology and equipment used in excavating and processing the ores. As during the Colonial Period, modern mining emphasized in its early stages the exporting of the ores to overseas markets. Those markets, however, while always interested to some degree in silver and gold (Godfrey 1992; Trapasso 1992), have come increasingly to stress the importation of copper, tin, nitrates, lead, zinc, and other minerals needed in the manufacture of modern industrial products (Warhurst 1999). To these has been

FIGURE 11.14

Leading industrial regions of modern Latin America and the Caribbean.

FIGURE 11.15

This truck and automobile repair shop is typical of the many small industries found in Latin American cities.

FIGURE 11.16

This glass blowing factory is another example of the small-scale urban manufacturing that is common to Latin America.

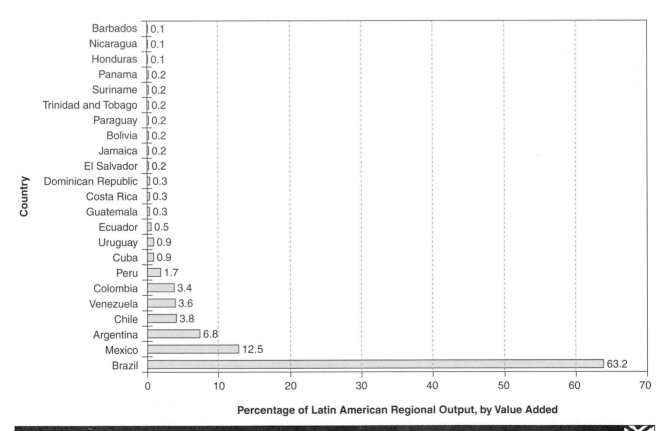

Percentage of Latin American Regional Output, by Value Added

FIGURE 11.17

Latin American manufacturing.

Source: Data compiled from *Britannica Book of the Year 2002* 2002, 832–837.

TABLE 11.3

Leading Latin American and Caribbean Manufacturers by Industrial Product Category, 1998

CATEGORY	Percentage of Latin American and Caribbean Output						
	BRAZIL	MEXICO	ARGENTINA	VENEZUELA	COLOMBIA	CHILE	PERU
Crude Steel	54.6	22.9	8.8	7.8	1.5	2.5	1.1
Pig Iron	76.4	13.8	6.1	—	0.6	2.6	0.5
Aluminum	55.0	11.5	8.1	24.3	—	—	—
Cement	31.0	27.1	6.3	6.9	7.6	3.4	3.8
Paper and paperboard	46.5	26.2	8.3	4.5	5.1	4.6	0.5
Automobiles	48.2	34.2	12.5	2.2	2.3	0.6	0.0

Source: Adapted from *Statistical Yearbook 1998* 2001, 460–497; Tuman and Morris 1998, 9.

FIGURE 11.18

Jamaica is the leading bauxite producing nation of Latin America and the Caribbean. These mines are situated in the northeastern interior of the island.

added, since World War II, a strong foreign demand for bauxite, iron ore, and ferroalloys, including manganese, chromium, and nickel (Figure 11.18). Much effort has also been expended in the search for fossil fuels, including petroleum and coal.

Recent decades have witnessed the discovery of a number of new ore deposits, many of which are among the largest and most productive in the world. These include the Venezuelan bauxite reserves at Los Pijiguaos and alluvial cassiterite, or tin ore, deposits in Rondônia and Amazonas and other portions of Brazil's upper Amazon Basin (Figure 11.19). The largest tin mine in

the world, the Pitinga, began operation in 1983 in Amazonas state some 300 kilometers northeast of Manaus. Brazilian tin output is now considerably greater than that of Bolivia, whose steadily declining production continues to originate from high-altitude Andean veins (Table 11.4).

Chile, which has dominated world copper production since the 1860s, has recently opened large new mines at Quebrada Blanca, Escondida, Andina, and Potrerillos in the northern and central Andes (Figure 11.20). Peru, now Latin America's second leading producer of copper, relies heavily on output from Cuajone and Toquepala, two huge open pit mines in the south that commenced operations in 1977 and 1960, respectively. Possibly the largest undeveloped copper reserves in the world are those at Cerro Colorado in the western Panamanian province of Chiriquí.

The northern Chilean and southern Peruvian Andes have also yielded significant iron ore deposits in recent decades. So, too, have Venezuela's Guiana Highlands. None of these, however, can compare in size to Brazil's massive Amazonian mines, which have been worked since 1985 at Carajás in central Pará state. Cauê Peak in southern Brazil has long supplied iron ore for the great Volta Redonda steel works in Rio de Janeiro state.

Major phosphate deposits have recently been opened in the Baja California peninsula of Mexico and along the coastal plain of northern Peru. A large find in Argentina's Río Colorado basin is also under development. A number of new ferroalloy operations have appeared in Minas Gerais and other portions of eastern Brazil, and Colombia's huge El Cerrejón coal strip mines have been producing since 1982.

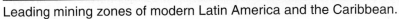

FIGURE 11.19

Leading mining zones of modern Latin America and the Caribbean.

TABLE 11.4

Leading Mineral Producing Nations of Latin America and the Caribbean

MINERAL	LEADING PRODUCER	SECOND LEADING PRODUCER	THIRD LEADING PRODUCER
Bauxite	Jamaica	Brazil	Suriname
Copper	Chile	Brazil	Peru
Diamonds	Brazil	Venezuela	Guyana
Gold	Brazil	Colombia	Chile
Iron ore	Brazil	Venezuela	Chile
Lead	Mexico	Brazil	Peru
Phosphate rock	Brazil	Mexico	Venezuela
Silver	Mexico	Peru	Chile
Tin	Brazil	Bolivia	Mexico
Zinc	Brazil	Peru	Mexico

FIGURE 11.20

Chile is the leading copper producing nation of Latin America and the Caribbean. One of the largest operations is the Chuquicamata mining site in the arid northern part of the country.

While the recent openings of these and other new operations have brought modest growth to Latin America's mining sector, it is important to note that mining now generates just 4 to 5 percent of the regional gross domestic product. This is a vast difference, indeed, from mining's preeminent position during the colonial era.

ENERGY RESERVES AND DEVELOPMENT

Because energy is required, in one form or another, to perform any kind of manufacturing or mining endeavor, we will now present a brief overview of its development in Latin America and the Caribbean. Commercial production of petroleum began in the early 1900s along the central Gulf coastal plain of Mexico and within the Maracaibo Basin of Venezuela. American and European interests were heavily involved in both nations, and most of the production was exported overseas. Subsequently, smaller oilfields have been developed in the Venezuelan and Colombian Llanos, in the eastern Amazonian regions of Ecuador, Peru, and Bolivia, in portions of Patagonia and Tierra del Fuego, and along the Brazilian and Argentine offshore continental shelves (see Figure 11.19). Exploration results have been disappointing to date in most of Brazil's Amazon Basin, mainland Chile, Paraguay, Central America and, with the exception of Trinidad and Tobago, the Caribbean Islands. Because Mexico's recently developed superfield in the Gulf of Campeche is geographically related to the older, adjoining coastal plain fields, and because three-fourths of Venezuela's output continues to originate in the Maracaibo Lowlands, it is correct to state that no new major discoveries of petroleum have occurred in Latin America and the Caribbean since the early twentieth century. While it is to be hoped that this situation will change in the future, the present reality is that over two-thirds of Latin America's total petroleum production comes from just two countries—Mexico and Venezuela (Figure 11.21). When we add to this the fact that Colombia and Mexico are the only nations to have discovered sizeable coal deposits, we begin to understand the sense of urgency felt by most Latin American and Caribbean peoples to develop alternative ways of meeting their growing energy needs.

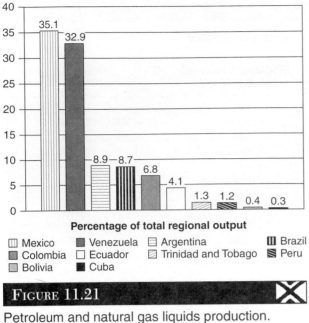

FIGURE 11.21

Petroleum and natural gas liquids production.

Source: Compiled from *Statistical Yearbook 1998* 2001, 628–633.

Hydroelectric Power Generation

The most significant alternative energy source utilized at present is hydroelectric power. As its name implies, hydropower is electricity produced by damming rivers and channeling the impounded water through turbo-generators. Because no fuels are consumed and no waste products are left behind, hydropower is often viewed as an environmentally friendly and economically low-cost source of electrical generation. There are, of course, environmental impacts from any human endeavor. In the case of hydroelectric dams, these often include severe siltation of the reservoir behind the dam. If unchecked, this can reduce greatly the water storage and electrical generating capacities of the complex itself. One of the most extreme cases of siltation yet documented is that of the Amaluza Reservoir behind Ecuador's Paute River hydroelectric complex. Deforestation and soil erosion were so severe in the surrounding highlands that the reservoir was virtually filled with sediment by 1992 and had to be dredged in order to remain operational just ten years after its completion (Hamilton 1993, 4). Numerous other Latin American reservoirs have silted up within twenty to twenty-five years.

Other negative environmental consequences of hydroelectric projects may include the loss of plant, fish, and animal species owing to the alteration of ecosystems situated downstream from the dam site and a greater occurrence of serious diseases (Smith et al. 1995). Barrow (1988, 68), for example, reported ex-panded populations of mosquitoes, snails, black flies, and rodents in the region impacted by the Tucuruí dam in the eastern Amazon. That, in turn, led to increased incidence of malaria, yellow fever, schistosomiasis (a liver fluke transmitted by snails), onchocerciasis (river blindness transmitted by black flies), leishmaniasis (ulcerating skin lesions transmitted by a protozoan), and South American sleeping sickness, or Chagas' Disease (chapter 13). Critics of large-scale hydroelectric projects also note that they often are associated with widespread blackouts or brownouts during the dry season and suggest that the power generated by the complex is actually very expensive if the reservoir is short-lived (ROSTLAC 1986).

These challenges notwithstanding, hydropower has proven to be very attractive throughout the region and has now become the leading source of electricity for many Latin American nations (Figure 11.22). While it would be impossible to note all of Latin America's hydroelectric complexes, a short list of the more prominent projects would include the Itaipú and Yacyretá dams on the Paraná River, the Paulo Afonso and Tres Marias plants on the São Francisco River, the Guri dam near Ciudad Guayana, Ecuador's Río Paute Project, and the Chicoasén, Angostura, and Malpaso dams within the Grijalva River Basin of southern Mexico (Figure 11.23). The massive Tucuruí dam, situated on the Tocantins River some 300 kilometers (185 miles) south of Belém, is but one of dozens of large hydroelectric complexes planned for Brazilian Amazonia (Biswas et al. 1999).

Renewable Organic Fuels

Two other important alternative energy sources in Latin America and the Caribbean are plants grown for the conversion of their biomass into ethyl alcohol (ethanol) and firewood. While neither of these strategies is new, both have received considerable attention from rural development personnel owing to their perceived potential as sustainable, low-cost energy sources. Unfortunately, the use of each has also involved tradeoffs that have limited the anticipated benefits.

The manufacturing of **ethanol** from sugar cane, cassava, and babacu palm has been pursued as a form of import substitution industrialization in Brazil since 1927, with a special urgency in the decades following the global oil shortages of the early 1970s. Advocates have encouraged the practice as a means of reducing costly petroleum imports, increasing Brazilian economic self-reliance, generating employment, and reducing regional income inequalities between the more industrialized South and the poverty-stricken sugar and ranching zones of the Northeast. Owing to government regulations, which tax gasoline-powered vehi-

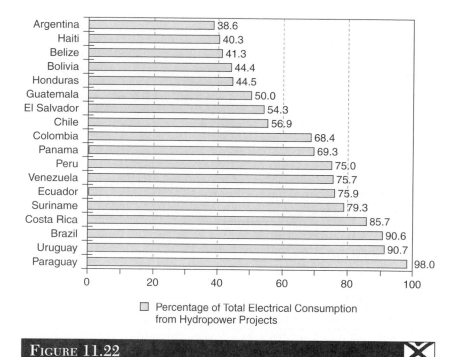

Country	Percentage
Argentina	38.6
Haiti	40.3
Belize	41.3
Bolivia	44.4
Honduras	44.5
Guatemala	50.0
El Salvador	54.3
Chile	56.9
Colombia	68.4
Panama	69.3
Peru	75.0
Venezuela	75.7
Ecuador	75.9
Suriname	79.3
Costa Rica	85.7
Brazil	90.6
Uruguay	90.7
Paraguay	98.0

☐ Percentage of Total Electrical Consumption
from Hydropower Projects

FIGURE 11.22

Reliance on hydroelectric power.

Sources: Compiled from *Britannica Book of the Year 2002* 2002, 838–843; *The Statesman's Year-Book 2002* 2001, various pages.

FIGURE 11.23

The Itaipú Dam on the Paraná River is the source of much of the electricity used in the industrial regions of southern Brazil, eastern Paraguay, and northern Argentina.

cles at more than twice the rate of vehicles that consume alcohol and which subsidize the costs of alcohol fuels, the program is an outward success, with the vast majority of Brazilian cars and trucks running either on a mixture of anhydrous alcohol and gasoline or on pure ethanol.

Negative side effects of the program have included its high cost, which critics have noted makes Brazilian alcohol fuels 35 to 50 percent more expensive than an equivalent amount of imported gasoline (Levinson 1987), and the expansion of sugar cane–driven *latifundios* in the Northeast and more especially in São Paulo state. This, in turn, has led to less land being planted to food crops and to accusations that the program has resulted in the subsidizing of fuel for wealthy car owners at the expense of the food supplies of the poor (Saint 1982). Environmentally, the program appears to have brought about a modest improvement in urban air quality, owing to reliance on the cleaner-burning alcohol fuels, but a deterioration of ground and surface water quality in the rural regions situated close to the distilleries.

Although considered by many to be old-fashioned and inefficient, the use of firewood as a fuel source continues to be extremely important in Latin America and the Caribbean, particularly among the rural peasantry who often convert it to charcoal for use in cooking foods (Figure 11.24). The most negative potential consequences of the practice, of course, are the deforestation of existing woodlands and the premature harvesting of reforested areas (Figures 11.25 and 11.26). Many of the reforested lands have been planted to fast-growing softwoods. While these monocultural stands

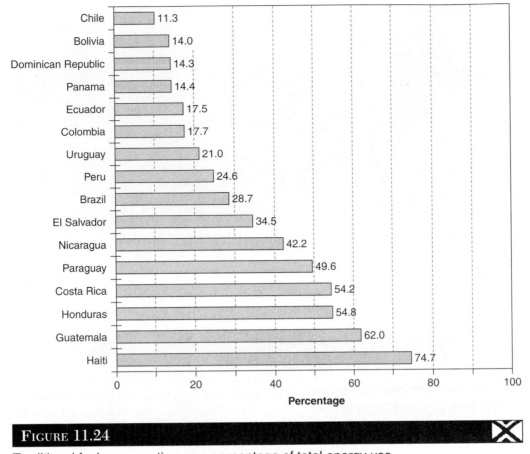

FIGURE 11.24

Traditional fuel consumption as a percentage of total energy use.
Source: Compiled from www.undp.org/hdr2001/indicator/indic_203_1_1.html

TOURISM

have the capacity to provide sustained timber harvests if properly managed, they are far less diverse in their genetic composition and economic products than the original tropical hardwood forests.

A final economic activity that is often categorized as an industry is tourism. Tourism can be defined and measured from a number of perspectives. Some countries include all visitors in their tourist counts, regardless of the length of stay. Most, however, distinguish between "tourists," defined as persons who stay at least one night but not more than one year in the country visited, and "excursionists," defined as guests who do not remain overnight in the country visited (Figure 11.27). Some nations even differentiate on the basis of whether the visitors arrived by air or by sea or whether they came for recreational as opposed to business purposes. Others base their counts on the type of accommodations chosen by the visitors. Almost lost oftentimes, in all the statistical hairsplitting, is the impact of the domestic tourist—that is, the visitor who arrives from another part of the same nation. While less glamorous to government officials than international arrivals, domestic tourism is a major industry in many regions (Theobald 1998; Meyer-Arendt 1990; Lea 1988).

International tourism exerts an enormous impact on some Latin American and Caribbean nations and yet plays a surprisingly small economic role in others (Table 11.5). The country that receives by far the greatest number of international tourists is Mexico, which was visited by almost 20 million persons in 1998. Brazil, Puerto Rico, Argentina, the Dominican Republic, Uruguay, Chile, the Bahamas, Cuba, and Jamaica each average over a million visitors annually. Although the total number of visitors is not as great in most Caribbean nations, many depend heavily upon tourist spending to provide jobs and to finance govern-

FIGURE 11.25

Highland Ecuadorian Indians collecting firewood.

FIGURE 11.26

Charcoal being transported on the backs of donkeys for sale as cooking fuel to urban consumers in Puebla, Mexico.

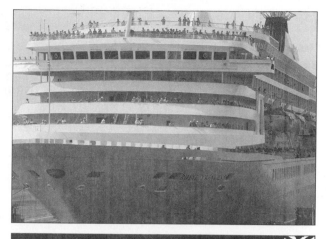

FIGURE 11.27

One of the fastest growing segments of the Caribbean tourist industry is cruise ships taking short, highly structured excursions to ports and other points of historical and cultural interest.

mental services. This is true now for Cuba as well, which since 1993 has attracted increased Western European and Canadian investment in the tourist sector (Honey 1999; Scarpaci 1998) (Figure 11.28). With the exception of Costa Rica, tourism contributes relatively little to the economies of most of the Central American countries, Suriname, Guyana, Bolivia, Paraguay, Ecuador, Peru, and Venezuela.

Considerable controversy exists over the merits of international tourism as a development model. Advocates cite its perceived economic benefits to the host nation, including the generation of badly needed jobs and foreign currency and the expansion of airports, roads, and public utilities (Aronsson 2000; Richards and Hall 2000; Archer 1984). In a growing number of countries, including Costa Rica, Ecuador (the Galapagos Islands), Argentina, and Brazil, an emphasis on environmentally sustainable, nature-based **ecotourism** has encouraged the establishment of national parks and biological preserves (Font and Buckley 2001; Fennell 1999; Schlüter 1998; Place 1988) (Figure 11.29). Costa Rica, for instance, has set aside 28 percent of its land in national parks and forests and in biological, wildlife, and indigenous reserves (Evans 1999). Some proponents go so far as to argue that international tourism has a beneficial impact on the preservation of local cultures by stimulating renewed interest in and development of local handicrafts, music, food, and dress.

Critics counter these claims by pointing out that much of the money spent by foreign tourists ends up in the hands of transnational hotel, advertising, and

TABLE 11.5

International Tourism to Latin America and the Caribbean, 1998

COUNTRY	TOTAL NUMBER OF TOURISTS (THOUSANDS)[a]	PERCENTAGE OF TOURISTS BY ORIGIN		TOTAL TOURIST RECEIPTS (MILLIONS OF U.S. DOLLARS)
		Americas	Europe	
Middle America				
Anguilla	43.9	79.0	18.2	58
Antigua and Barbuda	226.1	59.1	39.0	256
Aruba	647.4	91.9	7.6	715
Bahamas	1,540.0	87.3	7.7	1,408
Barbados	512.4	49.8	49.1	703
Belize	299.7	82.8	15.1	99
British Virgin Islands	279.1	80.2	7.7	230
Cayman Islands	404.2	90.4	8.6	450
Costa Rica	942.9	83.9	14.0	884
Cuba	1,415.8	42.0	56.0	1,571
Dominica	65.5	80.6	17.9	38
Dominican Republic[b]	2,334.5	34.5	45.6	2,142
El Salvador	541.9	91.9	5.0	125
Grenada	115.8	47.2	32.9	59
Guadeloupe	133.0	5.3	94.1	466
Guatemala	636.3	79.3	18.1	394
Haiti	146.8	89.5	9.3	57
Honduras	321.2	87.3	10.2	164
Jamaica	1,225.3	81.2	17.5	1,197
Martinique	548.8	14.1	85.2	415
Mexico[c]	19,809.5	97.9	1.8	7,897
Montserrat[b]	7.5	74.5	20.0	8
Netherlands Antilles[d]	270.6	62.1	37.5	749
Nicaragua	405.7	90.0	8.3	90
Panama	422.2	89.4	7.3	379
Puerto Rico	3,396.1	75.7	n.a.	2,233
St. Kitts and Nevis	93.2	82.7	16.3	76
St. Lucia	252.2	63.8	35.1	291
St. Vincent-Grenadines	67.2	68.8	30.2	74
Trinidad and Tobago	347.7	76.3	22.2	201
Turks and Caicos Islands	110.9	81.4	10.7	196
U.S. Virgin Islands	480.1	94.4	3.0	921
South America				
Argentina	2,969.8	86.0	11.5	2,936
Bolivia	420.5	64.1	32.3	174
Brazil	4,818.1	71.6	24.1	3,678
Chile	1,759.3	84.9	12.9	1,062
Colombia[d]	969.0	87.4	12.6	939
Ecuador	510.6	75.9	21.1	291
Guyana[d]	75.7	91.6	6.9	52
Paraguay	349.6	80.9	9.5	595
Peru	723.7	66.8	27.6	913
Suriname	54.6	10.1	84.4	44
Uruguay	2,324.0	77.9	4.1	695
Venezuela	685.4	53.4	44.0	961

[a]Excludes excursion visitors except as noted

[b]Includes excursionists

[c]Includes nationals of the country residing abroad

[d]Data is for 1997

Source: Compiled from *Statistical Yearbook 1998* 2001, 735–767.

FIGURE 11.28

Cuba is now aggressively promoting tourism as a means of attracting foreign investment. The greatest number of tourists come from Europe, the former Soviet Union, and Canada. Shown here is a resort in Holguín.

FIGURE 11.29

So-called ecotourism can take many forms. Shown here are cabins belonging to a "Jungle Inn" downstream from Iquitos, Peru, on a bank of the Marañón River. Visitors are made to feel that they are contributing to the preservation of nature by staying at the lodge.

transportation companies and that what resources remain are often squandered on unnecessary showcase projects that do little to benefit the local population (Southgate 1998; Bélisle 1983; Britton 1980). They note further that international tourist developments frequently lead to inflated land values and higher living costs in the impacted areas, which in turn often result in local food shortages brought on by agricultural land abandonment. Environmental degradation may occur through the loss of native species of plants and animals caused by the loss or contamination of native habitats (Hall and Page 1999; Savage 1993). Most troubling to many critics is the pattern of local people being hired primarily as poorly paid physical laborers. These conditions have led some to assert that international tourism is little more than a modified monocultural neoplantation, which perpetuates the social inequalities, cultural dilution, and economic dependence and underdevelopment of the colonial past (Weaver 1998 and 1988; Mowforth and Munt 1998; Sambrook, Kermath, and Thomas 1992).

Having stated the positions of both the advocates and opponents of international tourism, it must be noted that the political leaders of virtually every Latin American and Caribbean nation have consistently endeavored to promote it as a means of achieving accelerated economic growth. Those nations that have been least successful have generally been those most beset by internal political instability or tyranny and by poor health conditions.

SUMMARY

While it is clear that portions of Latin America and the Caribbean have achieved historically unprecedented levels of industrialization over the past century, it is important that we recognize that the manufacturing output of the region as a whole continues to be small by world standards and is likely to remain so in the foreseeable future. Japan alone, for instance, produces more manufactured goods than all the Latin American and Caribbean nations combined. The percentage of Latinos employed in manufacturing and mining remains relatively low compared to the more industrialized nations, and much of Latin America's manufacturing continues to be poorly paid, labor intensive, and of low technology. Vast areas remain without any industry whatsoever.

Having noted this, we must also stress that industrialization, where it has occurred, has brought increased levels of material living to many. Regrettably, it is also true that industrialization has been associated with systematic neglect of the rural countryside, with growing levels of foreign economic influence, and with the expansion of urban poverty. It is to the urban areas, where most Latin Americans now reside, that we next turn our attention.

KEY TERMS

SUGGESTED READINGS

Archer, Ewart. 1984. "Estimating the Relationship between Tourism and Economic Growth in Barbados." *Journal of Travel Research* 22:8–12.

Aronsson, Lars. 2000. *The Development of Sustainable Tourism.* London: Continuum.

Assadourian, Carlos Sempat. 1992. "The Colonial Economy: The Transfer of the European System of Production to New Spain and Peru." *Journal of Latin American Studies* 24 Quincentenary Supplement: 55–68.

Bakewell, Peter John. 1984. *Miners in Red Mountain: Indian Labor in Potosí, 1545–1650.* Albuquerque: University of New Mexico Press.

_____. 1971. *Silver Mining and Society in Colonial Mexico: Zacatecas 1546–1700.* Cambridge: Cambridge University Press.

Barham, Bradford L., and Oliver T. Coomes. 1996. *Prosperity's Promise: The Amazon Rubber Boom and Distorted Economic Development.* Boulder, Colo.: Westview Press.

Barrow, Chris. 1988. "The Impact of Hydroelectric Development on the Amazonian Environment: With Particular Reference to the Tucuruí Project." *Journal of Biogeography* 15:67–78.

Bebbington, Anthony. 2000. "Reencountering Development: Livelihood Transitions and Place Transformations in the Andes." *Annals of the Association of American Geographers* 90:495–520.

Bélisle, Francois J. 1983. "Tourism and Food Production in the Caribbean." *Annals of Tourism Research* 10:497–513.

Biswas, Asit K., et al., eds. 1999. *Management of Latin American River Basins: Amazon, Plata, and São Francisco.* Tokyo: United Nations University.

Blackman, Courtney N. 1991. "Tourism and Other Services in the Anglophone Caribbean." In *Small Country Development and International Labor Flows: Experiences in the Caribbean,* ed. Anthony P. Maingot, 53–81. Boulder, Colo.: Westview Press.

Brazil Trade and Industry, February 1986.

Britannica Book of the Year 2002. 2002. Chicago: Encyclopedia Britannica.

Britton, Robert. 1980. "Shortcomings of Third World Tourism." In *Dialectics of Third World Development,* eds. Ingolf Vogeler and Anthony de Souza, 241–248. Montclair, N.J.: Allanheld and Osmun.

Cabán, Pedro A. 1989. "Industrial Transformation and Labour Relations in Puerto Rico: From 'Operation Bootstrap' to the 1970s." *Journal of Latin American Studies.* 21:559–591.

Cardoso, F. H., and E. Faletto. 1979. *Dependency and Development in Latin America.* Berkeley: University of California Press.

Céspedes, Guillermo. 1976. *América Latina Colonial Hasta 1650.* México, D.F.: Secretaría de Educación Pública.

Cobb, Gwendolin B. 1949. "Supply and Transportation for the Potosí Mines, 1545–1640." *Hispanic American Historical Review* 29:25–45.

Conroy, Michael E. 1973. "Rejection of Growth Center Strategy in Latin American Regional Development Planning." *Land Economics* 49:371–380.

Corral, Carlos Montalvo. 2000. "Structural Determinants of Sustainability in the Maquiladora Industry on Mexico's Northern Border." In *Shared Space: Rethinking the U.S.-Mexico Border Environment,* ed. Lawrence A. Herzog, 313–335. La Jolla: University of California, San Diego Center for U.S.-Mexican Studies.

Craig, Alan K. 1984. "Foreign Exploitation of Natural Resources in Western South America: Peruvian Guano and Chilean Nitrate." In *Latin America: Case Studies,* eds. Richard G. Boehm and Sent Visser, 165–173. Dubuque, Iowa: Kendall Hunt.

Cravey, Altha J. 1998. *Women and Work in Mexico's Maquiladoras.* Lanham, Md.: Roman and Littlefield.

Dean, Warren. 1969. *The Industrialization of São Paulo, 1880–1945.* Austin: University of Texas Press.

Despres, Leo A. 1991. *Manaus: Social Life and Work in Brazil's Free Trade Zone.* Albany: State University of New York Press.

DeWind, Josh, and David H. Kinley, III. 1988. *Aiding Migration: The Impact of International Assistance on Haiti.* Boulder, Colo.: Westview Press.

Dore, Elizabeth. 1988. *The Peruvian Mining Industry: Growth, Stagnation, and Crisis.* Boulder, Colo.: Westview Press.

Evans, Sterling. 1999. *The Green Republic: A Conservation History of Costa Rica.* Austin: University of Texas Press.

Fennell, David A. 1999. *Ecotourism: An Introduction.* London: Routledge.

Font, X., and R. C. Buckley, eds. 2001. *Tourism Ecolabelling: Certification and Promotion of Sustainable Management.* Wallingford, U.K.: CABI Publishing.

Furtado, Celso. 1970. *Economic Development of Latin America.* Cambridge: Cambridge University Press.

Garner, Richard L. 1988. "Long-Term Silver Mining Trends in Spanish America: A Comparative Analysis of Peru and Mexico." *American Historical Review* 93:898–935.

Godfrey, Brian J. 1992. "Migration to the Gold-Mining Frontier in Brazilian Amazonica." *The Geographical Review* 82:458–469.

Grindle, Merilee S. 1986. *State and Countryside: Development Policy and Agrarian Politics in Latin America*. Baltimore: Johns Hopkins University Press.

Gwynne, Robert N. 1986. *Industrialization and Urbanization in Latin America*. Baltimore: Johns Hopkins University Press.

———. 1982. "Location Theory and the Centralization of Industry in Latin America." *Tijdschrift voor Economische en Sociale Geografie* 73:80–93.

Hall, C. Michael, and Stephen J. Page. 1999. *The Geography of Tourism and Recreation: Environment, Place, and Space*. 2nd ed. London: Routledge.

Hamilton, Roger. 1993. "Green Security for a Nation's Energy." *The IDB* 20 (January–February):3–12.

Harner, John. 1998. "Dependency and Development in the Copper Mining Region of Sonora, Mexico." *Yearbook, Conference of Latin Americanist Georgraphers* 24:17–30.

Honey, Martha. 1999. *Ecotourism and Sustainable Development: Who Owns Paradise?* Washington, D.C.: Island Press.

"Industry's Tiny Titans." 1998. *IDBAMERICA* 25(4):13.

Klein, H. F. 1987. *The Coal of El Cerrejón: Dependent Bargaining and Colombian Policy-Making.*. University Park: The Pennsylvania State University Press.

Kleinpenning, J. M. G. 1971a. "Objectives and Results of the Development Policy in North-East Brazil." *Tijdschrift voor Economische en Sociale Geografie* 62:271–284.

———. 1971b. "Road Building and Agricultural Colonization in the Amazon Basin." *Tijdschrift voor Economische en Sociale Geografie* 62:285–289.

Kopinak, Kathryn. 1996. *Desert Capitalism: Maquiladoras in North America's Western Industrial Corridor*. Tucson: University of Arizona Press.

Lea, John. 1988. *Tourism and Development in the Third World*. London: Routledge.

Levinson, Marc. 1987. "Alcohol Fuels Revisited: The Costs and Benefits of Energy Independence in Brazil." *Journal of Developing Areas* 21:243–258.

Long, Frank. 1983. "Industrialization and the Role of Industrial Development Corporations in a Caribbean Economy: A Study of Barbados 1960–80." *Inter-American Economic Affairs* 37 (Winter): 33–56.

Lord, J. Montague. 1992. "Latin America's Exports of Manufactured Goods." In *Economic and Social Progress in Latin America 1992 Report*, 191–279. Washington, D.C.: Inter-American Development Bank.

Lynch, John. 1992. "The Institutional Framework of Colonial Spanish America." *Journal of Latin American Studies* 24 Quincentenary Supplement: 69–81.

MacLachlan, Ian, and Adrian Guillermo Aguilar. 1998. "Maquiladora Myths: Locational and Structural Change in Mexico's Export Manufacturing Industry." *Professional Geographer* 50:315–331.

McCoy, James P. 1992. "Bauxite Processing in Jamaica and Guyana: An Extension to Aluminum Smelting." *World Development* 20:751–766.

Mejía, Oscar. 1988. "The Export Potential of Colombian Coal." *Natural Resources Forum* 12:345–352.

Meyer-Arendt, Klaus J. 1990. "Recreational Business Districts in Gulf of Mexico Seaside Resorts." *Journal of Cultural Geography* 11 (Fall/Winter):39–55.

Mowforth, Martin, and Ian Munt. 1998. *Tourism and Sustainability: New Tourism in the Third World*. London: Routledge.

Mullings, Beverley. 1999. "Sides of the Same Coin? Coping and Resistance among Jamaican Data-Entry Operators." *Annals of the Association of American Geographers* 89:290–311.

Peña, Devon G. 1997. *The Terror of the Machine: Technology, Work, Gender, and Ecology on the U.S.-Mexican Border*. Austin: University of Texas Press.

Place, Susan E. 1988. "The Impact of National Park Development on Tortuguero, Costa Rica." *Journal of Cultural Geography* 9 (Fall/Winter):37–52.

Pomeroy, Cheryl. 1988. "The Salt of Highland Ecuador: Precious Product of a Female Domain." *Ethnohistory* 35:131–160.

Portes, Alejandro, José Itzigsohn, and Carlos Dore-Cabral. 1994. "Urbanization in the Caribbean Basin: Social Change during the Years of Crisis." *Latin American Research Review* 29 (2):3–37.

Prebisch, Raúl. 1971. *Change and Development: Latin America's Great Task*. New York: Praeger.

Ratt, W. Dirk. 1981. *Revoltosos: Mexico's Rebels in the United States, 1903–1923*. College Station: Texas A & M University Press.

Richards, Greg, and Derek Hall. 2000. *Tourism and Sustainable Community Development*. London: Routledge.

Richardson, Harry W., and Margaret Richardson. 1975. "The Relevance of Growth Center Strategies to Latin America." *Economic Geography* 51:163–178.

Roberts, Susan. 1990. "Going Offshore: Finance and Development in the Caribbean." Paper presented at the annual meeting of the Association of American Geographers, Toronto, Canada, April.

ROSTLAC (Oficina Regional de Ciencia y Tecnología de la UNESCO para América Latina y el Caribe). 1986. *Agua, Vida, y Desarrollo*. 2 tomos. Montevideo, Uruguay: UNESCO.

Safa, Helen I. 1995. *The Myth of the Male Breadwinner: Women and Industrialization in the Caribbean*. Boulder, Colo.: Westview Press.

Saint, William S. 1982. "Farming for Energy: Social Options under Brazil's National Alcohol Programme." *World Development* 10:223–238.

Sambrook, Richard Alan, Brian M. Kermath, and Robert N. Thomas. 1992. "Seaside Resort Development in the Dominican Republic." *Journal of Cultural Geography* 12(2): 65–75.

Savage, Melissa. 1993. "Ecological Disturbance and Nature Tourism." *The Geographical Review* 83:290–300.

Savedoff, William D. 1995. *Wages, Labour, and Regional Development in Brazil*. Aldershot, England: Ashgate Publishing.

Scarpaci, Joseph L. 1998. "The Changing Face of Cuban Socialism: Tourism and Planning in the Post-Soviet Era." *Yearbook, Conference of Latin Americanist Geographers* 24:97–109.

Schlüter, Regina G. 1998. "Tourism Development: A Latin American Perspective." In *Global Tourism*, ed. William F. Theobald, 216–230. Oxford: Butterworth-Heinemann.

Sklair, Leslie. 1993. *Assembling for Development: The Maquila Industry in Mexico and the United States.* 2nd ed. San Diego: Center for U.S.-Mexican Studies, University of California at San Diego.

Smith, Nigel J. H., et al. 1995. *Amazonia: Resiliency and Dynamism of the Land and Its People.* Tokyo: United Nations University Press.

South, Robert B. 1990. "Transnational 'Maquiladora' Location." *Annals of the Association of American Geographers* 80:549–570.

Southgate, Douglas. 1998. *Tropical Forest Conservation: An Economic Assessment of the Alternatives in Latin America.* New York: Oxford University Press.

The Statesman's Year-Book 2002. 2001. New York: St. Martin's Press.

Statistical Yearbook 1998. 2001. New York: United Nations.

Sternberg, Rolf. 1984. "Hydroelectric Power in the Context of Brazilian Urban and Industrial Planning." In *Latin America: Case Studies,* eds. Richard G. Boehm and Sent Visser, 187–198. Dubuque, Iowa: Kendall/Hunt for the National Council for Geographic Education.

Theobald, William F. 1998. "The Meaning, Scope, and Measurement of Travel and Tourism." In *Global Tourism*, ed. William F. Theobald, 3–21. Oxford: Butterworth-Heinemann.

Thorp, Rosemary. 1992. "A Reappraisal of the Origins of Import-Substituting Industrialization 1930–1950." *Journal of Latin American Studies* 24 Quincentenary Supplement: 181–195.

Thoumi, Francisco E. 1991. "Economic Policy, Free Zones and Export Assembly Manufacturing in the Dominican Republic." In *Small Country Development and International Labor Flows: Experiences in the Caribbean*, ed.

Anthony P. Maingot, 167–182. Boulder, Colo.: Westview Press.

Trapasso, L. Michael. 1992. "Deforestation of the Amazon— A Brazilian Perspective." *Geojournal* 26:311–322.

Tuman, John P., and John T. Morris. 1998. "The Transformation of the American Automobile Industry." In *Transforming the Latin American Automobile Industry,* eds. John P. Tuman and John T. Morris, 3–25. Armonk, N.Y.: M. E. Sharpe.

Villarreal, Rene. 1990. "The Latin American Strategy of Import Substitution: Failure or Paradigm for the Region?" In *Manufacturing Miracles: Paths of Industrialization in Latin America and East Asia,* eds. Gary Gereffi and Donald L. Wyman, 292–320. Princeton, N.J.: Princeton University Press.

Warhurst, Alyson, ed. 1999. *Mining and the Environment: Case Studies from the Americas.* Ottawa: International Development Research Centre.

Weaver, David B. 1998. *Ecotourism in the Less Developed World.* Wallingford, U.K.: CAB International.

———. 1988. "The Evolution of a 'Plantation' Tourism Landscape on the Caribbean Island of Antigua." *Tijdschrift voor Economische en Sociale Geografie.* 79:319–331.

West, Robert C., ed. 1998. *Latin American Geography: Historical-Geographical Essays, 1941–1998.* Baton Rouge: Department of Geography and Anthropology Geoscience and Man, vol. 35.

———. 1993. *Sonora: Its Geographical Personality.* Austin: University of Texas Press.

Wilson, James R., and Cotton Mather. 1990. "Photo Essay: The Rio Grande Borderland." *Journal of Cultural Geography* 10 (Spring/Summer):66–98.

ELECTRONIC SOURCES

www.maquiladirectory.com/statistics/
www.solunet-infomex.com
www.state.gov/r/pa/ei/bgn5286.htm
www.undp.org/hdr2001/indicator/indic_203_1_1.html
www.worldbank.org/data/wdi2001/pdts/tab5_1.pdf

12

Urbanization, Population Growth, and Migration

One of the most momentous geographic transformations in the history of Latin America and the Caribbean has been the region's rapid evolution during the past century from a predominantly rural and agrarian society to one that is highly urbanized. Because this transformation has been associated to some extent with industrialization and the beginnings of social and political pluralism, some observers have suggested that the Latin American and Caribbean nations are simply following the same development path of many Western European and Anglo American peoples. These same observers reason that the continuing urbanization of the Latin American and Caribbean countries will eventually result in their becoming as "developed" as their European and Anglo American counterparts.

The problem with this logic is that many of the Latin American and Caribbean nations have now become as urbanized, in a statistical sense, as the European and Anglo American countries and yet remain much poorer economically and far less open in their social and political structures. Awareness of this fact has led other scholars to associate Latin American and Caribbean urbanization with economic and social underdevelopment.

To appreciate how Latin American and Caribbean urbanization has contributed to both the development and underdevelopment of the region, it is necessary to begin with a historical overview of the roles that cities have played in Latin America from the pre-Conquest era to the present. We will come to recognize urbanization as yet another expression of the intense social and economic divisions that have characterized the region since the coming of the Spanish and Portuguese *conquistadores* over five centuries ago.

NATIVE AMERICAN URBANIZATION

Urbanization, which can be defined as the construction of and residence within cities, was practiced to varying degrees by many pre-Hispanic populations but was totally absent among others. Hardoy (1975, 3) has noted that the proportion of pre-Columbian Latin American territory subjected to the control of urban centers was never greater than 5 percent and often was considerably less. Most native American cultures were predominantly rural in character, with the great majority of their populations either residing in small farming communities or belonging to seminomadic hunting and gathering societies.

While acknowledging that the majority of pre-Conquest indigenous peoples belonged to rural societies, we should also be aware that true urbanization did occur among some Amerindian cultures, and in size and splendor these cities were often equal or superior to those of the Old World at comparable times in history. By far the largest and most impressive of the pre-Columbian New World cities was the magnificent Aztec capital of Tenochtitlán, which, as we noted in chapter 7, is estimated to have supported a pre-Conquest population of 200,000 to 250,000 persons on a surface area of some thirteen to twenty square kilometers (five to eight square miles) (Griffin and Ford 1993; Garza and Schteingart 1978). Cortés and his fellow conquerors declared that the grandeur of Tenochtitlán's great plaza, broad avenues, monumental architecture, zoo, markets, and "floating" gardens, or *chinampas*, exceeded that of any city in Spain. Other true urban developments in south-central Mexico included Teotihuacán, whose population in the Valley of Mexico grew to between

85,000 and 200,000 in the fifth through seventh centuries A.D. (Millon 1966); Monte Albán, which supported some 45,000 to 50,000 persons in the Valley of Oaxaca between 200 and 700 A.D.; and Tajín, whose population reached 40,000 to 50,000 in the ninth through eleventh centuries A.D.

The presence of moderate-sized cities among the Maya is evidenced by the northern Guatemalan remains of Tikal, whose population peaked at around 45,000 in 550 A.D. (Butterworth and Chance 1981), and by the ruins of the smaller urban centers of Copán, Tulum, Calakmul, Palenque, Uxmal, and Chichén-Itzá (Figure 12.1). The largest urban development in ancient Peru appears to have been the community of Chan Chan. While lacking huge monumental structures, Chan Chan housed an estimated population of over 100,000 between 900 and 1463 A.D. in nine rectangular enclosures built over twenty square kilometers (eight square miles) of land along the arid coast of northern Peru (Butterworth and Chance 1981). Other, smaller urban centers that flourished at various times in Andean South America included Tiahuanaco on the Bolivian end of Lake Titicaca; Quito and Riobamba, Ecuador; and the Peruvian cities of Cajamarca, Pachacámac, and Tumbes (Chandler 1987). The Inca capital of Cuzco, which was considerably smaller than Tenochtitlán, was primarily an administrative center, with travel in and out of the city limited to authorized individuals.

Sanders and Webster (1988) have argued that pre-Columbian cities can be grouped according to their relative emphasis on ritual, administrative, and commercial functions. While differing in their individual character, all were supported by densely populated rural hinterlands, and the great majority were built on inland sites, reflecting the inward-looking focus of their occupants. One final trait shared by most of the cities occupied at the time of the Conquest was their near total destruction by the Spanish invaders, who in many cases began immediately to rebuild them to reflect a value system that was largely alien to the Amerindian peoples.

COLONIAL EUROPEAN SETTLEMENT

As we noted in chapter 5, the European settlers who colonized Latin America represented an Iberian culture that considered residence in an urban area essential to maintaining a refined, upper-class lifestyle. Because none of the Spaniards who came to the New World intended to live as peasants or social nobodies, the initial thrust of Hispanic colonization centered on the founding of towns and cities from which the conquerors could govern their rural subjects. So compelling was the need to create urban environments that at least 225 cities had been established within fifty years of the Conquest, and over 330 existed by the year 1630 (Céspedes 1976, 76) (Table 12.1).

In addition to having the opportunity to live as aristocratic *hidalgos* or *caballeros,* Spanish **vecinos,** or city residents, were also granted certain benefits that were unavailable to rural dwellers. These privileges included the right to govern themselves by electing their own municipal authorities, the right to petition the Crown for redress of grievances or for special favors or concessions, and the power to designate a municipal *procurador,* or purchaser, to obtain Spanish-regulated goods desired by the citizenry. Other advantages of urban residence were the protection and security afforded by the presence of the military and, for the devout, increased access to the clergy and Catholic sacraments. From the very beginning of the colonial era, then, cities embodied power, privilege, wealth, and rank. Although humble by modern standards, they stood in stark contrast to the obscurity and poverty that, from the Spanish point of view, enshrouded the intervening Indian hinterlands.

In a very real sense, the European-dominated cities of colonial Latin America functioned as predators of the rural zones, with the elites taking the agricultural produce, minerals, and other raw materials obtained from the countryside and investing the wealth generated by the sale of those products in the urban areas. Because this process led over time to increasing gaps in living levels between the city dwellers and rural folk, it can be argued that the process of urbanization has contributed to the chronic, long-term underdevelopment of Latin America and the Caribbean. In contrast, Anglo American and Western European cities are widely viewed as having contributed over time to the economic improvement of their adjoining rural districts and thus to the long-term development of their respective nations.

The predatory character of the colonial cities was facilitated by the Spanish Crown's policy of allowing each city to claim as its own territory all of the neighboring rural regions not controlled by surrounding cities. This consigned the peasantry to a serfdom whose sole purpose was to generate wealth to be consumed by the urban aristocracy. This practice of allocating rural territory to cities differed markedly from Anglo American law, which restricted the territorial holdings of urban centers to the land contained within their "city limits."

A third significant difference in the evolution of colonial Latin and Anglo American cities had to do with their spatial proximity to one another. In the English colonies of North America, settlement initially was clustered along the eastern Atlantic coastal plain and then gradually, over the course of several centuries, pushed westward. Physical distances between cities were thus relatively short and communication levels were high. This strengthened both the economic and political unification of the United States and Canada.

FIGURE 12.1

Major pre-Hispanic urban centers.

TABLE 12.1

Founding Dates of Leading Latin American and Caribbean Cities

DATE	CITY	DATE	CITY
1496	Santo Domingo, Dominican Republic	1548	Recife, Brazil
1511	San Juan, Puerto Rico		La Paz, Bolivia
1515	Havana, Cuba	1549	Salvador, Brazil
1519	Veracruz, Mexico	1550	Concepción, Chile
	Panama City, Panama		San Luís Potosí, Mexico
1521	Mexico City, Mexico	1554	Guanajuato, Mexico
1523	Cumaná, Venezuela		São Paulo, Brazil
1524	Maracaibo, Venezuela	1556	Valencia, Venezuela
	Guatemala City, Guatemala	1557	Cuenca, Ecuador
	San Salvador, El Salvador	1558	Mérida, Venezuela
	León, Nicaragua	1560	Santa Cruz, Bolivia
	Granada, Nicaragua		Durango, Mexico
1525	Santa Marta, Colombia	1561	Mendoza, Argentina
1526	Oaxaca, Mexico	1562	San Juan, Argentina
	San Cristóbal de las Casas, Mexico	1563	Barquisimeto, Venezuela
1531	Puebla, Mexico	1564	Cartago, Costa Rica
	Culiacán, Mexico	1565	Tucumán, Argentina
1532	São Vicente, Brazil		Rio de Janeiro, Brazil
1533	Cartagena, Colombia	1567	Caracas, Venezuela
	Cuzco, Peru	1572	Huancavelica, Peru
1534	Quito, Ecuador	1573	Santa Fé, Argentina
1535	Lima, Peru		Córdoba, Argentina
	Trujillo, Peru	1574	Cochabamba, Bolivia
	Pôrto Seguro, Brazil	1577	Saltillo, Mexico
1536	Buenos Aires, Argentina (1st)	1578	Tegucigalpa, Honduras
	Cali, Colombia	1580	Buenos Aires, Argentina (2nd)
	Ilhéus, Brazil	1582	Salta, Argentina
1537	Asunción, Paraguay	1585	Paraíba, Brazil
	Olinda, Brazil	1588	Corrientes, Argentina
1538	Bogotá, Colombia	1596	Monterrey, Mexico
	Guayaquil, Ecuador	1609	Fortaleza, Brazil
1540	Campeche, Mexico	1612	São Luís, Brazil
1541	Santiago, Chile	1616	Belém, Brazil
	Antigua, Guatemala	1675	Medellín, Colombia
1542	Guadalajara, Mexico	1711	Vila Rica, Brazil
	Mérida, Mexico	1726	Montevideo, Uruguay
	Comayagua, Honduras	1736	San José, Costa Rica
1544	La Serena, Chile	1742	Pôrto Alegre, Brazil
1546	Zacatecas, Mexico	1749	Port-au-Prince, Haiti
1547	Potosí, Bolivia		

Sources: Portes and Walton 1976; Hardoy 1975; Godfrey 1991; Gwynne 1986.

In contrast, the Spanish settlement of Latin America was characterized by the founding of hundreds of small urban centers within the first few decades following the Conquest (Figure 12.2). While the number of cities established was impressive, the territory over which they were scattered was so vast and rugged that most functioned in virtual isolation, particularly during the rainy season when weather conditions made transportation and communication between the cities almost impossible. Under these conditions, most of the larger cities evolved essentially as city-states, exploiting their rural possessions but seldom achieving the regional political and economic integration that served as the basis for Anglo American development.

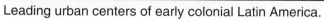

FIGURE 12.2

Leading urban centers of early colonial Latin America.

Colonial Urban Primacy

The city-state settlement form that emerged in colonial Latin America also marked the beginnings of a long-term pattern of urban primacy. When used to measure population levels, primacy refers to conditions under which one city grows more rapidly and eventually becomes much larger than others nearby. The existence of a **primate city** has been viewed by some development theorists, including growth pole advocates (chapter 11), as a factor that promotes economic development through a sort of spatial trickle-down effect. Other analysts have suggested that the condition exerts a negative influence on the development process by concentrating limited resources and political power in the hands of a relatively small elite, which neglects the development of the entire region or nation in favor of that of the primate city.

Prior to 1750, censuses of total urban populations were rarely undertaken for Latin American cities (Morse 1971), but accounting was made of the number of *vecinos*, or Spanish citizens. Although we know that some cities were more attractive to the Europeans for health and other reasons, it is nevertheless likely that the relative concentration of *vecino* population correlated positively with the total urban populations of the period. Using this criterion, it is apparent that Morris (1981, 87) was correct in noting that a high level of urban primacy emerged in the early colonial era. A number of regional capitals, including, for example, Mexico City, Quito, Panama City, and Santiago, Chile, served as home to half or more of the Spanish populations of their respective viceroyalties during the late sixteenth and early seventeenth centuries (Table 12.2).

TABLE 12.2

Urban Primacy of *Vecino* Populations in the Early Colonial Era

CITY	PERCENTAGE OF VICEROYALTY *VECINO* POPULATION	
	1580	1630
Mexico City	48.1	58.0
Lima	17.1	23.4
Bogotá	28.2	29.2
Quito	45.7	56.7
Panama City	69.3	46.7
Santiago, Chile	22.5	53.0
Santo Domingo	29.2	11.9
Guadalajara	13.7	22.2
Guatemala City	21.8	34.0

Source: Hardoy and Aranovich 1966, 185–193.

Functional Classification of Cities

By the late 1500s, Latin America's newly established cities were beginning to assume their individual characters and identities. While every city was unique and distinctive in some fashion, it is possible to place each of them in one or more functional categories (Table 12.3).

One of the most common categories was the **agricultural city,** whose primary purpose was the production and supplying of foodstuffs to the greater region that it served. Because most Spaniards refused to work the soil themselves, colonial agricultural cities almost invariably were established in the midst of preexisting indigenous populations, who could be forced to labor in the fields. Some of these cities, such as Cuzco and Bogotá, were erected literally on the ruins of earlier Indian settlements. Others, including Arequipa, Querétaro, Santiago, Chile, and the Nicaraguan cities of León and Granada, were founded within densely settled native rural zones. While the economies of most of the leading agricultural cities were focused on the growing of food crops and the production of meats, hides, and fibers, some such as Tucumán and Guadalajara devoted considerable land to breeding asses and mules for use in the mines of the distant highlands.

The second functional category of the early colonial settlements was the **mining city.** Because the mining cities' primary purpose was the provision of laborers and supplies to the mines themselves, these cities typically were situated at high elevations where the growing of foodstuffs was limited by low temperatures and aridity. The mountainous terrain on which the mining cities were built resulted in most of them being very densely settled, with narrow, winding streets and seemingly endless networks of steeply sloped footpaths and stairways connecting the upper and lower portions of the settlements (Figure 12.3). Although some of the mining centers, such as Potosí, Zacatecas, and Guanajuato, ranked for short periods of time among the largest and wealthiest cities of the New World, their cyclic boom-and-bust character and physical isolation limited their long-term growth and development.

TABLE 12.3

Functional Categories of Early Colonial Cities

CATEGORY	EXAMPLES
Agricultural cities	Tucumán, Argentina; Santiago, Chile
Mining cities	Potosí, Bolivia; Zacatecas, Mexico
Industrial cities	Quito, Ecuador; Puebla, Mexico
Commercial cities	Havana, Cuba; Mendoza, Argentina
Administrative cities	Mexico City, Mexico; Lima, Peru

Industrial cities, although often not as prominent as the agricultural and mining centers, also played a valuable role in the colonial economy. As was noted in chapter 11, early manufacturing in colonial Latin America centered on *obraje* production of inexpensive cotton and woolen textile products in areas where the native populations had sustained a strong industrial tradition prior to the Conquest. These included the Puebla, Tlaxcala, and Oaxaca basins of central and southern Mexico, and Otavalo, Quito, Cuenca, and Cajamarca in the Ecuadorian and Peruvian highlands. The securing of royal monopolies or subsidies earmarked for the manufacture of certain industrial products was another means through which many colonial cities came to specialize in the production of specific industrial goods. Beneficiaries of royal monopolies or subsidies included Puebla, which was favored with concessions to produce ceramic and silk products; Mexico City and Querétaro, which were given tobacco monopolies; and Havana and Guayaquil, which were authorized to build ships.

The fourth type of colonial city consisted of the **commercial settlements,** whose existence and economic vitality were linked directly to their location at key transportation points. Many of the leading commercial cities, including Havana, Santo Domingo, and San Juan in the Caribbean, and Veracruz, Cartagena,

Panama City, Guayaquil, and Callao, were seaports whose populations tended to fluctuate wildly with the arrival and departure of the Spanish *flota* (Figure 12.4). Other commercial cities owed their prosperity to their strategic positions along key overland transportation routes. Examples of the latter include the Andean foothill settlement of Mendoza, which controlled trade between Argentina and Chile, and Monterrey, which came to form an economic link between the cities of the Mexican Mesa Central and those of the United States.

The fifth and final functional category of Latin American colonial settlements was the **administrative centers,** many of which became the primate cities of their respective regions. The two largest and most influential administrative centers of the early colonial era were Mexico City and Lima, whose respective *vecino* populations stood at 15,000 and 9,500 in the year 1630 (Table 12.4). Secondary administrative cities, which also doubled as agricultural, industrial, or commercial settlements, included Buenos Aires, Asunción, Santiago, Quito, Bogotá, Caracas, Guatemala City, and Havana. The outward unification of church and state in colonial Latin America resulted in the administrative capitals also becoming ecclesiastical and educational centers. They thus attracted the most able and ambitious persons of both the secular and religious spheres and soon came to monopolize the resources of the regions they governed.

FIGURE 12.3

The narrow winding streets and dense, crowded settlement of Zacatecas, Mexico, are typical of Latin America's mining cities.

FIGURE 12.4

Panama City, situated along an inlet of the Pacific Ocean and at the principal transoceanic overland transportation route of colonial Latin America, has functioned since its founding as a commercial settlement.

TABLE 12.4

Urban *Vecino* Populations in 1580 and 1630

	VECINO POPULATION	
CITY	**1580**	**1630**
Mexico City	3,000	15,000
Lima	2,000	9,500
Potosí	400	4,000
Cuzco	NA	3,500
Quito	400	3,000
Bogotá	600	2,000
Havana	60	1,200
La Plata	100	1,100
Oruro	—	1,000
Guatemala City	500	1,000
Puebla	NA	1,000
El Callao	—	700
Santo Domingo	500	600
Guadalajara	150	600
Panama City	400	500
Tlaxcala	50	500
Santiago, Chile	375	500
Córdoba, Argentina	—	500

Sources: Hardoy and Aranovich 1966; 185–193; Sánchez-Albornoz 1974, 90–122.

Once established, the principal functions of most Latin American and Caribbean cities changed relatively little through the remainder of the colonial and the subsequent republican periods. As individual cities filled specific roles within the broader regions and nations to which they belonged, their inhabitants eventually came to acquire certain common values and culture traits that gave each city a collective personality. In the state of Veracruz, Mexico, for instance, the port of Veracruz city has historically acted as the regional commercial center, and its racially mixed *Jarocho* population came to be renowned throughout the country for its liberal, easygoing lifestyle. Xalapa, the state capital located 120 kilometers (74 miles) to the northwest in the foothills of the eastern escarpment, became the conservative administrative center of the region. Córdoba, also founded in the foothills some 96 kilometers (60 miles) south of Xalapa, emerged as the home of the ultraconservative agricultural aristocracy; while Orizaba, situated just 24 kilometers (15 miles) inland from Córdoba, grew into a liberal, blue-collar industrial city specializing in the manufacturing of textile products and beer. Although in the last half of the twentieth century modern transportation and mass communications began to dilute the traditional roles of some Hispanic cities, most continue to function largely as they have since their foundings hundreds of years ago (Bataillon 1982; Bromley 1979; Arreola and Curtis 1993). In contrast to the cultural sameness of the Anglo Ameri-

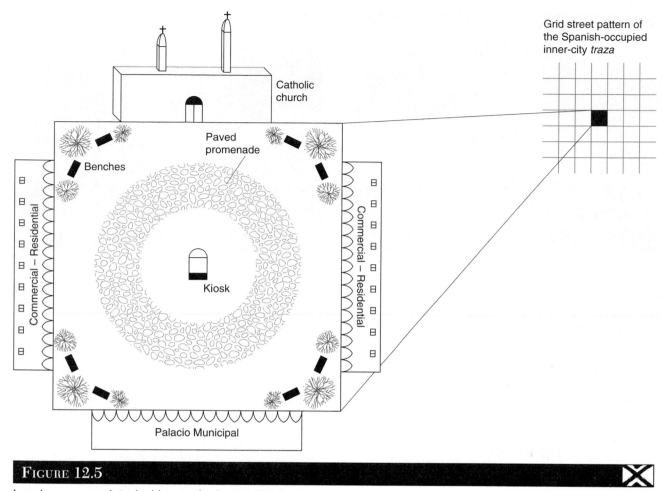

Figure 12.5

Landuses associated with a typical colonial plaza.

can urban landscape, one of the most rewarding and enjoyable aspects of geographic research in Latin America and the Caribbean is to come to understand, and experience, the distinctive group personality and culture of each individual city.

Morphology of the Colonial City

The first settlements established by the Spaniards on the island of Hispaniola and elsewhere in the New World were tiny, medieval-like fortress communities with narrow, twisting streets and paths. They were similar in design to many of the Castilian towns built by the Christians during the early centuries of the *Reconquista* (Butterworth and Chance 1981). In the decades preceding and following the founding of Santo Domingo in 1496, however, the growing influence of Italian Renaissance humanism led to a revival within the Spanish bureaucracy of the classical Greek and Roman grid town plan first described by Vitruvius (Stanislawski 1946 and 1947; Smith 1955). Santa Fé, founded by the Spanish Crown near Granada in 1491,

functioned as the base of operations for the final successful Christian siege of the Moors in southern Spain. Gade (1992, 472) has suggested that the grid street pattern incorporated into the design of Santa Fé may have served as an Iberian model for New World town planners. Whether or not this indeed was the case, it appears that the first New World city to be built entirely after the ancient Greek and Roman design was Mexico City, which was laid out by the Spanish surveyor Alonso García Bravo immediately following the destruction of Tenochtitlán. Mexico City's grid plan, which was first codified in 1523 and later incorporated into the Spanish New Town Ordinances of 1573, served as the inspiration for almost every settlement, great or small, built in the colonial period.

After stating, wishfully, that the site for each new settlement "must be one which is vacant and can be occupied without doing harm to the Indians and natives or with their free consent" (Nutall 1922, 250), the New Town Ordinances specified that the heart of the community should be a central rectangle or square called the **plaza mayor,** or *plaza real* (Figure 12.5). Designed

FIGURE 12.6

The colonial era building facing the plaza of Antigua, Guatemala, houses businesses and government offices on the ground floor and restaurants and residences of the elite on the second story.

FIGURE 12.7

Latin American urban architectural patterns, from the colonial period to the present, have commonly included buildings being situated close together, thick walls, high ceilings, and an absence of external lawn or green space. This street scene is of Guanajuato, Mexico.

initially to support horse racing and other equestrian activities sponsored by the leading *caballeros,* many plazas, or *zócalos* as they came to be called in Mexico, soon acquired a parklike atmosphere, with a central kiosk or bandstand set in the midst of paved walkways, benches, and ornamental trees and shrubs. Facing the plaza were the most important buildings of the community, including the government building, called the *ayuntamiento* or *palacio municipal;* the Catholic cathedral or principal church; and two- or three-story buildings housing cafes, restaurants, and government and business offices on the ground floor, with the residences of the urban elite above (Figure 12.6). Radiating out from the *plaza mayor* were streets laid out in a **grid pattern.** The blocks of land delimited thereby were either distributed to religious orders or subdivided into individual family plots called *solares,* which were given to the *vecinos.* Initially, vacant lots were common and most cities had a somewhat rural appearance, with *huertas,* or gardens, planted behind homes whose front walls typically bordered a narrow sidewalk set against the street. The Spanish preoccupation with rank and social standing was expressed in the relative location of the residences, with those situated closest to the plaza being the most prestigious and those farthest away the least. As the population of the city grew, the vacant plots, called *baldíos,* were allocated, and the population density of the city increased. Demand for space frequently reached the point where the side walls of one structure abutted those of neighboring buildings.

The architectural patterns of colonial homes varied somewhat by region in response to climatic variables, but most were modeled after those of southern Spain. Their principal features included thick, often stuccoed, walls and generous use of interior tile, high ceilings, and flat roofs (Figure 12.7). The living quarters of the wealthy generally were built around an inner courtyard, or patio, which was adorned with plants and other ornamentation and which afforded privacy and relaxation from the outside world (Figure 12.8). Tiny living quarters for the servants were set at the rear of the property next to the corrals where animals were penned. The residences of the poor, on the other hand, were usually small one- to three-room dwellings with little interior lighting, packed dirt floors, and minimal furnishings. Exterior windows were few for rich and poor dwellings alike, as the night air was believed to carry malaria, fevers, and other diseases and evil spirits. Those windows that did exist were generally grated to protect against unauthorized entry. Bathrooms and potable water were entirely absent, and sanitation conditions were very poor.

With the passage of time, some streets evolved into commercial strips, with the front portions of the homes converted into tiny stores and shops and the interior portions of the dwellings reserved for residential use. Often, enterprising citizens would endeavor to extend the fronts of their properties further out into the streets to gain additional commercial space, only to have the additions demolished on orders of the *cabildo,* or town council (Gakenheimer 1966). Other streets evolved into special showcase boulevards with elaborate landscaping and prosperous businesses.

FIGURE 12.8

The tranquil interior courtyard of an upper-middle-class home in Managua, Nicaragua.

The outer sections of town were occupied primarily by "civilized" Indians, who migrated to the city in large, unknown numbers in search of employment as household servants, manual laborers, and craftspeople. These outlying *barrios*, as they were called, also contained mills, tanneries, brick kilns, slaughterhouses, and other city-owned industries deemed by the Spaniards to be inappropriate inner city landuses (Chance 1976). The symmetrical grid street pattern of the *traza*, or Spanish-occupied inner city, tended to break down in the peripheral Indian *barrios*, although scattered, unpretentious secondary plazas were created to service the needs of the lower classes.

Ejido, or common grazing land, was usually set aside on the outskirts of the *barrios*, with individual land grants being available beyond the *ejido* (Figure 12.9). Those settlers who had served as mounted soldiers were entitled to a large estate known as a *caballería*. Spanish foot soldiers, or peons, were given smaller properties called *peonías*. As was noted earlier, the Spaniards could not, in theory, take any lands occupied by Indians without their "permission." This regulation was not respected in practice, of course, and any lands appropriated by the Spaniards because they were "vacant" had usually become so only because the previous Indian occupants had all been killed or had fled. The *caballerías* and *peonías* were to evolve over time into rural *haciendas* or *rancherías*, whose principal purpose was the bestowal of social standing on their absentee urban owners.

Settlement in Portuguese America

In contrast to the urban-based conquest of Spanish America, the Portuguese occupation of Brazil centered initially on rural estates whose primary function was the production of raw materials for export overseas. The principal rural settlement types were the sugar plantations, or *engenhos,* which came to dominate the fertile northeastern coastal plains, and the *fazendas,* or cattle ranches, which reigned in most of the semiarid interior backlands. Scattered small-scale mining camps were also established wherever tiny deposits of precious ores and gems were found, but none of these rivaled the Spanish mineral discoveries in Peru and Mexico until the emergence in the eighteenth century of Vila Rica and other gold and diamond centers of Minas Gerais.

The largest settlements in the early colonial era consisted of a string of small ports that stretched from São Vicente on the south to Belém on the north. Many of these communities served as administrative centers for the fifteen captaincies whose territories extended inland from the Atlantic coast. The three dominant ports of the period were Bahia (later Salvador), which functioned as the capital of the largely disunified Brazilian colony from its founding in 1549 until 1763; Olinda-Recife, which emerged as the leading city of the Northeast; and Rio de Janeiro, which controlled trade along the southern coast and later progressed to where it replaced Salvador as the capital of Brazil. While these and other ports eventually grew to rank among the largest cities in Latin America, they failed to attract the elite, rural upper class, so that Brazilian political power and social prestige were concentrated on the rural estates well into the eighteenth century. The first Portuguese viceroy to serve in Brazil, for example, wrote that Rio de Janeiro had "only friars, clerics, soldiers and beggars," and that "the noble men live in the country and are the ones which serve me" (Morse 1962, 333).

Once the Brazilian cities did begin to grow, the absence of a Portuguese master town plan resulted in most of the settlements being characterized by a haphazard mix of narrow, winding streets and lanes whose collective appearance was described by Smith (1955, 9) as one of "picturesque confusion." Despite the disarray, it is possible to identify four attributes shared by most of the early colonial Brazilian cities. The first was a location on a hilly, elevated site overlooking the Atlantic. The second was a two-tiered design that divided the settlement into a lower port and harbor zone, called the *cidade baixa,* or lower city, and an upper sector called the *cidade alta,* or upper city. The former typically contained the custom house, main market, and much lower-class housing, while the more heavily fortified upper city supported the principal administrative and religious edifices, monuments, and the residences of the elite. An excellent example of the upper-lower city pattern was Olinda-Recife (Figure 12.10).

Mid-Sixteenth Century

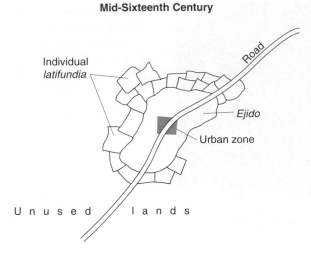

Mid-Seventeenth Century

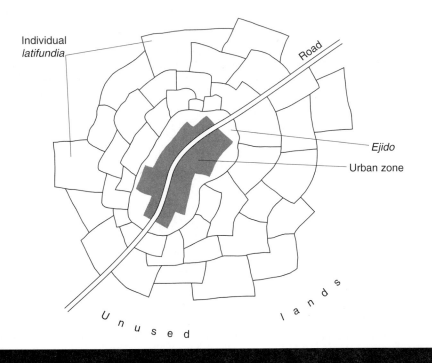

FIGURE 12.9

Generalized urban landuse of colonial Spanish settlements.
Source: Céspedes, 1976: 78.

A third common feature that appeared with the expansion of the ports was a tendency for much of the growth to occur close to the coast. This led to the emergence of a strip or linear settlement pattern, whose main road came to be called *o largo*. Local public squares were often established along the *largo*, giving it a somewhat clustered or polynuclear appearance. The fourth shared trait was the creation of a landscaped public walk or mall. These grounds, known as the *rossio*, were the Portuguese counterpart to the Spanish *alameda*.

FIGURE 12.10

Recife, or Pernambuco, as it was originally named, served in the early colonial era as the *cidade baixa,* or lower city, to the neighboring community of Olinda, which performed the functions of a *cidade alta,* or upper city. Today, Olinda (foreground) is a small community that has retained much of its colonial heritage while Recife (background) has grown into a major commercial center.

The only significant exception to the pattern of the dominant Brazilian colonial cities being ports was the community of São Paulo, founded by Jesuit colonists in 1554. From the very beginning, *Paulistas,* as the residents came to be called, turned their attention inland. By 1560, bands of able-bodied Paulistas were organizing themselves into quasi-pioneer groups called **ban-deiras** that penetrated ever deeper into the interior of the colony in search of slaves or precious metals. Together with the women and children who followed, the *bandeiras* became, as it were, mobile cities, stopping long enough along the way to clear some land and plant and harvest a crop before moving on. These settlements often evolved into permanent municipalities whose occupants adopted New World rather than European ways, eating native foods, sleeping in hammocks, and even speaking Tupi-Guaraní on the Brazilian frontier. The *bandeiras* thus emerged as a vigorous, indigenous settlement type, which was established widely in São Paulo, Minas Gerais, and Bahia and eventually as far away as Goías, Mato Grosso, and the Amazon Basin. As a result, São Paulo city came to control the interior of the colony, while the eastern, coastal regions fed into the scattered, Atlantic-oriented ports. As recently as the late nineteenth century, Brazil was a preeminently rural land with just three cities—Rio de Janeiro, Salvador, and Recife—having populations greater than 100,000.

Rural Settlements

While the European colonists were laying the foundations of an enduring urban value system during the early colonial era, they were also eager to promote a clustered, or nucleated, rural settlement pattern among the Indians they hoped to exploit. Where the natives had previously followed a nomadic or semi-nomadic lifestyle, the Jesuits and other religious orders often attempted to gather them into newly formed mission communities, called *aldeiamentos* in Brazil and *congregaciones* or *reducciones* in the Spanish-speaking territories. Some of the most active areas of mission settlement formation included eastern Paraguay and the neighboring regions of south-central Brazil, the floodplains of the Amazon River and its principal tributaries, and northern Mexico or New Spain, portions of which are now within the southwestern United States.

Most often, however, the Europeans chose to establish their cities and towns in the midst of relatively dense preexisting Indian populations and to allocate, through the *encomienda* and *repartimiento* systems, entire villages to the control of individual *conquistadores.* Because the Spanish and Portuguese masters rarely had any interest in learning the native languages and abhorred the thought of having to live among their subjects, these pre-Conquest Indian communities came to be governed through Spanish intermediaries, called *corregidores.* The *corregidores,* in turn, had little desire to become involved in the day-to-day affairs of the Indians, preferring instead to allow the *caciques, curacas,* and other traditional native leaders to continue to govern at the local level, so long as they fulfilled their labor and tribute quotas.

The autonomy of these indigenous settlements was further strengthened when, as part of the reform movement leading to the passage of the New Laws of 1542, the Spanish crown decided to recognize the right of the Indian communities to collectively hold title to their ancestral lands in perpetuity. These lands came to be called *tierras de resguardo,* or reservation lands. They were protected so strongly by Spanish law that regulations were enacted prohibiting Europeans not only from residing within the Indian settlements but even from spending more than two hours in one while "passing through" on a journey. In this manner, many of the pre-Conquest communities were able to preserve for centuries a mostly Indian way of life.

Having noted this, it is important also to recognize that most of the Indian settlements gradually adopted selected Hispanic traits (Gade 1992; Gade and Escobar 1982; Despres 1991). These traits often included the *alcalde-regidor,* or mayor-councilman, Spanish form of governance and the implementation of the grid street plan centered on a central plaza, complete with a small

FIGURE 12.11

The modest, attractive plaza or *zócalo* of San Buenaventura Nealtican, Mexico, is typical of those found in countless Latin American peasant villages whose founding predated the Spanish Conquest. Visible are the kiosk, benches, and shrubbery. The eighteenth century Catholic church, which continues to serve the community, faces the left side of the plaza.

kiosk, benches, and walkways (Figure 12.11). The Catholic church and town hall, however humble, also came to be situated around the plaza. One of the best indicators of the past or present prominence of an Indian community is the size, as gauged by the number of arches, of the local *palacio municipal*. Other Hispanic traits adopted by the Indians included the use of European crops, animals, and production techniques. European Catholicism, as noted in chapter 9, was also assimilated in folk form, as were selected aspects of Old World dress and music.

Even place names became hybridized (Gibson 1966). In the Puebla Valley of Mexico, for instance, rural communities that predate the Conquest commonly bear both Spanish and Indian names. Examples include San Buenaventura Nealtican, San Gabriel Ometoxtla, and Santa María Acuexcómac. Settlements founded by the Spaniards after the Conquest, such as San Nicolás de los Ranchos and the regional urban center of Puebla de los Angeles (later Puebla de Zaragoza), carry only Spanish names.

Ultimately, the process of cultural mixing resulted in most of the rural, pre-Colombian Indian settlements becoming peasant villages whose occupants formed "part-cultures," that is, mixed Indian and Hispanic societies. Living with one foot in the present and the other in the distant past, these villages, which survive by the thousands in the rural regions, provide excellent settings for studying the continuity and change that characterize Latin America today.

Late Colonial Population Levels

Having reviewed the functions and morphology of the early Spanish and Portuguese cities, as well as the transformation of most Indian communities to peasant villages, let us conclude our study of colonial settlement patterns by noting that on the eve of the nineteenth-century revolutionary wars, overall population levels were remarkably low. Morse's (1971) calculation of a total Latin American and Caribbean population of 22 million in the early 1800s is considerably lower than all but the most conservative pre-Colombian population estimates (Table 7.1). In other words, the population of Latin America and the Caribbean probably failed, in three centuries of colonial rule, to recover fully from the catastrophic Indian losses brought on by disease, malnourishment, and harsh treatment at the hands of the European masters. National population levels reflected this, with most of the soon-to-be independent nations having far fewer inhabitants than their largest cities do today (Table 12.5). Likewise, urban populations were in most cases exceedingly small, with the typical "major city" of the day supporting no more than 20,000 to 30,000 persons, including Indians, blacks, and *mestizos* (Table 12.6). The Hispanic urban bias notwithstanding, then, the vast majority of the Latin American and Caribbean peoples continued to live, as had their ancestors for millennia before, as rural agriculturists residing in tiny peasant villages or on isolated farmsteads, *haciendas*, or plantations (Figure 12.12).

THE REPUBLICAN PERIOD

While conditions varied somewhat from country to country, the political independence that ushered in the nineteenth century did little to alter the long-standing patterns of high birth rates and almost equally high death rates that had produced a modest but steady population growth rate in Latin America from the mid-1600s onward. By 1850, the total population of the region had expanded to 30 million (Hardoy 1975), an increase of over one-third from four to five decades earlier (Table 12.5). Much of this growth was concentrated in the capital cities, which were the major beneficiaries of government spending in the newly independent nations (Table 12.6). Typical of this pattern was Mexico City, which in the Díaz era received approximately 80 percent of all national outlays on streets, electricity, and water and sewer systems and most of the money spent on libraries and schools (Johns 1997, 15).

Population growth in Latin America and the Caribbean accelerated somewhat in the late 1800s, owing both to the economic growth spurred by the liberal trade policies followed in most nations and to

TABLE 12.5 ✖

National Populations in the Late Colonial and Mid-Republican Eras

COLONY/COUNTRY	TOTAL POPULATION (MILLIONS)	
	Late Colonial	Mid-Republican
Mexico	5.84	7.66
Brazil	2.39	7.21
Colombia	.83	2.24
Peru	1.08	1.89
Venezuela	.78	1.49
Bolivia	.55	1.37
Chile	.37	1.29
Cuba	.27	1.19
Argentina	.31	1.10
Haiti	.46	.94
Guatemala	.51	.85
Ecuador	.55	.82
Paraguay	.20	.50
El Salvador	.27	.39
Honduras	.13	.35
Nicaragua	.18	.30
Dominican Republic	.13	.20
Uruguay	.06	.13
Costa Rica	.06	.13

Sources: Burns 1980; Morse 1971; Sánchez-Albornoz 1974.

TABLE 12.6 ✖

Urban Populations in the Late Colonial and Mid-Republican Eras

CITY	POPULATION (THOUSANDS)	
	Late Colonial	Mid-Republican
Mexico City	137	210
Havana	51	197
Rio de Janeiro	65	186
Santiago, Chile	30	115
Salvador (Bahia)	70	108
Lima-Callao	64	107
Buenos Aires	38	91
Recife	25	86
Guadalajara	20	75
Puebla	68	74
Valparaíso	5	70
Guanajuato	71	69
Caracas	42	44
La Paz	22	44
Belém	11	41
Bogotá	28	40
Guatemala City	25	40
Santiago de Cuba	15	37
Fortaleza	10	35
Camagüey	16	31
Pôrto Alegre	6	30
São Paulo	24	26
Maracaibo	24	25
Mérida, Mexico	10	25
San Luís Potosí	10	25
Montevideo	7	25
Cuzco	32	24
Guayaquil	8	23
Arequipa	24	22
Ayacucho	26	20

Sources: Chandler 1987; Morse 1971; Morse 1975; Burns 1980; Moore 1978; Walton 1978; Butterworth and Chance 1981.

FIGURE 12.12

Leading urban centers of late colonial Latin America.

Case Study 12.1: Confederate Settlements in Latin America and the Caribbean

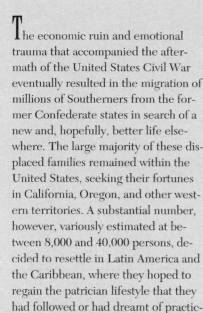

The economic ruin and emotional trauma that accompanied the aftermath of the United States Civil War eventually resulted in the migration of millions of Southerners from the former Confederate states in search of a new and, hopefully, better life elsewhere. The large majority of these displaced families remained within the United States, seeking their fortunes in California, Oregon, and other western territories. A substantial number, however, variously estimated at between 8,000 and 40,000 persons, decided to resettle in Latin America and the Caribbean, where they hoped to regain the patrician lifestyle that they had followed or had dreamt of practicing in the antebellum South.

The initial focus of post–Civil War Latin American Confederate settlement was Mexico, whose embattled monarch, Maximilian, encouraged several thousand Southerners to colonize portions of Veracruz and a number of other states in the hope that their eco-nomic and technological contributions would strengthen him in his campaigns against the forces of Juárez. Maximilian's death at the hands of a firing squad in 1867, together with widespread anti-American sentiment among the Mexican populace and questionable confederate leadership, led to a panicked abandonment of the country by the unfortunate Southerners. Their illusions shattered, most returned to the United States; but a few vowed to try again in Cuba, Venezuela, Guyana, or Brazil.

Brazil proved to be the principal Latin American Confederate destination, owing to its political stability under Emperor Dom Pedro II and to the fact that slavery, while in decline, was not outlawed until 1888. *Confederado* settlement was concentrated primarily in the state of São Paulo, with a number of communities, such as Americana, turning first to cotton farming and later to textile production. Other towns founded by the South-erners included Rio Doce along the middle Atlantic coast and the Amazonian river port of Santarém.

Most of the Confederate immigrants were Protestants. English, with a Southern accent, was spoken almost exclusively in their homes, and American symbols, such as the Fourth of July holiday and the Confederate flag, were honored on nostalgic occasions. The racial and religious tolerance of Brazil, together with the relatively strong educational backgrounds of the Confederates, soon resulted in the assimilation of most into the upper social class, where many took Brazilian spouses. Some of their children achieved success as farmers and industrialists, while others turned to dentistry, medicine, and the arts. Their descendants today consider themselves to be Brazilians in the fullest sense of the term, but they are proud of their mixed ancestry, which expresses itself, among other ways, in names such as Juan Carlos Calhoun.

Sources: Mahoney and Mahoney 1998; Dawsey and Dawsey 1995; Iwańska 1993; Harter 1985; Weaver 1952; Jefferson 1928.

the arrival of the railroad, telegraph, steam-powered ships, and other technological advances that opened many regions to foreign development and colonization. Foreign immigration, almost nonexistent during the long centuries of colonial rule, brought not only accelerated population growth but also much needed infusions of money and professional and industrial expertise to the Southern Cone countries of Argentina, Brazil, Chile, and Uruguay. The cities of São Paulo and Buenos Aires grew particularly rapidly from Italian and German immigration. Venezuela, Mexico, Peru, and Cuba also benefited, to lesser degrees. Most of the immigrants were from Europe, but Asians, religious migrants such as Mennonites and Mormons, and even disillusioned Southerners from the failed Confederacy joined in the action (Case Study 12.1). By 1900, the total number of inhabitants of Latin America and the Caribbean stood at 60 million. For the first time in 400 years the region had finally regained its likely pre-Conquest population levels (Figure 12.13).

THE TWENTIETH CENTURY

Although foreign immigration played an important role in the population growth of the Southern Cone countries in the late 1800s and early 1900s, its subsequent impact on the population of Latin America and the Caribbean has been relatively small when compared to the changes caused by fluctuations in the natural birth and death rates of the region. For any area or nation, the population growth rate is the difference between its birth and death rates, adjusted for net immigration or emigration. We will now examine the changes in these key demographic indicators over the past century, as portrayed in Table 12.7.

As the twentieth century dawned, birth rates in Latin America and the Caribbean averaged 43.8 for every 1,000 living persons, or 4.4 percent per year. They remained at 4.0 percent or higher per annum through the 1920s, and then dropped to 3.5 percent in the depression-struck 1930s before climbing back to the 3.7 percent range by the 1940s. Overall, then, birth rates

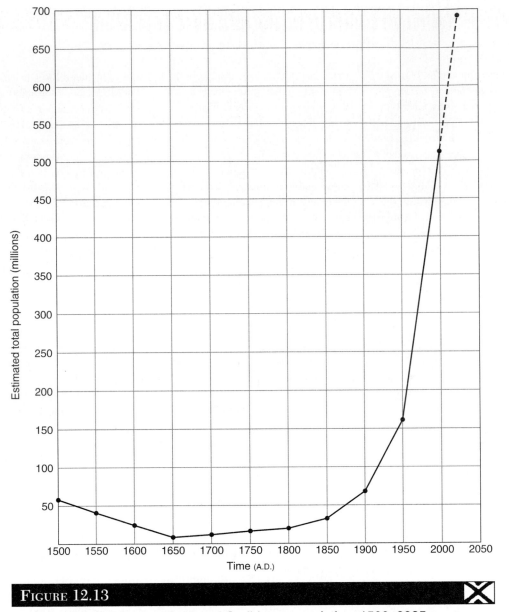

FIGURE 12.13

Total estimated Latin American and Caribbean population: 1500–2025.

declined little in the first half of the twentieth century. Setting aside the unquestionable influence of the Catholic Church in discouraging couples from limiting the size of their families, the sustained high birth rate of the early to mid–twentieth century can be further accounted for in the fact that the region remained overwhelmingly rural and in constant need of more laborers.

Meanwhile, the regional death rate was trending slowly but surely downward, dropping from 2.6 percent per year at the turn of the century to 1.5 percent annually in the 1940s. This consistent decline is attributable primarily to the steady diffusion of improved medical care and health practices, which brought about a drop in infant mortality rates. The overall result was that the annual population growth rate of the region grew from 1.8 percent in 1900 to 2.2 percent in the 1940s. By 1950, total regional population had grown to 159 million and the annual birth rate had returned to 3.9 percent, a level that was sustained through the 1960s as well. Continuing declines in the death rate resulted in annual population increases greater than 2.6 percent from the 1950s all the way through the 1970s. The 1980s and 1990s brought increased use of contraception among Latin American women and resultant declines in the birth rate, which lowered the annual growth rate to 1.7 percent. Fifty years of rapid growth, however, had left the total regional population at 519 million in 2000. Current estimates project the region's population will reach 697 million in the year 2025.

TABLE 12.7

Latin American and Caribbean Crude Birth, Death, and Population Growth Rates by Decade: 1900–2000

DECADE	CRUDE BIRTH RATE PER 1,000 POPULATION	CRUDE DEATH RATE PER 1,000 POPULATION	POPULATION GROWTH RATE PER 1,000 POPULATION
1900–1909	43.8	25.9	17.9
1910–1919	43.3	26.3	17.0
1920–1929	40.0	22.5	17.5
1930–1939	35.2	17.7	17.5
1940–1949	37.0	15.3	21.7
1950–1959	39.1	11.7	27.4
1960–1969	39.0	11.0	28.0
1970–1979	35.0	9.0	26.0
1980–1989	29.0	7.0	22.0
1990–1995	25.0	7.0	17.0
1995–2000	23.3	6.5	16.8

Sources: *Demographic Yearbook 1999* 2001; Sánchez-Albornoz 1974; Ruddle and Obermann 1972; *Economic and Social Progress in Latin America, 1980–1981 Report* and *1993 Report* 1981 and 1993; *World Development Report 1992* 1992.

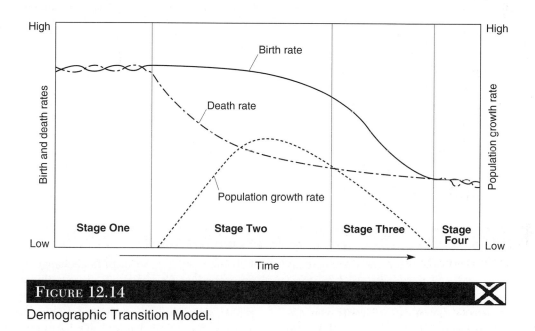

FIGURE 12.14

Demographic Transition Model.

The Demographic Transition Model

Whether or not the population will reach that level will depend, of course, on future demographic trends that are extremely difficult to forecast. Although it is clear that Latin America and the Caribbean experienced historically unprecedented population increases in the twentieth century (Figure 12.13), the long-term patterns also appear consistent with those predicted in the **Demographic Transition Model** (Figure 12.14). If this proves to be the case, we can anticipate one or two more generations of relatively high growth followed by a general long-term stabilizing of population levels. Let us

now examine the four stages of the model with an eye to their application to Latin America and the Caribbean.

The model is based on the collective experiences of the European, Asian, and Anglo American nations that have evolved, over the past several centuries, from predominantly rural to primarily urban societies. In Stage One, the traditional agrarian phase, high birth rates are balanced by equally high death rates, resulting in minimal population growth. While the history of each country and microregion is different, in most of Latin America and the Caribbean Stage One lasted through the early Republican Period of the mid–nineteenth century (Figure 12.13).

Stage Two of the model consists of the period when urbanization and the associated process of industrialization begin to increase rapidly. Because industrial technologies invariably are applied to improve medical care, Stage Two is characterized by a rapidly declining death rate. Yet, because the rural folk who have recently moved to the city in search of industrial employment retain their rural values, which include a desire for large families, Stage Two is also characterized by continuing high birth rates. The combination of high birth rates and greatly lowered death rates results in extremely rapid population growth during Stage Two. This was the condition of most Latin American and Caribbean nations from the late 1800s until the mid-1900s.

Stage Three of the model occurs only after the children and grandchildren of those persons who migrated from the rural to the urban zones have reached adulthood and are raising their own children. Because these new generations have been raised from birth in the city, their collective values are urban based, with the result that they choose to have smaller families than their parents. Stage Three is thus characterized by steadily falling birth rates and, consequently, greatly lowered overall population growth rates. On the basis of the trends portrayed in Table 12.7, it would appear that Latin America and the Caribbean as a whole entered Stage Three in the 1970s and 1980s. If this is correct, we can anticipate a continued slowing of the birth and growth rates in the decades ahead, although the total absolute population will likely continue to increase rapidly simply because the base size of the population has become so large. In other words, the 1.7 percent increase in the 2000 population of 519 million resulted in far more absolute growth than did the 2.7 percent increase in the 1950 population of 159 million. As Stage Three winds down, however, the birth rate becomes so low as to generate little overall population growth. This is not likely to occur for Latin America and the Caribbean as a whole until the middle of the twenty-first century, after the passing of the childbearing years of the children now being born in the cities to parents who migrated from the rural countryside. If and when the birth rate reaches this level, the region will then have entered into Stage Four, in which there is little or no long-term population growth owing to the fact that the birth and death rates are again in approximate balance, just as they were in Stage One. The principal difference, of course, is that the balance in Stage Four is maintained through much lower birth and death rates than those characteristic of Stage One.

At present, no one can say whether or not the Latin American and Caribbean nations will continue, as a group, to follow the trends predicted in the Demographic Transition Model. They do appear to have followed the trends up to this time. Indeed, by examining specific national data, one can place each individual nation, as well as larger Latin American and Caribbean subregions, at a certain point along the demographic transition continuum (Table 12.8). The Middle American nations, for example, average 2.4 percent annual growth, meaning that they are in the late second and early third stages of the Demographic Transition Model. The Andean and Guiana countries, with regional growth rates averaging 2.0 and 1.8 percent, respectively, are in the mid-third stage, while the Southern Cone (1.4 percent) and Caribbean Island nations (1.1 percent) fall primarily within the late third stage (Figure 12.15).

Once we recognize where a given country or subregion is presently situated along the timeline of the model, we should be able to predict future demographic trends with a reasonable degree of confidence. Those countries, such as Guatemala, Nicaragua, Honduras, and Paraguay with annual rates of increase of 2.7 percent or higher are still in the middle of Stage Two, meaning that they are likely to experience relatively rapid population growth for several generations to come. On the other hand, Cuba, Uruguay, Argentina, Chile, Brazil, and most of the Caribbean island states are clearly past their peak natural increase periods and may actually approach zero population growth within another generation or two. Overall population growth rates for Latin America and the Caribbean as a whole peaked in the 1950s and 1960s and have now begun the gradual decline characteristic of regions in the late second and early third stages of the Model. While these recent patterns could unexpectedly change, the rate of natural increase likely will continue to decline steadily in the decades to come.

Urbanization

Urbanization, and the attitudinal changes that urbanization promotes toward family size, is ultimately most responsible for the slowing of the birth and population growth rates. So let us now analyze how Latin America and the Caribbean have evolved from predominantly rural to overwhelmingly urban regions. In reviewing the long-term urbanization trends depicted in Table 12.9, we learn that as recently as 1850, 94 percent of the population continued to live in peasant villages or in other isolated rural settings. Then, starting in the last half of the nineteenth century, there began a pattern of slow but steady increases in urban residence, which was associated largely with the expansion of industrial activity. By 1910, one in eight Latin Americans lived in a city, and by 1950 the ratio had grown to two in five. Then, beginning in the 1950s, the earlier trickle of urbanization suddenly seemed to become a torrent, with annual

TABLE 12.8

National Birth, Death, and Natural Increase Rates: Early Twenty-First Century

COUNTRY	BIRTH RATE*	DEATH RATE*	NATURAL INCREASE RATE*
Middle America			
Mexico	2.6	0.5	2.1
Guatemala	3.6	0.7	2.9
Belize	2.9	0.6	2.3
Honduras	3.3	0.6	2.8
El Salvador	3.0	0.7	2.3
Nicaragua	3.4	0.5	2.8
Costa Rica	2.1	0.4	1.7
Panama	2.3	0.4	1.9
Caribbean			
Antigua and Barbuda	2.2	0.6	1.6
Aruba	1.7	0.6	1.1
Barbados	1.5	0.8	0.6
Cuba	1.2	0.7	0.5
Dominica	1.6	0.8	0.8
Dominican Republic	2.6	0.5	2.1
Grenada	1.9	0.7	1.2
Guadeloupe	1.7	0.6	1.2
Haiti	3.3	1.5	1.7
Jamaica	2.0	0.5	1.5
Martinique	1.4	0.6	0.8
Netherlands Antilles	1.4	0.6	0.7
Puerto Rico	1.5	0.7	0.8
St. Kitts and Nevis	1.9	0.9	1.0
St. Lucia	1.8	0.6	1.2
St. Vincent and the Grenadines	2.0	0.8	1.2
Trinidad and Tobago	1.4	0.8	0.7
Andean South America			
Venezuela	2.4	0.5	1.9
Colombia	2.2	0.6	1.7
Ecuador	2.8	0.6	2.2
Peru	2.6	0.7	2.0
Bolivia	3.2	0.9	2.3
Southern South America			
Chile	1.8	0.6	1.2
Argentina	1.9	0.8	1.1
Paraguay	3.1	0.5	2.7
Uruguay	1.6	1.0	0.7
Brazil	2.0	0.7	1.3
Guianas			
Guyana	2.4	0.8	1.5
Suriname	2.4	0.7	1.7
French Guiana	2.6	0.5	2.1
Non–Latin American Comparisons			
United States	1.5	0.9	0.6
Canada	1.1	0.8	0.3
United Kingdom	1.1	1.0	0.1
Spain	0.9	0.9	0.0
Portugal	1.1	1.1	0.1
Japan	1.0	0.8	0.2

*Percentage

Note: All figures are rounded to the nearest tenth of a point.

Sources: Population Reference Bureau, *2002 World Population Data Sheet:* www.prb.org/Content/Navigation Menu/Other_reports/2000–2002/sheet1.html; *Demographic Yearbook 1999* 2001; *The Statesman's Yearbook 2002* 2001.

FIGURE 12.15

Regional patterns of population natural increase.

TABLE 12.9

Growth of Urbanization, 1850–2000

NATION	PERCENTAGE OF POPULATION RESIDING IN URBAN AREAS				
	1850	1910	1950	1970	2000
Middle America					
Mexico	7.9	10.8	42.6	59.0	74.4
Belize	—	—	—	—	46.5
Guatemala	5.0	8.9	24.9	35.7	40.4
Honduras	3.0	6.3	28.5	28.9	46.9
El Salvador	3.0	8.8	36.6	39.4	46.6
Nicaragua	3.0	—	35.0	47.0	64.7
Costa Rica	15.0	7.9	33.8	37.9	51.9
Panama	—	13.4	36.0	47.6	57.7
Caribbean					
Cuba	—	28.0	44.1	60.2	75.3
Dominican Republic	5.0	3.1	23.7	40.3	65.0
Haiti	3.0	8.2	14.8	19.8	35.7
Jamaica	—	—	24.3	41.5	56.1
Puerto Rico	—	—	59.5	58.3	75.2
Andean South America					
Venezuela	7.0	9.0	53.8	72.4	87.4
Colombia	3.0	7.3	34.0	57.2	74.9
Ecuador	6.0	12.0	28.4	39.5	62.4
Peru	5.9	5.4	43.9	57.4	72.8
Bolivia	4.0	9.2	33.5	40.8	64.8
Southern South America					
Chile	5.9	24.2	60.0	75.2	84.6
Argentina	12.0	28.4	61.4	78.4	89.4
Paraguay	4.0	17.7	32.7	37.1	56.0
Uruguay	13.0	26.0	30.0	82.1	91.3
Brazil	7.0	9.8	36.1	55.8	81.3
Guianas					
Guyana	—	—	25.6	29.4	38.2
Suriname	—	—	16.2	45.9	52.2
Latin America	6.3	12.8	39.3	57.3	75.4

Note: Definitions of what population level is required in order for a settlement to be classified as an urban area have varied widely over time and from one country to another.

Sources: *Economic and Social Progress in Latin America 1998–1999 Report* 1998; Wilkie 1978; *Statistical Yearbook,* various years; *Demographic Yearbook,* various years; World Bank World Development Indicators 2002: http://devdata.worldbank.org/data-query/

increases of almost 1 percent per year being common over the following four decades. So rapid and complete was this transformation that by 2000 three of every four Latinos had become a city dweller. A region that from time immemorial had been oriented toward the land was suddenly ranked among the most urbanized on earth (Figure 12.16).

As with the rates of natural population increase, there are clear regional patterns in the degree of urbanization of the Latin American and Caribbean nations (Figure 12.17). The populations of Uruguay (Latin America's most urbanized country), Argentina, Venezuela, Chile, and Brazil are each now more than 80 percent urban. As noted previously, these countries, with the exception of Venezuela, also have some of the lowest rates of natural population increase. On the other end of the spectrum, the populations of Haiti, Guyana, and many of the Central American nations are still mostly rural and also rank among the most rapidly growing. Increased urbanization thus appears to be associated over time with decreased rates of natural population increase. Several of the Andean populations, as well as those of Mexico and Cuba, are now 70 percent or more urbanized, meaning that their urban transformations are largely complete.

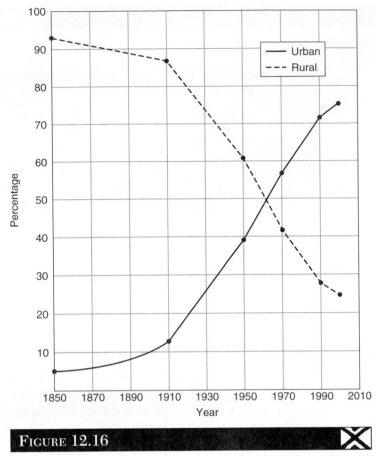

Urbanization of the Latin American and Caribbean population: 1850–2000.

City Size in the Modern Era

As the Latin American and Caribbean elites have continued to allocate almost all of their national financial resources to the traditional urban centers where they and their associates reside, tens upon tens of millions of rural peasants have migrated to the cities in search of a better life for themselves and their children. This migration has been so great and so rapid that the populations of many cities have now reached levels that would have been unthinkable as recently as a generation or two ago.

Because attempts to determine the true populations of the Latin American and the Caribbean cities are often hampered, even today, by inaccurate census counts, changing definitions of what constitutes the spatial limits of a given city, and occasional political sensitivities relative to the true size of the larger cities, all urban population figures should be regarded as educated approximations, at best. Note, for instance, the widely varying population estimates for the cities shown in Table 12.10, where São Paulo's population is listed as high as 18.5 million and as low as 9.9 million. Similar fluctuations occur with most other major met-

ropolitan areas. Such variations are often caused by the failure to include major suburban populations in the grand total.

Regardless of the actual populations, which in most cases we will likely never know, these figures clearly portray a growth that has spiraled almost out of control. One result is that Latin America, by every account, now contains two of the world's five largest metropolitan areas—Mexico City and São Paulo—and two others—Buenos Aires and Rio de Janeiro—rank among the twenty largest. Such enormous cities are increasingly being referred to in the academic literature as "global cities" (Scott 2001; Simmonds and Hack 2000; Rojas and Daughters 1998), or "megacities" (Butler, Pick, and Hettrick 2001; Drakakis-Smith 2000; Gilbert 1996b).

A second consequence of the metropolitan growth has been the emergence, in the past twenty to thirty years, of incipient megalopoli. A **megalopolis** is a highly urbanized region that includes two or more large, neighboring cities that were originally spatially separated but whose suburbs and satellite communities have grown to form a densely settled, although not

FIGURE 12.17

Regional patterns of urbanization.

TABLE 12.10

Estimates of Metropolitan Area Populations in the Early Twenty-First Century

CITY	SOURCES, WITH ESTIMATED POPULATIONS IN THOUSANDS		
	World Gazetteer	Statesman's Yearbook	Demographic Yearbook
Mexico City	20,965	16,674	15,048
São Paulo	18,505	16,583	9,928
Buenos Aires	12,917	11,802	11,298
Rio de Janeiro	11,247	10,192	5,584
Bogotá	7,798	6,276	6,545
Lima	7,604	6,465	6,321
Santiago, Chile	5,637	5,641	4,739
Belo Horizonte	4,414	3,803	2,124
Guadalajara	3,847	3,462	2,870
Pôrto Alegre	3,646	3,247	1,306
Caracas	3,478	3,127	2,784
Monterrey	3,468	3,022	2,563
Recife	3,440	3,087	1,368
Salvador	3,174	2,709	2,274
Medellín	2,994	1,958	2,767
Fortaleza	2,976	2,583	2,056
Curitiba	2,835	1,476	1,550
Cali	2,801	2,111	2,143
Santo Domingo	2,800	3,523	2,135
Guayaquil	2,686	2,205	2,118
Havana	2,682	2,192	n.a.
Guatemala City	2,565	3,119	1,007
Puebla	2,540	1,562	1,266
Brasília	2,089	1,821	1,923

Sources: *The World Gazetteer 2002:* www.world-gazetteer.com/st/statn.htm; *The Statesman's Yearbook 2002* 2001; *Demographic Yearbook 1999* 2001.

necessarily continuously built up, urban corridor. The concept was developed originally by the French geographer Jean Gottmann (1961), who applied the term to the Washington D.C.–New York City–Boston Atlantic coastal region of the northeastern United States. Urban geographers and demographers now recognize that megalopoli are forming throughout the world. In Latin America and the Caribbean, these include Toluca–Cuernavaca–Mexico City–Puebla on the Mesa Central, the Niterói–Rio de Janeiro–Santos–São Paulo–Campinas axis of southern Brazil, and the La Plata–Buenos Aires–Montevideo–San Nicolás–Rosario corridor of Argentina's and Uruguay's lower Paraná–Río de la Plata basin (Figure 12.18). Other regions that may emerge as megalopoli in the future include the Caracas–Maracay–Valencia basins of northern Venezuela; Colombia's middle and lower Cauca River Valley, which contains Medellín, Manizales, and Cali; and the Central Valley of Chile from Valparaíso and Viña del Mar southward through Santiago to Rancagua.

Urban Primacy

The accelerated growth over the past century of the leading cities of Latin America and the Caribbean has reinforced long-term primacy patterns in most nations. As we noted previously in Table 12.2, primacy levels, as measured in terms of Latin America's Hispanic population, were high throughout the colonial era. Morris (1981) has documented the continuation of those levels through the Republican Period and their general tendency to increase in the twentieth century.

There has been and continues to be considerable disagreement among Latin Americanist scholars over the best way to measure urban primacy. Jefferson (1939) relied mostly on a three-city primacy ratio in which he compared the relative population sizes of the three largest cities of a given country. Garza and Schteingart (1978) have utilized a two-city ratio or index, while Boswell (1989) and Portes and Walton (1976) employed a four-city index. Perhaps the clearest

FIGURE 12.18

Leading urban centers and emerging megalopoli: early twenty-first century.

TABLE 12.11

Urban Primacy in the Early Twenty-First Century

COUNTRY	LARGEST URBAN AGGLOMERATION (LUA)	NATIONAL POPULATION (MILLIONS)	LUA PERCENTAGE OF TOTAL URBAN POPULATION	LUA PERCENTAGE OF TOTAL NATIONAL POPULATION
Suriname	Paramaribo	0.4	88.4	63.9
Guyana	Georgetown	0.7	86.4	31.5
Costa Rica	San José	3.8	75.5	39.2
Haiti	Port-au-Prince	7.1	68.2	24.3
Puerto Rico	San Juan	4.0	66.2	49.8
Panama	Panama City	2.9	61.9	35.7
El Salvador	San Salvador	6.4	59.1	27.5
Uruguay	Montevideo	3.4	55.5	50.7
Dominican Republic	Santo Domingo	8.7	49.5	32.2
Guatemala	Guatemala City	13.3	47.7	19.3
Jamaica	Kingston	2.7	45.1	25.3
Honduras	Tegucigalpa	6.6	44.3	20.8
Paraguay	Asunción	5.9	43.3	24.3
Chile	Santiago	15.5	43.0	36.4
Nicaragua	Managua	5.0	41.5	26.8
Argentina	Buenos Aires	37.8	38.2	34.2
Peru	Lima	28.0	37.3	27.2
Ecuador	Guayaquil	13.5	32.0	19.9
Cuba	Havana	11.2	31.6	23.8
Bolivia	La Paz	8.5	28.0	18.1
Mexico	Mexico City	103.4	27.3	20.3
Colombia	Bogotá	41.0	25.4	19.0
Venezuela	Caracas	24.3	16.4	14.3
Brazil	São Paulo	176.0	12.9	10.5
Trinidad and Tobago	Port of Spain	1.2	5.5	4.0
Non–Latin American Comparisons				
United States	New York City	281.6	10.0	7.5
Canada	Toronto	30.8	22.8	17.8
United Kingdom	London	59.7	20.9	18.8
France	Paris	58.9	25.9	19.2
Japan	Tokyo	126.9	31.4	24.5

Note: The "Largest Urban Agglomeration" is defined as the area within the city limits plus the suburban fringe or thickly settled territory lying outside but adjacent to the city.

Sources: U.S. Census Bureau International Programs Center, 2002: www.census.gov/cgi-bin/ipc/idbrank.pl; The World Gazetteer 2002: www.world-gazetteer.com/st/statn.htm; *The Statesman's Yearbook 2002* 2001.

way to express urban primacy is simply to calculate the percentage of the total urban population of a nation or region that resides within the largest urban agglomeration.

Table 12.11, which utilizes this latter approach, reveals a number of significant trends related to Latin American and Caribbean primacy. The first is that most nations have very high levels of primacy. Indeed, three-fifths of the countries have over 40 percent of their total urban population living in the primate city. A second trend is that Latin America's most populous countries—Brazil, Mexico, Colombia, Argentina, Peru, and Venezuela—have much lower primacy levels, as a group, than do the less populous nations. A third pattern is that the primate cities are usually also the capital cities of their respective nations. Finally, the primate cities have become so large that most now contain between a fifth to a third of their entire national populations. This has reinforced the age-old Hispanic patterns of concentration of wealth and power in the principal urban centers and neglect of the rural areas and smaller provincial cities.

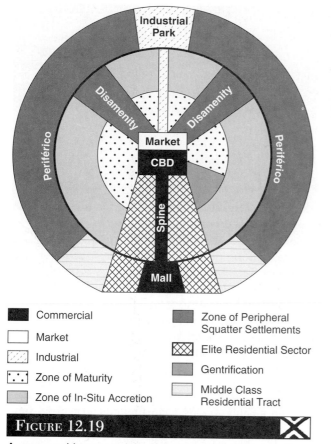

Commercial
Market
Industrial
Zone of Maturity
Zone of In-Situ Accretion

Zone of Peripheral Squatter Settlements
Elite Residential Sector
Gentrification
Middle Class Residential Tract

FIGURE 12.19

A new and improved model of Latin American city structure.

Source: Adapted with permission from Larry R. Ford, "A New and Improved Model of Latin American City Structure," *The Geographical Review*, Vol. 86, p. 438. Copyright 1996 The American Geographical Society.

MORPHOLOGY OF THE MODERN CITY

A number of observers have commented correctly on the widespread absence of enforced landuse planning and zoning in Latin American and Caribbean cities. However, it would be erroneous to assume that population growth in the modern era has left the Hispanic urban center as a chaotic, formless, ungoverned settlement. In reality, the modern city consists of a number of spatially and socioeconomically distinct landuse zones, a generalized model of which was first presented in 1980 by Griffin and Ford and subsequently updated by Ford in 1996 (Figure 12.19).

The Ford model of current Latin American city structure recognizes that one of the principal changes flowing from population growth and technological advancement has been the replacement of a portion of the old inner city elite residential–market district with a dynamic Central Business District (CBD), which houses the offices of the leading domestic and, in some cases,

FIGURE 12.20

The central business district of São Paulo, Brazil, one of the world's largest cities.

FIGURE 12.21

These inner city tenement housing units are found four blocks from the downtown plaza of Lima, Peru, and were created from a property that formerly housed an upper-class family. The signs hanging from the balcony and over the door state that the tenants are on strike and fighting eviction.

foreign commercial firms and corporations. The size and physical appearance of the CBD varies, of course, from one city to another, but those of many of the larger communities, such as Mexico City, Caracas, and São Paulo, are now comparable to their Anglo American counterparts (Figure 12.20). One result of the development of the CBD has been that many elites have abandoned portions of the old inner city living space in favor of new upper-class commercial and residential districts bordering either side of a showcase, automobile-oriented thoroughfare or "spine" (Figure 12.21). The thoroughfare is now likely to lead to a suburban retail mall, or edge city, whose upscale establishments offer stiff competition to traditional inner city businesses (Figure 12.22).

FIGURE 12.22

The entrance to a suburban retail mall in Aguascalientes, Mexico.

FIGURE 12.23

The restoration of Barrio Amón, a historic inner-city district of San José, Costa Rica, is but one of countless examples of urban renewal now underway in Latin America and the Caribbean. The building shown in this image has been converted into a hotel and restaurant that caters largely to international tourists.

A second outer city development of recent origin may be an industrial park that houses space-extensive factories and warehousing operations. The industrial park is generally situated along an industrial transportation corridor that is served by a highway and rail line and, occasionally, a river or inlet of the sea. Increasingly, the edge city malls and industrial parks are being linked by a modern, rapid transit perimeter road called the *periférico*, whose purpose is to relieve some of the chronic traffic congestion characteristic of the inner city districts. New middle class housing districts are also springing up near the edge city malls and industrial parks, as well as along more favored sections of the *periférico* itself.

With the exception of occasional corridors of "disamenity," where physical decay is accelerating and public services are limited, the Ford model suggests that the remaining portions of the modern Latin American city consist of three concentric landuse rings whose economic value and relative social standing decrease with increasing distance from the central city. It should be noted that this sequence is exactly the opposite of the Anglo American city, which is characterized by inner city decay and suburban affluence.

The innermost of the Latin American landuse rings is a "zone of maturity," which comprises older, well-established, middle class neighborhoods, picturesque plazas, occasional historic districts (labeled "gentrification" on the model), and high levels of public services (Figure 12.23) (Case Study 12.2). It grades eventually into a "zone of *in-situ* accretion," which consists mostly of newer, modestly priced homes and businesses, many of which seem forever to be under construction owing to the presence of unfinished upper stories and rooms (Figure 12.24). Finally, the outermost ring now comprises vast expanses of squatter settlements or shantytowns whose functions are similar in many respects to those of the colonial Indian *barrios*. Because these peripheral neighborhoods receive minimal financial support from the elite-dominated municipal governments, levels of public services, such as potable water, sewage disposal, street paving and drainage, garbage disposal, and even electricity, are generally inadequate.

While they are not cartographically feasible to depict in the model, Ford also acknowledges in his study the intensely mixed landuse patterns that are characteristic of the modern Latin American city. The model does not apply perfectly, of course, to every Latin American city. Godfrey (1991), for example, in stressing the differences between the cities of Latin America's Spanish and Portuguese realms, has noted that the elites of Brazil's coastal cities have recently shown a strong preference for beachfront residences, even if they are situated far from the inner city CBD. These and other limitations notwithstanding, the model is valuable in its ability to capture the broad outlines of landuse within the Latin American city. Because these uses are constantly evolving, let us now briefly analyze two areas subject to intense change: the inner city and the peripheral squatter settlements.

The Inner City: Zone of Commercial High-Rises and Low Income Tenements

We have noted previously that one result of the progressive abandonment of the inner city by the traditional elites has been the conversion of portions of the urban core to an upscale Central Business District

Case Study 12.2:   Social Expressions of the Latin American Plaza

The plazas, or town squares, that are found from the smallest to largest settlements of Latin America can serve a number of different functions or roles. These may include remaining as a largely unimproved site where animals are grazed and soccer matches are held; the hosting of religious and political gatherings and rallies; and the location of daily or periodic markets where food, clothing, medicines and other essential goods and services are bought and sold.

As important as these activities are, perhaps the most significant function of the plaza is that of facilitating social interaction. When visiting a traditional Latin American town or city, one of the most enjoyable and instructive experiences available to the traveler is to sit on a bench in the plaza and observe the activities of the people.

One of the principal activities associated with the plaza is upper class adult male bonding. True to the traditional gender role of the *macho,* whose sphere of action is in *la calle,* or the public domain, rather than the home, the men will gather daily at their favorite cafe or restaurant, where they will spend hours discussing politics and business while taking their noonday meal, perusing the newspaper, and watching a nearby game of dominoes or chess.

Another form of social interaction focuses on the Catholic church that fronts almost every Latin American plaza. If one watches carefully, there is generally an almost constant trickle of the faithful who come to attend Mass

or, often, simply to worship in solitude. Most of the communicants are older women, many of whom come dressed in black with a small grandchild in tow.

A third form of social discourse centers on the benches set around the outer edge of the plaza itself. Here people of all stations of life may come to meet a friend or simply to experience the unique combination of public exposure and privacy offered in the plaza. In many smaller, traditional communities, certain people come to have unspoken claim to a spot on a specific bench at given hours of the day. Access to personal space on plaza benches can thus become an expression of social standing and acceptance. One American geographer, who had lived in a small Honduran city for several extended periods of time while conducting field research, reported that he knew that he had become a fully accepted member of the community when he arrived at the usual time to find his customary place on his plaza bench occupied by a newcomer. Without saying a word, the other townspeople seated nearby quickly communicated to the stranger that he was seated in the wrong spot and had better move at once.

The crowning social activity of the plaza is the gathering of the townspeople on a Friday, Saturday, or Sunday evening. As darkness sets in, the plaza will fill with people of all ages and backgrounds who come to be entertained and to interact with one another. Entire families will often dine

out at the nearby restaurants and, if a theater is within walking distance, watch a movie before blending into the mass of dimly lit faces. Others will purchase food from street vendors.

As the evening progresses, the town band or orchestra will then serenade the public as the teenaged youth engage in the age-old custom of promenading. This typically entails the girls walking hand in hand around the kiosk in one direction and the boys circling in the other. As they pass, the girls and boys both appear to ignore each other at first, but then will begin to exchange longer and longer glances with each ensuing pass. Finally, if a boy can summon up enough courage, he will position himself to where he can be next to the girl of his choice on the following pass. As she approaches, still holding hands with her girlfriend, her male suitor will reach out and attempt to hook her arm in his. As their arms touch, the moment of decision arrives for the young woman. If she would prefer to spend the remainder of the evening with the boy, she will let go of her girlfriend's hand and turn to walk with him. If, however, she does not wish to be courted by him, she will keep hold of her girlfriend's hand and the young man will be forced to withdraw his arm as they move apart. All the while, parents and grandparents relax and visit while pretending not to be watching their teenagers.

Eventually, the lateness of the hour sees the plaza emptying. The social activities will begin again tomorrow.

Recommended readings: Arreola and Curtis 1993; Nelson 1963; Gade 1976; Elbow 1975; Crowley 1987; Richardson 1982.

whose clients are housed in modern high-rise office buildings. While these structures are often pointed to with pride by local authorities as evidence of the "progress" and "modernization" being realized by the nation, they tend to obscure the fact that the Latin American and Caribbean inner city is also a zone of deepening poverty and physical squalor.

Although some of the inner city poor can now be seen living in flimsy shanties or sleeping in the open on the streets and sidewalks, most reside in old colonial era buildings that were formerly the homes and offices of the elite but have recently been converted by their absentee owners into tenement or apartment projects. The tenements, variously called **vecindades,** *tugurios, callejones,*

FIGURE 12.24

The zone of *in-situ* accretion found in Latin American cities is characterized by modest-sized homes and business buildings that seem to be constantly receiving new additions. This view is of Quito, Ecuador.

FIGURE 12.25

Inner-city tenement, or *vecindad,* in downtown Mexico City, just a few blocks from the main plaza.

inquilinatos, solares, and *conventillos,* are in especially high demand among recently arrived rural migrants whose poverty and unfamiliarity with the city leave them few residential alternatives. Typically the buildings are subdivided into dimly lit one-room apartments, each of which may house a nuclear or extended family with as many as ten to twelve members (Figure 12.25). Population densities thus rise to almost unthinkable levels, as buildings that formerly contained a single well-to-do family of six to eight persons now support 100 to 200 or more lower class occupants. Keeling (1996, 101) has written of a Buenos Aires mansion that once belonged to the widow of an Argentine president that was subsequently occupied by forty families. Access to electricity in the inner city tenements frequently depends on dangerous, illegally installed power lines, and use of toilets and clothes-washing facilities is often limited to a shared interior communal patio.

The living levels of the inner city poor have often been made worse by the enactment of socially well intentioned but economically misguided municipal rent control laws that set rents at such low levels that the building owners cannot make a profit after deducting expenses for maintenance. Many consequently decide that their only option is to do nothing at all to maintain the properties, allowing them instead to continue to deteriorate until they are finally condemned by zoning officials who then permit the buildings to be demolished and replaced by high income housing or office space (Zaaijer 1991). The problem, of course, is that the poor lose both ways: either through the ongoing worsening of their living conditions or through the loss of affordable inner city housing.

Occasionally one reads of an inner city neighborhood, such as the fictitiously named "El Centro" downtown Mexico City district studied by Eckstein (1990), which is able to halt, at least temporarily, the seemingly inevitable process of inner city landuse change. These cases appear to be exceptions to the norm, however, which owe their success largely to short-term external factors. One reason El Centro residents were able to preserve and even improve their levels of living down through the 1980s was the Mexican debt crisis, which limited the abilities of the elites to finance the demolition of old buildings and the construction of new ones in the city.

The end result is that the Latin American inner city has now become a zone of sharp contrasts, where decaying tenements often stand in the shadows of soaring skyscrapers and rundown busses fight for street positions with chauffeured limousines. Traffic congestion, street litter, and air pollution are all mounting, and the sense of community that formerly characterized the district has now largely been lost.

Peripheral Squatter Settlements

A second area currently experiencing rapid change is the outer edge of the city where **shantytowns** are spreading out onto land that was formerly considered too distant or physically too marginal to be developed. Shantytowns, or **squatter settlements** as they are often called, are, as their names suggest, communities of the poor lower class who erect tiny, makeshift

FIGURE 12.26

The homes of squatter settlements are often constructed of discarded materials salvaged by the occupants. This dwelling is on the outskirts of Ibagué, Colombia.

TABLE 12.12

Selected National Names for Shantytowns and Squatter Settlements

NATION	NAME(S)
Mexico	*ciudades perdidas, jacales, fraccionamientos clandestinos, colonias paracaidistas, colonias proletarias*
Guatemala	*barrios de invasíon*
Panama	*barriadas brujas*
Venezuela	*ranchos, cerros, barrios, quebradas, pueblos marginales*
Colombia	*tugurios, barrios de invasión, barrios piratas, barrios clandestinos*
Ecuador	*barrios populares, suburbios*
Peru	*barriadas, pueblos jóvenes*
Chile	*callampas*
Argentina	*villas miserias, villas de emergencia*
Uruguay	*cantegriles*
Brazil	*favelas, invasões*
Haiti	*bidonvilles*

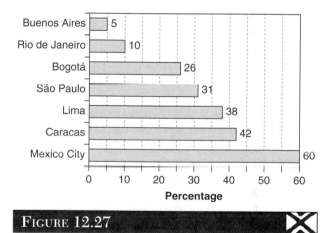

FIGURE 12.27

Percentage of urban populations residing in squatter settlements.

Source: Adapted from Gilbert 1996a: 75.

dwellings or shanties on land to which they do not hold title (Figure 12.26). Viewed historically, shantytowns likely have existed in Latin American and Caribbean cities from the time of the European Conquest. Many of the poor Indian and mixed-race occupants of the outlying *barrios* of the colonial city, for instance, were squatters in the sense that they never owned the land on which they built their houses.

Urban squatting, then, is certainly nothing new; nor, as the extensive array of national names for squatter settlements attests (Table 12.12), it is limited to one or two parts of the region. What is historically unprecedented, however, is the explosive growth of shantytowns that has been fueled in the post–World War II era by the combination of high national population growth rates and massive rural to urban migration. Whereas in most of the major Latin American cities at the turn of the century approximately 10 to 15 percent of the population resided in squatter settlements (Stann 1975; Pearse 1961), now it is not uncommon to find that 25 to 60 percent of the inhabitants of a given community live in the outer shantytowns (Figure 12.27). One consequence of the expansion of the squatter settlements is their emergence as the principal centers of new building construction in most cities. De Soto (1990, 76) has noted, for instance, that 90 percent of all edifices erected in Lima in recent years have been placed on squatters' land.

Scholarly and official government policy responses to the presence of the burgeoning squatter settlements have fluctuated greatly over the past half century. When the shantytowns were first studied in the 1950s and early 1960s, they were widely perceived as insidious urban cancers whose delinquent populations lived in perpetual squalor. Characteristic of the prevailing sentiments of that era was an autobiographical account of life in the *favelas* of São Paulo, written by a woman named Carolina Maria de Jesus, with the assistance of a journalist. In her best-selling book, entitled *Child of the Dark* (1962), de Jesus

portrayed a life where "the only thing that does not exist is friendship" (p. 21) and concluded that the only solution was to "tear down the *favelas*" (p. 25). Her sentiments were widely shared by government planners of the day, most of whom proposed bulldozing the shantytowns and replacing them with large public sector housing units.

By the late 1960s and 1970s, however, scholars such as Mangin (1967) and Portes (1972), were stressing that squatter settlements were a rational and, indeed, a desirable low cost, self-help response of the honest, hard-working lower class. More recent assessments have offered both negative and positive perspectives. Representative of the former is Eckstein (1990), who wrote of life in the "squatter settlement of despair." In contrast, Mangurian (1997) has described an "urban renaissance" centering on the physical upgrading of the *favelas* of Rio de Janeiro and their increased integration with the economic and social life of the city as a whole.

Regardless of whether a given study emphasizes the positive or the negative aspects of the Latin American and Caribbean shantytowns, clearly many Anglo Americans have fallen victim to a number of popular but inaccurate misconceptions and stereotypes regarding the nature of these communities and their inhabitants (Flinn and Converse 1970; Stokes 1991). These include: (1) the perception that shantytowns are spatially disorganized and socially ungoverned communities that have evolved without any planning or administration; (2) the belief that the squatters themselves are composed almost entirely of recent migrants who have just stepped off a bus from the rural countryside; and (3) the idea that the residents of shantytowns are mostly unemployed and unproductive members of society with strong criminal tendencies.

Field research has demonstrated repeatedly that none of these perceptions is accurate. Contrary to the notion that all shantytowns are disorganized and ungoverned, for example, we find that the founding of many squatter settlements is the result of a carefully orchestrated plan carried out with the support of high-level government officials (Case Study 12.3). Similarly, researchers have consistently reported that the overwhelming majority of squatters are longtime residents of the city who used *personalismo* to arrange for their place in the new community. Finally, the upper class suspicion that most shantytown residents are unemployed troublemakers vanishes upon a personal visit to the community. There one finds that, except for its poverty and the resultant lack of some material advantages such as indoor plumbing and sanitation facilities, the shantytown differs little from any other neighborhood of the city. Streets are lined with tiny stores, shops, and restaurants, a never-ending stream of busses carries adults to work, and children seem to be everywhere, laughing and playing all kinds of games (Figure 12.28).

Urban Population Densities

One consequence of the close living conditions in both inner city tenements and peripheral squatter settlements is that population densities within Latin American and Caribbean cities are generally higher than in the urban centers of Anglo America. The typical large Latin American city, for example, presently counts 3,000 to 7,000 persons per square kilometer (7,800 to 18,200 per square mile). In contrast, most cities in the United States and Canada have population densities of 700 to 2,000 persons per square kilometer (1,800 to 5,200 per square mile) (Table 12.13).

The high population densities have led, in turn, to severe traffic congestion in the mid- and large-sized cities. Mexico City, for instance, is estimated to experience about 40 million person journeys per day, some 5 million of which are made on one of the world's most efficient subway systems (Figueroa 1996) (Figure 12.29).

NEW CITIES

As we noted earlier in the chapter, most of the prominent cities of Latin America and the Caribbean were founded in the early colonial era and are now hundreds of years old. Cities and towns have been and continue to be founded in modern times, however, as circumstances warrant. Although many of these are just tiny frontier settlements of a few hundred inhabitants or less, a few were planned from the beginning to serve as specialized commercial centers or as national or regional capitals. With generous government funding, these have grown into major cities whose existence symbolizes to a large degree the geographical aspirations of the nations they represent. One common form of the specialized commercial centers developed in recent years has been the coastal beach resort designed to attract international tourists. One of the most successful of these has been Cancún, which was developed by Mexican government planners on the northeast coast of Yucatán.

By far the most grandiose and publicized of Latin America's new capital cities is Brasília, Brazil, which was founded in 1960 on a central plateau site in eastern Goías, more than 960 kilometers (600 miles) inland from Rio de Janeiro and São Paulo. While Brasília is justifiably honored the world over for its sweeping design and futuristic, monumental architecture, which were conceived by Lúcio Costa and Oscar Niemeyer, respectively, its existence represents the ongoing realization of the centuries-old Brazilian dream of occupying its western Amazonian territory (Freitas Marcondes 1959; Crist 1963; Epstein 1973) (Figure 12.30). Critics of Brasília have accused it over the years of being socially sterile and physically rigid. This perception was promoted primarily by Brazilian elites and government leaders. The latter were so hesitant to leave the pleasures of Rio de Janeiro during Brasília's early years that most maintained two offices: an unofficial one in Rio that they used Friday through Monday and an official one in Brasília to which they commuted by air on Tuesday mornings and abandoned on Thursday afternoons (Griffin and Ford 1993, 258). In fairness, it must be acknowledged that all cities require time to achieve a pleasing personality and ambiance of their own, and there can be no question today of Brasília's long-term contributions to the development of central and western Brazil.

Case Study 12.3:   Planned Squatter Settlement Invasions

Every squatter settlement, although unique and distinct in one way or another, falls within one of two broad categories: spontaneous or planned. Spontaneous squatter settlements are those that grow little by little as one or more families establish themselves at their own initiative on land that they do not own and then are joined over time by other squatters. The establishment of a planned squatter settlement, on the other hand, entails advance site selection and considerable preparation by the organizers prior to an *en masse* "invasion" by the squatters.

One type of person known to have secretly encouraged planned squatter invasions is the high-level politician who uses the illegal occupation of a parcel of land as a means of rewarding his supporters and/or punishing his rivals. In deciding which parcel of land should be invaded, the public official may consider the political sympathies of the present landowner and whether the future development of the parcel will advance or threaten the ambitions of the politician. A second type of individual who may be inclined to sponsor a planned squatter settlement is the owner of a piece of land that is worth little in its present state. Such an owner may secretly arrange for the invasion of his own property and then pretend to be an innocent victim in the hope that the squatters will eventually succeed in persuading the municipal government to purchase the land on their behalf at inflated prices from the poor, victimized landowner. Yet a third type of person disposed to organized planned invasions is the professional squatter who earns his living by selling the plots he reserves for

himself within a new settlement and by forcing other would-be squatters to pay him for his "assistance" in helping them secure their plots within the new development.

Regardless of whether a politician, the present landowner, or the career squatter organizes the invasion, the actual occupation of the land is a carefully staged event designed to arouse public sympathy and support for the squatters and their actions. To accomplish this aim, the organizers recruit potential squatters from among persons already known to be honorable, hardworking, law abiding members of the community. Criminal elements and recent migrants whose character is unknown to the organizers stand a much lower chance of being selected to participate. While squatter selection proceeds, the target site is thoroughly scouted out with an eye to where streets, public buildings, and residential lots will be placed. Those approved for participation in the invasion begin to stockpile essential building supplies, and the organizers alert sympathetic members of the media to their plans so that favorable publicity is assured. They also purchase a national flag, which is displayed prominently at the invasion site in the hope of convincing both the government and the public to support the squatters because they are such good, loyal citizens.

When the appointed day of the invasion finally arrives, the squatters load their belongings and construction materials onto trucks and busses and await the coming of nightfall, after which they converge on the designated site. In the course of just a few hours, streets are marked off and individual

lots partitioned. Working feverishly throughout the night, the squatters then erect flimsy, makeshift shacks out of whatever materials they were able to secure, and set up house inside.

As morning dawns, neighbors awake to see a new settlement, complete with streets and homes, where none existed the night before. The designated newspaper, radio, and television reporters arrive as scheduled and find the national flag fluttering in the breeze and possibly a patriotic slogan painted on the side of a nearby hill or building. Photographs are taken and interviews completed, and by afternoon the media is full of heartwarming accounts about how some of the city's poorest but most loyal citizens have finally obtained the hope of a better life. Frequently the organizers adopt, as a finishing touch, an extra optimistic name for the community, such as "El Futuro" (The Future), "El Paraíso," (Paradise), or "La Esperanza" (Hope).

The following days are busy as resident associations are formed to work for the securing of titles to the newly occupied plots and for the installation of public utilities and bus services. Security measures are also implemented, and schools, churches, soccer fields, and community centers planned. Although occupation of the land is free, monthly dues are collected by the associations. As the settlement becomes more established over time, the residents are frequently able to improve and enlarge the size of their homes as their incomes increase. The number of commercial enterprises also expands, and renting and subleasing of homes become common.

Sources: De la Cadena 2000; Matos Mar 1961; Pérez Perdomo and Nikken 1982; Mangin 1967; Moser 1982; Lutz 1970; Roberts 1992.

FIGURE 12.28

Children of a San José, Costa Rica, squatter settlement.

FIGURE 12.29

Without its subway system, Mexico City's traffic congestion and air pollution levels would be even worse than they already are.

TABLE 12.13

Population Densities of Latin American and Non–Latin American Cities

CITY	PERSONS PER SQUARE MILE	PERSONS PER SQUARE KILOMETER
Latin American		
Guayaquil	28,424	10,972
Salvador	18,823	7,266
Barranquilla	17,531	6,767
São Paulo	17,228	6,650
Belo Horizonte	16,426	6,341
Caracas	11,818	4,562
Rio de Janeiro	11,518	4,446
Bogotá	10,106	3,901
Cali	9,748	3,763
Curitiba	9,407	3,631
San Salvador	8,860	3,420
Havana	7,831	3,023
Non–Latin American		
Chicago	8,326	3,214
Detroit	5,401	2,085
Atlanta	3,878	1,497
Houston	3,702	1,429
Los Angeles	3,062	1,182
Toronto	1,943	750
New York City	1,886	728
Denver	1,798	694

Source: Derived from data in *Demographic Yearbook 1999* 2001.

FIGURE 12.30

Brasília is known worldwide for the stunning architecture of its principal buildings. Shown here is the National Palace.

A second new capital city established in modern times is Belmopan, Belize, which was planted on its present site some eighty kilometers (fifty miles) inland from the Caribbean Sea following the near total destruction of Belize City by hurricane in October of 1961. Unfortunately for Belmopan, Belize City was rebuilt and, like Rio de Janeiro in Brazil, continues to be the principal port and emotional heart of the country.

A third nation that has seriously considered moving its capital in modern times is Argentina, which has long looked southward to Patagonia, Tierra del Fuego, and even Antarctica as untapped frontiers (Child 1986; Caviedes 1984). In 1983, then Argentine president Raúl Alfonsín introduced a proposal to move the national capital from Buenos Aires to Viedma, a small port and the capital of Río Negro province in northern Patagonia. Alfonsín argued that the move would strengthen Argentine claims to the Falkland (Malvinas) Islands and Antarctica, hasten the economic development of Patagonia, and reduce the excessive size and influence of Buenos Aires. A number of reports and plans related to the project have since been generated by government officials, but there presently appears to be little public support for the concept.

MIGRATION

The final topic that we will address in our study of population growth and urbanization is the migration or relocation of many Latin American and Caribbean peoples from one place of residence to another. While migration has always been a part of human existence, its scholarly study in modern times can be traced to the works of the nineteenth-century British demographer E. G. Ravenstein, who on two occasions (1885; 1889) published studies in the *Journal of the Royal Statistical Society* entitled "The Laws of Migration." In these articles, Ravenstein discussed a number of characteristics or patterns of human migration and noted that decisions to move or relocate almost always result from the interplay of **push and pull factors.**

Persons who are happy and satisfied with their circumstances in life are not likely to seek a new place of residence. If they come to believe, however, that their present life is not good enough and is unlikely to ever improve, they will begin to feel as though they are being pushed out of their present home. In the process, they will have become, either knowingly or unknowingly, prime migration candidates. The ultimate destinations of those who do eventually choose to leave will then be determined by their mental images of the quality of life in other places, whose perceived positive attributes serve to pull the migrant toward his or her destination.

While weather, climate, political and social conditions, access to health care, and proximity to friends and relatives can all function as either push or pull factors for certain individuals, research has shown that most Latin American and Caribbean migrants, be they permanent or short-term, are motivated primarily by a desire to improve their economic circumstances (Bebbington 2000; De Souza 1998). Ugalde, Bean, and Cárdenas (1979), for example, found that 78 percent of those persons migrating to the United States from the Dominican Republic cited unemployment in the home region, the search for higher incomes, or a desire to upgrade their education (and subsequently their future earnings) as the principal reason for leaving. Similar findings have been reported by Duany (1992) for migrants entering Puerto Rico from the Dominican Republic and Cuba and by Reichert and Massey (1979) and Rubenstein (1992), both of whom analyzed Mexican migration to the United States.

Migration thus serves both as an economic and an emotional survival mechanism. It occurs most often when two nations or regions of widely differing circumstances coexist in close spatial proximity. We will now review specific Latin American and Caribbean migration flows from the perspectives of three settings: those flows occurring between regions within the same country, those that entail resettlement between two or more Latin American or Caribbean countries, and those that center on international movements involving European and Anglo American nations.

Interregional Flows

Although international resettlement has attracted more publicity, the movement of people from one region to another within a given nation has long been the leading form of Latin American and Caribbean migration. Chief among interregional flows is **rural to urban migration.** While precise counts are impossible to come by, in all likelihood the number of rural to urban migrants over the past century in Latin America and the Caribbean has exceeded 150 million, with no end in sight. Mexico City and São Paulo, the region's two largest urban agglomerations, are each believed to be currently receiving about 300,000 rural migrants annually (Villa and Rodríguez 1996). Other large-scale interregional migrations include the colonization of the Amazon and Orinoco river lowlands by Andean highlanders (Crist and Nissly 1973; Eidt 1962; Bromley 1972) and the resettlement of millions of Brazilians from the drought-prone Northeast to the industrialized South and westward into Amazonia (Ozorio de Almeida 1992; Godfrey 1990; Gauthier 1975) (Figure 12.31).

Historically, most rural to urban migrants have been married men who sent money, or remittances, back home to their families until they could eventually be reunited in their new place of residence. More recently, there has been an increased flow of young, single women, for whom there are growing opportunities to obtain work as domestics and workers in *maquiladora* and other factory settings (Radcliffe 1990).

A lesser-known form of interregional migration is that of individuals forced by armed conflict, internal strife, or systematic violation of human rights to move from their home region to another within their country of residence. These **Internally Displaced Persons** (IDPs) are estimated currently to number approximately 2.5 million for Latin America as a whole, almost all of whom are found in Colombia (*World Refugee Survey 2002*). Unlike refugees, who benefit from an established system of international protection and support, IDPs have no legal basis for appealing to the international community for assistance, owing to the fact that they have not migrated across national borders.

Migration between Latin American and Caribbean Countries

The movement of people from one nation to another within Latin America and the Caribbean has been dictated in recent years primarily by economic opportunity in the receiving nation and the perceived absence of that opportunity in the donor country. The leading migrant destination overall has been Argentina, whose factories and farms have at various times attracted hundreds of thousands of workers each from Chile, Bolivia, Paraguay, and Uruguay. Similarly, Venezuelan petroleum and mining operations have functioned as a magnet to nearby Colombians, Ecuadorians, and Trinidadians. Other large intra–Latin America and Caribbean movements have included Haitians to the Dominican Republic, Dominicans and "down islanders" from the Lesser Antilles to Puerto Rico, Peruvians to Chile, and Central Americans to Mexico.

Migration to Europe and Anglo America

The migration of Latin American and Caribbean nationals to Europe and Anglo America has centered on three recipient nations. The first, France, has long supported the resettlement of modest numbers of Blacks from French Guiana, Guadeloupe, and Martinique. Because these persons hold French citizenship, they have benefited from considerable assistance in securing housing and employment on the continent (Condon and Ogden 1991).

Like France, Great Britain also turned to its Caribbean populations as a source of inexpensive laborers needed to rebuild the nation's infrastructure after World War II. For a time, English-speaking West Indian Blacks were well received and their numbers rose steadily. By the late 1950s, however, public sentiment was turning against the continued presence of the immigrants, whom some considered to be filling jobs needed by unemployed native Britons. This led Parliament in 1961 to pass the British Commonwealth Immigrants Act, which not only made entry into the United Kingdom more difficult but even went so far as to require the forced deportation of some West Indians already residing within the country (North and Whitehead 1991). Critics have charged that the legislation was racially motivated, while British leaders have insisted that it is simply a reflection of changing labor needs within the nation.

By far the leading destination of Latin American and Caribbean overseas migrants in recent decades has been the United States, whose high levels of living, stable political environment, and porous borders have proven to be an irresistible combination to millions of

FIGURE 12.31

Principal Latin American and Caribbean migration flows.

TABLE 12.14

Foreign-Born and Illegal Alien Resident Population of the United States: 2001

SOURCE COUNTRY	TOTAL POPULATION (THOUSANDS)	
	Foreign-Born	Illegal Alien
Latin America	14,477	
Mexico	7,841	2,700
Caribbean	2,813	
Cuba	952	...
Dominican Republic	692	75
Jamaica	411	50
Haiti	385	105
Trinidad and Tobago	173	50
Barbados	54	...
Grenada	42	...
Other Caribbean	104	...
Central America	1,948	
El Salvador	765	335
Guatemala	327	165
Honduras	250	90
Nicaragua	245	70
Costa Rica	77	...
Panama	69	...
Belize	59	...
Not elsewhere classified	156	...
South America	1,875	
Colombia	435	65
Peru	328	30
Ecuador	281	55
Guyana	202	...
Brazil	160	...
Venezuela	126	...
Argentina	89	...
Chile	83	...
Uruguay	73	...
Bolivia	44	...
Paraguay	n.a.	...
Not elsewhere classified	54	...

Sources: Foreign-Born=U.S. Census Bureau: www.census.gov/population/socdemo/foreign/ppl-145/tab03-4.pdf. Illegal Alien=U.S. Immigration and Naturalization Service: www.ins.usdoj.gov/graphics/aboutins/statistics/illegalalien/illegal.pdf.

legal and illegal immigrants. Mexico has been the leading donor country, but Cuba, El Salvador, the Dominican Republic, Colombia, Jamaica, Haiti, and many other nations have all sent significant numbers of migrants northward (Table 12.14).

Although we will never know for sure the precise numbers, there is universal agreement among scholars and government officials that illegal Latin American and Caribbean migrants to the United States have greatly outnumbered their legal counterparts. In recent years, for example, the United States Border Patrol in the Southwest Border Region has apprehended from 60,000 to 120,000 illegal migrants monthly, and the number who are not apprehended is likely many times greater (U.S. Immigration and Naturalization Service: www.ins.usdoj.gov/). Occasionally, the sheer volume of illegal aliens and the costs of trying to apprehend and deport them have prompted Congress to offer amnesty to the migrants. This occurred most recently in the 1986 Immigration Reform and Control Act, which permitted illegal aliens who had resided continuously in the United States since January 1,

1982, as well as workers who had spent a minimum of ninety days in perishable crop agriculture during 1986, to apply for and receive permanent legal status. Over 3.1 million persons had applied for citizenship under this legislation by August of 1990 (Papademetriou 1991, 315).

In other instances, the United States government has authorized "temporary" resident status for illegal aliens whose home countries have been severely damaged by natural disasters and which are judged to be incapable of effectively assimilating large numbers of repatriated citizens. This happened in late 1998 when, in the aftermath of Hurricane Mitch, the U.S. Immigration and Naturalization Service announced that undocumented Honduran and Nicaraguan nationals already residing in the country would be allowed to remain at least until July 2000 without being deported if they registered with the federal government.

Critics have long maintained that illegal immigrants take jobs that would otherwise be filled by American citizens, are responsible for much crime and social unrest within the country, and cost the federal government huge sums of money in the form of subsidized public services. These claims have been countered, however, by research that suggests that undocumented aliens, as a group, are a significant economic asset to the United States. Bustamante (1977), North and Houstoun (1976), and Villalpando (1975), for instance, each found that approximately three-fourths of all illegal immigrants paid social security and federal income taxes, while less than 1.0 percent had received welfare benefits and fewer than 4.0 percent had children enrolled in public schools. Defenders of undocumented immigrants argue, further, that the migrants tend to fill low-paying manual labor positions that most American citizens are unwilling to accept, and that the illegal aliens are not likely to engage in socially deviant behavior owing to their fear of being apprehended and deported.

Regardless of the relative economic and social costs of illegal immigration, the long-term persistence and creativity that potential Latin American and Caribbean migrants often demonstrate in pursuit of their goal of entering the United States speaks volumes about the perceived quality of life differences between the neighboring regions. Members of one Dominican family studied by Garrison and Weiss (1979), for example, used each of the following legal or illegal strategies to reunite themselves in the United States: arranging in the Dominican Republic to work legally in the United States; one family member already legally residing in the United States sponsoring the admission of another; contracting for a large sum of money a "marriage of convenience" to a legal American resident who later divorces the newly arrived immigrant; overstaying a tourist visa; sending one person to the United States on documents "borrowed" from another; use of an illegally purchased passport; and crossing international borders with no papers at all.

POLITICAL REFUGEES

Beyond its regular legal immigration quotas, the United States government has, from time to time, allowed the entry of additional Latin American and Caribbean migrants who are classified as **political refugees.** The process of determining who qualifies as a refugee and who does not has come under increasing scrutiny in recent years, particularly with respect to would-be Caribbean immigrants.

The principal controversy has centered on Cuban applicants, who have been virtually assured of admission, and the Haitian boat people, who have been routinely turned back or deported. Critics have charged that American refugee policy is racist, owing to the fact that most of the Cuban immigrants are relatively light-skinned while the Haitians are largely Blacks or *mulattos*. Government policy defenders have countered that the Haitian migrants are economically motivated, while the Cubans are fleeing Castro's political repression. The key weakness in both arguments is the unspoken assumption that immigrants leave for only one reason. The reality is that human beings generally have multiple motivations for their actions, which are often based on both political and economic considerations.

THE BRAIN DRAIN

A second form of recent specialized Latin American and Caribbean immigration to the United States has been the flight of large numbers of highly educated and skilled professional and technical personnel, who are often frustrated by the relatively low levels of compensation and the antiquated equipment available to them in their native countries. The seeds of this discontent may be sown unintentionally when bright young foreign students come to pursue advanced degrees at American, Canadian, or European universities, where they are given access to modern equipment and libraries and exposed to persons living at high levels of material consumption. On completing their studies and returning home, the new graduates frequently find that local economic and/or political constraints make it virtually impossible for them to work and live as they desire. Professionals trained in Latin American and Caribbean universities are also subject to the same frustrations as their American-trained colleagues. If their disenchantment becomes strong enough that they choose to leave, they become a part of the so-called **brain drain** now occurring between the technologically less developed and more developed nations.

The extent of the brain drain for Latin America and the Caribbean as a whole is difficult to measure, but surveys of individual regions and nations suggest that it is widespread. Thomas-Hope (1999), for instance, reported that over half of all immigrants to the United States from the English-speaking Caribbean nations were employed as executives or skilled professional and technical persons prior to their departure. Escobar-Navia (1991) reported that over half of the emigrants who leave Colombia for the United States have more than twelve years of schooling, and Muñiz (1991) stated that in 1989 a minimum of 300 leading Argentine researchers and 15,000 professionals were living abroad. The exodus of nurses from Jamaica has been so great in recent years that the Ministry of Health has been forced to close whole hospital wards and reduce the services offered in many facilities (*World Development Report* 1993).

The consequences of the brain drain, which has been and continues to be encouraged by United States immigration priorities that favor the admission of professional and technically skilled persons, are most serious. In addition to losing the desperately needed services of these individuals, the Latin American and Caribbean nations also forfeit the time and money invested in their training. In 1980, for example, the estimated 20,595 professionals native to the Andean countries residing in the United States represented a loss of U.S. $617,850,000 in training expenses alone (Escobar-Navia 1991, 224). Such losses, multiplied many times over, have led some observers, such as Chaney (1979, 206), to describe the brain drain as a "reverse aid program favoring the more developed countries."

Remittances

While the brain drain clearly represents a net economic loss to the Latin American and Caribbean nations, we should recognize that international migration generally results in considerable financial gain to the donor regions. One way in which this gain is realized is through a reduction of local unemployment levels, which would be higher than they already are if the migrants were to remain in their native countries.

The second significant economic benefit to the donor regions is the money, or **remittances,** sent back home by migrants to support their loved ones. Remittance levels vary greatly by country, with those that send the greatest number of migrants tending to benefit the most (Figure 12.32). In many cases, remittances constitute one of the principal sources of national income, often ranking higher than other more traditional economic activities. Remittances received in 1999, for instance, were equal in Mexico to all revenues from tourism and in Peru to the value of exports of fish products. In the Dominican Republic they were triple the value of all agricultural exports, and in El Salvador they were nine times the value of official development assistance (Inter-American Development Bank 2001). Needless to say, the loss of remittances, should it ever occur, would inflict great hardship on millions of Latin American and Caribbean families.

CONCLUSIONS: ARE LATIN AMERICA AND THE CARIBBEAN OVERPOPULATED?

As we complete our study of urbanization and population growth, let us consider briefly two interrelated questions. The first is whether the Latin American and Caribbean nations have become overpopulated, and the second is to what extent a correlation exists between population densities and levels of poverty or wealth.

The first thing we must do, when asking these questions, is to recognize that the term "overpopulation" can mean different things to different people (Lappé, Collins, and Rosset 1998). To some, it is simply a measure of population densities, with any region or nation having more than a certain number of people per square mile being defined as overpopulated. The value of this approach is limited by the difficulty of deciding exactly what density constitutes overpopulation and also by the realization that not all areas are equally endowed in the productive potentials of their natural environments.

This dilemma has led some scholars to attempt to define overpopulation as a certain population density per arable, or farmable, unit of land. While there is some merit in this second approach, it suffers from the lack of a satisfactory definition of "arable." Mountain slopes, for example, may normally be defined as nonarable, yet if the local peasantry has no other recourse, they will surely farm the mountains.

A third factor complicating an attempt to define overpopulation is the level of technology available to the people. The same plot of ground that may yield very poor harvests without fertilization can be made much more productive, for instance, with regular applications of fertilizer. Similarly, regions with large industrial capacities can comfortably support many more people than regions that have similar physical environments but are largely unindustrialized. Overpopulation is therefore a difficult condition to objectively define because a nation with a relatively high population density

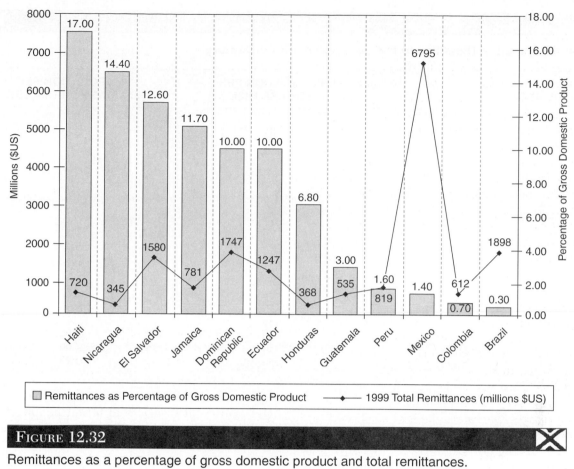

Remittances as a percentage of gross domestic product and total remittances.
Source: Inter-American Development Bank 2001.

actually may be able to economically better support its people than one with a lower population density.

This principle is illustrated well by the data found in Table 12.15, where we learn that the population densities of most of the Middle and South American countries are far lower than those of the Netherlands, Japan, the United Kingdom, Germany, and other more prosperous European and Anglo American nations. The lack of a correlation between population densities and income levels (which will be addressed in detail in chapter 13) is further evidenced by the fact that many of the poorest Latin American nations, including Nicaragua, Honduras, Bolivia, and Paraguay, also have some of the lowest population densities. Likewise, the Caribbean Islands with the highest population densities—Barbados, Puerto Rico, Martinique, and Aruba—also rank among the most prosperous.

It is thus debatable whether Latin America and the Caribbean are overpopulated. All observers do agree on three things: (1) overall population levels are far higher than they likely have ever been before; (2) most Latin American and Caribbean countries have only light to moderate population densities when compared to other world nations; and (3) population densities vary widely from place to place within the region, with the Caribbean Island realm being the most densely settled and the Southern Cone countries, Brazil, and the Guianas the least. Future Latin American and Caribbean income levels will reflect not only population densities but also the intensity of landuse and the utilization of industrial technologies. It is to the analysis of these income levels, and their impacts on the quality of human life, that we turn our attention in the concluding chapter.

TABLE 12.15

National Population Densities in the Early Twenty-First Century

COUNTRY/AREA	POPULATION PER SQUARE MILE	POPULATION PER SQUARE KILOMETER
Caribbean		
Antigua and Barbuda	394	152
Aruba	939	363
Barbados	1,666	643
Cayman Islands	356	173
Cuba	254	98
Dominica	242	93
Dominican Republic	466	180
Grenada	671	259
Guadeloupe	660	255
Haiti	668	258
Jamaica	633	245
Martinique	994	384
Netherlands Antilles	691	267
Puerto Rico	1,152	445
St. Kitts and Nevis	384	148
St. Lucia	673	260
St. Vincent and the Grenadines	776	300
Trinidad and Tobago	588	227
Middle America		
Mexico	137	53
Guatemala	317	122
Belize	33	13
Honduras	152	59
El Salvador	782	302
Nicaragua	101	39
Costa Rica	196	76
Panama	97	37
Andean South America		
Venezuela	69	27
Colombia	93	36
Ecuador	129	50
Peru	56	22
Bolivia	20	8
Southern South America		
Chile	55	21
Argentina	35	14
Paraguay	38	15
Uruguay	47	18
Brazil	54	21
Guianas		
Guyana	8	3
Suriname	7	3
French Guiana	4	2
Non-Latin American Comparisons		
United States	78	30
Canada	8	3
France	271	105
Germany	604	233
Portugal	277	107
Spain	206	79
United Kingdom	634	245
Switzerland	458	177
Netherlands	1,019	393
Japan	871	336

Sources: Modified from *The Statesman's Yearbook 2002* 2001; U.S. Census Bureau International Programs Center, 2002: www.census.gov/cgi-bin/ipc/idbrank.pl; *The World Almanac and Book of Facts 2001* 2001.

KEY TERMS

vecinos 306	*plaza mayor* 313	push and pull factors 341
primate city 310	grid pattern 314	rural to urban migration 342
agricultural city 310	*bandeiras* 317	Internally Displaced Persons 342
mining city 310	Demographic Transition Model 323	political refugees 345
industrial city 311	megalopolis 328	brain drain 346
commercial settlements 311	*vecindades* 335	remittances 346
administrative centers 311	shantytowns or squatter settlements 336	

SUGGESTED READINGS

Arreola, Daniel D., and James R. Curtis. 1993. *The Mexican Border Cities: Landscape Anatomy and Place Personality.* Tucson: University of Arizona Press.

Bataillon, Claude. 1982. *Las Regiónes Geográficas en México.* 6a ed. Trans. Florentino M. Torner. México, D. F.: Siglo Veintiuno Editores, S. A.

Bebbington, Anthony. 2000. "Reencountering Development: Livelihood Transitions and Place Transformations in the Andes." *Annals of the Association of American Geographers* 90: 495–520.

Boswell, Thomas D. 1989. "Population and Political Geography of the Present-Day West Indies." In *Middle America: Its Lands and Peoples,* 3rd ed., eds. Robert C. West and John P. Augelli, 103–127. Englewood Cliffs, N.J.: Prentice-Hall.

Bromley, R. J. 1972. "Agricultural Colonization in the Upper Amazon Basin: The Impact of Oil Discoveries." *Tijdschrift voor Economische en Sociale Geografie* 63:278–294.

Bromley, Rosemary D. F. 1979. "The Functions and Development of 'Colonial' Towns: Urban Change in the Central Highlands of Ecuador, 1698–1940." *Institute of British Geographers Transactions* n.s. 4:30–43.

Burns, E. Bradford. 1980. *The Poverty of Progress: Latin America in the Nineteenth Century.* Berkeley: University of California Press.

Bustamente, Jorge A. 1977. "Undocumented Immigration from Mexico: Research Report." *International Migration Review* 11:149–177.

Butler, Edgar W., James B. Pick, and W. James Hettrick. 2001. *Mexico and Mexico City in the World Economy.* Boulder, Colo.: Westview.

Butterworth, Douglas, and John K. Chance. 1981. *Latin American Urbanization.* Cambridge: Cambridge University Press.

Caviedes, Cesar. 1984. *The Southern Cone: Realities of the Authoritarian State in South America.* Totowa: Rowman and Allanheld.

Céspedes, Guillermo. 1976. *América Latina Colonial Hasta 1650.* México, D. F.: Secretaría de Educación Pública.

Chance, John K. 1976. "The Urban Indian in Colonial Oaxaca." *American Ethnologist* 3:603–632.

Chandler, Tertius. 1987. *Four Thousand Years of Urban Growth: An Historical Census.* Lewiston, N.Y.: St. David's University Press.

Chaney, Elsa M. 1979. "The World Economy and Contemporary Migration." *International Migration Review* 13:204–212.

Child, Jack. 1986. "Antarctica and Argentine Geopolitical Thinking." In *Yearbook 1986,* ed. David L. Clawson, 11–16. Muncie, Ind.: Conference of Latin Americanist Geographers.

Condon, Stephanie A., and Philip E. Ogden. 1991. "Emigration from the French Caribbean: The Origins of an Organized Migration." *International Journal of Urban and Regional Research* 15:505–523.

Crist, Raymond E. 1963. "Why Move a Capital: Brasília's Origins." *Americas* 15:13–17.

Crist, Raymond E., and Charles M. Nissly. 1973. *East from the Andes.* Gainesville: University of Florida Press.

Crowley, William K. 1987. "La Gran Plaza, Monterrey's Grand Attempt at Urban Renewal." In *Yearbook 1987,* ed. Martha A. Works, 36–44. Auburn, Ala.: Conference of Latin Americanist Geographers.

Dawsey, Cyrus B., and James M. Dawsey, eds. 1995. *The Confederados: Old South Immigrants in Brazil.* Tuscaloosa: University of Alabama Press.

de Jesus, Carolina Maria. 1962. *Child of the Dark: The Diary of Carolina Maria de Jesus.* Trans. by David St. Clair. New York: New American Library.

De la Cadena, Marisol 2000. *Indigenous Mestizos: The Politics of Race and Culture in Cuzco, Peru, 1919–1991.* Durham, N.C.: Duke University Press.

de Soto, Hernando. 1990. "Finding the Path Out of Latin American Poverty, Debt and Drugs: The Informal Revolution." In *Latin America: How New Administrations Will Meet the Challenge,* 73–79. Washington, D.C.: Inter-American Development Bank.

De Souza, Roger-Mark. 1998. "The Spell of Cascadura: West Indian Return Migration." In *Globalization and Neoliberalism: The Caribbean Context,* ed. Thomas Klak, 227–253. Lanham, Md.: Rowman and Littlefield.

Demographic Yearbook 1999. 2001. New York: United Nations.

Despres, Leo A. 1991. *Manaus: Social Life and Work in Brazil's Free Trade Zone.* Albany: State University of New York Press.

Drakakis-Smith, David. 2000. *Third World Cities.* 2nd ed. London: Routledge.

Duany, Jorge. 1992. "Caribbean Migration to Puerto Rico. A Comparison of Cubans and Dominicans." *International Migration Review* 26:46–66.

Eckstein, Susan. 1990. "Urbanization Revisited: Inner-City Slum of Hope and Squatter Settlement of Despair." *World Development* 18:165–181.

Economic and Social Progress in Latin America 1980–1981 Report, 1993 Report, and *1998–1999 Report.* 1981, 1993, 1998. Washington, D.C.: Inter-American Development Bank.

Eidt, Robert C. 1962. "Pioneer Settlement in Eastern Peru." *Annals of the Association of American Geographers* 52:255–278.

Elbow, Gary S. 1975. "The Plaza and the Park: Factors in the Differentiation of Guatemalan Town Squares." *Growth and Change* 6:14–18.

Epstein, David G. 1973. *Brasília, Plan and Reality.* Berkeley: University of California Press.

Escobar-Navia, R. 1991. "South-North Migration in the Western Hemisphere." *International Migration* 29:223–230.

Figueroa, Oscar. 1996. "A Hundred Million Journeys a Day: The Management of Transport in Latin America's Mega-cities." In *The Mega-city in Latin America,* ed. Alan Gilbert. Tokyo: United Nations University Press.

Flinn, W. L., and J. W. Converse. 1970. "Eight Assumptions Concerning Rural-Urban Migration in Colombia: A Three-Shantytowns Test." *Land Economics* 46:456–466.

Ford, Larry R. 1996 "A New and Improved Model of Latin American City Structure." *The Geographical Review* 86:437–440.

Freitas Marcondes, J. V. 1959. "Brasília, The New Capital of Brazil." *Mississippi Quarterly* 12:157–167.

Gade, Daniel W. 1992. "Landscape, System, and Identity in the Post-Conquest Andes." *Annals of the Association of American Geographers* 82:444–459.

———. 1976. "The Latin American Central Plaza as Functional Space." In *Latin America: Search for Geographic Explanations,* ed. Robert J. Tata, 16–23. Muncie, Ind.: Conference of Latin Americanist Geographers.

Gade, Daniel W., and Mario Escobar. 1982. "Village Settlement and the Colonial Legacy in Southern Peru." *The Geographical Review* 72:430–449.

Gakenheimer, Ralph A. 1966. "Decisions of Cabildo on Urban Physical Structure: 16th Century Peru." In *The Urbanization Process in America from its Origins to the Present Day,* eds. Jorge E. Hardoy and Richard P. Schaedel, 241–260. Buenos Aires: Editorial del Instituto, Instituto Torcuato Di Tella.

Garrison, Vivian, and Carol I. Weiss. 1979. "Dominican Family Networks and United States Immigration Policy: A Case Study." *International Migration Review* 13:264–283.

Garza, Gustavo, and Martha Schteingart. 1978. "Mexico City: The Emerging Megalopolis." In *Latin American Urban Research,* Vol. 6, eds. Wayne A. Cornelius and Robert V. Kemper, 51–85. Beverly Hills: Sage Publications.

Gauthier, Howard L. 1975. "Migration Theory and the Brazilian Experience." *Revista Geográfica del Instituto Panamericano de Geografía e Historia* 82:51–62.

Gibson, Charles. 1966. "Spanish-Indian Institutions and Colonial Urbanism in New Spain." In *The Urbanization Process in America from its Origins to the Present Day,* eds. Jorge E. Hardoy and Richard Schaedel, 225–239. Buenos Aires: Editorial del Instituto, Instituto Torcuato Di Tella.

Gilbert, Alan G. 1996a. "Land, Housing, and Infrastructure in Latin America's Major Cities." In *The Mega-city in Latin America,* ed. Alan G. Gilbert, 73–109. Tokyo: United Nations University Press.

———. 1996b. *The Mega-city in Latin America.* Tokyo: United Nations University Press.

Gilbert, Alan G., and Peter J. Sollis. 1979. Migration to Small Latin American Cities: A Critique of the Concept of 'Fill-In' Migration." *Tijdschrift Voor Economische en Sociale Geografie* 70(2):110–113.

Godfrey, Brian J. 1991. "Modernizing the Brazilian City." *The Geographical Review* 81 (January): 18–34.

———. 1990. "Boom Towns of the Amazon." *The Geographical Review* 80:103–117.

Gottmann, Jean. 1961. *Megalopolis: The Urbanized Northeastern Seaboard of the United States.* New York: The Twentieth Century Fund.

Griffin, Ernst, and Larry Ford. 1980. "A Model of Latin American City Structure." *The Geographical Review* 70:397–422.

———. 1993. "Cities of Latin America." In *Cities of the World: World Regional Urban Development,* 2nd ed., eds. Stanley D. Brunn and Jack F. Williams, 225–265. New York: Harper Collins College Publishers.

Gwynne, Robert N. 1986. *Industrialization and Urbanization in Latin America.* Baltimore: The Johns Hopkins University Press.

Hardoy, Jorge E. 1975. "Two Thousand Years of Latin American Urbanization." In *Urbanization in Latin America: Approaches and Issues,* ed. Jorge E. Hardoy, 3–55. Garden City, N.J.: Anchor Press.

Hardoy, Jorge E., con Carmen Aranovich. 1966. "Escalos y Funciones Urbanos en América Hispánica Hacia el Año 1600-Primeras Conclusiones." In *The Urbanization Process in America from its Origins to the Present Day,* eds. Jorge E. Hardoy and Richard P. Schaedel, 171–208. Buenos Aires: Editorial del Instituto, Instituto Torcuato Di Tella.

Harter, Eugene C. 1985. *The Lost Colony of the Confederacy.* Jackson: University Press of Mississippi.

Inter-American Development Bank. 2001. "Remittances as a Development Tool: A Regional Conference." Statistical overview data for a conference held in Washington, D.C. on May 17–18, 2001; accessible on the Internet at: www.iadb.org.

Iwańska, Alicja. 1993. *British American Loyalists in Canada and U.S. Southern Confederates in Brazil: Exiles from the United States.* Lewiston, N.Y.: Edwin Mellen Press.

Jefferson, Mark. 1928. "An American Colony in Brazil." *The Geographical Review* 18:226–231.

———. 1939. "The Law of the Primate City." *The Geographical Review* 29:226–232.

Johns, Michael. 1997. *The City of Mexico in the Age of Díaz.* Austin: University of Texas Press.

Keeling, David J. 1996. *Buenos Aires: Global Dreams, Local Crises.* Chichester: John Wiley & Sons.

Lappé, Frances Moore, Joseph Collins, and Peter Rosset. 1998. *World Hunger: 12 Myths,* 2nd ed. New York: Grove Press.

Lutz, T. M. 1970. *Self-help Neighborhood Organization, Political Orientations of Urban Squatters in Latin America: Contrasting Patterns from Case Studies in Panama City, Guayaquil and Lima.* Ph.D. diss., Georgetown University, Washington, D.C.

Mahoney, Harry Thayer, and Marjorie Locke Mahoney. 1998. *Mexico and the Confederacy 1860–1867.* San Francisco: Austin and Winfield.

Mangin, William. 1967. "Latin American Squatter Settlements: A Problem and a Solution." *Latin American Research Review* 2:65–98.

Mangurian, David. 1997. "A New Future for Rio's Favelas." *IDB Extra.* Washington, D.C.: Inter-American Development Bank.

Matos Mar, José. 1961. "Migration and Urbanization: The Barriadas of Lima: An Example of Integration into Urban Life." In *Urbanization in Latin America,* ed. Philip N. Hauser, 170–190. New York: International Documents Service of Columbia University Press.

Millon, Rene. 1966. "Urbanization at Teotihuacán: The Teotihuacán Mapping Project." In *The Urbanization Process in America from its Origins to the Present Day,* eds. Jorge E. Hardoy and Richard P. Schaedel, 105–120. Buenos Aires: Editorial del Instituto, Instituto Torcuato Di Tella.

Moore, Richard J. 1978. "Urban Problems and Policy Responses for Metropolitan Guayaquil." In *Latin American Urban Research,* Vol. 6, eds. Wayne A. Cornelius and Robert V. Kemper, 181–203. Beverly Hills: Sage Publications.

Morris, Arthur S. 1981. *Latin America: Economic Development and Regional Differentiation.* Totowa, N.J.: Barnes & Noble.

Morse, Richard M. 1962. "Some Characteristics of Latin American Urban History." *American Historical Review* 67:317–338.

———. 1975. "The Development of Urban Systems in the Americas in the Nineteenth Century." *Journal of Interamerican Studies and World Affairs* 17:4–26.

Morse, Richard M., ed. 1971. *The Urban Development of Latin America 1750–1920.* Stanford: Stanford University Center for Latin American Studies.

Moser, Caroline O. N. 1982. "A Home of One's Own: Squatter Housing Strategies in Guayaquil, Ecuador." In *Urbanization in Latin America: Critical Approaches to the Analysis of Urban Issues,* ed. Alan Gilbert, 1982. New York: John Wiley & Sons.

Muñiz, C. M. 1991. "The Emigration of Argentine Professionals and Scientists." *International Migration* 29:231–239.

Nelson, Howard J. 1963. "Townscapes of Mexico: An Example of the Regional Variation of Townscapes." *Economic Geography* 39:74–83.

North, D. S., and M. F. Houstoun. 1976. *The Characteristics and Role of Illegal Aliens in the U.S. Labor Market: An Exploratory Study.* Washington, D.C.: Linton & Company.

North, David S., and Judy A. Whitehead. 1991. "Policy Recommendations for Improving the Utilization of Emigrant Resources in Eastern Caribbean Nations." In *Small Country Development and International Labor Flows: Experiences in the Caribbean,* ed. Anthony P. Maingot, 1–52. Boulder, Colo.: Westview Press.

Nutall, Zelia. 1922. "Royal Ordinances Concerning the Laying Out of New Towns (July 3, 1573)." *Hispanic American Historical Review* 5:249–254.

Ozorio de Almeida, Anna Luiza. 1992. *The Colonization of the Amazon.* Austin: University of Texas Press.

Papademetriou, D. G. 1991. "South-North Migration in the Western Hemisphere and U.S. Responses." *International Migration* 29:291–316.

Pearse, Andrew. 1961. "Some Characteristics of Urbanization in the City of Rio de Janeiro." In *Urbanization in Latin America,* ed. Philip N. Hauser, 191–205. New York: International Documents Service of Columbia University Press.

Pérez Perdomo, Rogelio, and Pedro Nikken. 1982. "The Law and Home Ownership in the *Barrios* of Caracas." In *Urbanization in Latin America: Critical Approaches to the Analysis of Urban Issues,* ed. Alan Gilbert, 205–229. New York: John Wiley & Sons.

Population Reference Bureau. 2002. *2002 World Population Data Sheet.* Washington, D.C.: Population Reference Bureau.

Portes, Alejandro. 1972. "Rationality in the Slum: An Essay on Interpretive Sociology." *Comparative Studies in Society and History* 14:268–286.

Portes, Alejandro, and John Walton. 1976. *Urban Latin America: The Political Condition from Above and Below.* Austin: University of Texas Press.

Radcliffe, Sarah A. 1990. "Between Hearth and Labor Market: The Recruitment of Peasant Women in the Andes." *International Migration Review* 24:229–249.

Ravenstein, E. G. 1885. "The Laws of Migration." *Journal of the Royal Statistical Society* 48:167–227.

———. 1889. "The Laws of Migration." *Journal of the Royal Statistical Society* 52:241–305.

Reichert, Josh, and Douglas S. Massey. 1979. "Patterns of U.S. Migration from a Mexican Sending Community: A Comparison of Legal and Illegal Migrants." *International Migration Review* 13:599–623.

Richardson, Miles. 1982. "Being-in-the Market versus Being-in-the-Plaza: Material Culture and the Construction of Social Reality in Spanish America." *American Ethnologist* 9:421–436.

Roberts, J. Timmons. 1992. "Squatters and Urban Growth in Amazonia." *The Geographical Review* 82:441–457.

Rojas, Eduardo, and Robert Daughters, eds. 1998. *La Ciudad en el Siglo XXI: Experiencias Exitosas en Gestión del Desarrollo Urbano en América Latina.* Washington, D.C.: Banco Interamericano de Desarrollo.

Rubenstein, H. 1992. "Migration, Development and Remittances in Rural Mexico." *International Migration* 30:127–153.

Ruddle, Kenneth, and Donald Obermann, eds. 1972. *Statistical Abstract of Latin America 1971.* Los Angeles: University of California at Los Angeles Latin American Center.

Sánchez-Albornoz, Nicolás. 1974. *The Population of Latin America: A History.* trans. W. A. R. Richardson. Berkeley: University of California Press.

Sanders, William T., and David Webster. 1988. "The Mesoamerican Urban Tradition." *American Anthropologist* 90:521–546.

Scott, Allen J., ed. 2001. *Global City-Regions: Trends, Theory, Policy.* Oxford: Oxford University Press.

Simmonds, Roger, and Gary Hack, eds. 2000. *Global City Regions: Their Emerging Forms.* London: Spon Press.

Smith, Robert C. 1955. "Colonial Towns of Spanish and Portuguese America." *Journal of the Society of Architectural Historians* 14 (December): 3–12.

Stanislawski, Dan. 1947. "Early Spanish Town Planning in the New World." *The Geographical Review* 37:94–105.

———. 1946. "The Origin and Spread of the Grid-Pattern Town." *The Geographical Review* 36:105–120.

Stann, E. Jeffrey. 1975. "Transportation and Urbanization in Caracas, 1891–1936." *Journal of Interamerican Studies and World Affairs* 17:82–100.

The Statesman's Yearbook 2002. 2001. New York: St. Martin's Press.

Statistical Yearbook 1995. 1997. New York: United Nations.

Stokes, Susan C. 1991. "Politics and Latin America's Urban Poor: Reflections from a Lima Shantytown." *Latin American Research Review* 26(2):75–101.

Thomas-Hope, Elizabeth. 1999. "Emigration Dynamics in the Anglophone Caribbean." In *Emigration Dynamics in the Developing Countries, volume III: Mexico, Central America, and the Caribbean,* ed. Reginald Appleyard, 232–284. Aldershot, England: Ashgate.

Ugalde, Antonio, Frank D. Bean, and Gilbert Cárdenas. 1979. "International Migration from the Dominican Republic: Findings from a National Survey." *International Migration Review* 13:235–263.

Villa, Miguel, and Jorge Rodríguez. 1996. "Demographic Trends in Latin America's Metropolises, 1950–1990. In *The Mega-city in Latin America,* ed. Alan G. Gilbert, 25–52. Tokyo: United Nations University Press.

Villalpando, V. 1975. *A Study of the Impact of Illegal Aliens on the County of San Diego on Specific Socioeconomic Areas.* San Diego, Calif.: San Diego County Human Resources Agency.

Walton, John. 1978. "Guadalajara: Creating the Divided City." In *Latin American Urban Research,* Vol. 6, eds. Wayne A. Cornelius and Robert V. Kemper, 25–50. Beverly Hills: Sage Publications.

Weaver, Blanche Henry Clark. 1952. "Confederate Immigrants to Brazil." *Journal of Southern History* 18:446–448.

Wilkie, James W., ed. 1978. *Statistical Abstract of Latin America.* Los Angeles: University of California at Los Angeles Latin American Center Publications.

The World Almanac and Book of Facts 2001. 2001. New York: Primedia Reference.

World Development Report 1992, 1993, and *1995.* 1992, 1993, 1995. New York: Oxford University Press for the World Bank.

Zaaijer, Mirjam. 1991. "Quito." *Cities* 8:87–92.

ELECTRONIC SOURCES

U.S. Census Bureau: www.census.gov/population/socdemo/foreign/ppl-145/tab03-4.pdf

U.S. Census Bureau International Programs Center, 2002: www.census.gov/cgi-bin/ipc/idbrank.pl

U.S. Immigration and Naturalization Service: www.ins.usdoj.gov/graphics/aboutins/statistics/

World Bank World Development Indicators 2002: http://devdata.worldbank.org/data-query/

The World Gazetteer 2002: www.world-gazetteer.com/st/statn.htm

World Population Data Sheet: www.prb.org/Content/Navigation Menu/Other_reports/

World Refugee Survey 2002: www.refugees.org/downloads/wrs02/wrs02_table 5.pdf

13

Development and Health

The term "development" has come to mean different things to different people. Government officials, both within and outside of Latin America and the Caribbean, frequently define development in terms of a composite measurement of key economic indicators. According to this reasoning, the higher the per capita economic output of a country or region, the more "developed" or "advanced" its people are. In addition to per capita gross product, other commonly used economic indicators of development or underdevelopment include the proportion of the population with access to electricity, potable water, and other physical amenities, as well as consumption levels of automobiles, television sets, and similar consumer goods.

Although few, if any, observers would question the usefulness of these variables as indicators of the economic aspects of development, most social scientists and humanists would argue that full development, like life itself, consists of much more than the acquisition and consumption of material goods. According to this perspective, the attainment of a developed state is the process of achieving, as fully as possible, one's innate human potential (Simpson 1994; Seers 1972; Goulet 1971). This potential has numerous noneconomic, as well as economic, expressions. De Souza and Porter (1974, 3–4), for example, listed the following nine "objectives of development":

1. A healthful, balanced diet in all seasons
2. Adequate medical care throughout life
3. Environmental sanitation and control of disease
4. Labor opportunities of sufficient variety to enable individuals to develop their varied talents
5. Adequate opportunities for learning useful skills and developing the mind

6. Safety of person and freedom of conscience, including religious belief
7. Adequate housing
8. Systems of economic production that are in balance with the environment
9. Political and social equality

Other variables might be added, but de Souza's and Porter's list is sufficient to make the point that development is a multifaceted process that entails interdependent economic and noneconomic expressions. The presence of an egalitarian social structure, for instance, certainly increases the likelihood that the youth of a given nation will receive a good education, which in turn will improve their access to adequate housing, health care, nutrition, and other economic manifestations of development.

As we come to understand the all-encompassing nature of the development process, we begin to appreciate also the difficulties in comparing levels of development between different nations or regions. Although there likely never will be a completely satisfactory method of measuring development, one of the most balanced assessments devised to date is the Human Development Index (HDI), which is published annually by the United Nations Development Programme. Recognizing that many noneconomic aspects of development are difficult, if not impossible, to measure and also that economic variables tend to be interrelated, the authors of the HDI have chosen to incorporate just three key criteria: life expectancy, educational attainment, and adjusted real income. Although care should be taken not to associate a high or low national ranking with any notion of inherent inferiority or superiority, many analysts use the HDI to

TABLE 13.1

Human Development Index Rankings of Latin American and Caribbean Nations, 2002

COUNTRY	RANKING*	COUNTRY	RANKING*
Barbados	31	El Salvador	104
Argentina	34	Bolivia	114
Chile	38	Honduras	116
Uruguay	40	Nicaragua	118
Bahamas	41	Guatemala	120
Costa Rica	43	Haiti	146
St. Kitts and Nevis	44		
Trinidad and Tobago	50	**NON–LATIN AMERICAN AND CARIBBEAN COMPARISONS**	
Antigua and Barbuda	52		
Mexico	54	Norway	1
Cuba	55	Sweden	2
Panama	57	Canada	3
Belize	58	United States	6
Dominica	61	Japan	9
St. Lucia	66	France	12
Colombia	68	United Kingdom	13
Venezuela	69	Germany	17
Brazil	73	Italy	20
Suriname	74	Spain	21
Peru	82	Portugal	28
Grenada	83	Russian Federation	60
Jamaica	86	Philippines	77
Paraguay	90	China	96
St. Vincent and the Grenadines	91	India	124
Ecuador	93	Nigeria	148
Dominican Republic	94	Sierra Leone	173
Guyana	103		

*1 = most developed; 173 = least developed

Source: Compiled from Human Development Report 2002.

determine whether or not overall human conditions are improving over time in a given country. The HDI also provides a sense of the relative or comparative well-being of the citizens of different nations, both within Latin America and the Caribbean and within the world generally (Table 13.1 and Figure 13.1).

In analyzing the HDI rankings, it is apparent that the Latin American and Caribbean peoples, as a group, are presently experiencing a mostly middle to moderately low level of social and economic development. While their overall standing is clearly not in the upper stratum, which is dominated by Western European and Anglo American nations, neither is it in the lowest echelon of countries, the majority of which are in Africa or Asia.

In the remainder of this chapter, we will first assess the economic manifestations of Latin American and Caribbean underdevelopment and then its cultural dimensions. Because social class structure, political behav-

ior, education, and racial and gender issues have been addressed in previous chapters, we will center our treatment of cultural expressions on health and nutrition.

ECONOMIC DEVELOPMENT AND UNDERDEVELOPMENT DURING THE COLONIAL AND REPUBLICAN PERIODS

If we define development in a technological sense as the utilization of mechanized industrial processes to generate wealth, it is apparent that the Latin American and Caribbean nations were largely underdeveloped on the eve of the early nineteenth-century revolutionary wars. While limited mechanization had been introduced into the mining, textile, and food processing sectors, these activities existed only as relatively small and isolated enterprises in the principal urban centers. Separating the industrial enclaves were vast rural hinterlands, the great

FIGURE 13.1

Human development index rankings of Latin American and Caribbean nations, by quartiles of world rankings: 2002.

Source: Data from Human Development Report, 2002.

majority of whose residents lived and died in deep and seemingly never-ending poverty.

With their attainment of independence in the early 1800s, most of the Latin American and Caribbean nations were opened for the first time to non-Hispanic foreign investment and to the attendant influences of the international financial community. Much of the money that entered the region during the republican era came from British, French, American, and German investors, who, as we have noted in previous chapters, became increasingly prominent in the mining, agricultural, and transportation sectors. While these private investors were concentrating their efforts on the acquisition of minerals, foods, fibers, and other raw materials destined for overseas processing and consumption, the home governments of these entrepreneurs were themselves becoming increasingly active in promoting the sale of European and American industrial equipment and finished goods to the Latin American and Caribbean peoples. The principal mechanisms used to accomplish these sales included the issuing of bonds and the granting of loans and other forms of "foreign aid," which required Latin American and Caribbean officials to spend the money received from their benefactors on the purchase of industrial commodities manufactured in the nation granting the assistance. The inexperienced and sometimes self-serving Latin American governments were thus unwittingly drawn from two directions into the turbulent world financial markets, whose alternating cycles of expansion and contraction have considerably influenced the economic health of the Latin American and Caribbean peoples over the past two centuries.

The International Financial Cycle

The cycle that repeatedly has impacted the region begins with a surge of foreign loans and investments in the less developed nations during a period of surplus international capital accumulation. Because the foreign financiers are under considerable pressure to realize large returns on their investments within a short period of time, many succumb to the temptation to invest in speculative projects or in countries of low or marginal credit worthiness. The rapid influx of foreign money leads initially to a short period of accelerated economic growth that fuels inflation. With inflation soon raging, the government is eventually forced to devalue its currency, which leads to a reduction of national purchasing power. Soon, the borrower nations find their economies slipping into recession. In an ill-fated attempt to head off economic collapse, they endeavor to borrow yet more money in order to make the scheduled payments on their existing loans. By this time, or shortly thereafter, however, the world financial markets have entered into the next period of economic downturn and are unwilling to loan as much

money as before. Unable to meet the obligations on their existing debt, the Third World countries are then forced to declare some form of moratorium on their loan repayments, which action places them, in effect, in economic default. Faced with choosing between losing all their investment or just a part of it, the foreign financial institutions are then forced to renegotiate the terms of the loans, which in turn allows the borrower nations an opportunity to experience some degree of economic recovery before the entire process begins anew.

Recurring Economic Crises

The cycles of international economic expansion and contraction that have plagued Latin American and Caribbean development efforts throughout the nineteenth and twentieth centuries have each followed the general sequence of events outlined above. The principal differences between the recurring cycles have not been in their characteristics but rather in their size and scope, with each succeeding round having proven, to date, to be larger and its impacts more harmful than those that preceded it. The crises of the early independence era involved relatively small sums of money, and the ensuing recessions consequently had little impact on the lives of the masses, most of whom, as we have seen, lived a virtually subsistence existence scarcely touched by the outside world. By the 1920s, however, several of the Latin American nations were beginning to industrialize, and great amounts of money were being borrowed. Most of that money came from the United States, which, in the aftermath of the opening of the Panama Canal and its victory in World War I, had emerged as the most powerful economic and political force in the Western Hemisphere. The wild, speculative spirit of the 1920s was followed, of course, by the worldwide economic crash and depression of the 1930s. Although none of the Latin American and Caribbean nations repudiated its foreign financial obligations at this time, only Argentina, the Dominican Republic, and Haiti continued to make payments on their debts (Drake 1994, xiv–xv). One of the reasons default was so widespread was the staggering amount of debt being carried, with national foreign debts exceeding exports by ratios ranging from 4.8 to 8.6 throughout the region (Meier 1989). Much of the debt was not retired until the 1950s and then only at substantially discounted terms (Portes 1990).

The Latin American debt crisis of the 1930s and 1940s eventually gave way, in the 1950s and 1960s, to a period of relative stability and prosperity that was marked by yet another upsurge in loan activity. The leading source of funding during this period was bilateral government-to-government grants. The large majority of these came from the United States, whose generosity, as manifested in the much-publicized **Alliance**

for Progress assistance program, was largely motivated by a fear that Communism was spreading among the impoverished masses. A second major source of foreign capital during this time was the Soviet Union, which began pouring the equivalent of 5 to 6 billion dollars annually into Cuba following Castro's overthrow of the Batista regime. Loans from private banks were of little consequence, constituting only 1.0 to 3.0 percent of all borrowing in most years (Gwynne 1986).

As the economic growth of the preceding two decades continued into the early 1970s, it may have seemed to some that the centuries-old boom and bust international economic cycle had at last been broken. Then, in 1973, the world was suddenly plunged into crisis by the decision of the Organization of Petroleum Exporting Countries (OPEC) to raise oil prices to record-high levels. As hundreds of billions of dollars flowed into the coffers of the OPEC states, their leaders began depositing huge sums with the major private banks of Europe and Anglo America. These institutions then sought out, and in many cases actually created, ways that they could loan their newly gained oil monies and thereby profit from the interest to be paid by the borrowers. Because the economies of the Anglo American and European nations were fast falling into recession, the banks were unable to loan the amounts of capital they desired to their traditional customers. The banks then turned, somewhat in haste, to the less-industrialized, Third World nations as potential clients. Simultaneously, the Latin American and Caribbean countries were becoming desperate for funds needed to pay for the imported petroleum products required to sustain their industrial development. There thus began again the now all too familiar cycle of excessive loaning by foreign commercial interests to the Latin American and Caribbean nations, many of which had poor prospects of ever being able to repay the loans. Ironically, it was those countries whose economies appeared to be the strongest—Brazil, Mexico, Argentina, and Venezuela—that attracted the most attention from the international banking community and that ultimately became the most indebted.

THE LOST DECADE

As during Latin America's previous debt cycles, the rapid economic growth that had been made possible by the wild borrowing of the mid to late 1970s ended when loans became more expensive in the early 1980s. One of the leading causes of the increased cost of money was the United States government's massive, largely defense-related borrowing under the Reagan administration. The Latin American and Caribbean nations, already burdened by low levels of domestic savings, high public sector costs, widespread corruption and mismanagement of government resources, and ar-

tificially high currency exchange rates, found themselves having to pay higher and higher interest rates on their loans. Simultaneously, demand for their exports was dropping as the world entered yet another recession. Unable to make the payments on their existing loans, the Hispanic leaders felt compelled to borrow more and more money in a futile, ill-fated attempt to keep their creditors at bay. We will now analyze, first, the extent to which indebtedness has increased over the past few decades, and second, the effects of the indebtedness on other regional economic conditions.

Foreign Indebtedness

We have noted previously that debt loads among the Latin American and Caribbean nations have fluctuated widely throughout the nineteenth and twentieth centuries. One of the periods of greatest indebtedness occurred in the aftermath of the Great Depression, the effects of which were not overcome in many Hispanic countries until the 1950s. The 1960s were a decade of relatively low levels of indebtedness, with the result that in 1970 the external debt as a ratio of the gross domestic product stood in the relatively manageable 5.0 to 20.0 percent range for every Latin American and Caribbean nation except Bolivia and Chile (Table 13.2).

The borrowing frenzy that followed the OPEC price hikes of the 1970s brought the Latin American and Caribbean states once again to an extremely precarious financial position. By 1989, for instance, five countries had debts that exceeded a full year's economic output and another ten had indebtedness of between 50.0 and 90.0 percent of their gross domestic product (GDP). The 1990s brought welcomed reductions in the debt loads of many nations but overall indebtedness levels remain high, with year 2000 data showing all but seven countries continuing to owe the equivalent of 30.0 percent or more of GDP (Table 13.2). One of the consequences of the return of historically high levels of indebtedness has been runaway inflation.

Inflation

As we analyze the inflation data contained in Table 13.3, several patterns and trends emerge. The first is that the decade of the 1960s was characterized by very low levels of inflation. Of the twenty-six reporting nations, for example, seventeen, or two-thirds, experienced an average annual inflation rate of less than 5.0 percent. Further evidence that inflation was firmly under control during the 1960s came from the fact that only four states—Uruguay, Brazil, Chile, and Argentina—experienced annual rates averaging 12.0 percent or more. In other words, with the exception of the Southern Cone nations, the 1960s were a period of remarkably stable consumer prices.

TABLE 13.2

Total External Debt and Debt to Gross Domestic Product Ratios, 1970–2000

COUNTRY	MILLIONS U.S. DOLLARS			AS PERCENTAGE OF GDP		
	1970	1989	2000	1970	1989	2000
Argentina	1,872	59,615	154,961	7.5	77.8	54.5
Belize	—	153	484	—	39.8	62.6
Bolivia	477	4,269	2,747	54.9	86.6	33.1
Brazil	3,680	111,088	223,841	8.0	26.9	37.7
Chile	2,066	19,227	34,859	24.0	63.9	49.4
Colombia	1,250	17,194	33,485	18.1	42.7	40.2
Costa Rica	134	3,678	4,483	13.8	80.5	28.1
Dominican Republic	215	3,866	4,341	14.7	56.8	22.2
Ecuador	209	10,587	13,143	13.3	104.3	96.6
El Salvador	88	2,142	3,761	8.6	36.2	28.5
Guatemala	106	2,866	4,326	5.7	29.9	22.7
Guyana	—	1,882	842	—	402.9	118.1
Haiti	40	865	691	10.0	—	17.5
Honduras	90	3,634	3,110	13.0	62.2	52.5
Jamaica	129	4,546	4,341	10.3	108.9	56.3
Mexico	3,228	104,442	157,038	9.8	42.1	27.1
Nicaragua	146	9,020	5,545	19.4	842.1	231.3
Panama	193	5,660	7,285	18.9	115.7	72.7
Paraguay	98	1,989	2,950	16.7	52.8	39.2
Peru	898	16,331	28,411	14.8	35.5	53.1
Suriname	—	126	249	—	33.7	29.4
Trinidad and Tobago	78	2,511	2,553	9.7	49.5	33.1
Uruguay	267	4,415	8,204	11.0	55.7	40.9
Venezuela	729	33,170	38,744	6.7	75.6	32.0

Sources: Compiled from World Bank World Development Indicators database, 2000; *World Development Report 1978* 1978, 96–97; *Annual Report 1998*, 1999, 147; *Economic and Social Progress in Latin America 1998–1999 Report* 1998, 247.

Although the 1970s was a decade of increasing levels of inflation largely associated with the economic aftershocks of the petroleum price hikes, price increases were still relatively manageable, with all but Chile, Argentina, and Uruguay reporting annual rates of under 36.0 percent. On the other hand, the 1980s were an unmitigated fiscal disaster in many Latin American and Caribbean nations, with inflation rates tracking the unprecedented increases in foreign debt. So uncontrolled did inflation become that in five countries—Nicaragua, Argentina, Brazil, Peru, and Bolivia—prices increased more than 200 percent per year over the course of the decade (Table 13.3). Some of the more infamous individual national rates of inflation that occurred during this period were Bolivia's 11,804.8 percent increase in 1985, Nicaragua's 14,295.3 percent jump in 1988, and Peru's 3,398.7 percent climb in 1989. The hyperinflation of the 1980s was followed by a period of declining inflation levels in the 1990s.

With the value of their deposits falling literally by the hour, the elite upper class began converting their savings into more stable foreign currencies and transferring their wealth to overseas financial institutions. The shifting of funds to more secure foreign locations, a process known as **capital flight,** became so widespread that the citizens of several of Latin America's most indebted nations came to have almost as much or more money on deposit overseas as their governments owed to foreign creditors. By 1989, for instance, when Mexico's foreign debt was hovering near U.S. $100 billion, Mexican citizens had an estimated U.S. $84 billion on deposit in foreign accounts. Similar patterns were evident in Argentina, Brazil, and Peru, but none of these could compare to Venezuela, whose foreign debt total of some U.S. $29 billion was only half the amount of Venezuelan funds held overseas (Pauly, Thomas, and Evans 1989).

Currency Devaluation

Another consequence of the chronic hyperinflation of the 1980s was loss of purchasing power of the Latin American and Caribbean currencies relative to the United States

TABLE 13.3

Latin American and Caribbean Inflation, 1961–2000

COUNTRY	ANNUAL AVERAGE GROWTH OF CONSUMER PRICE INDEXES (in percent)				
	1961–70	1971–80	1981–90	1990–99	2000
Argentina	21.4	141.5	437.7	10.6	0.8
Belize	—	—	4.1	2.3	1.1
Bolivia	5.6	19.6	220.0	9.3	3.8
Brazil	46.2	35.6	337.0	253.5	8.2
Chile	27.1	174.3	20.3	9.7	4.1
Colombia	11.1	21.1	23.7	21.7	13.6
Costa Rica	2.5	10.9	25.6	16.2	6.5
Dominican Republic	2.1	10.4	24.6	9.0	7.8
Ecuador	4.4	12.5	36.4	34.5	105.9
El Salvador	0.7	10.7	19.0	9.4	3.9
Guatemala	0.8	9.5	14.4	10.7	5.7
Guyana	2.3	10.2	31.0	6.4	6.6
Haiti	2.9	10.7	6.7	23.2	14.6
Honduras	2.2	8.1	7.8	19.5	8.7
Jamaica	4.2	18.1	14.8	26.1	10.9
Mexico	2.8	16.6	65.2	19.9	12.0
Nicaragua	1.7	19.6	618.9	35.1	13.0
Panama	1.3	7.1	1.8	1.1	2.1
Paraguay	3.4	13.1	21.7	13.8	8.9
Peru	9.7	30.3	332.0	31.6	3.7
Suriname	4.2	9.8	12.9	88.0	73.8
Trinidad and Tobago	3.1	13.0	11.0	5.9	5.7
Uruguay	47.8	64.0	60.3	38.2	3.6
Venezuela	1.0	8.5	23.3	51.8	27.6

Sources: Compiled from World Development Indicators database, 2000; Human Development Report 2001; *Annual Report 1986* 1987, 117; *Annual Report 1995* 1996, 111; *Annual Report 1998* 1999, 148.

dollar, the Japanese yen, and the stronger western European currencies. By late 1986, for example, one American dollar was valued at 1.923 million Bolivian pesos. Bus fares in La Paz had climbed to 200,000 pesos, and a hamburger cost 3 million pesos. Many other countries in the region experienced similar crises at one time or another during the 1980s and 1990s (Table 13.4).

As national currencies became devalued beyond recovery, the most common governmental response was simply to replace the old currency with a new one whose value was a thousand or a million times greater than its predecessor. On January 1, 1987, for instance, the Bolivian government replaced the peso with the boliviano, which initially traded at 1.93 bolivianos to the dollar. Many other Latin American countries have followed a similar path in recent years, with some having changed currencies more than once. Brazil, for example, changed currencies seven times between 1976 and 1993, during which period inflation reached a compounded 200 billion percent (Epstein 1995)! Needless to say, while periodically dropping three or six zeroes

TABLE 13.4

Currency Devaluation in Venezuela, 1988–2001

YEAR	BOLIVARES PER U.S. DOLLAR
1988	14.50
1989	34.68
1990	46.90
1991	56.82
1992	68.38
1993	90.83
1994	148.50
1995	176.84
1996	417.33
1997	488.63
1998	547.56
1999	605.72
2000	679.96
2001	699.70

Sources: CIA World Factbook 2001; *Economic and Social Progress in Latin America 1998–1999 Report* 1998, 280.

from the currency clearly simplifies the handling of commercial transactions, it does nothing of itself to slow inflation and the attendant loss of purchasing power.

Economic Stagnation and Human Suffering

With foreign debt totals low and inflation under control, the 1960s were a decade of considerable overall improvement in the living levels of most of the Latin American and Caribbean peoples (Table 13.5). The per capita gross domestic product (GDP) of the region rose from U.S. $1,497 in 1960 to U.S. $1,982 in 1970, an impressive gain of 5.4 percent per year. At the national level, every country except Haiti showed an increase, and hope and confidence in a brighter future were the order of the day.

The optimism of the 1960s was followed by a sense of uncertainty and foreboding in the 1970s. On the positive side, the per capita GDP of the region grew at an annual rate of 5.6 percent to U.S. $2,754 in 1980. Much of this growth, however, was achieved through large foreign borrowings, and inflation was becoming increasingly burdensome to the average family.

As challenging as these conditions were at the time, they proved to be mere forerunners of the extreme economic hardships endured by most Latinos in the 1980s. Driven by the almost unchecked growth in foreign indebtedness and inflation, per capita GDP actually decreased in twenty of twenty-six reporting nations. Because individual and family living levels regressed rather than progressed in most of the Latin American and the Caribbean countries during the 1980s, the period has come to be called the **Lost Decade.**

TABLE 13.5

Gross Domestic Product per Capita, 1960–2000

COUNTRY	IN 1990 U.S. DOLLARS				
	1960	1970	1980	1990	2000[a] (PPP[b])
Argentina	4,220	5,449	5,942	4,742	7,480 (12,050)
Belize	—	—	1,702	2,125	2,890 (5,240)
Bolivia	686	936	1,181	888	990 (2,360)
Brazil	980	1,396	2,708	2,539	3,590 (7,300)
Chile	1,777	2,138	2,353	2,623	4,590 (9,100)
Colombia	716	872	1,195	1,417	2,030 (6,060)
Costa Rica	1,205	1,537	1,986	1,865	3,830 (7,980)
Dominican Republic	503	600	927	896	2,120 (5,710)
Ecuador	568	705	1,324	1,227	1,190 (2,910)
El Salvador	985	1,228	1,219	1,026	2,000 (4,410)
Guatemala	615	795	1,044	857	1,700 (3,770)
Guyana	633	690	703	471	860 (3,670)
Haiti	288	262	352	314	510 (1,470)
Honduras	403	503	636	585	860 (2,400)
Jamaica	1,277	1,859	1,483	1,633	2,760 (3,440)
Mexico	1,556	2,223	3,162	2,971	5,110 (8,790)
Nicaragua	950	1,330	979	645	460 (2,080)
Panama	1,153	1,852	2,386	2,089	3,280 (5,680)
Paraguay	713	839	1,458	1,445	1,440 (4,450)
Peru	1,204	2,155	2,356	1,670	2,080 (4,660)
Suriname	504	636	1,066	824	1,790 (3,480)
Trinidad and Tobago	3,148	4,272	6,598	4,259	5,160 (8,220)
Uruguay	2,324	2,434	3,183	3,099	6,080 (8,880)
Venezuela	2,624	3,244	3,274	2,830	4,310 (5,740)
Latin America and the Caribbean	1,497	1,982	2,754	2,450	3,690 (6,860)

[a]2000 data is Gross National Income (GNI), calculated using the World Bank Atlas method. Gross National Income has replaced Gross Domestic Product (GDP) and Gross National Product (GNP) as the World Bank's estimate of national income.

[b]Purchasing Power Parity (PPP) is a new measure that adjusts GNI data to reflect cost of living differences among countries. The per capita GNI of each country is converted to "international dollars," thereby facilitating a more accurate comparison of living levels between nations. For comparative purposes, the year 2000 per capita GNI PPP of the United States of America was $34,100.

Sources: Compiled from World Bank World Development Indicators database, 2000; *World Population Data Sheet 2002; Annual Report 1998* 1999, 146.

The suffering experienced during the Lost Decade can be expressed with both collective and individual data. One of the most telling statistics at the aggregate level of analysis is that the total number of persons in Latin America and the Caribbean living below the poverty line increased from 120 million in 1980 to between 163 million and 170 million in 1989 (Bonilla 1990, 215–216; Porter 1990, 2). The latter figure represented two of every five Latinos, a ratio that has not changed in the ensuing years. A second revealing statistic is that real disposable income dropped by an average of 20 percent, meaning that the typical household lost a fifth of its purchasing power over the decade (Iglesias 1990).

At the personal and family level, Selby (1991) noted that larger household kinship support networks were formed. Several related nuclear families often moved into a single home or apartment in order to save on the costs of housing, utilities, food, and other necessities. Klak (1992) documented, further, a downward movement in housing quality, with increasing numbers of families being forced to reside in low-cost squatter dwellings. Table 13.6 portrays the effects of the downward spiral of living levels by tracing increases in the costs of food, clothing, and rental shelter from 1986 to 1995 in selected Latin American and Caribbean countries. Note that the cost of each of these necessities in Colombia, for example, was approximately five to seven times greater in 1995 than in 1986. In some nations, conditions were even worse, while in others they were slightly better. Regardless of the precise degree of deterioration experienced in each locale, in the ultimate analysis the 1980s and early 1990s were, for most Latinos, a most difficult period.

FISCAL RESTRUCTURING AND AUSTERITY

When the great lending and borrowing spree began in the 1970s, little concern was expressed either by the international financial community or by Hispanic and Caribbean officials about the capacities of the Latin American and Caribbean nations to repay their debts. After all, went the reasoning, the borrowed funds would be invested in sound economic development projects

TABLE 13.6

Cost of Food, Clothing, and Rent Indices for Selected Latin American and Caribbean Nations[†]

COUNTRY	FOOD 1986	FOOD 1991	FOOD 1995	CLOTHING 1986	CLOTHING 1991	CLOTHING 1995	RENT 1986	RENT 1991	RENT 1995
Barbados	79.1	104.8	106.1	100.9	102.6	106.3[*]	87.1	117.8	171.6[*]
Colombia	37.7	129.9	279.9	40.8	126.8	211.8[*]	39.4	129.6	265.7[*]
Costa Rica	52.0	126.0	196.7[*]	58.2	126.6	172.0[*]	53.2	114.6	155.6[*]
Chile	46.8	125.8	195.6	53.1	118.8	154.2	50.7	120.7	189.5
Ecuador	17.0	148.6	490.4	20.7	150.9	367.5[*]	37.5	152.5	514.5[*]
El Salvador	38.6	117.9	206.1	66.0	107.6	148.7	71.8	110.1	129.6
Guatemala	45.3	132.3	198.3	60.2	126.7	168.4	47.1	120.3	198.7
Honduras	64.5	143.7	281.1	67.1	142.2	232.1	74.3	—	228.4
Jamaica	58.1	154.8	554.3	64.4	142.3	466.8[*]	84.6	—	228.4
Martinique	90.1	103.0	112.8	88.0	103.0	106.5[*]	80.0	103.5	117.6[*]
Mexico	13.7	120.1	208.7	16.3	111.2	190.6	10.2	130.9	180.4[*]
Paraguay	37.0	120.2	194.9[*]	41.8	126.5	180.2[*]	39.1	125.8	176.8[*]
Peru	0.4	448.2	1,483.4	5.5	459.0	—	24.8	1,064.6	—
Suriname	40.4	118.7	7,828.4	52.1	138.0	5,427.5	93.7	103.2	1,657.1
Uruguay	10.0	185.4	893.9	9.1	215.5	956.1	6.9	107.6	1,661.1
Venezuela	15.2	137.7	585.3	29.5	125.6	299.7[*]	44.5	138.0	364.5[*]
NON–LATIN AMERICAN COMPARISONS									
Canada	86.4	104.8	109.2	85.3	109.5	112.4	84.6	103.4	112.1
France	89.2	103.0	105.8	87.3	103.3	107.3	80.4	105.0	121.5
Japan	94.3	104.8	106.1	89.4	104.7	106.1	89.8	102.9	112.7
Spain	180.1	103.5	120.7	77.1	105.3	121.9	80.2	105.4	121.3
United States	82.6	103.6	112.7	85.3	103.7	106.4	85.7	103.5	114.0

[†]1990 = 100; [*]1994 data

Source: Compiled from *Year Book of Labor Statistics 1996* 1996, 876–896.

that would soon more than pay for themselves. Viewed with the benefit of hindsight, this somewhat cavalier attitude included two problems: first, its failure to factor in the almost unlimited capacity of many Hispanic and Caribbean government leaders and bureaucrats to embezzle and otherwise misuse the borrowed funds, and second, the economic downturn experienced by most of the Anglo American and Western European nations in the 1980s. The latter greatly reduced overseas demand for Latin American and Caribbean products and, by extension, the revenues needed to repay the foreign loans.

With economic conditions thus deteriorating steadily, a number of the more heavily indebted Latin American nations, including Mexico and Brazil, reached the point where they felt they no longer could afford to make the payments due on their loans. This led them to declare, on a nation by nation basis, indefinite suspensions or moratoriums on their scheduled debt payments.

Faced with the prospect of having a number of the United States' largest banks fail and thereby threaten the entire world financial system, the Reagan administration began to search with newfound urgency for solutions to what was now labeled the **debt crisis.** The first proposal to come forth, in 1985, was named the **Baker Plan** after its chief architect, U.S. Secretary of the Treasury James Baker. While recognizing the need for internal structural reforms among the borrower countries, the heart of the Baker Plan was the strategy to simply buy time by loaning even more money to the debt-stricken Hispanic and Caribbean nations and to hope that economic conditions in both the lending and borrowing countries would improve to the point where the debts could then be paid. The Baker Plan failed miserably, because the last thing both the indebted nations and the international banks wanted was to increase the already unmanageable debt loads, and because there was never any effort to provide debt relief.

As the cry for a solution that would allow for meaningful debt relief mounted among the borrower states (Bacha 1987; Blackman 1989), a second proposal, named the **Brady Plan** after the Bush administration Treasury Secretary Nicholas Brady, was introduced in March 1989. In contrast to its predecessor, the Brady Plan generally was well received in the Latin American and Caribbean nations and contributed positively to the economic recovery of the region.

Much of the success achieved through the Brady Plan and later programs has been attributable to their flexible approaches, which tailor a specific combination of debt relief measures to the needs of each individual country. These measures include debt-equity swaps or privatization, debt-for-debt swaps, debt buybacks, debt-for-education swaps, and debt-for-nature swaps. Under the terms of the debt-equity swaps, private firms, both foreign and domestic, are allowed to purchase enterprises, such as airlines, steel mills, and utility companies, that were formerly owned by a national government. The new private owner promises, for its part, to assume the past debts of the firm it is purchasing, to invest additional monies in the company being acquired, and to pay a purchase price to the seller government. While the **privatization** process has resulted in freeing some countries from billions of dollars worth of debt (see Table 11.1), its drawbacks include increased foreign economic influence in the host nation and the possibility that a large proportion of the future profits of the acquired firm will be exported to overseas shareholders (Manzetti 1999; Rogozinski 1998; Ramamurti 1992).

Debt-for-debt swaps entail, in effect, a debtor country trading a larger amount of its debt for a smaller amount held by the lender. Debt buybacks allow the debtor country to sell at a discounted rate some of its debt to a second party in return for guaranteeing payment on the remaining debt. Debt-for-education and debt-for-nature swaps center on foreign universities or conservation groups purchasing a given amount of debt at a discounted rate in exchange for a fund being established in the host country to support either the advanced education of local students abroad or the establishment of protected biospheres or preserves.

Although the debt relief provided by the Brady Plan was most needed, it is important to note that by 1992 the total amount of Latin American and Caribbean foreign indebtedness exceeded the previous pre–Brady Plan high level registered in 1987 (*Economic and Social Progress in Latin America 1993 Report* 1993, 301). The success of the Brady and subsequent plans in promoting renewed economic growth, then, is to be found not only in the short-run debt relief that they provide but also in a controversial set of austerity measures that they impose on the debtor nations. These measures, often referred to collectively as **conditionality,** have been implemented primarily through the International Monetary Fund (IMF) and its companion organization, the World Bank. While varying slightly in their specific details from country to country, each conditionality package has required the privatization of government-owned enterprises, currency exchange rate deregulation, trade liberalization, tax reform, and increased accountability, or "transparency," of expenditures of public funds. In order to further reduce government expenditures, the IMF austerity programs have also required the removal of long-cherished subsidies of basic food commodities, utilities, and public transportation as well as the suppression of wages.

The intent of each of the measures has been to strengthen the fiscal position of the debtor government by lowering expenses and/or increasing revenues. To the

Case Study 13.1:   IMF Austerity Measures and Civil Unrest in Latin America and the Caribbean

With levels of foreign indebtedness spiraling out of control in many Latin American and Caribbean nations during the 1970s and 1980s, government officials in a number of the most highly indebted countries responded to the problem by ordering the national treasury to print more money. The effect of issuing more money without achieving a corresponding increase in national economic output, however, is invariably to lessen the value of the currency, which soon leads to soaring inflation and an equivalent decline in the living levels of the people.

Realizing that deteriorating economic conditions can easily result in a loss of political support, the government leaders endeavored to lessen the hardships brought on by the mounting inflation by subsidizing the costs of basic necessities such as food and public transportation. The long-term effect of the subsidies, however, was to increase the national debt even further, thereby leading to ever higher rates of inflation and continuing slip-

page in the living levels of the lower and middle classes.

This vicious downward economic cycle eventually forced the leaders of most of the Latin American and Caribbean nations to submit to a package of International Monetary Fund–imposed austerity measures that included the reduction or removal of government food and fuel subsidies. In August of 1990, Peruvians awoke one morning to learn that food prices had tripled since the day before and that a gallon of gasoline, which had been selling for the equivalent of U.S. seven cents, now cost the equivalent of U.S. two dollars. Lima's privately owned bus lines immediately ceased their runs and looting of stores and marketplaces erupted throughout the city. The government responded by ordering the army to patrol the main streets, and armed clashes broke out between soldiers wearing ski masks to avoid being recognized and rioting civilians desperate for food. Other countries experienced similar civil unrest. In Argentina, rioting by hungry working class looters

in May of 1989 led to the arrest of over 1,500 people in Rosario, Tucumán, and Buenos Aires. In Venezuela, food riots took the lives of over 300 persons in February of 1989. Similar disturbances in the Dominican Republic, which left 60 civilians dead, 200 wounded, and 4,300 arrested, prompted the nation's planning minister to say: "It is not that we are unwilling to put our own house in order. It is that we want to keep our house and not let it go up in flames" (Pastor 1987, 259).

After a decade of relative calm, civil unrest began to reemerge in many countries in the early twenty-first century, fueled by a growing perception that privatization and other economic reforms had enriched corrupt public officials and foreign corporations while failing to alleviate poverty. In addition to street demonstrations, the recent wave of protests has spawned a new generation of leftist politicians who portray the austerity measures as violations of national sovereignty and promise help to the poor through a return to statist economic policies.

extent that these neoliberal reform programs have been implemented, their fiscal goals generally have been realized (Thorp 1998; García-Rodríguez 1990; Hojman 1994). The reluctance of many Latin American and Caribbean governments to submit fully to all the IMF terms, however, has stemmed from the painful costs conditionality initially imposes on the members of the lower socioeconomic classes, who are forced to try to cope with huge price increases while having their wages repressed.

For instance, during an IMF-imposed fiscal adjustment period, it has not been unusual for the masses to go to sleep one night thinking that their budgets were reasonably adequate to meet their needs and then to arise the following morning to learn that the prices of bread, tortillas, beans, rice and other basic necessities have doubled or tripled since the night before. Bus fares and water and electricity bills are equally volatile (Gilbert 1990). The anger and frustration and sense of hopelessness that have been spawned by repeated price hikes have ignited, in some nations, riots and other forms of

mass protest that have threatened the survival of the government itself (Weaver 2000) (Case Study 13.1). For these reasons, the political leadership of several Latin American and Caribbean nations has resisted implementation of the conditionality measures.

THE FOUND DECADE

Those countries whose leaders adopted conditionality, however, emerged in the early to mid-1990s with greatly improved economic conditions and renewed social tranquility (Table 13.5; Case Study 13.2). The gross national income of Latin America and the Caribbean as a whole, for instance, increased an average of 3.3 percent per year from 1990 to 2000 (Figure 13.2). Equally important to the well-being of the lower socioeconomic classes has been a greatly lowered rate of inflation which, by 2000, had dropped to an average of 5.2 percent (Table 13.3). The increased economic confidence has also reversed the capital flight of the 1980s, with the

Case Study 13.2:   Structural Reform and Argentine Economic and Political Turmoil

One of the Latin American countries that embraced economic reform most enthusiastically in the late twentieth century was Argentina. Although long recognized as a nation possessing extensive natural resources and great economic potential, Argentina was in a state of fiscal and social crisis by 1989. Inflation was causing the prices of goods and services to grow at more than 200 percent per month, the government operated countless deficit-producing industries, and corruption was notoriously widespread. Tax collections were in a shambles, with approximately half of all wage earners either underreporting their income or simply refusing to pay in the expectation that the government would never track them down. At one point, the Central Bank actually ran out of currency. Foreign trade and investment had dwindled owing to endless bureaucratic delays and expenses, and the national telephone company was so inept that the average waiting period to have a phone line installed in a private home was twenty years. Conditions were so chaotic that the country's outgoing president, Raul Alfonsín, voluntarily resigned from office six months before the scheduled expiration of his term.

Alfonsín's successor, Carlos Saul Menem, immediately introduced a radical economic restructuring program that included widespread privatization of government-held companies. Foreign trade was promoted by slashing import duties and regulations. Over 6,000 new federal tax agents were hired, and negotiations with Brazil, Uruguay, and Paraguay to form an economic common market were intensified. Of great importance, also, was the passage of a new Convertibility Law, which attacked inflation by

establishing a one-to-one exchange rate between the Argentine peso and the American dollar and by requiring the Central Bank to have reserves of gold, dollars, and dollar-denominated bonds equal to the amount of the nation's circulating currency. The foreign debt was also reduced and rescheduled through the Brady Plan.

The initial impact of the restructuring measures was generally positive. Argentine economic output surged, and foreign investment reached all-time record high levels. The hyperinflation of the previous decade was overcome to such a degree that prices rose by only 1.6 percent from 1996 to 1998 and actually declined by 1.2 percent in 1999.

Unfortunately, however, the economic growth of the 1990s masked a failure to fully implement the needed reforms. The tax collection system continued to be plagued by high evasion rates; rampant corruption persisted; and the provincial governments routinely overspent their revenues. The national government performed no better, paying for long-cherished subsidies with borrowed funds and proceeds derived from the sale of government-owned enterprises. Once privatization revenues began to decline, however, the fiscal deficits grew to unsustainable levels. When Menem's successor, Fernando De la Rua, belatedly attempted to implement tax increases and spending cuts in an effort to stave off impending economic collapse, he received almost no support from the powerful provincial governors, and street rioting and demonstrations broke out among those who feared the loss of government subsidies.

The social unrest became so severe that De la Rua was forced to resign

from office in December 2001. There followed an extremely chaotic period during which Argentina went through three additional presidents in fifteen days before Eduardo Duhalde, an old-style Peronist who had long opposed free-market reforms, formed a caretaker government. Under Duhalde, Argentina suspended payments on its foreign debt and abandoned the Convertibility Law, which led, in turn, to a severe devaluation of its currency. Access to funds deposited in personal checking and savings bank accounts was also restricted, causing further public unrest. By 2002, unemployment had climbed to an unprecedented 21.5 percent and half of all Argentines were living below the poverty line.

Although some suggest that Argentina's economic and social turmoil is attributable to the implementation of externally imposed fiscal austerity measures, others believe that the root of the crisis is the failure of the Argentine people to embrace the reforms required for the country to live within its collective means. Regardless of these different perspectives, almost all would agree that the Argentine people have been the victims of failed leadership. The leaders of Argentina have failed both to enforce strict accounting of governmental resources and to instill fiscal discipline among those they have appointed to serve. For its part, the leadership of the International Monetary Fund has also failed to require compliance with the terms of conditionality and has continued to channel additional funds to Argentina despite its failure to sustain needed reforms. As a result, Latin America's historically most prosperous nation is now in danger of losing both its middle class and its fragile democracy to the ever-worsening downward economic spiral.

FIGURE 13.2

Gross national income purchasing power parity per capita: 2000.

late 1990s averaging $50 to $80 billion dollars returning to the region annually. The overall effect of the Brady Plan and later programs, with their debt relief packages and internal austerity and structural reform programs, has been to breathe new economic life into a region whose output was in serious decline only a few years earlier. Some observers went so far as to label the 1990s the **Found Decade.** Whether or not this is the best appellation, it is important to recognize that the economies of most of the Latin American and Caribbean nations continue to be relatively small and fragile when compared to those of the more industrialized Anglo American and Western European countries. It is the growing recognition of the development difficulties facing states with small economies and the potential benefits of economic cooperation in an increasingly interconnected world that has inspired the current interest in multinational economic unions.

REGIONAL AND SUBREGIONAL ECONOMIC UNIONS

The economic integration of Latin America and the Caribbean historically has been hindered both by severe physical isolation and by enduring cultural and political divisions. Interregional trade among the pre-Conquest Indians was limited primarily to the exchange of products within the Aztec, Mayan, and Incan empires, and much of that was in the form of forced tribute that did little to alter the predominantly subsistence character of the masses. Coercion also shaped colonial trade patterns, with the Spanish and Portuguese mercantile systems serving to divide, rather than to unify, the economies of the Hispanic colonies. Even the achievement of independence did little to bind the Latin American and Caribbean nations to one another, owing largely to the perpetuation of long-standing rivalries and to the emergence of new conflicting territorial claims that resulted, as often as not, in the outbreak of armed hostilities among neighboring states (chapter 6). As recently as the 1950s, regional and subregional economic cooperation was virtually nonexistent, and international trade with the United States, the western European countries, and Japan far exceeded, in both volume and value of cargo, the flow of commodities among the Latin American and Caribbean nations.

A modest beginning of regional economic cooperation was finally achieved in 1960 with the formation of the **Latin American Free Trade Association,** whose name was changed under the 1980 Treaty of Montevideo to the **Latin American Integration Association** (LAIA). The principal objective of the organization, whose initial membership included Argentina, Bolivia, Brazil, Chile, Colombia, Ecuador, Mexico, Paraguay, Peru, Uruguay, and Venezuela, was to increase trade and economic integration among the member states. This was to be accomplished through the progressive lowering and eventual removal of duties or tariffs on products shipped between two or more of the participating countries. A second multinational entity, the **Latin American Economic System,** or SELA, was formed by twenty-five nations in 1975 to provide additional coordination of and support for the economic integration of the region.

The fortunes of LAIA, SELA, and other related organizations that have been formed over the past three to four decades have reflected the broader economic currents to which the region has been subjected (Instituto para la Integración de América Latina 1990). The relative prosperity of the 1960s and early 1970s, for instance, brought the establishment of a number of subregional groupings and modest gains in levels of trade and economic cooperation. These advances were largely lost, however, in the turbulence of the late 1970s and 1980s when a new wave of protectionism emerged in most countries. The economic rebound of the 1990s witnessed additional progress toward regional economic integration, including the formation of two new subregional groupings and the rejuvenation of those formed previously. Given their actual and potential importance, we will now review the history and success to date of each of these subregional organizations.

Central American Common Market

With their small domestic markets and modest natural resource endowments, Central American nations have long needed economic integration. Appropriately, therefore, the first subregional economic union to be organized among the Latin American and Caribbean countries was the Central American Common Market (CACM), which came into existence in 1960 with Guatemala, Honduras, El Salvador, and Nicaragua as founding members. Costa Rica and Panama subsequently joined (Figure 13.3).

Considerable progress toward economic integration was realized during the early years of CACM's existence, and the leaders of the participating nations even went so far as to assign exclusive production rights of given industrial commodities to specific members (Irvin 1988). This ultimately contributed to feelings of jealousy by the Honduran and Nicaraguan populations, who felt that El Salvador and Guatemala were unfairly favored in their industrial allotments. These disagreements, together with the reemergence of armed conflicts between and within the member states, led to a weakening of the CACM, which virtually ceased to function in the 1980s. Progress in resolving the wars had resulted, by the early 1990s, in renewed desires to revitalize the CACM. By the early

FIGURE 13.3

Regional economic unions and trading groups.

twenty-first century, the CACM had become a more cohesive entity and had negotiated trade agreements with the United States and the European Union.

Andean Community

The second subregional economic organization to form in Latin America was the Andean Group, which was established through the Cartagena Agreement in 1969 with Colombia, Ecuador, Peru, Bolivia, and Chile as members. Venezuela subsequently joined in 1973, and Chile withdrew in 1976. The principal impediment to trade between members over the years has been the high costs of transporting goods between the mountain-dominated nations. This has resulted in the maintenance of a mostly bilateral, rather than regional, pattern of economic exchange within the group. Bolivia, for instance, has traditionally sent the bulk of its exports to Peru, Colombia to Venezuela, Ecuador to Peru, and both Peru and Venezuela to Colombia. The recent expansion of trade between Bolivia and Brazil has further diminished Bolivia's ties to its Andean neighbors. Nevertheless, a free trade zone among the member states was agreed to in 1992, and the new Andean Integration System became operational in 1997. Current priorities center on reaching a free trade accord with MERCOSUR. Advocates of such a step believe that it would lead to the formation of a South American Free Trade Area (SAFTA) which could eventually be merged with the North American Free Trade Association (NAFTA) to form the Free Trade Area of the Americas (FTAA) (Carranza 2000).

Caribbean Community

The third subregional economic union to be formed was the Caribbean Community (CARICOM), which began operations in 1973 with Barbados, Guyana, Jamaica, and Trinidad and Tobago as charter members. These were subsequently joined by other English-speaking island states, Suriname, and Belize, whose leaders saw the establishment of a Caribbean common market with a common external tariff as working to their mutual advantage. Haiti was admitted as a provisional member in 1997, bringing the number of community members to fifteen. In addition to having difficulty obtaining agreement among such a large and diverse membership base, the CARICOM states have also been challenged in recent years by the need to redefine their objectives and strategies in the face of changing geopolitical and economic conditions abroad (Klak 1998). Trade agreements have been signed with the European Union, and the United States recently extended special trade and tax incentives to the region through the Trade and Development Act of 2000, the latest in a series of trade preferences that began with the passage in 1983 of the **Caribbean Basin Initiative.**

TABLE 13.7

Gross Domestic Products of Regional Economic Unions

ECONOMIC UNION	GDP, YEAR 2000 U.S. $ BILLIONS	PERCENTAGE OF COMBINED GDP
North American Free Trade Association	11,099.8	89.7
Southern Cone Common Market	907.7	7.3
Andean Community	277.2	2.2
Central American Common Market	66.3	0.5
Caribbean Community	31.1	0.3

Source: Calculated from World Bank World Development Indicators database.

Southern Cone Common Market

In 1990, Argentina and Brazil announced an agreement to eliminate all barriers to the free movement between them of goods, services, and labor by 1995. The accords were later extended in 1991 to include Uruguay and Paraguay, at which time the Southern Cone Common Market (MERCOSUR in Spanish and MERCOSUL in Portuguese) came into formal existence. In 1996, Bolivia and Chile negotiated free trade agreements with MERCOSUR, and Bolivia was also granted associate membership status.

MERCOSUR has emerged over the past decade as the dominant South American economic union. This progress is attributable to the steady pace of economic integration and also to the large economies of the member countries, which collectively generate approximately 80 percent of South America's gross domestic product (Table 13.7). MERCOSUR is currently negotiating a free trade area with the European Union, with which it has half again more trade than does the United States.

North American Free Trade Association

The most recent and economically the largest subregional union to affect Latin America is the **North American Free Trade Association** (NAFTA), which was accepted by the presidents of Mexico, the United States, and Canada in December of 1992 and went into effect in January of 1994 (Figure 13.4). Prior to its adoption, NAFTA was criticized severely by opposition groups in each member country, who feared the possible loss of jobs and accelerated environmental degra-

FIGURE 13.4

This Xalapa, Veracruz, hotel flies the flags of the members of the North American Free Trade Association: the United States, Mexico, and Canada.

dation such as that associated with the *maquiladora* industrial corridor (chapter 11). Proponents argued, however, that it created a mutually advantageous relationship between Mexico and its Anglo American neighbors to the north. One of the first consequences of NAFTA was a rapid growth in Mexican imports of American products, which threatened to increase the Mexican trade deficit to the point of exhausting the country's monetary reserves. The government's decision to severely devalue the new peso in late 1994 led to a decline of foreign investment and credit, which threatened a collapse of the entire economy. Aside from these short-term problems, which were magnified by the deteriorating political climate, Mexico's decision to align itself with the United States represented, as Weintraub (1991, 10) noted, a break of "historic proportions" from its previous adversarial political relationship with its Anglo neighbor. This alignment was clearly expressed, for better or for worse, in the response to the crisis by the United States, which arranged for a $20 billion international rescue package, a move that linked even more tightly the economic fortunes of the two countries.

Free Trade Area of the Americas

Much time must pass before final judgment can be made on the success or failure of any of the economic unions discussed above. Seen collectively, the developments of the past decades do suggest that there is presently greater momentum toward the economic integration of the Latin American and Caribbean nations than at any time in history. Some observers view the regional trade groups as forerunners of a pan-American economic union that will eventually encom-

pass the entire Western Hemisphere. This concept, which was promoted by the administration of George H. Bush through its Enterprise for the Americas Initiative (Porter 1990), was approved by the leaders of thirty-four nations, who in December of 1994 agreed to negotiate a **Free Trade Area of the Americas** by the year 2005. The commitment was subsequently renewed through the signing of the Quebec Declaration in April 2001. Although the realization of this goal is far from certain, it is clear that the process of integration currently underway throughout Latin America and the Caribbean is of great economic and political significance.

HEALTH AND NUTRITIONAL DEVELOPMENT IN LATIN AMERICA AND THE CARIBBEAN

Having reviewed the more purely economic expressions of Latin American and Caribbean development, we will now analyze the quality of human life as it is manifested in the varying levels of health and nutrition found throughout the region. Access to adequate levels of clean water, food, and health care is one of the most basic of all human needs, and its realization constitutes one of the most reliable measures of human development.

One of the greatest obstacles to objectively comparing levels of health between two or more nations or regions is that the diseases that most afflict one area often tend to be different from those that most trouble another. This is attributable both to cultural differences between the study populations and to the presence of different physical environments. It is futile to argue, for instance, that a region characterized by a high incidence of malaria and schistosomiasis is more or less healthy than one whose occupants experience elevated levels of cancer and tuberculosis.

Because one nation or region may have a higher incidence of one disease and a lower incidence of another, most observers believe that the best overall indicator of human health is life expectancy. In this regard we find that the Latin American and Caribbean peoples are healthier than the inhabitants of the other less industrialized world areas but not as healthy as the inhabitants of the high income countries (Table 13.8). As would be expected, average life expectancies vary widely by nation within the region, from a low of 53.2 years in Haiti to a high of 77.5 years in Costa Rica (Figure 13.5). While the overall life expectancy of the Latin American and Caribbean peoples has steadily improved over time, from 49 years in 1950 to 70 years in 2000, much work remains to be done to improve health conditions in many areas. The most essential and fundamental of these tasks is the provision of clean water and adequate sanitation facilities.

TABLE 13.8

Life Expectancy at Birth by World Region

REGION	LIFE EXPECTANCY IN YEARS	
	1960	2001
Sub-Saharan Africa	40	49
South Asia	44	63
Arab States	47	66
East Asia and the Pacific	48	69
Latin America and the Caribbean	56	70
High Income Countries	70	78
World	**51**	**67**

Sources: Human Development Report 2001; Bellamy 1997, 98–99.

TABLE 13.9

Proportion of Population with Access to Safe Drinking Water, 1970–2000

COUNTRY	PERCENTAGE	
	1970	2000
Argentina	56	79
Bahamas	65	96
Barbados	98	100
Belize	67	76
Bolivia	33	79
Brazil	55	87
Chile	56	94
Colombia	63	91
Costa Rica	74	98
Dominican Republic	37	79
Ecuador	34	71
El Salvador	40	74
Guatemala	38	92
Guyana	75	94
Haiti	12	46
Honduras	34	90
Jamaica	62	71
Mexico	54	86
Nicaragua	35	79
Panama	69	87
Paraguay	11	79
Peru	35	77
Suriname	97	95
Trinidad and Tobago	96	86
Uruguay	92	98
Venezuela	75	84

Sources: World Bank World Development Indicators database, 2000; *Annual Report 1998* 1999, 151.

Provision of Clean Water and Sewage Treatment

One of the most significant advances in Latin American and Caribbean health care in recent decades has been the expansion of potable water services from approximately 66 million beneficiaries in 1960 to 440 million in 2000 (World Bank World Development Indicators database 2000; Douglas 1990). Progress within the individual countries has been equally impressive, in a statistical sense, with many nations having achieved 50 to 100 percent increases over the past twenty-five to thirty years in the proportion of the population with access to safe water (Table 13.9).

These gains notwithstanding, a number of circumstances lead most authorities to conclude that unclean water supplies and untreated sewage will continue to constitute the most serious health challenges facing the region for many years to come. The first of these circumstances is that approximately 15 percent of the population still lacks access to potable water. A second concern centers on the criteria used by many Latin American countries to define "safe" or "improved" water and its access. Water that has been filtered to remove particles of dirt and other minute solids may appear visually to be "clean," for example, and yet contain any number of viruses, bacteria, and other disease-causing organisms. If such water is drunk directly, or if it is used to irrigate vegetable crops that are later consumed without being properly disinfected, serious disease may occur. Many government officials and scholars also question the appropriateness of the Pan American Health Organization's criterion for "access,"

which is defined as including all persons living within 200 meters of a public water outlet.

The third reason health officials continue to express great concern about the water supplies of most Latin American and Caribbean nations is the direct link between contaminated water and a number of serious diseases, including diarrhea, intestinal worm infestation, eye and skin infection, schistosomiasis, typhoid, and cholera (Case Study 13.3). While the occurrence of diarrhea is generally not viewed by the public with the same degree of alarm as the other illnesses, it is actually the most threatening of all these diseases, accounting, through dehydration and malnourishment, for approximately one-third of all infant and early childhood deaths (Williams 1987, 4).

FIGURE 13.5

Life expectancy at birth.

Source: Data from World Bank World Development Indicators database, 2000.

Case Study 13.3:   The Latin American Cholera Epidemic

Cholera is the most lethal of all diarrheal diseases: it can kill its victims within ten hours of exposure to the bacterium, which may be present in water and foods that have come into contact with untreated human feces. Symptoms include intense stomach and intestinal pain, uncontrolled diarrhea, profuse vomiting, fever, and chills. Death, which is most common among children, comes primarily from dehydration and/or respiratory failure, both of which are related to the rapid loss of body fluids that may reduce the victim's weight by as much as 10 percent in a single night. In a large majority of cases, the disease's symptoms are relatively mild and the victim survives, but fatality rates can be as high as 50 percent among more vulnerable populations.

The disease was almost eradicated worldwide in the 1950s but managed to survive in some of the more isolated and poverty-stricken parts of the Old World tropics. Until its recent reappearance in South America, the illness had not been reported in the Western Hemisphere since 1895, and natural resistance levels among the Hispanic peoples were low.

The outbreak of cholera in Latin America began in Peru in 1991. Most victims were residents of the *barriadas* of Lima and other coastal cities, where the presence of raw, untreated sewage combined with widespread consumption of foods and drinks purchased from street vendors to create an ideal setting for the spread of the cholera bacillus. From Peru, the disease soon spread northward into Ecuador, Colombia, Venezuela, Central America, and Mexico and southward and eastward into Chile and Brazil. Over 325,000 cases and more than 3,000 deaths were reported in Peru alone in 1991, and world health authorities feared for a time that the number of cases might soon reach into the millions among South America's urban poor. Fortunately, this did not occur, and the number of Peruvian cases had fallen by 1993 to less than 40,000. Much of the credit for containing what could have been a plague of almost unthinkable proportions goes to a massive educational campaign launched in the affected regions, which stressed the importance of consuming only boiled or chlorinated water and cooked foods and the adoption of improved personal hygiene habits (Case Study Figure 13.1). The lives of hundreds of thousands of victims were also saved through the administration of oral rehydration therapy.

Although the spread of the disease appears to have been arrested, its very appearance, after almost a century of absence, came to symbolize what Carlyle Guerra de Macedo, director of the Pan American Health Organization, aptly described as the "profound crisis in living conditions" that confronts tens of millions of Latinos on a daily basis ("Cholera's Message" 1991, 3). As these conditions are ameliorated, possibly cholera will once again be eliminated from the region.

CASE STUDY FIGURE 13.1

Poster mounted on the wall of a rural school in the uplands of central Panama teaching children how to avoid contracting cholera.

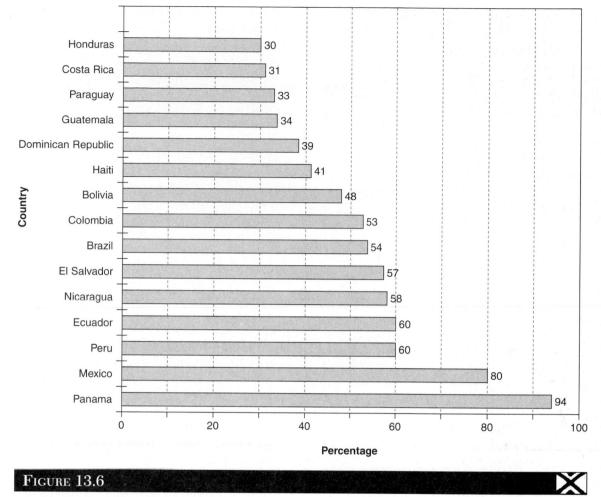

FIGURE 13.6

Oral Rehydration Therapy use rate.

Source: Data from Human Development Report, 2001.

One of the most significant developments in the treatment of childhood dehydration was the discovery in the United States in the 1950s that an oral solution consisting of eight teaspoons of sugar, one teaspoon of salt, and a quart of water acts to rapidly rehydrate victims of diarrhea, many of whom would otherwise die. The treatment, which came to be called **Oral Rehydration Therapy,** or ORT, was so simple and novel that the medical establishment in most Third World countries was initially reluctant to embrace it, preferring instead conventional intravenous treatments. Its effectiveness, low cost, and suitability for use in even the most remote and isolated locations eventually resulted in generous financial support from the United Nations Children's Fund (UNICEF) and the United States Agency for International Development (USAID), both of which shifted monies from the construction of treated water systems to ORT. By 2000, the percentage of diarrhea cases in children under 5 years of age treated with oral rehydration therapy varied from a high of 94 percent in Panama to a low of 30 percent in Honduras (Figure 13.6).

Critics of the UNICEF and USAID funding priorities have suggested that ORT has become a "silver bullet" that provides the development agencies with dramatic statistical accomplishments in the war on diarrhea by placing a greater emphasis on short-term treatment rather than long-term prevention (Douglas 1990, 7). Agency defenders counter that much money is being invested in the construction of water and sewerage treatment systems, but that their installation and subsequent utilization requires many years to complete and that immediate action must be taken to save as many lives as possible. Regardless of these differences, there is widespread consensus among development specialists that unsafe water supplies continue to constitute one of the leading impediments to improving the health of the Latin American and Caribbean peoples.

Food and Nutrition

One of the most visible effects of prolonged poverty is a shortage of food and/or the absence of a balanced diet in the lives of the inhabitants of a region. While the Latin American and the Caribbean nations, as a group, are certainly not the poorest countries on earth, nevertheless, severe poverty is widespread and tens, if not hundreds, of millions of persons live out their lives in either an undernourished or malnourished state.

Undernourishment can be defined as the condition of not having enough food in one's diet. Its most common outward symptom is chronic, or long-term, low body weight. In extreme cases, victims lose most of their muscle tissue and become physically emaciated and mentally lethargic or listless. In milder cases, victims are underweight to varying degrees and experience much hunger but continue their daily activities.

The simplest way to gauge the extent of undernourishment in a given population is to measure average per capita daily caloric intake. Table 13.10 has this data for two time periods: 1970 and 1997. As we analyze the statistics, at least three important conclusions can be drawn. The first is that levels of caloric intake vary widely from one nation to another, with the diets of some countries now averaging 3,000 or more calories of food per person per day and others averaging less than 2,300 (Figure 13.7). The second point is the significant gains achieved in most nations over the past three decades. Finally, we learn from the regional comparisons found at the end of the table that Latin American and Caribbean caloric consumption compares favorably to intake levels among the world's other less developed nations but lags considerably behind levels in the high income countries.

Malnourishment is the state of consuming enough food but lacking a balanced diet. One common form of malnourishment is a deficiency of essential vitamins and minerals. Although these substances do not contain calories, their extended absence in the human diet is likely to lead to the development of serious disease. Vitamin A deficiency is associated, for example, with eye disease and blindness, a shortage of Vitamin B with beriberi and pellagra, Vitamin C with scurvy, and Vitamin D with rickets. Similarly, a deficiency of iron leads to anemia, and insufficient quantities of iodine can cause goiter, mental retardation, and neuromuscular disorders (Henschen 1967).

A second form of malnourishment is protein deficiency. Some food sources, such as meat, fish, eggs, dairy products, nuts, wheat, and beans and other leguminous plants, are relatively high in protein. Many other foods, including maize or corn, rice, bananas and plantains, and most tropical tubers such as cassava, sweet potatoes, yams, and taro, are high in carbohydrates but relatively low in protein. If a person's diet is dominated by carbohydrates or starches, he or she may eat large quantities of food and even become obese, yet suffer from protein-based malnutrition. Such individuals are likely to have low levels of physical stamina and, ultimately, lower economic productivity. One of the classic descriptions of this condition comes from a United States Peace Corps volunteer named Moritz Thomsen (1969, 260), who, after having experienced protein malnourishment while living in a village along Ecuador's northern Pacific coast, wrote that "there are just so many miles to a gallon of bananas—not one foot more."

In addition to low energy levels, protein-based malnutrition is also associated with diarrheal and respiratory diseases. Severely malnourished children are further subject to stunting, the failure to grow to a height within the normal age range. Tragically, reductions in body size of infants and children can persist into adulthood (Stini 1982). Malnourishment also reduces one's overall resistance to disease, thereby increasing susceptibility to such nondietary diseases as malaria, tuberculosis, pneumonia, and measles. Because the presence of malnourished and undernourished individuals is often viewed with shame and embarrassment by government authorities and civic leaders, it has become customary for Latin American and Caribbean medical personnel not to list dietary-related conditions as the cause of death, even when they are the principal underlying factors. Instead, other, more socially and culturally acceptable diseases are entered on the civil registries. Nevertheless, competent observers estimate that malnourishment and undernourishment together are responsible for the deaths of more than 700,000 children in Latin America and the Caribbean each year (Cordovez 1990, 12). Protein consumption levels are thus an excellent indicator of the true quality of diets in a given country (Table 13.10).

Two of the most effective ways to increase dietary protein intake among the young are to promote proper prenatal and postnatal nutrition of mothers and to encourage breastfeeding or nursing of their babies. In addition to providing all the nutrients needed by the infant during the first four months of life (Latham 1979), human milk contains antiinfective and immunologically active substances that reduce the risk of the infants contracting infectious disease (Gordon 1979; Cole 1979). Recent studies in southern Brazil have concluded that babies that are not breastfed are three times more likely to die from respiratory illness and eighteen times more likely to die from diarrhea than infants that are breastfed (*World Development Report 1993* 1993, 78). Part of the reason for findings such as these is that nursing children are more likely to get enough food to eat than nonbreastfed infants. Another contributing factor is that

TABLE 13.10

Caloric and Protein Intake per Capita per Day, 1970–1997

COUNTRY/REGION	CALORIES		GRAMS OF PROTEIN	
	1970	1997	1970	1997
Argentina	3347	3093	106	95
Bahamas	2600	2443	78	78
Barbados	2854	3176	77	92
Belize	2266	2907	57	65
Bolivia	1998	2174	50	57
Brazil	2409	2974	60	76
Chile	2637	2796	69	77
Colombia	1938	2597	45	63
Costa Rica	2370	2649	58	68
Cuba	2640	2480	68	52
Dominica	2051	3059	50	86
Dominican Republic	2003	2288	44	50
Ecuador	2188	2679	52	59
El Salvador	1830	2562	47	64
Grenada	2251	2768	58	67
Guatemala	2097	2339	57	61
Guyana	2281	2530	57	69
Haiti	1944	1869	45	41
Honduras	2155	2403	56	58
Jamaica	2538	2553	67	63
Mexico	2706	3097	70	83
Nicaragua	2338	2186	72	49
Panama	2257	2430	57	65
Paraguay	2589	2566	73	77
Peru	2198	2302	55	60
St. Kitts and Nevis	1989	2771	44	75
St. Lucia	2008	2734	51	80
St. Vincent	2331	2472	53	65
Suriname	2225	2665	56	65
Trinidad and Tobago	2486	2661	64	59
Uruguay	3045	2816	93	84
Venezuela	2352	2321	59	59
High Income Countries	3041	3412	92	105
Arab States	2225	2930	60	79
Latin America	2474	2798	65	73
East Asia	2050	2906	49	78
South Asia	2103	2467	52	59
Sub-Saharan Africa	2271	2237	55	53
World	2358	2791	62	74

Source: Derived from Human Development Report 2001.

mothers who feed their babies a prepackaged formula may be forced to mix contaminated water with the powdered formula. Other consequences of breast-feeding include enhanced mother-child psychological bonding, substantial reductions in family food expenditures, and, in many but not all cases, an inhibitory effect on female reproductive functions that increases

control over the spacing of the birth of the subsequent child (Lawrence 1985).

Given the clear advantages of breastfeeding, one might expect the practice to be almost universal among Latin American and Caribbean mothers. Recent data suggest, however, that only one in four infants are now exclusively breastfed from ages 0 to 3 months

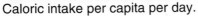

FIGURE 13.7

Caloric intake per capita per day.

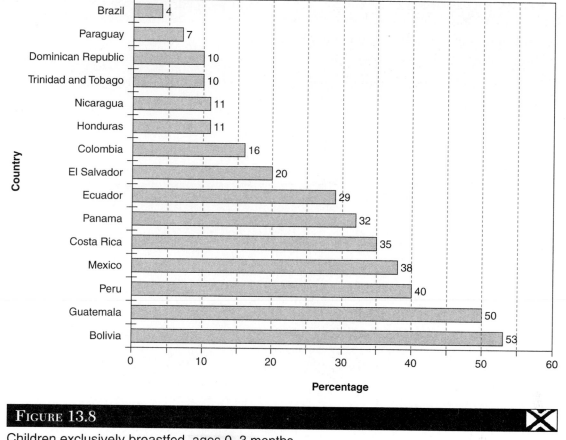

FIGURE 13.8

Children exclusively breastfed, ages 0–3 months.
Source: Derived from Bellamy 1997, 82–83.

(Figure 13.8). These data reflect campaigns by a number of multinational firms in the 1960s and 1970s to depict breastfeeding as culturally backward and unprogressive and the changing lifestyles of increasing numbers of Hispanic women who are employed outside the home.

Access to Health Care

Access to adequate health care is a fourth general indicator that, together with the measurement of life expectancy, provision of clean water, and availability of adequate nutrition, can be used to evaluate the overall levels of human health in a nation or region. Two of the most useful indicators of health care accessibility are the number of persons per physician and the number of persons per hospital bed. These data, which are presented in Table 13.11, confirm that great differences in health care accessibility are to be found within Latin America and the Caribbean. In Uruguay, for instance, there is 1 physician for every 271 persons. In contrast, in Honduras, the ratio is 1 doctor for every

10,000 persons. Similar patterns are evident in the data on the availability of hospital beds.

While we recognize that the health care delivery systems of some countries, such as the Southern Cone nations, are considerably more developed than those of others (Scarpaci 1988), we should realize also that use of health care facilities is contingent on more than their spatial proximity to the potential consumers. For example, Annis (1981) found that in the rural highlands of western Guatemala the large majority of the population had reasonably good access to health services and yet rarely sought care at the existing facilities. He concluded that this reluctance was caused not by the inaccessibility of the medical services nor by the cultural incompatibility of Western medical practices with native folkways, but rather by the presence of poorly trained personnel and underequipped clinics. In other words, the people failed to come not because of long travel distances but because health care delivery systems failed to cure. This may help to account for the simultaneous reliance by many Latinos both on professionally trained medical practitioners and on traditional healers, including

TABLE 13.11

Health Infrastructure in Latin America and the Caribbean

COUNTRY	POPULATION PER PHYSICIAN (Persons) 1998	POPULATION PER HOSPITAL BED (Persons) 1998
Argentina	346	455
Belize	8,482	345
Bolivia	1,929	714
Brazil	714	278
Chile	1,493	455
Colombia	1,111	714
Costa Rica	1,192	588
Dominican Republic	1,173	1,111
Ecuador	753	625
El Salvador	1,429	667
Guatemala	3,333	909
Guyana	3,226	279
Haiti	2,500	1,251
Honduras	10,000	1,000
Jamaica	2,000	476
Mexico	814	1,250
Nicaragua	1,429	833
Panama	821	370
Paraguay	1,258	833
Peru	969	400
Suriname	1,208	232
Trinidad and Tobago	1,429	312
Uruguay	271	435
Venezuela	491	417

Source: *Annual Report 1998* 1999, 150–151.

shamen (*curanderos*), midwives, bonesetters, and herbalists (Huber and Sandstrom 2001). In a related vein, Weil (1992, 43) has noted that many Latin American nations have more health practitioners than can be employed. The proportion of recent medical school graduates hired by the national health service system of Chile, for instance, fell from 80 percent in the mid-1970s to approximately 25 percent in 1982. Mexico had over 18,000 unemployed physicians in 1985.

One of the reasons many Latinos have been, and continue to be, reluctant to seek care from trained physicians and nurses is the native American and Iberian traditions of self-medication. **Automedicación,** as the practice is known in Spanish, is extremely widespread, accounting for 25 to 90 percent of all initial treatments, depending on the nature of the symptoms (Browner 1989; Logan 1983). Deciding how to treat a given condition is typically a very social process, which involves securing advice from numerous family members and friends. Two of the most trusted and most frequently consulted of these are generally the corner pharmacist and the market vendor of medicinal herbs.

One reason that these individuals are initially consulted more often than doctors or nurses is that their advice is free and considered authoritative. A second reason, however, is the lax regulation of prescription drugs, which are routinely dispensed without a doctor's prescription in Latin American pharmacies (Price 1989). Thus, most Latinos consider access to a pharmacy far more important than access to physicians and hospitals, although the latter are utilized in cases of serious illness.

HEALTH AND THE PHYSICAL-CULTURAL ENVIRONMENT

In addition to the general health issues discussed above, many Latin American and Caribbean peoples face disease threats specific to the physical environment within which they reside. While overall levels of health throughout the region have never been better and most of the population enjoys relatively good health, significant health challenges remain. We will now discuss briefly the risks associated with four key environments: tropical lowlands, highlands, urban centers, and rural regions.

FIGURE 13.9

Malaria cases per 100,000 persons.

Source: Data derived from Human Development Report 2002.

Tropical Lowlands

The year-round heat of the Latin American and Caribbean lowlands has provided habitats for a number of disease-bearing organisms whose populations are either absent or greatly reduced in colder environments. Foremost in impact among these vectors is the mosquito, which is responsible for the transmission of three life-threatening diseases: malaria, yellow fever, and dengue fever.

Malaria is caused by a protozoan parasite or plasmodium that lives in the body of the *Anopheles* mosquito. Although mosquitoes can tolerate cool temperatures, as is evidenced by their presence in great numbers in Arctic summers, their populations decrease dramatically in areas that are subject to frost. Furthermore, the malaria plasmodium requires summer temperatures of 59 to 60° F or higher in order to develop (Henschen 1967, 141–145). For these reasons, malaria is of little consequence in the *tierra fria* and *tierra helada* altitudinal life zones. On the other hand, its presence in the *tierra caliente* is so widespread, and its effects so deadly or debilitating, that it is widely acknowledged as having contributed more to the underdevelopment of the lowland tropics than any other physical factor (Figure 13.9). Malaria is present throughout the year in the Latin American and Caribbean lowlands, but its incidence tends to peak at the end of the dry season when local water bodies are at their lowest levels and are most stagnant. The coming of the rains brings some relief as the rising streams and rivers cleanse backwater areas of mosquito eggs and larvae (Smith 1982, 99–127). The accelerated deforestation of many Latin American and Caribbean lowland zones has contributed to higher incidence rates owing to the increased flooding that follows the felling of the trees (Weil 1981). Malaria was formerly treated with Chloroquine, but a number of resistant strains have emerged in recent years. Sulphur-based treatments are effective but prohibitively expensive to the masses, many of whom experience the intense fevers and chills of the disease over and over again in the course of their lives.

Yellow fever, which was formerly known widely as the "black vomit," is a viral disease transmitted primarily by the *Aedes aegypti* mosquito and consists of two forms. The first is passed from human to human and is therefore common in cities and towns. It was this urban-based yellow fever that was most responsible for the French failure to complete the Panama Canal (Figure 13.10). The second form, called sylvan or jungle yellow fever, is hosted by monkeys and transmitted by at least five species of mosquitoes, including the *Haemogogus* (Cueto 1992). It is less common than the first form owing to the ongoing loss of monkey habitats and to the fact that the monkeys that do survive in the wild tend to reside in the upper stories of the rain forest and have little contact with humans. Yellow fever can be prevented by a vaccine, but the high cost of the treatment has not only precluded eradication but has permitted its reintroduction in recent years into a number of urban areas that had been free of its effects over the past half century (Knouss 1992, 15–16).

Dengue fever is a mosquito-borne viral disease whose symptoms include high fever, intense headache, diarrhea, and joint pain. The hemorrhagic form causes leakage in the blood vessels of its victims, sending them into shock if they do not receive prompt medical care, and proves fatal in approximately 5 percent of all cases (Centers for Disease Control and Prevention 2002). The illness is known to have existed for at least two centuries but had been much reduced in the Latin American tropics by the 1970s as a consequence of an *Aedes aegypti* mosquito eradication campaign organized by the Pan American Health Organization. The relative success of the campaign, together with its high cost, led to the discontinuation of the control program, which, in turn, resulted in the reemergence of dengue fever as a major threat to human health in the 1980s. Over 650,000 cases were reported in the Americas in 2001, and the disease has become a serious concern once again in the Amazonian and Caribbean lowlands (Figure 13.11).

FIGURE 13.10

This plaque, honoring the Cuban doctor and scientist Carlos J. Finlay, has been erected on the site of the colonial-era fortress that guarded the entrance to Panama City. Finlay's 1881 discovery that yellow fever is transmitted by mosquitoes made possible the understanding that subsequently led to the construction of the Panama Canal. It also made possible an increased utilization of Latin America's humid lowlands.

FIGURE 13.11 ✕

Distribution of dengue fever.

Source: Adapted from Centers for Disease Control and Prevention 2002.

Legend for Figure 13.11:
- Areas infested with *Aedes aegypti*
- Areas with *Aedes aegypti* and dengue epidemic activity

FIGURE 13.12 ✕

Distribution of Chagas' Disease.

Source: Adapted from Mangurian 1995.

Legend for Figure 13.12:
- Areas affected by Chagas' Disease

A fourth illness of the Latin American lowlands, *Leishmaniasis braziliensis,* is a protozoal disease that frequently leads to the development of facial lesions and ulcers, the loss of facial cartilage, infection of the larynx and vocal cords, and ultimately death. It is transmitted from wild animals and domestic dog hosts to humans by the *Phlebotomus* sand fly. Because the flies do not tolerate the colder temperatures that occur above 2,400 meters elevation, the disease is limited to the low- and mid-altitudes, from Yucatán southward through Central America and into the Amazon Basin. Gade (1979) has suggested that leishmaniasis was largely responsible for the failure of the pre-Colombian Inca Empire to conquer the humid coca-producing lowlands east of the Andes.

American trypanosomiasis, or Chagas' Disease, is one of the most widespread, and yet least understood, of the deadly illnesses of the Latin American and Caribbean lowlands, with 16 to 18 million persons believed to be infected (Centers for Disease Control and Prevention 2002; Haddock 1979) (Figure 13.12). The disease is caused by the protozoa *Trypanosoma cruzi,* which is carried by lice and a number of triatomid insects of the *Reduviidae* family. Both wild and domestic animals, including rats, dogs, armadillos, and opossums, can serve as hosts. Although the disease may be spread through blood transfusions and breastfeeding,

the most common form of transmission is through insects, which often nest in great numbers in the air spaces of thatch and daub and wattle walls and ceilings and bite the human victim at night. The bite, which is intended to draw blood, creates a puncture wound through which contaminated insect feces enter the body. In other instances, infection occurs when the sleeping person inhales feces falling from the ceiling. Once present, the disease progresses through three distinct phases. The initial, or acute stage, is characterized by swollen eyelids, high fever, and the enlargement of the heart, liver, spleen, and other internal organs. The acute stage may be fatal in young children, who often develop meningoencephalitis, but most victims survive. The second stage is one of latency, which can last anywhere from ten to twenty years. It is followed by the chronic stage, which is associated with a rapid weakening of the heart and increased susceptibility to other diseases. Because the initial symptoms of Chagas' Disease are of short duration and because many victims in the chronic stage succumb first to other maladies, the incidence of American trypanosomiasis has been, until recently, widely underestimated. It is now recognized as a leading killer in the *tierra caliente,* especially among the rural and urban poor whose dwellings are not treated regularly with insecticide (Figure 13.13).

FIGURE 13.13

Buildings constructed of thatch ceilings and daub and wattle walls, such as this home in Barranquilla, Colombia, frequently host reduviid insects capable of transmitting Chagas' Disease.

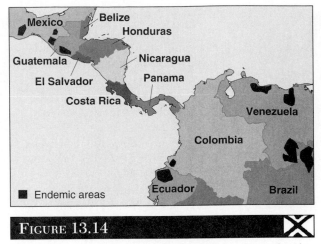

FIGURE 13.14

Remaining areas where river blindness is endemic.
Source: Adapted from "River Blindness in Retreat" 1997.

River blindness, or onchocerciasis, is associated with several species of the *Simulium* genus of black fly, which nests near rivers and transmits the larvae of a parasitic worm through its bite. Once present in the human body, the larvae cause intense itching, followed by general weakness, and eventually blindness. The prevalence of the disease fluctuates with black fly populations, with an estimated 100,000 to 215,000 persons being affected as recently as the 1980s in southern Mexico, Guatemala, Ecuador, Colombia, Venezuela, and northern Brazil ("New Offensive Against River Blindness" 1994; Vachon 1993; Weil 1981) (Figure 13.14). The 1990s have brought considerable progress in the campaign against the disease through use of the drug *invermectin,* which is effective in controlling larval populations within the human body. The adult worm can be killed only through insecticidal spraying.

Another group of illnesses endemic to the Latin American lowlands consists of the helminth, or worm, diseases. These include *Necator americanus,* a parasitical hookworm that enters humans through the skin as they walk with either the feet or legs exposed to contaminated ground or water, and *Dracunculus medinensis,* or Guinea worm, a three-foot-long nematode that lives in human muscle tissue (Henschen 1967, 146–156). Other helminth diseases common to the lowlands are *Ascaris lumbricoides,* which is a roundworm infestation, and a variety of tapeworm maladies.

A final serious helminth disease, schistosomiasis, is caused by suckerworms that utilize snails as intermediate hosts. The eggs of the worm hatch best in shallow, slow-moving or stagnant, well-lit water bodies at temperatures between 50° and 86° F (Weil and Kvale 1985; Haddock 1981). Once hatched, the free-swimming larvae, called miracidia, enter the snail and reside there for approxi-

mately one month before leaving the snail in a modified form, called cercariae. Given the opportunity, the cercariae then attach themselves to humans bathing, swimming, or working in the water and, within seconds, penetrate the skin and enter the circulatory system. Passing through the heart and lungs, the cercariae eventually cluster in the liver. There the worms mature and pair before moving into the intestinal tract and depositing eggs, many of which are then passed through defecation back into the water bodies where the cycle begins anew.

Schistosomiasis victims experience chronic weakness, diarrhea, anemia and, in advanced stages, lung and liver failure. The number of cases in Latin America and the Caribbean is unknown but is thought to be rising rapidly. At least 14 million persons are believed to be infected, primarily in Brazil, where the disease has established itself in and around many of the major cities of the Northeast (World Health Organization 2002; Barrow 1988). Secondary concentrations are found in northcentral Venezuela, the Lesser Antilles, and Hispaniola. The recent expansion of the disease is related, in part, to increases in the number of hydroelectric dams and reservoirs and to the growth of irrigated agriculture where flooded lands provide additional snail habitats. Control efforts to date have centered on largely ineffective attempts to limit snail populations through the use of molluscicides and on equally ineffective educational campaigns designed to persuade people to avoid contact with contaminated water. Meaningful progress in limiting the disease ultimately will depend in large measure on the provision of clean potable water and sanitation systems.

Highlands

The special health challenges confronting residents of the Latin American highlands primarily involve the effects of oxygen deficiency, or **hypoxia,** and cold on body growth and function. Atmospheric oxygen levels, like temperatures, decrease steadily with increasing elevations. One of the most common evidences of this principle is the tendency of most lowland visitors to highland communities to experience shortness of breath, dizziness, nausea, headaches, and a marked reduction of physical stamina following their arrival at their upland destination. This condition, called *soroche* or mountain sickness in Andean South America, generally passes after a week or two, but some individuals never adjust and find themselves unable to function normally at high elevations.

Native highlanders, on the other hand, possess certain physiological adaptations that enable them to live and work comfortably at elevations as high as 15,000 to 16,000 feet above sea level, where oxygen levels may be only half or less of those of the coastal lowlands. These adaptations, which contribute in varying ways to increased pulmonary and cardiovascular capacity, include larger than average chests, lungs, and hearts, greater than average bone marrow and red blood cell production, and elevated basal metabolisms (Salzano and Callegari-Jacques 1988; Frisancho 1978; Buskirk 1978; Little and Hanna 1978).

These physiological adaptations notwithstanding, the highlanders experience higher incidences of respiratory illness, including tuberculosis, bronchitis, and pneumonia, than do their lowland counterparts. On the positive side, mosquito-borne diseases are absent, and gastrointestinal, helminth, and cardiovascular disease rates are significantly lower than those experienced by residents of the lowlands (Baker 1978; Little and Baker 1976).

Urban Centers

Residence in a city does not, in and of itself, automatically condemn a person to live in a less healthful environment than a rural dweller. Each community, be it urban or rural, differs from others in size, culture, technological resources, and physical geography, and thus is subject to varying types and degrees of environmental degradation. Having noted this, we must still recognize that densely settled regions are more likely, simply by virtue of the number of people they support, to experience higher levels of certain environmental stresses than do rural zones (Price 1994; Harpham 1994; Case Study 13.4). These may include, but are not limited to, air pollution, water pollution, solid waste pollution, emotional disorders, soil pollution, visual pollution, noise pollution, thermal pollution, and radioactive contamination (Herman 1989) (Figure 13.15). Because water-related illnesses have been treated previously in this chapter, we will focus briefly here on two other significant urban health issues, air pollution and emotional disorders.

Air pollutants can come from a number of sources, including industrial wastes, automobile and other vehicular emissions, and heating and cooling system contaminants (Figure 13.16). When mixed indiscriminately by nature with sunlight, dust particles, and other naturally occurring substances, these materials may combine to form toxic compounds that impact human health in very negative ways. The author of a study of the incidence of respiratory diseases in São Paulo, for example, gathered data over an eleven-year period on sulfur dioxide and dust particulate levels at thirty-nine monitoring stations situated throughout the city (Ribeiro Sobral 1989). Three areas were then selected for detailed analysis of respiratory illness patterns among 12- and 13-year-old children. The findings led to two conclusions: that prevalence rates are generally higher in areas with higher pollution levels, and that young people living in economically poor high-pollution neighborhoods are at the greatest risk of all (Table 13.12). The latter finding suggests that poverty, especially as it expresses itself in limited access to adequate clothing and shelter, increases susceptibility to respiratory disease.

FIGURE 13.15

Solid waste disposal is rapidly becoming a major environmental concern in Latin America's urban centers. Citizens of Heredia, Costa Rica, are encouraged to throw their waste into curbside public trash receptacles that are painted with the phrase "clean city, cultured city."

Case Study 13.4: Environment and Health in Mexico City

Residents of Mexico City, Latin America's most populous urban agglomeration, are subject to a number of serious health problems associated with the contamination of the valley's physical environment and lax regulatory procedures. One source of great concern to civil authorities is the estimated 25,000 tons of garbage and other solid wastes generated each day. Approximately two-thirds of the refuse is collected by the sanitation department and transported to huge open dumps, where it is searched by thousands of poor people who rely on scavenging to provide a meager income. Many of these individuals actually reside in tiny shacks erected on top of the garbage mounds themselves. The remaining third of the metropolitan area's solid waste goes uncollected and is left to rot on the streets before being consumed by rats and other vermin.

A second environmental challenge is an emerging problem of groundwater contamination. As the water table is progressively lowered in response to increased pumping, tests are revealing troubling increases in salinity, ammonium, and fecal coliform levels in water obtained from wells in many parts of the city.

As severe as the capital's solid waste and groundwater contamination

problems are, they pale in significance to the city's air pollution woes. Some 4 million cars, busses, and cargo trucks, many of them operating with old and inadequately maintained diesel engines, combine with 35,000 factories to generate 3 to 4 million tons of air pollutants annually. Because the valley is surrounded by high mountains on three sides and subject to frequent atmospheric temperature

inversions that trap the contaminants at the lower elevations, the city is enveloped throughout most of the year with a thick yellow-gray haze that often reduces visibility to a few blocks or less (Case Study Figure 13.2). The smog buildup is particularly severe in the dry winter months when little or no rain falls to cleanse the atmosphere. The situation has become so serious that pollutant levels routinely

CASE STUDY FIGURE 13.2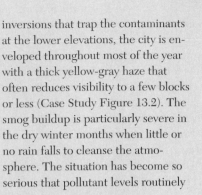

The buildup of air pollutants is so severe in Mexico City that a heavy blanket of smog appears to envelop the basin during most of the dry season. Visibility within the smog often is reduced to a few blocks and the beautiful twin volcanoes of Popocatépetl and Ixtaccíhuatl, which stand at the eastern edge of the valley, are obscured almost entirely from view.

FIGURE 13.16

Diesel engine-powered vehicles, such as this bus in Tegucigalpa, Honduras, often lack adequate emission-control equipment and contribute to high levels of air pollution and human respiratory illnesses in many Latin American cities.

Case Study 13.4—continued

reach 400 percent of maximum safe levels established by the World Health Organization. Because 55 percent of schoolchildren have unsafe levels of lead in their blood, many multinational corporations and foreign embassies offer hardship salary supplements to employees stationed in the capital. The benefits of physical exercise are considered by many to be outweighed by increased risk to the lungs and heart. Over 80 percent of children treated at state-run hospitals suffer from bronchitis, asthma, and other respiratory ailments. Eye infections and skin rashes are also common. One of the most troubling forms of atmospheric particulates is dried human fecal matter, which contains salmonella, streptococcus, staphylococcus, and other bacteria that can cause serious infections.

Municipal and national government authorities have been slow and inconsistent in their responses to the problems. Unleaded gasoline has been available only since 1991, and vehicle emission testing has been of limited impact. The "No Driving Today" program, which requires that each automobile be left unused one business day per week, was implemented in 1989. Its potential benefits have been partly negated, however, by many commuters

purchasing an additional, older car to drive on the assigned off-day rather than taking public transportation. One positive development has been a requirement that all new cars sold after 1990 be equipped with a catalytic converter. The government also closed a huge, heavily polluting oil refinery in 1991, and millions of trees have been planted in and around the city. Many

of these trees, unfortunately, have been damaged by the city's extraordinarily high concentrations of atmospheric ozone (Case Study Figure 13.3). Despite these and other official policies, enforcement of the regulations has been inconsistent, and little progress has been achieved in lowering the city's air pollution levels, which are among the highest in the world.

CASE STUDY FIGURE 13.3

A sign posted near small, struggling shrubs in downtown Mexico City asks passersby to "Please, let us live. Thank you, the plants." In reality, air pollutants and the lack of systematic care are far greater threats to urban vegetation than is human abuse.

Sources: Ezcurra et al. 1999; Drakakis-Smith 2000; Pick and Butler 1997; Onursal and Gautam 1997; Collins and Scott 1993.

Incidence of mental or emotional disorders is often reported to be higher in urban areas than in rural zones, although it is possible that this pattern reflects increased access to professional treatment in cities rather than more frequent occurrence of mental illness (Rotondo 1961). Possible causes of the high levels of emotional stress found among Latin American and Caribbean urban residents include overcrowded living conditions, increased risk of becoming a victim of child or adult violence, accidents, geographic dislocation of rural to urban migrants, burdens of double workloads carried by women, and long travel times to and from work (Reichenheim and Harpham 1991; Caccia Bava 1990; Curto de Casas 1994). Commuters in the largest cities often spend as much as four hours per day in travel.

Rural Regions

The health risks of rural residents in Latin America and the Caribbean are associated not so much with exposure to contaminants and stresses as with the lack of access to potable water, sanitation, and health care, including the services of doctors, nurses, and hospitals (Table 13.13). Childhood malnutrition is also more common in rural than in urban regions, owing to the historic neglect of the agrarian countryside. Most telling are the mortality rates of children under age 5, which in most nations are significantly higher in rural areas than in the cities (Table 13.14).

TABLE 13.12

Prevalence Rates of Symptoms of Respiratory Illness in São Paulo City, Brazil

ILLNESS/SYMPTOM	PERCENTAGE RESPONDENTS POSITIVE			
	Low-Pollution Area	Medium-Pollution Area	Prosperous High-Pollution Area	Poor High-Pollution Area
Cough with colds	55.7	55.6	65.5	8.1
Phlegm with colds	56.4	55.6	72.4	90.5
Congested chest (weeks per year)	10.7	15.7	24.1	52.4
Wheezing with shortness of breath	9.3	17.6	20.7	38.1
Chest illness before the age of 2 years	5.0	9.3	9.7	28.6
OTHER ILLNESSES				
Measles	32.1	49.1	43.4	76.2
Bronchitis	25.7	23.1	17.2	42.9
Asthma	4.3	3.7	5.5	19.0
Pneumonia	5.7	10.2	12.4	14.3
Whooping cough	20.0	15.7	22.1	33.3
Ear infections	34.3	58.3	60.0	52.4
Heart disease	0.0	2.8	4.8	9.5
Allergy	16.4	23.1	32.4	33.3

Source: Modified from Ribeiro Sobral 1989, 960 and 963.

TABLE 13.13

Rural versus Urban Health Indicators

COUNTRY	PERCENTAGE OF POPULATION WITH ACCESS TO					
	SAFE WATER		ADEQUATE SANITATION		HEALTH SERVICES	
	Rural	Urban	Rural	Urban	Rural	Urban
Haiti	23	37	16	42	39	—
Bolivia	36	87	32	72	52	77
Guatemala	49	87	52	72	—	—
Brazil	69	85	4	55	—	—
Nicaragua	29	84	34	77	60	100
Peru	18	75	25	58	—	—
Dominican Republic	—	80	83	76	67	84
Ecuador	49	80	49	95	20	70
El Salvador	46	85	65	91	—	—
Honduras	79	96	78	97	55	86
Colombia	56	97	56	97	72	86
Paraguay	10	70	14	65	38	90
Mexico	57	92	32	85	—	—
Argentina	29	77	37	73	21	80
Venezuela	75	80	30	64	—	—
Uruguay	5	85	65	60	—	—
Trinidad and Tobago	91	99	98	99	99	100
Costa Rica	92	100	70	95	—	—
Cuba	69	96	82	95	—	—

Source: Bellamy 1997, 84–85.

TABLE 13.14

Rural versus Urban Mortality Rates, Age 4 and Under

COUNTRY	DEATHS PER 1,000 POPULATION	
	Rural	**Urban**
Bolivia	168	114
Brazil	121	88
Colombia	33	36
Ecuador	112	65
Guatemala	130	99
Mexico	104	32
Paraguay	38	43
Peru	131	67
Trinidad and Tobago	30	41

Source: Behrman 1993, 246.

SUMMARY

As we conclude our discussion of health and development in Latin America and the Caribbean, several key points need to be stressed. The first is that every physical and cultural environment, be it tropical lowlands, highlands, urban centers, or rural regions, has certain unique health risks. However, no one environment is inherently more or less "healthy," or suited to human occupation, than another.

The second lesson is that poverty, rather than the physical environment, is the greatest of all health threats. From the insect- and worm-transmitted diseases of the lowlands to the respiratory illnesses of the highlands, it is money, with the associated capacity to secure safe housing, clean drinking water, qualified medical care, and adequate food, that is most necessary to prevent or treat any malady. The pivotal place of economic resources in health care highlights further the importance of using these monies wisely; presently, an estimated 25 to 33 percent are misspent (Knouss 1992).

A third fundamental point is that great progress has been achieved in the twentieth century in understanding and treating disease and in raising the overall living standards of the peoples of the region. While the illnesses described in this chapter do exist, a balanced perspective requires that we recognize that the overall incidence levels of disease are lower and the quality of treatment is better than ever before. The effects of the Lost Decade notwithstanding, most Latin American and Caribbean peoples have access to better medical care, are healthier, and live longer than did their ancestors.

A fourth conclusion is that education plays a critical role in the advancement of Latin American and Caribbean health care and that women are the preeminent participants in that process. Women exercise the greatest influence in the sanitation and health care behaviors of most Hispanic families (Menéndez 1982; Division of Environmental Health 1986); so it is not surprising to learn that children of mothers with no education are up to three times more likely to die in their youth than those whose mothers have completed seven to nine years of schooling (Sánchez and Eberwine 1994, 6). To be most effective, then, health care strategies must include significant educational components that will upgrade the knowledge and skills of women, who are the principal health care providers of the region.

KEY TERMS

SUGGESTED READINGS

Annis, Sheldon. 1981. "Physical Access and Utilization of Health Services in Rural Guatemala." *Social Science and Medicine* 15D:515–523.

Annual Report 1986, 1994, 1995, and *1998.* 1987, 1995, 1996, 1999. Washington, D.C.: Inter-American Development Bank.

Bacha, Edmar L. 1987. "IMF Conditionality: Conceptual Problems and Policy Alternatives." *World Development* 15:1457–1467.

Baker, Paul T., ed. 1978. *The Biology of High Altitude Peoples.* Cambridge: Cambridge University Press.

Barrow, Chris. 1988. "The Impact of Hydroelectric Development on the Amazonian Environment: With Particular Reference to the Tucuruí Project." *Journal of Biogeography* 15:67–78.

Behrman, Jere R. 1993. "Investing in Human Resources." In *Economic and Social Progress in Latin America 1993 Report,* 187–255. Washington, D.C.: The Johns Hopkins University Press for the Inter-American Development Bank.

Bellamy, Carol. 1997. *The State of the World's Children 1997.* New York: Oxford University Press.

Blackman, Courtney N. 1989. "The International Debt Crisis in Third World Perspective." *Caribbean Affairs* 2:1–8.

Bonilla, Elssy. 1990. "Working Women in Latin America." In *Economic and Social Progress in Latin America 1990 Report,* 207–256. Washington, D.C.: Inter-American Development Bank.

Browner, C. H. 1989. "Women, Household, and Health in Latin America." *Social Science and Medicine:* 461–473.

Buskirk, E. R. 1978. "Work Capacity of High-Altitude Natives." In *The Biology of High Altitude Peoples,* ed. Paul T. Baker, 173–187. Cambridge: Cambridge University Press.

Caccia Bava, Silvio. 1990. "Urban Policies for Social Transformation: The Case of São Paulo." *Cities* 7:60–64.

Carranza, Mario Esteban. 2000. *South America Free Trade Area or Free Trade Area of the Americas? Open Regionalism and the Future of Regional Economic Integration in South America.* Aldershot, England: Ashgate.

"Cholera's Message." 1991. *The IDB* 18 (July–August): 3.

Cole, Elizabeth. 1979. "Breastfeeding: A Critique of the Literature." In *Breastfeeding and Food Policy in a Hungry World,* ed. Dana Rafael, 137–145. New York: Academic Press.

Collins, Charles O., and Steven L. Scott. 1993. "Air Pollution in the Valley of Mexico." *The Geographical Review* 83:119–133.

Cordovez, Diego. 1990. "New Ideas for Progress: From Debt to Renewed Growth." In *Latin America: How New Administrations Will Meet the Challenge,* 11–13. Washington, D.C.: Inter-American Development Bank.

Cueto, Marcos. 1992. "Sanitation from Above: Yellow Fever and Foreign Intervention in Peru, 1919–1922." *Hispanic American Historical Review* 72:1–22.

Curto de Casas, Susana Isabel. 1994. "Health Care in Latin America." In *Health and Development,* eds. David R. Phillips and Yola Verhasselt, 234–248. London: Routledge.

de Souza, Anthony R., and Philip W. Porter. 1974. *The Underdevelopment and Modernization of the Third World.* Washington D.C.: Association of American Geographers Commission on College Geography Resource Paper No. 28.

Division of Environmental Health. 1986. *Women, Water and Sanitation.* Geneva: World Health Organization.

Douglas, David. 1990. "In the Vessel's Wake: The U.N. Water Decade and Its Legacy." *Grassroots Development* 14(2):3–11.

Drakakis-Smith, David. 2000. *Third World Cities.* 2nd ed. London: Routledge.

Drake, Paul W. 1994. "Introduction: The Political Economy of Foreign Advisers and Lenders in Latin America." In *Money Doctors, Foreign Debts, and Economic Reforms in Latin America from the 1890s to the Present,* ed. Paul W. Drake, xi–xxxiii. Wilmington, Del.: Scholarly Resources Inc.

Economic and Social Progress in Latin America, 1992 Report, 1993 Report, and *1998–1999 Report.* 1992, 1993, 1998. Washington, D.C.: Inter-American Development Bank.

Epstein, Jack. 1995. "Where Carrots Cost Twice as Much as Chicken." *The Christian Science Monitor,* April 21: 6.

Ezcurra, Exequiel, et al., eds. 1999. *The Basin of Mexico: Critical Environmental Issues and Sustainability.* Tokyo: United Nations University Press.

Frisancho, A. R. 1978. "Human Growth and Development Among High-Altitude Populations." In *The Biology of High Altitude Peoples,* ed. Paul T. Baker, 117–171. Cambridge: Cambridge University Press.

Gade, Daniel W. 1979. "Inca and Colonial Settlement, Coca Cultivation and Endemic Disease in the Tropical Forest." *Journal of Historical Geography* 5:263–279.

García Rodríguez, Enrique. 1990. "Turning Around the Bolivian Economy and the Challenges for Future Development." In *Latin America: How New Administrations Will Meet the Challenge,* 89–96. Washington, D.C.: Inter-American Development Bank.

Gilbert, Alan. 1990. "The Provision of Public Services and the Debt Crisis in Latin America: The Case of Bogotá." *Economic Geography* 66:349–361.

Gordon, Karen A. 1979. "Infant Health from a Public Health Perspective." In *Breastfeeding and Food Policy in a Hungry World,* ed. Dana Rafael, 155–161. New York: Academic Press.

Goulet, Denis A. 1971. *The Cruel Choice.* New York: Atheneum.

Gwynne, Robert N. 1986. *Industrialization and Urbanization in Latin America.* Baltimore: The Johns Hopkins University Press.

Haddock, Kenneth C. 1981. "Control of Schistosomiasis: The Puerto Rican Experience." *Social Science and Medicine* 15D:501–514.

———. 1979. "Disease and Development in the Tropics: A Review of Chagas' Disease." *Social Science and Medicine* 13D:53–60.

Harpham, Trudy. 1994. "Cities and Health in the Third World." In *Health and Development*, eds. David R. Phillips and Yola Verhasselt, 111–121. London: Routledge.

Henschen, Folke. 1967. *The History and Geography of Diseases*. Trans. Joan Tate. New York: Delacorte Press.

Herman, Mauricio. 1989. "The Urban Cleanup." *The IDB* 17 (April–May): 7–9.

Hojman, David E. 1994. "The Political Economy of Recent Conversions to Market Economies in Latin America." *Journal of Latin American Studies* 26:191–219.

Huber, Brad R., and Alan R. Sandstrom, eds. 2001. *Mesoamerican Healers*. Austin: University of Texas Press.

Iglesias, Enrique V. 1990. "Latin America: New Approaches for the 1990s. How New Administrations Will Meet the Challenges." In *Latin America: How New Administrations Will Meet the Challenges*, 1–5. Washington, D.C.: Inter-American Development Bank.

Instituto para la Integración de América Latina. 1990. *El Proceso de Integración en América Latina en 1989*. Buenos Aires: Banco Interamericano de Desarrollo.

Irvin, George. 1988. "ECLAC and the Political Economy of the Central American Common Market." *Latin American Research Review* 23(3):7–29.

Klak, Thomas, ed. 1998. *Globalization and Neoliberalism: The Caribbean Context*. Lanham, Md.: Rowman and Littlefield.

———. 1992. Recession, the State and Working-Class Shelter: A Comparison of Quito and Guayaquil During the 1980s." *Tijdschrift voor Economische en Sociale Geografie* 83:120–137.

Knouss, Robert F. 1992. "The Health Situation in Latin America and the Caribbean: An Overview." In *Health and Health Care in Latin America During the Lost Decade: Insights for the 1990s*, eds. Connie Weil and Joseph L. Scarpaci, 9–29. Minneapolis: Minnesota Latin American Series No. 3 and Ames: Iowa International Papers Nos. 5–8.

Latham, Michael C. 1979. "International Perspectives on Weaning Foods: The Economic and Other Implications of Bottle Feeding and the Use of Manufactured Weaning Foods." In *Breastfeeding and Food Policy in a Hungry World*, ed. Dana Rafael, 119–127. New York: Academic Press.

Lawrence, Ruth. 1985. *Breastfeeding: A Guide for the Medical Profession*. St. Louis: C. V. Mosby.

Little, Michael A., and J. M. Hanna. 1978. "The Responses of High-Altitude Populations to Cold and Other Stresses." In *The Biology of High Altitude Peoples*, ed. Paul T. Baker, 251–298. Cambridge: Cambridge University Press.

Little, Michael A., and Paul T. Baker. 1976. "Environmental Adaptations in Perspectives." In *Man in the Andes*, eds. Paul T. Baker and Michael A. Little, 405–428. Stroudsburg, Pa.: Dowden, Hutchinson & Ross.

Logan, Kathleen. 1983. "The Role of Pharmacists and Over the Counter Medications in the Health Care System of a Mexican City." *Medical Anthropology* 7:68–89.

Mangurian, David. 1995. "Fight Against Chagas' Disease." *The IDB* 22(8):12.

Manzetti, Luigi. 1999. *Privatization South American Style*. Oxford: Oxford University Press.

Meier, Gerald M. 1989. "Misconceptions about External Debt." In *Economic Development and World Debt*, eds. H. W. Singer and Soumitra Sharma, 27–32. New York: St. Martin's Press.

Menéndez, E. L. 1982. "Automedicación, Reproducción Social y Terapeutica, y Medios de Comunicación Masiva." In *Medios de Comunicación Masiva, Reprodución Familiar y Formas de Medicina "Popular"*, ed. E. L. Menéndez, 29. México, D. F.: Centro de Investigaciones y Estudios Superiores en Antropología Social. Cuadernos de la Casa Chata No. 57.

"New Offensive Against River Blindness." 1994. *The IDB* 21 (November): 10.

Onursal, Bekir, and Surhid P. Gautam. 1997. *Vehicular Air Pollution: Experiences from Seven Latin American Urban Centers*. Washington, D.C.: The World Bank.

Pastor, Manual, Jr. 1987. "The Effects of IMF Programs in the Third World: Debate and Evidence from Latin America." *World Development* 15:249–262.

Pauly, David, Rich Thomas, and Judith Evans. 1989. "The Dirty Little Debt Secret." *Newsweek*, 17 April: 46.

Pick, James B., and Edgar W. Butler. 1997. *Mexico Megacity*. Boulder, Colo.: Westview Press.

Porter, Roger B. 1990. "The Enterprise for the Americas Initiative: A New Approach to Economic Growth." *Journal of Interamerican Studies and World Affairs* 32 (Winter): 1–12.

Portes, Richard. 1990. "Development vs. Debt: Past and Future." In *Latin America: How New Administrations Will Meet the Challenge*, 40–45. Washington, D.C.: Inter-American Development Bank.

Price, Laurie J. 1989. "In the Shadow of Biomedicine: Self Medication in Two Ecuadorian Pharmacies." *Social Science and Medicine* 28:905–915.

Price, Marie. 1994. "Ecopolitics and Environmental Nongovernmental Organizations in Latin America." *The Geographical Review* 84:42–58.

Ramamurti, Ravi. 1992. "The Impact of Privatization on the Latin American Debt Problem." *Journal of Interamerican Studies and World Affairs* 34 (Summer): 93–125.

Reichenheim, M., and T. Harpham. 1991. "Maternal Mental Health in a Squatter Settlement of Rio de Janeiro." *British Journal of Psychiatry* 159:683–690.

Ribeiro Sobral, Helena. 1989. "Air Pollution and Respiratory Diseases in Children in São Paulo, Brazil." *Social Science and Medicine* 29:959–964.

"River Blindness in Retreat." 1997. *The IDB* 24(7): 12–13.

Rogozinski, Jacques. 1998. *High Price for Change: Privatization in Mexico*. Washington, D.C.: Inter-American Development Bank.

Rotondo, H. 1961. "Psychological and Mental Health Problems of Urbanization Based on Case Studies in Peru." In *Urbanization in Latin America*, ed. Philip M. Hauser, 249–257. New York: International Documents Service of Columbia University Press.

Salzano, Francisco M., and Sidia M. Callegari-Jacques. 1988. *South American Indians: A Case Study of Evolution*. Oxford: Clarendon Press.

Sánchez, Jane, and Donna Eberwine. 1994. "Patient and Practitioner." In *IDB Extra: Investing in Women,* 6–7. Washington, D.C.: Inter-American Development Bank.

Scarpaci, Joseph L. 1988. *Primary Medical Care in Chile: Accessibility under Military Rule.* Pittsburgh: University of Pittsburgh Press.

Seers, Dudley. 1972. "What Are We Trying to Measure?" *The Journal of Development Studies* 8:21–36.

Selby, Henry A. 1991. "The Oaxacan Urban Household and the Crisis." *Urban Anthropology* 20:87–98.

Simpson, E. S. 1994. *The Developing World: An Introduction.* 2nd ed. Essex, England: Longman.

Smith, Nigel J. H. 1982. *Rainforest Corridors: The Transamazon Colonization Scheme.* Berkeley: University of California Press.

Stini, William A. 1982. "The Interaction between Environment and Nutrition." *Mountain Research and Development* 2:281–288.

Thomsen, Moritz. 1969. *Living Poor: A Peace Corps Chronicle.* Seattle: University of Washington Press.

Thorp, Rosemary. 1998. *Progress, Poverty, and Exclusion: An Economic History of Latin America in the 20th Century.* Washington, D.C.: Inter-American Development Bank and the European Union.

Vachon, Michael. 1993. "Onchocerciasis in Chiapas, Mexico." *The Geographical Review* 83:141–149.

Warren, Kenneth S., and A. F. Mahmoud Adel, eds. 1990. *Tropical and Geographical Medicine.* 2nd ed. New York: McGraw-Hill.

Weaver, Frederick Stirton. 2000. *Latin America in the World Economy: Mercantile Colonialism to Global Capitalism.* Boulder, Colo.: Westview Press.

Weil, Connie. 1981. "Health Problems Associated with Agricultural Colonization in Latin America." *Social Science and Medicine* 15D:449–461.

———. 1992. "Medical Geographic Research in Latin America and the Caribbean in the 1980s and Beyond." In *Benchmark 1990,* ed. Tom L. Martinson, 223–230. Auburn, Ala.: Conference of Latin Americanist Geographers.

Weil, Connie, and Katherine M. Kvale. 1985. "Current Research on Geographical Aspects of Schistosomiasis." *The Geographical Review* 75:186–216.

Weintraub, Sidney. 1991. "The New U.S. Economic Initiative Toward Latin America." *Journal of Interamerican Studies and World Affairs* 33 (Spring): 1–18.

Williams, Glen. 1987. *A Simple Solution: How Oral Rehydration Is Averting Child Death from Diarrheal Dehydration.* New York: United Nations Children's Fund.

World Development Report 1978 and 1993. 1978, 1993. New York: Oxford University Press for the World Bank.

World Population Data Sheet 2002. 2002. Washington, D.C.: Population Reference Bureau.

Year Book of Labour Statistics 1996. 1996. Geneva: International Labour Organization.

ELECTRONIC SOURCES

Centers for Disease Control and Prevention: www.cdc.gov/ncidod

CIA World Factbook: www.odci/cia/publications/factbook/menugeo.html

Human Development Report: www.undp.org/hdr

World Bank World Development Indicators database: http://devdata.worldbank.org/data-query/

World Health Organization: www.who.int/ctd/

GLOSSARY OF FOREIGN TERMS

A

alameda Grove of trees; also a public walk or mall

alcalde Municipal mayor

Altiplano Largest of the Andean intermontane plateaus, encompassing portions of Bolivia and Peru

amigo de carne Special friend

aparcería Sharecropping

Area Canalera Panama Canal Zone

ayllu Basic unit of Inca society, generally comprising a number of extended families

B

bajo Deeply cut valley of the Patagonian Plateau

bañado Shallow depression containing the channels of a braided stream whose locations shift from year to year

bandeiras Colonial era slave-hunting bands from São Paulo state

barrio Urban neighborhood or district

bolsón Basin characterized by interior drainage

bracero program Arrangement from 1942 to 1964 between Mexico and the United States wherein Mexican agricultural laborers were allowed to work legally in the United States on a temporary basis

C

caatinga Vegetation of the Brazilian backlands consisting of a mixture of palms, low deciduous trees, and cacti

caballería Large colonial estate granted to a mounted soldier

caballero Horseman and/or gentleman

cabildo Municipal council

cafetal Coffee farm of the northern Andean highlands

Callejón Andino Narrow elongated intermontane valley of the Ecuadorian Andes that includes ten distinct basins

calpulli Small community of Aztec-controlled Indians

camino real Royal highway of the colonial era

campo cerrado Form of Brazilian savanna vegetation characterized by scattered deciduous trees and grasses

cantina Neighborhood bar

capitanias Portuguese land grants of the colonial era that became the basis of many of the modern Brazilian states

carguero system Traditional Catholic lay religious hierarchy

carreta Animal-drawn cart

Castellano Spanish language or, more literally, the language of Castile

caudillo Military or political strongman

cenote Sinkhole, associated primarily with the limestone strata of the Yucatán Peninsula

central Steam-powered sugarcane mill

chapada Plateau (Brazil)

chinampa Form of raised field agriculture practiced in the Valley of Mexico

Chiquitos Plateau Moderately elevated region separating the Amazon and Paraná river basins

cholo Peruvian who attempts to change his racial status from Indian to *mestizo*

chubasco Hurricane that occurs in Pacific waters west of Mexico and northern Central America

cidade alta Upper sector portion of a Brazilian city

cidade baixa Lower sector portion of a Brazilian city

cimarrón Maroon, or escaped slave, of the colonial era

colonos Colombian peasants, many of whom became caught up in *La Violencia;* also indentured laborers of southern Brazil, Uruguay, and Argentina

compadrazgo Roman Catholic system of godparenthood

corregidor Spaniard appointed by the Crown to cleanse the *curaca* system

criollo American-born person of European descent; also used to refer to New World breeds of cattle

curacas Local Indian chiefs who functioned as low- to mid-level administrators within the Inca empire and the subsequent Spanish colonial era

curandero Folk or faith healer; shaman

D

descamisados Shirtless ones; displaced Argentine farm workers who benefited temporarily from government-subsidized industrial jobs during the first Perón era

diezmo Literally, a 10 percent tithe given to the Church; also a tax assessed by the colonial government on individuals or businesses

dignidad de la persona An individual's inherent self-worth or personal dignity

distinguido Distinctive or conspicuous; a nonconformist

doctrinero Colonial Catholic indoctrinator employed to prepare Indians for baptism

dón/doña Male/female honorary title of respect

E

ejido Common grazing land, usually on the outskirts of a Spanish settlement; also communally farmed land belonging to Mexican Indian villages; their Peruvian counterparts are called *comunidades indígenenas or comunidades campesinas*

encomendero Spanish colonial lord who received an *encomienda*; later an honorary title

encomienda Grant of Indians to a Spanish lord in the early colonial period; later, a landed estate

estancia Large cattle-grazing estate

estero Low depression of the eastern Pampa; also called *cañado*

evangélicos Charismatic Protestants of the Pentecostal tradition

F

favela Portuguese term for squatter settlement or shantytown

fazenda Brazilian agrarian estate

finca Small farm of the Central American highlands, frequently producing coffee

flota Convoy of Spanish ships utilized to transport gold and silver from the New World

friaje Antarctic air mass that brings relatively cool and dry atmospheric conditions to the western Amazon Basin; called *friagem* in Brazil

fútbol Soccer

G

gachupín Iberian-born Spaniard living in the New World; more literally, a spur-wearer; also called *peninsular*

garúa Thick fog of Pacific origin that forms along the southern Peruvian and northern Chilean coasts between June and October

gaucho Cowboy of the *estancias* of the Pampa

Gran Chaco Arid interior lowland region of northern Argentina and southern Paraguay

Gran Pantanal Inland swamp of South America that feeds the Paraguay River

guano Nitrogen-rich organic fertilizer comprised of seabird droppings

H

hacendero Owner of a *hacienda*

hacienda Large, landed estate used for cattle grazing; found primarily in the interior highlands; also called *fundo* in Chile

hidalgo Person of upper-class standing and prestigious parentage

hombre de confianza Man in whom confidence and trust are placed

huacas Sacred places or shrines of the Inca peoples

I

indigenismo Social and political movement that glorifies American Indian heritage

Indios mingados Free Indians hired to work the mines in colonial Peru and Bolivia; also called *mingas*

ingenio Water- or wind-powered sugarcane mill; also called *engenho* in Brazil

L

Ladino Guatemalan person of Hispanic culture

lahar Wall of mud released on a mountain slope by an earthquake or volcanic eruption

latifundio Large, landed estate

latifundista Owner of a *latifundio*

limpieza de la sangre Roman Catholic doctrine of the colonial era that excluded from the priesthood any individual who had non-Catholic or heretical ancestors among his previous four generations

Lingua Geral Portuguese-modified Tupinambá language that became widely spoken in colonial Brazil

Llanos Orinoco River Plains of Venezuela and Colombia

Llanos de Mojos Lowland region of northern Bolivia and western Brazil drained by the Beni and Mamoré rivers

loma Peruvian coastal vegetation association consisting of short grasses and shrubby plants that derive a portion of their moisture requirements from *garúa* fog droplets

M

machete Long metal knife used to perform work or to engage in physical conflict

machismo Culture trait of male dominance

maquiladora Foreign-owned industrial plant that hires native workers at low wages to assemble imported manufactured items that are then returned for sale to a foreign market

Marianismo Idealized Latin American female gender role

Marranos Outwardly Christianized Jews in medieval Iberia and colonial Latin America, many of whom continued to secretly practice Judaism

mestizos Persons of mixed European and Indian ancestry; also called *mamelucos* or *caboclos* in Brazil

milpa Field of maize; also called *conuco, roza,* and *chacra*

minga Cooperative, reciprocal labor practiced within the Inca Empire

minifundio Small landholding

minifundista Owner of a *minifundio*

mita Communal labor performed by Inca villagers

mita minera Literally, a turn in the mine; tribute of labor for the silver mines required of Bolivian and Peruvian colonial Indian villages

mitayos Indians fulfilling the labor requirements of the *mita minera*

montaña Densely forested foothills of the eastern Peruvian Andes

monte Steppe vegetation of western Patagonia characterized by low trees, shrubs, and grasses

mordida A bribe; literally, "the bite"

mulattos Persons of mixed European and African ancestry

N

naborías Free Indian workers of the colonial silver mines of Mexico

négritude Social and political movement that glorifies African-American heritage

niños de la calle Street children; also called *gamines*

norte Arctic air mass that brings colder winter temperatures to Mexico and northern Central America

O

obraje Colonial textile factory

Oriente Lower eastern slopes of the Ecuadorian Andes drained by the Amazon River and its tributaries

P

palacio municipal Municipal government building; also called *ayuntamiento*

Pampa Fertile plain of eastern Argentina

pampero Antarctic cold front that sweeps northward across the Pampa, and often beyond into southern Brazil

páramo Treeless vegetation of the high, humid Andes

pardos Individuals of mixed African and non-African ancestry

pasto duro Native, tough pasture grasses of the Pampa

pasto tierno Introduced tender pastures of the Pampa

patrón Upper-class master and/or benefactor

peonía Small estate granted to a colonial foot soldier

personalismo One's personal sphere of influence

pet kotoob Raised field agriculture practiced by the pre-Conquest Maya Indians

plaza Town square, usually containing the main Catholic church and the *palacio municipal*; also called *zócalo*

pulque Fermented drink made from the juice of the maguey plant

puna High plateau of the central Andes; also, treeless vegetation of the high, arid Andes

Q

quinto real A 20 percent tax assessed by the Spanish government on goods and services exchanged within the mercantile system

R

ranchería Small rural community

reales de minas Royal mines or mining centers

reducciones Nucleated settlements into which many Indians were forced by the Spaniards; also called **resguardos** or **congregaciones** in Spanish and *aldeiamentos* in Portuguese

regidor Municipal councilman

repartimiento Tribute or services required of colonial Indian villages by Spanish lords

reparto de efectos System that imposed forced purchases of unneeded goods on the Indians of colonial Peru by the Spanish Crown

S

salar Salt flat

Santería African folk Catholicism centered in Cuba and Puerto Rico

segundos Second and subsequent sons who are ineligible to inherit the family estate

sertão Severely eroded backlands of Northeastern Brazil

servicio personal Drafted Indian mining labor in colonial Mexico

sierra Mountain range; also called *serra* in Brazil

siesta Nap or period of rest generally taken in the early afternoon

solar Individual lot within a city

soroche Highland oxygen deficiency syndrome characterized by shortness of breath, dizziness, nausea, headaches, and reduction of physical stamina

sukacollo Raised field agriculture practiced on the *Altiplano*

T

tener cuello/palanca Exercising personal connections and influence; also called *jeito* in Brazil

tepuis Flat-topped mountains of the Guiana Highlands

tierra caliente Lowest altitudinal life zone; literally, "hot land"

tierra fría Next-to-highest altitudinal life zone; literally, "cold land"

tierra helada Highest altitudinal life zone; literally, "frozen land"

tierra templada Next-to-lowest altitudinal life zone; literally, "temperate land"

tinterillismo/papeleo Expressions for bureaucratic excess; literally, "ink-bottleism" or "paper shuffling"

trapiche Animal-powered sugarcane mill

traza Heart of the central city, characterized by grid street pattern

tumbeiros Portuguese ships that transported slaves from the Old World to the Americas

V

vecindad Inner city tenement; also called *tugurio, callejón, solar, and conventillo*

vecinos Residents of Spanish-American cities

veta Ore-bearing vein

W

waru waru Raised field agriculture practiced on the *Altiplano*

Y

yanaconas Free Indians, many of whom became skilled craftsmen during the colonial period

Yungas Subtropical lowland region of northern Bolivia

Z

zafra Sugarcane harvest season

zambos Persons of mixed African and American Indian ancestry

CREDITS

All photographs supplied by David L. Clawson unless noted below. All line art illustrations are courtesy of Carto-Graphics.

CHAPTER 1
1.1: Charles M. Nissly; 1.3: Charles M. Nissly; 1.4: Robert L. Layton; 1.5: Alexander Coles; 1.7: Charles M. Nissly; 1.9: John W. Dewitt.

CHAPTER 2
2.3: Charles M. Nissly; 2.5: Oscar H. Horst; 2.6: Oscar H. Horst; 2.7: Peter H. Yaukey; 2.8: Peter H. Yaukey; 2.9: Robert L. Layton; 2.10: Peter H. Yaukey; 2.11: Charles M. Nissly; 2.12: Charles M. Nissly; 2.13: Charles M. Nissly; 2.14: Robert L. Layton; 2.18: Charles M. Nissly; 2.19: Christoph Stadel; 2.20: Oscar H. Horst; 2.21: Oscar H. Horst; 2.22: Charles M. Nissly; 2.23: Oscar H. Horst; 2.24: Charles M. Nissly; 2.25: Charles M. Nissly; 2.26: Charles M. Nissly; 2.27: Christoph Stadel; 2.29: Delwin L. Clawson; 2.30: Peter H. Yaukey; 2.36: Robert L. Layton; 2.45: Charles M. Nissly; 2.46: Oscar H. Horst.

CHAPTER 3
3.9: Christoph Stadel; 3.10: Oscar H. Horst; 3.11: Charles M. Nissly.

CHAPTER 4
4.4: Charles M. Nissly; 4.5: Charles M. Nissly; 4.6: Peter H. Yaukey; 4.7: John W. Dewitt; 4.9: Kally Marie Ray Squires; 4.11: Robert L. Layton; 4.12: Charles M. Nissly; 4.13: Oscar H. Horst; 4.15: Robert L. Layton; 4.16: Robert L. Layton; 4.17: Oscar H. Horst; 4.18: Charles M. Nissly; 4.19: Peter H. Yaukey.

CHAPTER 5
5.9: Charles M. Nissly; 5.10: Charles M. Nissly; 5.17: Courtesy of Dr. Gonzalo Palacios, Embassy of Venezuela, Washington, D.C.; 5.18: Charles M. Nissly.

CHAPTER 6
6.9: Charles M. Nissly; 6.10: Courtesy of Pablo V. Alacantara, Consulate of the Dominican Republic, New Orleans, Louisiana; 6.11: Courtesy of the Laboratorio Fotografico, Universidad de Puerto Rico, Recinto de Rio Piedras, Sistema de Bibliotecas, San Juan, Puerto Rico; 6.12: Courtesy of Dr. Gonzalo Palacios, Embassy of Venezuela, Washington, D.C.; 6.14: Courtesy of the Embassy of Argentina, Washington, D. C.; 6.15: Courtesy of Juan Macris, Consulate of Peru, Houston, Texas; 6.16: Courtesy of Ambassador Andrea Petricevic, Embassy of Bolivia, Washington, D.C.; 6.17: Courtesy of Isabel Allende and the Fundacion Salvador Allende, Santiago de Chile; 6.18: Courtesy of Hoy Magazine, Santiago de Chile; 6.19: Courtesy of the Embassy of Uruguay, Washington, D.C.; 6.20: Courtesy of Editora Abril, São Paulo, Brazil; 6.21: Courtesy of Editora Abril, São Paulo, Brazil; 6.22: Courtesy of Ambassador Odeen Ishmael, Embassy of the Republic of Guyana, and the Guyana Information Services, Washington, D.C.

CHAPTER 7
7.1: Christoph Stadel; 7.8: Delwin L. Clawson; 7.9: Charles M. Nissly; 7.10: Christoph Stadel; 7.11: Charles M. Nissly; 7.12: Charles M. Nissly; 7.13; 7.18: Christoph Stadel; 7.19: Christoph Stadel; 7.27: Delwin L. Clawson; 7.28: Cyrus B. Dawsey.

CHAPTER 8
8.5: Christoph Stadel; 8.6: Robert L. Layton; 8.8: Charles M. Nissly; 8.18: John W. Dewitt; 8.19: Oscar H. Horst; 8.20: Charles M. Nissly; 8.23: John W. Dewitt.

CHAPTER 9
9.2: Kally Marie Ray Squires; 9.4: Oscar H. Horst; 9.6: Christoph Stadel; 9.8: Oscar H. Horst; 9.19: John W. Dewitt; 9.21: Charles M. Nissly.

CHAPTER 10
10.1: Charles M. Nissly; 10.5: Robert L. Layton; 10.8: Delwin L. Clawson; 10.12: Christoph Stadel; 10.19: Charles M. Nissly; 10.25: John W. Dewitt; 10.26: John W. Dewitt; 10.29: Christoph Stadel; 10.30: Christoph Stadel.

CHAPTER 11
11.2: Christoph Stadel; 11.4: Christoph Stadel; 11.6: Robert L. Layton; 11.8: Robert L. Layton; 11.10: Charles M. Nissly; 11.20: Christoph Stadel; 11.23: Christoph Stadel; 11.25: Christoph Stadel; 11.28: Charles M. Nissly; 11.29: Robert L. Layton.

CHAPTER 12
12.10: Christoph Stadel; 12.20: Christoph Stadel; 12.21: John W. Dewitt; 12.26: Delwin L. Clawson; 12.30: Charles M. Nissly.

CHAPTER 13
13.13: Delwin L. Clawson.

INDEX

·························